MATHEMATICAL AND CONCEPTUAL
FOUNDATIONS OF 20TH-CENTURY PHYSICS

NORTH-HOLLAND MATHEMATICS STUDIES 100
Notas de Matemática (100)

Editor: Leopoldo Nachbin

Centro Brasileiro de Pesquisas Fisicas
and University of Rochester

NORTH-HOLLAND – AMSTERDAM • NEW YORK • OXFORD

MATHEMATICAL AND CONCEPTUAL FOUNDATIONS OF 20TH-CENTURY PHYSICS

Gérard G. EMCH

Departments of Mathematics
and of Physics
University of Rochester
Rochester, N.Y.,
U.S.A.

NORTH-HOLLAND – AMSTERDAM • NEW YORK • OXFORD

ISBN: 0 444 87585 9

Fisrt edition: 1984
Second printing: 1986

Publishers:
ELSEVIER SCIENCE PUBLISHERS B.V.
P.O. BOX 1991
1000 BZ AMSTERDAM
THE NETHERLANDS

Sole distributors for the U.S.A. and Canada:
ELSEVIER SCIENCE PUBLISHING COMPANY, INC.
52 VANDERBILT AVENUE
NEW YORK, N.Y. 10017
U.S.A

Library of Congress Cataloging in Publication Data

Emch, Gérard G.
Mathematical and conceptual foundations of 20th-century physics.

(North-Holland mathematics studies ; 100)
Bibliography: p.
Includes index.
1. Mathematical physics. 2. Physics. 3. Relativity (Physics) 4. Quantum theory. I. Title. II. Series.
QC20.E43 1984 530.1 84-14741
ISBN 0-444-87585-9

PRINTED IN THE NETHERLANDS

CONTENTS

PART III. QUANTUM THEORY

Chapter Interdependence

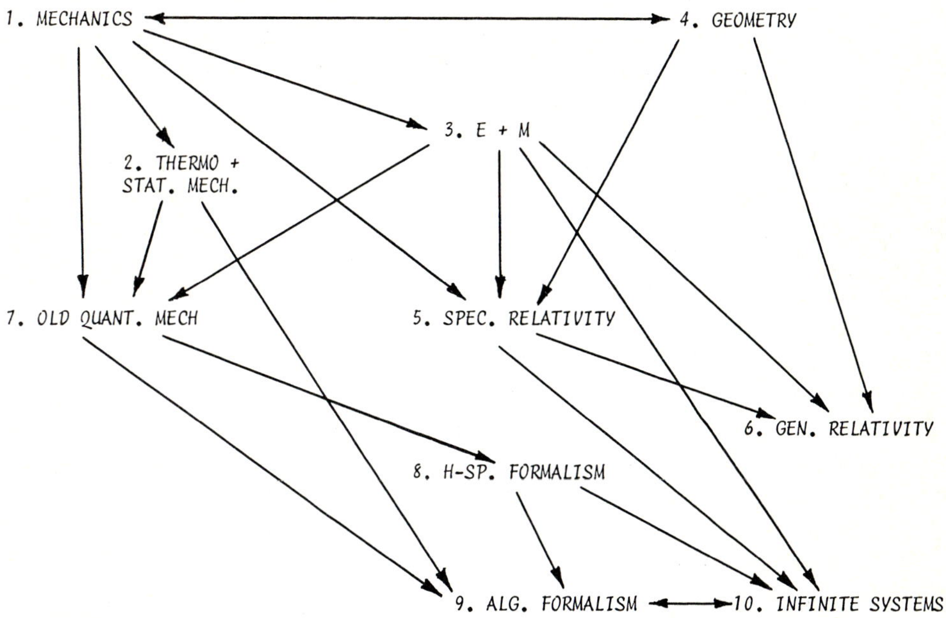

The arrows indicate logical dependences. As much as possible in two dimensions, the lay-out has been organized in such a manner that, along most paths, the mathematical analytic sophistication increases from left to right, while the physical synthetic conceptualization increases from top to bottom of the diagramme.

PREFACE

This book is primarily intended for Mathematicians with no, or little, background in Physics; much emphasis is laid on the interlocked historical development of mathematical and physical ideas.

For didactic purposes, the book is divided in three parts. Parts II & III can be covered in one semester each, as part of an introductory survey course on Applied Mathematics in the Graduate mathematics curriculum. Part I is elementary enough to be assigned for independent readings. The mathematical level of discourse should present no problem for a beginning graduate student in an American mathematics department. If this student were to consider the book for self-study, (s)he might find useful to have had, or to take concurrently, one-semester introductory courses in Functional Analysis and in Differential Geometry: these are two amongst the main tools of today's Mathematical Physics; they, however, are not prerequisites: all terms appearing in the text, mathematical as well as physical, are defined. What is required from the reader is the curiosity and the breadth necessary to undertake repeated crossings of the bridges which exist between Mathematics and Physics, two disciplines traditionally considered as closely allied fields; (s)he will follow in this journey the steps of such luminaries as Archimedes, Fermat, Newton, Euler, the Bernoullis, Lagrange, Laplace, Fourier, Cauchy, Gauss, Poincaré, Cartan, Hilbert, Weyl and von Neumann, not to mention the names of many Mathematicians or Mathematical Physicists now active.

For the mathematical audience thus defined, this book will present the main ideas and fundamental concepts of 20th-century Physics, with special attention to the concurrent mathematical developments. This century has indeed been marked by two conceptual revolutions from which Mathematics drew considerable impetus: the theory of Relativity and Quantum theory, both owing much to the seminal ideas of Einstein, one of the great geniuses of all times.

As no intellectual revolution can be properly understood without some knowledge of the paradigms prevalent at the time of its inception, Part I provides a survey of Classical Physics, which we divided in three chapters: Mechanics, Thermodynamics and Statistical Mechanics, and Electromagnetism. This study provides opportunities to place in perspective the successive advents of Calculus, of Probability and Statistics, of Differential and Symplectic Geometry, and of classical Functional Analysis.

Relativity is presented in Part II of this book, and Quantum Theory in Part III. The motivation provided by physical problems in the development of mathematical disciplines such as, for instance, pseudo-Riemannian Geometries, Hilbert Spaces and Operator Algebras, are emphasized.

Aside from the primary aim of this book, which is to present a unified mathematical account of the conceptual foundations of 20th-century Physics, under a single cover and in a form suitable for use in a survey course in Applied Mathematics, it is hoped that the book will also serve another function, namely that various parts of the work will be excerpted, and incorporated in separate courses pertaining to the Pure Mathematics curriculum, to provide illustrative examples, further motivations, and testimony to the unity of the Mathematical Sciences.

Finally, the author hopes that this book will help mathematicians broaden their exchanges with physicists, and with philosophers and historians of science.

Acknowledgements

Some long time ago, Leopoldo Nachbin impressed upon me that I should take some "rest", and write this book. I have had the advantage of being able to discuss both text and tune with him over the several years I have tested a patience that, by now, must be proverbial. This one-hundredth volume of the *Notas de Mathemática* is a testimony to his perseverance.

The materials presented here have been taught in courses I gave at the University of Rochester; the Ecole Polytechnique Fédérale à Lausanne; the Universität Tübingen; the Universidade do São Paulo; the Université de Paris VII; and the Virginia Polytechnic Institute and State University. I wish to acknowledge with thanks the stimulating comments I received from my colleagues at these Institutions: William Eberlein, Nicolas Gisin, Christian Günther, Lawrence Helfer, Richard Lavine and Malcolm Savedoff; Philippe Choquard and Philippe Martin; Peter Kramer, Burckhard Kümmerer, Alfred Rieckers and Wolfgang Schröder; Moysés Nussenzveig and Kalyan Sinha; Jacqueline Bertrand, Raymond Jancel and Guy Rideau; George Hagedorn, Gerhard Hegerfeldt and Paul Zweifel.

I would like to put on record the generosity of the leave of absence policy of the University of Rochester which also allowed me to spend sizable stretches of time with colleagues at the Zentrum für interdisziplinäre Forschung der Universität Bielefeld; the Université de Genève; Harvard University; the Mathematical Centre at the University of Warwick; the Centre de Physique théorique du CNRS à Marseille; and the University of Pennsylvania. For arranging the other head of these bridges, I am grateful to Ludwig Streit; Jean-Pierre Eckmann and Constantin Piron; George Mackey and Shlomo Sternberg; Klaus Schmidt; Daniel Kastler and Madeleine Sirugue-Colin; and Dick Kadison.

The always supportive attitude of Antoinette Emch-Dériaz over the twenty-five years of our association took a new turn in the present venture; sometimes at the cost of delays in her own historical researches, she tried to inform me away from the naturally whiggish views scientists tend to have on matters of intellectual history. For one, I appreciate her unrelenting efforts, even if the reader will soon realize that I still have a long way to tread on the road to a proper historiographical method. If anything is right here on that account, it is probably hers.

My students Stephan De Bièvre, Sung-Pyo Hong and Orietta Protti have responded with juvenile enthusiasm to the suggestion that they read first drafts of this book; they have helped clear many ambiguities.

Joan Robinson produced the work on AMS-TeX with extraordinary celerity, more than professional conscience, and remarkable resources of ingenuity. Arnie Pizer, the expert TeX tamer on our Faculty, saw to it that the beast would roar in harmony with a true mathematician's taste.

CHAPTER 1. MECHANICS

SYNOPSIS

There is a good deal of truth in A. N. Whitehead's caveat: "... in a memoir one's whole trouble is with the first chapter, or even the first page. For it is there, at the very outset, where the author will probably be found to slip in his assumptions. Further, the trouble is not with what the author does say, but with what he does not say. Also, it is not with what he knows he has assumed, but with what he has unconsciously assumed". Still, the author must make the plunge somewhere, swim as he may in whatever murky currents he encounters, reserving all the while his right to come back, stir and plumb for new depths those calm expanses he once thought had been properly charted. We make our plunge in classical mechanics.

By the turn of the Twentieth Century, Hamilton mechanics was the widely accepted form of mechanics, the one to be used when a correspondence was to be found with the new quantum mechanics. For instance, the first formula on the first page of the first chapter—"Principes de la dynamique"—of Poincaré's *Leçons de mécanique céleste* (1905) is the system of first order differential equations of Hamilton, which we write:

$$(1) \qquad \frac{dq^i}{dt} = \frac{\partial H}{\partial p^i} \quad ; \frac{dp^i}{dt} = -\frac{\partial H}{\partial q^i} \quad i = 1, 2, \ldots, d$$

The system of equations reflects the presence of a mathematically interesting underlying structure, namely that of a 2d-dimensional symplectic manifold, i.e. a differentiable manifold $\mathbf{M}$ endowed with a symplectic form ω which, by Darboux theorem, can be written (at least locally):

$$(2) \qquad \omega = \sum_{i=1}^{d} dp^i \wedge dq^i$$

In many applications $\mathbf{M}$ is T^*M, the cotangent bundle of a manifold M, called the "configuration space"; and ω is the canonical form $\omega = d\theta$ [where θ is the 1-form defined, for every $p \in T^*M$ and every $\xi \in T(T^*M)_p$, by $\theta(\xi) = p(\pi_*\xi)$ with π the projection $\pi : T^*M \to M$]. When $M = \mathbb{R}^{3N}$, $\mathbf{M}$ can simply be identified with the usual "phase space" $\mathbb{R}^{6N} = \{(q^1, \ldots q^{3N}; p^1, \ldots p^{3N}\}$; the general form is nevertheless necessary, as soon as holonomic constraints are considered.

To every smooth (e.g. C^∞ —) function F on $\mathbf{M}$ one can then associate the vector field ξ_F defined by

$$\xi_F \rfloor \omega = -dF \tag{3}$$

from which one defines in turn the Poisson bracket

$$\{F, G\} = -\omega(\xi_F, \xi_G)\,; \tag{4a}$$

in local coordinates this becomes

$$\{F, G\} = \sum_{i=1}^{d} \frac{\partial F}{\partial q^i} \cdot \frac{\partial G}{\partial p^i} - \frac{\partial F}{\partial p^i} \cdot \frac{\partial G}{\partial q^i} \tag{4b}$$

Note that one obtains for the coordinate functions q^i and p^j:

$$\left.\begin{array}{r} \{q^i, q^j\} = 0 = \{p^i, p^j\} \\ \{p^i, q^j\} = -\delta^{i,j} 1 \end{array}\right\} \quad \forall i, j = 1, \ldots, d \tag{4c}$$

Note further that the Hamilton equations (1) now appear as the equations for the integral curves of the vector field ξ_H associated to a special function H, the Hamilton function. In the particular case of a system of N particles moving freely in the ordinary one-particle configuration space $\mathbb{R}^3$, except for their mutual (velocity independent) interactions, H takes the form

$$H = T + V \tag{5a}$$

with

$$T = \sum_{n=1}^{N} \frac{1}{2m_n} |\mathbf{p}_n|^2 \tag{5b}$$

$$V = V(\mathbf{q}_1, \ldots, \mathbf{q}_N) \tag{5c}$$

In this case, Equations (1) reduce to:

$$\dot{q}_n^i = \frac{1}{m_n} p_n^i \quad ; \quad \dot{p}_n^i = -\frac{\partial V}{\partial q_n^i} \tag{6a}$$

so that

$$\mathbf{F}_n = m_n \ddot{\mathbf{q}}_n \quad \text{with} \quad F_n{}^i = -\frac{\partial V}{\partial q_n^i} \tag{6b}$$

To understand the facts of experience lying behind the interpretation of $\mathbf{F}_n$ as the "force" acting on the n-th particle, m_n as its "mass", $\mathbf{p}_n$ as its "momentum"; of T as the "kinetic energy" of the system, V as its "potential energy", and H as its "total energy", it is helpful to back up in history to what should be called Euler's mechanics. It is traditional and convenient to divide classical mechanics into three parts, the first being associated to Newton's *Philosophiae Naturalis Principia Mathematica* (1687), the second to Lagrange's *Mécanique Analytique* (1788), and the third to Hamilton's *General Method in Dynamics* (1834, 1835) and Jacobi's *Vorlesungen über Dynamics* (published by Clebsch in 1866).

Our three sections' headings perforce have to use this discrete nomenclature—Newtonian, Lagrangean and Hamiltonian mechanics—which artificially parcels out a continuous development starting before Galileo and continuing after Poincaré. We will also allude to some of the problems of current interest: however "classical" it may be called, mechanics is still a field of active research where new mathematics and, thus, new physical understandings are in the process of being developed.

Section 1. NEWTONIAN FORMULATION

Mechanics is a science of experience, an experience apprehended by observation, controlled by experiment, and comprehended by theory.

One of the established sources of experience that played an important role in Newton's formulation of rational mechanics and of universal gravitation was observational astronomy, culminating in Kepler's three laws of planetary motion, published in Prague, the first two in his *Astronomia Nova* (1609) and the third in his *De Harmonice Mundi* (1619). These laws are: (1) *each planet moves in an elliptical orbit, with the sun at one focus of the ellipse;* (2) *the focal radius from the sum to a planet sweeps equal areas of space in equal intervals of time;* (3) *the square of the sideral periods of the planets are proportional to the cube of their mean distance to the sun,* a statement which we transcribe as:

$$A^3 = k\,T^2 \tag{1}$$

where T is the period of the planet on its orbit, and A is the semi-major axis of its elliptical orbit, i.e. the average of the distances between the sun and the planet at its aphelion and perihelion; as we shall see below, the proportionality factor k can be computed in term of the gravitational constant of Newton's theory of universal gravitation.

The use of controlled experiments was persuasively advocated by Galileo in his *Discorsi e Dimostrazioni Matematiche intorno à due nuove Scienze attenti alla Mecanica & i Movimenti Locali* (1638). Whether Galileo was faithfully reporting the results of experiments he had himself actually performed is not terribly relevant, although we should mention that doubts were raised on that account, already by some of his contemporaries, e.g. Mersenne in France. The main point we want to make here is that Galileo was eloquently expounding, as a reflection of contemporary concerns, that the primary role of controlled experiments is to discriminate between opposing theories. Hence, theory has to come first, to organize observational experience and to inform the sensible planning of experiments.

The rational mechanics of the 17th century, however, did not happen as a sudden reaction to some hitherto unchallenged misconceptions inherited from antiquity; it was instead slowly brought into focus through the perceptive work of the medieval critics of Greek science. These now obscure scholars started the programme that would eventually replace the qualitative notions, that emerge from common experience of motion, by qualitative statements on: the geometric description of motion (*kinematics*); the analysis of its causes or manifestations (*dynamics*); the notions of force, torque, pressure, stress and their effects; the concepts of mass and inertia; the derived quantities we now call linear and angular momentum, kinetic and potential energy, work, power and action.

One of the achievements of medieval science was the logical distinction between the computational geometric questions of kinematics and the conceptually more involved problems of dynamics. Between 1328 and 1350, the scholars of Merton College in Oxford succeeded indeed in formulating clear enough ideas of instantaneous velocities and accelerations, allowing them to state a rule to the effect that in a rectilinear uniformly accelerated motion, the space $(x - x_0)$ travelled in an interval of time $(t - t_0)$ is given by:

$$(x - x_0) = \frac{1}{2}(v + v_0)(t - t_0) \tag{2}$$

where v_0 (resp. v) is the velocity at time t_0 (resp. t). Similarly dim, but sound, ideas on the change of rate of changes can be found in *De Uniformitate et Difformitate Intensionum* (1350) and *Tractatus de Latitudinibus Formarum* (n.d.) by Nichole Oresme in France, who can be credited with a geometrical proof of the "*Merton rule*" (2) akin, again in modern language, to the computation of the trapezoidal area under the graph of v as a function of t. The fact that v is here a linear function of t did obviously help since calculus was still a long way ahead. Indeed, it took some three centuries until enough power and confidence were built into the mathematical apparatus to allow the elegance and economy of concepts on which Newton erected his *Principia* (1687). What can be regarded as the first, even if incomplete, axiomatic formulation of mechanics holds indeed in the following three laws, stated in the beginning of the *Principia*: (1) *every body continues in its state of rest, or of uniform motion along a straight line, unless it is compelled to change that state by forces impressed upon it;* (2) *the change of motion is proportional to the motive force impressed, and it takes place in the direction of the straight line in which that force is impressed;* (3) *to every action there is always an opposite and equal reaction; or, the mutual actions of two bodies upon each other are always equal and opposed in direction.* Newton complements these three laws with a few "definitions", among which it is necessary to quote here the following ones: (a) "the quantity of matter is the measure of the same, arising from its density and bulk conjunctly ... it is this quantity that I mean ... under the name of body or mass. And the same is known by the weight of each body, for it it proportional to the weight ... "; (b) "the quantity of motion is the measure of the same, arising from the velocity and quantity of matter conjunctly ... "; (c) "an impressed force is an action exerted upon a body, in order to change its state, either of rest, or of moving uniformly along a straight line. This force consists in the action only; and remains no longer in the body, when the action is over. For a body maintains every new state it acquires, by its *vis inertiae* [force of inactivity] only".

These laws and definitions call for some immediate comments.

Firstly, the reader will have noticed that Newton assumed an acquaintance with kinematics. He stated explicitly—in a separate commentary, or scholium—that he does "not define time, space, place [i.e. volume occupied by a body] and motion [in particular velocities, absolute or relative, and acceleration], as being known to all." One might add that, although Newton refrained from using explicitly the notation of the fluxions in the *Principia,* the notions and methods of infinitesimal and integral calculus nonetheless pervade implicitly his exposition throughout. If he tried to hide this, as some authors would have it, the subterfuge would be so transparent to the modern eye, that little (if any) insight would result from trying to hide the obvious in the present survey: our aim is primarily to outline the substratum of concepts on which 20th-century physics was erected.

Secondly, we should point out that the definition of mass in Newton's *Principia* requires some elaboration. Mach (1883a) noticed that Newton's third law gives an empirical mean to determine the ratio of two masses; he started with the empirical definition: "all those bodies are bodies of equal mass, which, mutually acting on each other, produce in each other equal and opposite accelerations"; on that basis, Mach then argued that this empirical determination of the equality of two masses can be extended to an empirical determination of the ratio of two arbitrary masses, and that this relation is transitive. It might be of some incidental interest to notice that Mach (1883b) used a similar reasoning to reach an empirical definition of the quantity of electricity, i.e. electric charge (see Chapter Three). Of more immediate interest here, we must notice that Newton, in his definition of mass, identifies the *inertial mass* (resistance to changes of linear motion) and the *gravitational mass* (as measured through the weight of the body). This identification is by no means trivial; it was to be elevated, by Einstein, to the status of a fundamental *principle of equivalence,* according to which no external static homogeneous gravitational field can be detected in a laboratory in free fall in this field, since both the observer and his measuring apparatus will respond to the field with the same acceleration. Newton himself felt it necessary to state explicitly that he did verify the identification between these two notions of mass by experiments "very accurately made" on pendulums of equal lengths, but different compositions, and that he found no detectable differences in their period. This result was subsequently verified with increasing precision. At the beginning of the 20th century, as a result of delicate balancing experiments conducted by Eötvös, the difference between the ratio of the inertial and gravitational masses of wood and platinum was known to be less than one part in 10^9; the method was further improved by Dicke et al. (1964) to reach an experimental coincidence with 10^{-11} between aluminum and gold. We can, therefore, build with some confidence on a theory that identifies inertial mass and gravitational mass, as Newton did, followed on that account by Einstein.

We now continue our analysis of Newton's theoretical set-up with a third remark, the purpose of which is to emphasize that Newton's originality, at the beginning of the *Principia*, was mostly in his determination to write down basic axioms from which the theory could proceed deductively by mathematical reasoning, and his ability to extract concise axiomatic statements from the mist of ideas that had accumulated on mechanics, ideas some of which were correct, others were irretrievably wrong-headed, but most were vague, or blurred by extraneous circumstances. For instance, Buridan (c. 1300–c. 1360) had discussed a quantity that he called *impetus* which, once imparted to the motion of a body, would continue with the motion until it is destroyed by some external agency, such as the resistance of the air to the motion of a projectile. However, Buridan does not properly distinguish between the nature of air-resistance and of gravity, nor does he seem to recognize the unique role of linear motion. Newton's definition (b) of the *quantity of motion* is crisp; without hesitation, we can write it

$$p = m\,v \tag{3}$$

and then note that Newton's second law of motion can be now transcribed to read:

$$F = \frac{d}{dt}(m\,v) \tag{4a}$$

or, upon assuming that the quantity of matter, as measured by the mass m, is independent of the state of motion of the body:

$$F = ma \quad \text{with} \quad a \equiv \frac{dv}{dt} = \frac{d^2x}{dt^2} \equiv \ddot{x} \tag{4b}$$

which, after the work Euler did in the mid-18th century, has become the textbook form of Newton's second law. Anticipating further the post-Newtonian developments into the 18th century, we can already remark here that if it happens, as is the case for gravitational forces, that there exists a function

$V : x \in \mathbb{R}^3 \mapsto V(x) \in \mathbb{R}$ such that $F = -\text{grad}\, V$, i.e. in modern language, if the force derives from a potential, Newton's second law takes the form

$$\dot{p}_k = -\partial_k V \quad (k = 1, 2, 3) \tag{4c}$$

These formal manipulations of Newton's second law, while legitimate and useful, may nevertheless divert attention from the fact that this second law subsumes a wealth of unconscious assumptions and idealizations. We shall discuss later the question of whether Newton's formulation suffices or not for establishing the equations of motion of bodies that are more complicated than systems of point masses. Our next remark is directed, rather, to the motivation behind the concept of force. Newton's contribution suggests, at least implicitly by the use he makes of it, that *force is a primary concept.* It is nevertheless of interest to understand how it came about that forces should be represented by vectors—or by vector fields, as is the case for gravitation—even before they could be fitted so snugly into Newton's second law. In this context, it is useful to remember that dynamics not only assumes kinematics, but also contains statics as a particular case. *Statics* is circumscribed, in modern language, as the science concerned with the conditions under which an array of forces (and torques), acting on material bodies—e.g. particles, rigid or elastic bodies, fluids, or assemblages thereof—results in an equilibrium state, i.e. a state of rest or, more generally, a state characterized by the absence of accelerated motions. Devoting here a few lines to statics seems justified on account of the following three circumstances: (i) elementary situations come more easily under experimental scrutiny; (ii) the theoretical analysis of these situations does not require, at the start, the full epistemological apparatus necessary to distinguish all the fine threads running through the fabrics of the shared intuitions stored under the name of experience; (iii) the historical development of statics, especially in the 18th century, but also earlier and later well into the 20th century, shows that its purview is broad enough to provide essential clues into the general case of dynamics.

In particular, the mathematical nature and rules of composition of forces as vectors attached to different points of space, came to light in the 16th-century study of statics, although one finds a forerunner in Archimedes' law of levers. We find for instance that in *De Beghinselen der Weeghconst* [The Elements of the Art of Weighing] (1586), Stevin starts with a discussion of various systems of levers, and then begins his theory of the inclined plane with the following result: "Proposition XIX. Given a triangle, whose plane is at right angle to the horizon, with its base parallel thereto, while on each of the other sides there shall be a rolling sphere, of equal weight to one another: as the right side of the triangle is to the left side, so is the apparent weight of the sphere on the left side to the apparent weight of the sphere on the right side." Although Stevin does not seem to care giving a clear-cut definition of what he means by the term "apparent weight", the meaning springs out from the context, namely the "proof" he proposes, and the "corollaries" he derives from his proposition. Indeed his reasoning amounts to showing that if two spheres are of such unequal weights that their apparent weights would maintain them in equilibrium were they to be linked by a (weightless, inextendible) string, then their apparent weight would be what we call today the tension in the string or, more abstractly, the components of the weights of the spheres along a direction parallel to the inclined plane on which they are placed. From the variants of this problem, which Stevin analyses using various combinations of strings attached to hanging weights, to pull first on a sphere on an inclined plane, and then in subsequent examples, to pull on freely hanging rigid bodies of arbitrary shapes, it becomes clear that Stevin had recognized that "apparent weights"—i.e. forces—are characterized by their magnitude, their direction, and the point of the body on which they act. Moreover, the nature of his argument is such that he has been credited for having arrived at the *law of the parallelogramme of forces*, i.e. the law of addition of vectors. It should, nonetheless, be noted that Stevin's proof of his proposition XIX is based on what he presents as a reductio ad absurdum involving the absence of perpetual motion; the reasoning, however, was formalized by Varignon in his *Nouvelle Mécanique ou Statique* (1725) where he shows that the law of the parallelogramme of forces can be obtained as an application of the principle of virtual work. In that treatise moreover, Varignon reproduces a geometric argument, he had presented to the Paris Academy in 1687, to the effect that the law of the parallelogramme of forces implies the *law of levers*. This law states that if n forces F_k $(k = 1, 2, \ldots, n)$ act at n points x_k $(k = 1, 2, \ldots, n)$ of a rigid system whose motions are constrained to be rotations around a fixed point x_0, then the equilibrium condition is:

(5) $$M = 0 \quad \text{with} \quad M \equiv \sum_{k=1}^{n} (x_k - x_0) \wedge F_k \,.$$

$M_k \equiv (x_k - x_0) \wedge F_k$ is called the *moment* (or torque) *of the force* F_k *with respect to* x_0, and M is the total torque applied to the constrained system. To maintain this constraint, namely that x_0 be fixed, a force

(6) $$F_0 = -\sum_{k=1}^{n} F_k$$

has to be exerted at the point x_0. Hence the conditions of equilibrium of such a rigid system are

(7a) $$F = 0 \quad (\text{ with } \quad F \equiv \sum_{k=0}^{n} F_k \,)$$

(7b) $$M = 0 \quad (\text{ with } \quad M \equiv \sum_{k=0}^{n} x_k \wedge F_k \,)$$

Note that, because of (7a), we can replace x_k in (7b) by $\xi_k = x_k + \xi$ with ξ arbitrary.

Remark further that equations of the form (7) can also be obtained dynamically when one considers a system of n masses m_k, concentrated at the points x_k $(k = 1, 2, \ldots, n)$, and submitted to external forces F_k and two-body forces F_{kl} satisfying the so-called "strong Newton third law":

(8) $$F_{lk} = -F_{kl} \quad \text{and} \quad (x_k - x_l) \wedge F_{kl} = 0 \,.$$

Newton's second law reads then

(9) $$m\ddot{x}_k = F_k + \Sigma_l\, F_{kl}$$

Upon defining the *total linear momentum* of the system

(10a) $$P \equiv \Sigma_k\, p_k = \Sigma_k\, m_k\, \dot{x}_k$$

its *total angular momentum* (or moment of momentum)

(10b) $$L \equiv \Sigma_k\, x_k \wedge p_k = \Sigma_k\, m_k\, x_k \wedge \dot{x}_k$$

the *total external force* applied to the system

(11a) $$F \equiv \Sigma_k\, F_k$$

and the *total external torque* applied to the system

(11b) $$M \equiv \Sigma_k\, x_k \wedge F_k$$

we compute

(12a) $$\dot{P} = F$$
(12b) $$\dot{L} = M$$

where we used (9), *and* (8) to conclude that

(13a) $$\Sigma_{kl}\, F_{kl} = 0$$

(13b) $$\Sigma_{kl}\, x_k \wedge F_{kl} = \frac{1}{2}\Sigma_{kl}\, \{x_k \wedge F_{kl} + x_l \wedge F_{lk}\} = \frac{1}{2}\Sigma_{kl}(x_k - x_l) \wedge F_{kl} = 0$$

We thus see that the conditions $\dot{P} = 0$ and $\dot{L} = 0$ correspond respectively to equilibrium conditions of the form (7a) and (7b). Note that the expressions F (see 11a) and M (see 11b), which now enter these equilibrium conditions, still depend *only* on the external forces applied to the system; the same remark evidently holds for the two equations (12a & b), so that these equations can be regarded as expressing the response of the system to external agencies. We should emphasize that the two equations (12)—and hence the two corresponding equilibrium conditions—are logically independent; equation (12a) expresses that the total linear momentum is conserved exactly when the total applied force vanishes, while equation (12b) expresses that the total angular momentum is conserved exactly when the total applied torque vanishes: $\dot{P} = 0$ is a statement about translational invariance, while $\dot{L} = 0$ is a statement about rotational invariance.

Before going any deeper into the consequences of the above remark for the foundations of the mechanics of deformable materials, we should mention that Newton met his greatest success with his mechanics of the solar system and his theory of universal gravitation, both expounded in Book III of the *Principia*. The first logical step, which is not original with Newton, was to recognize that it is the same kind of force that keeps the planets moving around the sun, the moon circling around the earth, and causes the proverbial apple to fall. What is original with Newton is to have quantified this idea, and done so correctly.

To get the theory started, we should first agree on the idealization consisting in temporarily assuming that, for planetary motion, the dimensions of the celestial bodies are small enough, compared to the distances they travel, to justify regarding them, in zeroth-order approximation, as point masses; as we shall see, Newton himself went further than that, but let us nevertheless start here.

Secondly, if the mass m_0 of the sun is much larger than the mass m of the planet whose motion we want to describe, we can also neglect the motion of the sun and regard it as being at rest; as should become apparent in the sequel this restrictive assumption can easily be disposed of, provided one considers only two-body systems: a planet and the sun. We denote by x the position of the planet with respect to the sun.

Thirdly, one should agree that in view of Newton's third law, relation (8) is a reasonable assumption for the forces between two point masses.

Under these circumstances, the force of attraction the sun exerts on a planet is directed along the line joining the planet to the sun, and its magnitude is $F = F(r)$ where $r = |x|$; i.e. it is a *central force*. In view of Newton's second law, the acceleration of the planet is directed straight toward the sun, and is of magnitude $a = a(r)$. Note then that the spherical symmetry of the field of forces around the sun implies that the angular momentum of the planet is conserved in the course of its motion. This statement is equivalent (see def. 10b) to Kepler's second law of planetary motion.

The question then is to determine the form of the function $a : r \in \mathbb{R}^+ \mapsto a(r) \in \mathbb{R}$. It seems that several members of the Royal Society in London—among them Halley, who was to exert a most stimulating influence on the writing and publication of the *Principia*—had realized that *if* the motion of the planets were circular, Kepler's third law would imply $a(r) = K r^{-2}$. This is indeed an elementary result in kinematics. With v denoting the tangential velocity of the planet on its circular orbit of radius r, and with $T = (2\pi r/v)$ denoting its period, we obtain readily:

$$a = \frac{v^2}{r} = (2\pi)^2 \cdot \frac{r}{T^2} \tag{14}$$

Kepler's third law (1) reads, for planets moving on circular orbits: $r^3 = k\,T^2$, where k is the same constant for all planets, irrespective of their mass or their distance to the sun. Hence (14) reads then

$$a = (2\pi)^2\, k \cdot r^{-2} \tag{15}$$

Newton took that hint seriously, and proved that in a central field of forces, of magnitude given by:

$$F(r) = G\, m_0\, m\, r^{-2} \tag{16}$$

a point-like planet of mass m must describe an elliptic orbit with one of its foci at $r = 0$, which is Kepler's first law. Newton further showed that the semi-major axis A of such an elliptic orbit, and its period T, satisfy the relation

$$A^3 = (2\pi)^{-2} G\, m_0\, T^2 \tag{17}$$

which is Kepler's third law. Newton, however, was not satisfied with this derivation of the laws of planetary motion, and he apparently tried, long and hard, to move away from the above mentioned zeroth-order approximation. He finally managed to convince himself, by a reasoning which can now easily be straightened out, that as a consequence of (16) the gravitational effect of a spherical body, centered at $x = 0$ and of local mass density

$$\rho(x) = \begin{cases} \rho(r) & \text{when } \; r \equiv |x| \leq x_0 \\ 0 & \text{otherwise} \end{cases} \tag{18}$$

acting on a mass m situated at x with $r \equiv |x| > r_0$, is the same as that of a point-mass of magnitude

$$m_0 = \int d^3x \;\; \rho(x) \tag{19}$$

situated at $x = 0$. He then extended this result to the mutual attraction of two spherical bodies, both satisfying the symmetry assumption (18), thus vindicating the use of what we called a zeroth-order approximation. Moreover, on the strength of this remark, Newton could then present another result, that he had guessed previously, and thus show how universal his theory of gravitation really was. Indeed he could now use (14) to compute the centripetal acceleration of the moon, due to the attraction of the earth, knowing its period (29.5 days), its distance from the earth (about 60 times the earth radius) and the radius of the earth; upon using then (15), he could compare his evaluation of the centripetal acceleration of the moon, and the easily measured acceleration g experienced by ordinary bodies falling on the earth. The numerical agreement turned out to be satisfactory enough to justify calling G in (16) the *universal gravitation constant.* Note, however, that one had to wait until the end of the 18th century for Cavendish to be able to carry out the first absolute measurement of G in the laboratory, from the torque produced by the minute attraction of two pairs of masses. Since g and G are related by the formula $g = Gmr^{-2}$, where r is the radius of the earth, and m its mass, Cavendish presented his results as a measurement of m—he spoke of a measurement of the mean mass density of the earth. One could then introduce this value of G in (17), and obtain the value m_0 of the mass of the sun.

At this point, the reader should notice that we only considered so far what is called, in modern language, the solution of the *equations of motion* for a *test particle* in an *external field of forces* on which it has no effect: a single planet moving in the gravitational field of the sun, the moon circling the earth, the fall of an ordinary body on the surface of the earth. For the two-body problem, where the masses involved are of comparable magnitude, one can still get through by separating out the motion of the center of mass, i.e. a mass $m = \Sigma_i m_i$ concentrated at $x_0 = \Sigma_i m_i x_i / \Sigma_j m_j$.

Serious difficulties, of a completely different magnitude, are however immediately involved in the study of the true n-body problem, in which one aims to describe the motions of n massive bodies interacting with one another, even if one limits one's attention to two-body interactions, as is for instance the case for the dynamics of the sun-earth-moon system, or the study of the simultaneous motions of all the planets around the sun under their mutual gravitational interactions ... not to speak of the theory of the tides, which would involve, in addition, an understanding of fluid flows. Quantitative approximation methods were devised, and series expansions were developed for the n-body problem, right through the end of the 19th century. These methods, however, were found lacking when it came to the general study of stability problems. A new era was to be opened by Poincaré (1892), who introduced qualitative global methods that are still the object of much research activity today, see e.g. Moser (1968, 1973, 1975).

Another problem of mechanics is to describe the deformable bodies of finite extension, familiar from every-day experience. Here too, we are faced with a long history; in this history, Newton's original attempts, mostly in Book II of the *Principia,* are only an incomplete, often misdirected and certainly non-rigourous, intermediary step.

One of the very first, and most instructive, achievements in this area was obtained in a problem pertaining to statics, namely the determination of the equilibrium figure of a perfectly flexible, inextendible chain of constant linear mass density, freely hanging from two posts, in a homogeneous, i.e. constant and parallel, gravitational field of forces.

Although one can find drawings of the solution in Leonardo da Vinci, who attacked the problem by replacing the continuous chain by a discrete model, and evidently did not solve the problem analytically, Galileo thought that the equilibrium figure was a parabola. If we have to believe a letter from his brother Johann, it was still under the impression that the solution was an algebraic curve that Jackob Bernoulli, in 1690, proposed this problem as a challenge, which was successfully and independently met by Huygens, Leibniz and Johann Bernoulli: their solutions were published in the same issue of *Acta Eruditorum* (1691). The figure of equilibrium is that curve, of given length, the center of gravity of which is the lowest. The idea behind this answer is the notion of contact force: the action of any part of the chain on a neighbouring part is equivalent to a tangential force acting at their common boundary point. The differential equation of the solution curve is then:

$$\frac{dy}{dx} = \frac{s}{a} \tag{20}$$

where y (resp. x) is the vertical (resp. horizontal) coordinate of a point on the curve, s is the length of the curve up to this point, as measured from the lowest point on the curve, and a is a constant (having to do with the total length of the chain between the two posts from which it hangs). The figure of equilibrium is, therefore, a transcendental curve, namely Leibniz' *catenary*:

$$y = a \quad \cosh\,(x/a) \tag{21}$$

In his famous *Methodus Inveniendi Lineas Curvas Maximi Minimive Proprietate Gaudentes* (1744), Euler showed that his curve generates the minimal surface of revolution, around the x-axis; i.e. it is the solution of a purely geometric, variational problem, namely to minimize the area

$$S = \int_{x_1}^{x_2} dx\; L(y, y') \quad \text{with} \quad L(y,\eta) = 2\pi\; y(1+\eta^2)^{\frac{1}{2}} \tag{22}$$

a general problem Euler had reduced (see Section Two below) to that of solving the differential equation:

$$\frac{\partial L}{\partial y} - \frac{d}{dx}\left(\frac{\partial L}{\partial y'}\right) = 0\ . \tag{23}$$

In the meantime, the response to Jackob Bernoulli's challenge had been reviewed several times, and one knew also how to treat the case where the linear density varies along the chain; more importantly, it had become clear that the solution could be obtained independently by any one of the following methods: (i) minimizing the total potential energy; (ii) using the principle of virtual work; (iii) balancing separately the tangential and the normal forces acting at every point of the chain; (iv) similarly balancing the forces in any two different directions; (v) balancing the moments of the forces. The interest of this remark is to be found along two different lines of thought. First, it focus our attention on the principle of virtual work and the nascent variational methods that were to play an essential role in the formulation of mechanics discussed in our next section. Second, it raises the question of the putative equivalence of two conditions, namely the balancing of forces, and the balancing of torques. Euler thought over this question for much of his life until he finally realized, in 1776, that in a general theory of deformable bodies made up of arbitrary materials—fluid, flexible, elastic or rigid—the two conditions are not equivalent: equations (12a) and (12b) [or at equilibrium, equations (7a) and (7b)] are independent, and both conditions must in general be assumed. Euler's approach to the mechanics of continuous media was to enclose an arbitrary part ω of the body under study by a smooth, closed, orientable, fictive surface $\partial\omega$, and to impose (12a) and (12b) with:

$$\text{(24a)} \qquad P(\omega) \equiv \int_\omega dx \;\; \rho\, \dot{x}$$

$$\text{(24b)} \qquad L(\omega) \equiv \int_\omega dx \;\; \rho\, x \wedge \dot{x}$$

and, following the formulation in which Cauchy proposed in 1823 to capture the ideas of his 18th-century predecessors (the Bernoullis and Euler, Parent and Coulomb) as well as of his contemporary Fresnel,

$$\text{(24c)} \qquad F(\omega) \equiv \int_\omega dx \;\; f \; + \int_{\partial\omega} d\sigma \; \tau$$

$$\text{(24d)} \qquad M(\omega) \equiv \int_\omega dx \;\; x \wedge f \; + \int_{\partial\omega} d\sigma \; x \wedge \tau$$

where ρ is the mass-density of the body; f is the (smooth) vector field of *body forces* acting on ω; τ is the integrable vector field of *contact forces* describing the action on $\partial\omega$ of the part of the body that is outside ω : τ is usually expressed (Cauchy theorem!) through the so-called *stress-tensor* field T defined by

$$\tau = T \cdot n \quad \text{i.e.} \quad \tau^j = T^{jk} n_k$$

with $j, k = 1, 2, 3$ (the convention of summation over repeated indices is used here), and n denoting the unit outward normal to the surface element $d\sigma$. The pair of conditions (12) is then equivalent to a pair of equations, which we shall refer to as the *Euler equations of motion*, namely:

$$\text{(25a)} \qquad \operatorname{div}\; T + f = \rho\, \ddot{x}$$

$$\text{(25b)} \qquad T^t = T \quad \text{i.e.} \quad T^{kj} = T^{jk}$$

These equations of motion must be further supplemented by constitutive equations that describe the nature of the material considered. For instance, a non-viscous fluid is characterized by the fact that τ is normal to $d\sigma$, i.e. by the fact that there are no tangential stresses; in this case, T is described by a scalar field p, the *pressure*, via:

$$T = -p \begin{pmatrix} 1 & 0 & 0 \\ 0 & 1 & 0 \\ 0 & 0 & 1 \end{pmatrix} \tag{26}$$

and (25a) reduces to the *Euler hydrodynamic equation*

$$-\text{grad}\, p + f = \rho\, \ddot{x} \tag{27}$$

From this equation one derives easily, for instance, the relation between the speed v of a steady flow of incompressible fluid through a vertical tube placed in the gravitational field of the earth, and the pressure p it exercises on the walls of the tube at the altitude h, namely:

$$p + \frac{1}{2}\rho\, v^2 + \rho\, g\, h = \text{ constant} \tag{28}$$

This result is known as *Bernoulli's theorem*, in honour of still another member of the family, Daniel, who had indeed published an argument to this effect in his *Hydrodynamica* (1738). Cauchy's accomplishment, almost a century later, was to produce a concise and general rational mechanics of deformable continuous bodies, namely the above equations (25), that finally fulfills Newton's programme.

Section 2. LAGRANGEAN FORMULATION

This section is devoted to a presentation of the variational formulation of rational mechanics that is usually associated with Lagrange's *Mécanique analytique* (1788). Lagrange's systematic and synthetic approach indeed contributed significantly to this formulation, for the motivation of which he gives credit to: (i) Johann Bernoulli, through his influence on the work of Varignon in the beginning of the 18th century; (ii) the principles stated around 1740–1743 by Maupertuis and d'Alembert; and (iii) the ideas developed around the same time by Euler. Poisson (1809) and Hamilton (1834, 1835) reformulated and generalized Lagrange's work to bring it to the form in which we know it today. The philosophical idea behind this multiplicity of attempts was, as Euler put it, that "the fabric of the universe is most perfect ... nothing at all takes place ... in which some rule of maximum or minimum does not appear". While qualitative statements of this sort permeated much of the development of mechanics since antiquity, the contribution of the 18th-century mathematical physicists was to address squarely the question of a sharp definition of the quantities the extrema of which had to be found, and to develop the mathematical methods allowing to solve the resulting variational problem.

To approach the first of these problems realistically, rather than axiomatically, it might be helpful to recall the following fact of experience. If two projectiles, say of the same mass $m = m_1 = m_2$, are thrown into a head-on collision, with initial velocities $v_1 = -v_2 = v$, then in the absence of other external forces the velocities w_1 and w_2 of the projectiles after the collision are again opposed, i.e. $w_1 = -w_2 = -w$, since by the law of conservation of momentum

$$m_1 v_1 + m_2 v_2 = m_1 w_1 + m_2 w_2 \tag{1}$$

but the magnitude w can be anywhere between 0 (if the projectiles are, say, blobs of clay) and v (if they are instead, say, perfect billard balls). The latter case, referred to by saying that one has an elastic collision, is characterized by another conservation law, namely

$$\frac{1}{2} m_2 {v_1}^2 + \frac{1}{2} m_2 {v_2}^2 = \frac{1}{2} m_1 {w_1}^2 + \frac{1}{2} m_2 {w_2}^2 \tag{2}$$

For a system of n point particles of mass m_k and velocity $v_k (k = 1, 2, \ldots, n)$, the quantity

$$T \equiv \frac{1}{2} \sum_{k=1}^{n} m_k \, {v_k}^2 \tag{3}$$

was introduced, up the factor $\frac{1}{2}$ which it acquired later in the 19th century, by Leibniz under the name of *vis viva*, and it is called today the *kinetic energy* of the system. It is the most immediately perceived and quantifiable form of energy.

In the primitive form of *scattering theory* described by (1) and (2), one limits one's attention to the situations *before* and *after* the collision, but avoids making any statement on what might happen *during* the collision itself. A closer look would reveal that one cannot assume the collision to be instantaneous, unless one is willing—and we are not—to accept that, at the time of the collision, the forces exerted by each projectile on the other are infinite. It appears more reasonable, although still admittedly vague, to imagine that in the course of the collision process the kinetic energy is absorbed by the projectiles, stored in some other form of energy, and then restored to the projectiles again in the form of kinetic energy, as they bounce back; the situation presented by the blobs of clay alluded to earlier would then represent a case where the storage and restitution has noticeably gone astray: energy has been absorbed by the system, or dissipated away, e.g. in the form of heat. In this chapter we shall restrict our attention to those idealized cases where such things do not happen, and thus first study only conservative systems.

Consider now a system of n point particles, of masses m_k $(k = 1, 2, \ldots, n)$, acted upon by forces F_k that derive from a time-independent potential $V : (x_1, \ldots, x_n) \in \mathbb{R}^{3n} \mapsto V(x_1, \ldots, x_n) \in \mathbb{R}$; we thus have

$$F_k \equiv -\mathrm{grad}_k\, V\,. \tag{4}$$

Let now $x_k : t \in \mathbb{R} \mapsto x_k(t)$ be the trajectory of the k-th particle. Newton's second law

$$\frac{d}{dt}(m_k v_k) - F_k \equiv 0 \quad \text{with} \quad v_k \equiv \frac{d}{dt}x_k \tag{5}$$

implies

$$\sum_{k=1}^{n} \left\{ \frac{d}{dt}(m_k v_k) \cdot v_k - F_k \cdot \frac{d}{dt}x_k \right\} = 0 \tag{6}$$

i.e.

$$\frac{d}{dt}\{T(v_1, \ldots, v_n) + V(x_1, \ldots, x_n)\} = 0 \tag{7}$$

Hence what is gained (resp. lost) in kinetic energy T in the course of the motion is lost (resp. gained) in V. For this reason V is called the *potential energy* of the system, and the constant of the motion

$$E \equiv T + V \tag{8}$$

is called the *total energy* ; under the same circumstances

$$dW \equiv \sum_{k=1}^{n} F_k \cdot dx_k \quad \text{and} \quad W|_{t=0} = 0 \tag{9}$$

defines the *work* W done by the forces causing the motion of the system between the times $t = 0$ and $t = t_0$.

While hindsight allows one to isolate, as forerunners of a variational formulation of the equilibrium laws of the levers and of the inclined planes, some of the arguments in Jordanus de Nemore's *De ratione ponderis* (early 13th century) taken over, towards the end of the 16th century, by Galileo and by Stevin, still another century had to pass before the appearance of the first formulation of a variational principle that was mathematically non-trivial, i.e. one that would generate unexpected solutions of old and new problems, as well as a new mathematical methodology. This was Bernoulli's treatment of the catenary (see Section One) as summarized by Euler: up to a constant multiplicative factor, (1.1.22) can be interpreted as the potential energy of a homogeneous linear chain, hanging in the uniform gravitational field of the earth. Indeed, upon using the same parameterization of the chain as in Section One,

$$V(s) = y(x(s))\, g\, \rho\, ds \tag{10}$$

is the potential energy of a chain-element ds, situated in $y(x(s))$ and of weight $g\,\rho\, ds$, with

$$ds^2 = dx^2 + dy^2 \quad \text{i.e.} \quad ds = \sqrt{1 + (dy/dx)^2}\, dx \tag{11}$$

so that the potential energy of the whole chain, hanging from the two points $y(x_0)$ and $y(x_1)$ is indeed:

$$V = g\,\rho \int_{x_0}^{x_1} dx\; y(x)\sqrt{1 + (dy/dx)^2} \tag{12}$$

The catenary is the curve $y : x \in [x_0, x_1] \mapsto y(x) \in \mathbb{R}$, of length ℓ, that minimizes (12), subject to the end conditions $y(x_0) = y_0$ and $y(x_1) = y_1$; its analytic form is given by (1.1.21), where the parametrization (x, y) is chosen so that $(0, a)$ is the lowest point of the chain, and a is determined by ℓ, x_0, y_0, x_1, y_1.

The general class of problems, of which the above variational problem is an archetype, can be formulated as follows. Given a smooth function $L : (\xi, \eta, t) \in \mathbb{R}^n \times \mathbb{R}^n \times \mathbb{R} \mapsto L(\xi, \eta, t) \in \mathbb{R}$, find a smooth function $\gamma : t \in \mathbb{R} \mapsto \gamma(t) \in \mathbb{R}^n$ with $\gamma(t_0) = \gamma_0$ and $\gamma(t_1) = \gamma_1$ that makes extremal the functional

$$S(\gamma) = \int_{t_0}^{t_1} dt \;\; L(\gamma, \dot{\gamma}, t) \tag{13}$$

where $\dot{\gamma}$ denotes the derivative of γ. We shall have the occasion, later on, to demand a slight generalization of this problem, where $\mathbb{R}^n \times \mathbb{R}^n$ is replaced by the tangent bundle TM of a n-dimensional differentiable manifold M. For the time being however, we shall first go in the opposite direction and consider instead the particular case $n = 1$ with $M = \mathbb{R}$. For every smooth variation $h : t \in \mathbb{R} \mapsto h(t) \in \mathbb{R}$, with $|h| < \epsilon$ and $|dh/dt| < \epsilon$, we form:

$$\begin{aligned} & S(\gamma + h) - S(\gamma) = \\ & \int_{t_0}^{t_1} dt\{L(\gamma + h, \dot{\gamma} + \dot{h}, t) - L(\gamma, h, t)\} = \\ & \int_{t_0}^{t_1} dt\{\frac{\partial L}{\partial \gamma} h + \frac{\partial L}{\partial \dot{\gamma}} \dot{h}\} + 0(h^2) \end{aligned} \tag{14}$$

Upon integrating by parts, we receive:

$$\begin{aligned} S(\gamma + h) - S(\gamma) &= \sigma(\gamma, h) + 0(h^2) \quad \text{with} \\ \sigma(\gamma, h) &= \Big(\frac{\partial L}{\partial \dot{\gamma}} h\Big)\Big|_{t_0}^{t_1} + \int_{t_0}^{t_1} dt \Big\{\frac{\partial L}{\partial \gamma} - \frac{d}{dt}\Big(\frac{\partial L}{\partial \dot{\gamma}}\Big)\Big\} h \end{aligned} \tag{15}$$

A smooth function $\gamma : t \in \mathbb{R} \mapsto \gamma(t) \in \mathbb{R}$ is said to be an *extremal* of S if and only if $\sigma(\gamma, h) = 0$ for all variations h. From (15) we obtain that, if γ is an extremal of S, subject to the conditions $\gamma(t_0) = \gamma_0$ and $\gamma(t_1) = \gamma_1$, i.e. $h(t_0) = 0 = h(t_1)$, then

$$\frac{\partial L}{\partial \gamma} - \frac{d}{dt}\left(\frac{\partial L}{\partial \dot{\gamma}}\right) = 0 \tag{16}$$

which is the Euler equation (1.1.23). Its generalization from one to n dimensions, namely

$$\frac{\partial L}{\partial q^k} - \frac{d}{dt}\left(\frac{\partial L}{\partial \dot{q}^k}\right) = 0 \quad k = 1, 2, \ldots, n \tag{17}$$

is referred to as the *Euler-Lagrange* system of differential equations; in this expression, $\{q^1, q^2, \ldots, q^n\}$ is an arbitrary system of (local) coordinates in $\mathbb{R}^n$ (or more generally in the differentiable manifold M).

While this will not cause any serious problem in the simple situations to be discussed in this section, we should, nevertheless, remark here that the Euler-Lagrange equations only identify local extrema in the space of smooth functions over which S is defined: a solution of this system of second-order differential equations is not necessarily a global extremum, and the equations moreover do not specify the kind (e.g. minimum, maximum, etc.) of stationary point a given solution may turn out to be. Detailed information of this sort would require a study of the so-called second variation, and of the concept of conjugate points, a mathematical development started in the 19th century by Jacobi and Weierstrass.

The applications considered, in the motivation we presented so far for the Euler equation, were to geometry and to statics. In dynamics, the problem that first led to the formulation of a variational principle was the famous *brachistochrone* challenge proposed by Johann Bernoulli in 1696. The problem is to determine the path that a massive particle will travel in the shortest time, between two given points (x_0, y_0), (x_1, y_1), in a homogeneous gravitational field parallel to the y-axis. The problem is thus to minimize

$$S = \int_{(x_0,y_0)}^{(x_1,y_1)} ds \;\; n(x, y) \tag{18a}$$

where

$$ds^2 = dx^2 + dy^2 \quad \text{i.e.} \quad ds = \sqrt{1 + (dy/dx)^2}\, dx \tag{18b}$$

and $n(x, y)^{-1} \equiv v(x, y)$ is the speed of the particle, determined from the conservation law (8) by:

$$\frac{1}{2}m\, v(x, y)^2 + mg\, y = \frac{1}{2}m\, v(x_0, y_0)^2 + mg\, y_0 \tag{18c}$$

One has thus to minimize

$$(19)\qquad S = \frac{1}{\sqrt{2g}} \int_{x_0}^{x_1} dx\{(\alpha - y)^{-1}[1 + (\frac{dy}{dx})^2]\}^{\frac{1}{2}}$$
$$\text{with}\quad \alpha \equiv y_0 + {v_0}^2/2g$$

in which we recognize again a variational problem of the type (13). The brachistochrone problem was correctly solved by Newton, Leibniz, de L'Hospital, Jackob and Johann Bernoulli, and their solutions—an arc of cycloid—appeared in the same issue of *Acta Eruditorum* (1697). Jackob Bernoulli's solution was the most modern, while Johann Bernoulli recognized the similarity between this problem and an even older variational problem: the Fermat principle according to which light rays follow the path of shortest time between two points; the geometric solution to that problem for two contiguous homogeneous media of different index of refraction n was well-known to be given by Snell law of refraction, a fact Bernoulli used to solve (18) by imagining that the space travelled by the particle was made up of infinitesimally thin horizontal layers of refractive index $n(x,y) = \{2g(y_0 - y) + {v_0}^2\}^{-1/2}$; thus Bernoulli ended up solving another purely geometrical problem, that we would now describe as the computation of a geodesic in a Riemann manifold where the metric is given by

$$(20)\qquad ds^2 = n(x,y)^2\,(dx^2 + dy^2)$$

Having these results at hand, including Euler's equation (16), Lagrangee set out to formalize the motivations outlined in the first paragraph of this section. He did characterize the motions of what he called a mechanical system as those trajectories that are extremals of the "action"

$$(21)\qquad S = \int_{t_0}^{t_1} dt\; T$$

where T is the kinetic energy of the system and is subject to the constraint that $E \equiv T + V$ be constant, with V denoting the potential energy of the system. From this postulate, Lagrange derived the equations

$$(22)\qquad \frac{d}{dt}\left(\frac{\partial T}{\partial \dot{q}^k}\right) + \frac{\partial V}{\partial q^k} = 0$$

Note that for a system of n particles where all the forces derive from a potential $V(q^1, \ldots, q^n)$, Newton's second law of motion follows from (22).

Poisson (1809) proposed another reading of (22): he introduced the function

$$(23)\qquad L \equiv T - V$$

from which he constructed the *action*

$$S(\gamma) = \int_{t_0}^{t_1} dt\, L(\gamma, \dot{\gamma}) \tag{24}$$

and then defined the motion as those smooth curves γ that are extremals of the action (24). For "conservative systems" where $E \equiv T + V$ is a constant, $L = 2T - E$ implies that the two variational problems (21) and (24) are indeed equivalent. Hamilton (1834, 1835) noticed, however, that the formulation (24) has the advantage that the solution of this variational problem does not require explicitly the auxiliary condition that $E \equiv T + V$ be constant; he, therefore, generalized further what he called the Lagrange function L, introduced in (24) by Poisson, and allowed it to depend on time explicitly, in addition to its implicit dependence on t through γ and $\dot{\gamma}$ as was the case in Poisson's formulation. As a result, the dynamical variational problem takes the general mathematical form (13): we saw how to derive from this the Euler-Lagrange equations (17), which are now to be interpreted as the equations of motion of the system; the equations (22) thus appear as a particular case in this more general formulation.

Since the Euler-Lagrange equations (17) are obtained from the variational problem (13), it is clear that they are equally valid in every smooth coordinate-system one cares to choose; we briefly indicate how one may take advantage of this flexibility.

With Poisson (1809), we introduce for every coordinate q^k the *conjugate variable*

$$p_k \equiv \frac{\partial L}{\partial \dot{q}^k} \tag{25}$$

Notice that

$$H \equiv p_k \dot{q}^k - L \tag{26}$$

(where the convention of summation over repeated indices is used) satisfies, because of (17):

$$\frac{d}{dt} H = 0 \quad \text{if and only if} \quad \frac{\partial L}{\partial t} = 0 \tag{27}$$

which, in the Newtonian case, reduces to (7). Notice further, as an immediate consequence of (17), that

$$\frac{d}{dt} p_k = 0 \quad \text{if and only if} \quad \frac{\partial L}{\partial q^k} = 0 \tag{28}$$

i.e. p_k is a constant of the motion whenever L does not depend explicitly on q^k. The historical prototype of this situation is the case of a particle moving in a plane, subject to a central field of forces; we indeed have then, in the polar system of coordinate (r, φ) on $\mathbb{R}^2 \setminus \{0\}$:

$$(29) \qquad L \equiv T - V = \frac{1}{2} m v^2 - V(r) = \frac{1}{2} m(\dot{r}^2 + r^2 \dot{\varphi}^2) - V(r)$$

so that $(\partial L / \partial \varphi) = 0$, and thus

$$(30) \qquad \frac{d}{dt} p_\varphi = 0 \quad \text{with} \quad p_\varphi \equiv \frac{\partial L}{\partial \dot{\varphi}} = m r^2 \dot{\varphi}$$

which is Kepler's second law of planetary motion. For this historical reason, a coordinate q^k that satisfies (28) is sometimes called a cyclic coordinate, or even more misleadingly, an "angle variable". The corresponding constant of the motion p_k can then be introduced as a parameter in the other Euler-Lagrange equations, thus reducing the number of coupled differential equations one has to consider when solving the dynamical problem at hand. The geometrical meaning of this procedure will appear more clearly in the next section.

Another advantage of the freedom one has to choose the coordinate system in which one expresses the Euler-Lagrange equations appears in problems where constraints are imposed on the system, for instance by requesting that the representative point $(x^1, \ldots, x^{3n})$ of a system of n particles be restricted to belong to a m-dimensional smooth manifold $M \subset \mathbb{R}^{3n}$; the case $(m, n) = (2, 1)$ is the familiar situation where a single particle is restricted to moving on a smooth surface in $\mathbb{R}^3$; notice, incidentally, that the corresponding restriction of the kinetic energy T, to the manifold M, equips the latter with a Riemann metric. More sophisticated problems, such as the study of the motion of a bead along a vertical circle that rotates with constant angular velocity around its vertical diameter, or the motion of rigid bodies, e.g. a spinning top, can all be first approached in this way. To this day, these problems have been an inexhaustible source of mathematical fun and insight; we unfortunately have to restrain ourselves from entering into this type of problem here; see, however, Arnold (1978) and Abraham & Marsden (1978).

In closing this section, we should point out that the description of a motion through a system of second order differential equations does not determine the Lagrange function L: even if it exists, it is generally not unique: for instance, for $(x, \dot{x}) \in \mathbb{R} \times \mathbb{R} \setminus \{0, 0\}$, the Lagrange functions

$$(31a) \qquad L_1 = \frac{1}{2}(\dot{x}^2 - x^2)$$

$$(31b) \qquad L_2 = \frac{\dot{x}}{x} \arctan\left(\frac{\dot{x}}{x}\right) - \frac{1}{2} \ell n\{\dot{x}^2 + x^2\}$$

respectively attribute x to the conjugate variable

$$(32a) \qquad p_1 = \dot{x}$$

$$(32b) \qquad p_2 = \frac{1}{x} \arctan\left(\frac{\dot{x}}{x}\right)$$

and lead to the same equation of motion, namely that of the harmonic oscillator

$$\ddot{x} + x = 0 \,. \tag{33}$$

This example can moreover be generalized to illustrate the fact that $(\partial L/\partial t) = 0$ is not a characteristic that can be associated without further ado to a given equation of motion. Consider, for instance, the Lagrange function

$$L_1(x, \dot{x}, t) = \frac{1}{2}\, e^{2\epsilon t}(\dot{x}^2 - x^2) \tag{34}$$

with $0 < \epsilon < 1$, leading to the differential equation of the damped harmonic oscillator

$$\ddot{x} + 2\epsilon\dot{x} + x = 0 \tag{35}$$

A straightforward computation shows indeed that this same equation of motion can also be obtained from a Lagrange function L_2 satisfying $(\partial L_2/\partial t) = 0$, namely:

$$\begin{aligned} &L_2(x, \dot{x}) = \omega^2\, U \arctan U - \frac{\omega^2}{2} \ell n\, [x^2(1 + U^2)] \\ &\text{with} \quad \omega^2 = 1 - \epsilon^2 \quad \text{and} \quad U = \frac{1}{\omega}\Big(\frac{\dot{x}}{x} + \epsilon\Big) \end{aligned} \tag{36}$$

With the help of the conjugate variable

$$p_2 \equiv \frac{\partial L_2}{\partial \dot{x}} = \frac{\omega}{x} \arctan\, U \tag{37}$$

one can construct, upon using (26), the constant of the motion

$$H_2 = -\omega\epsilon \arctan\, U + \frac{\omega^2}{2} \ell n\, [x^2(1 + U^2)]. \tag{38}$$

For an up-to-date entry into the mathematical theory and the physical circumstances behind and around this phenomenon, see e.g. Della Riccia (1982) and references quoted therein.

3. HAMILTONIAN FORMULATION

Poisson's reformulation of Lagrange's formalism can be exploited to replace the Euler-Lagrange system (1.2.17) of n second-order differential equations by a mathematically equivalent system of $2n$ first-order differential equations that allow a conveniently simpler description of the motion. In the Hamiltonian formulation, this is achieved by introducing a function, whose differential describes these equations of motion, and which is defined on a space that is different from the space on which the Lagrange function is defined. As we shall see in Chapter Eight, this formalism and its building blocks—canonically conjugate variables, Hamilton function and Poisson bracket—played a seminal role in the inception of a consistent mathematical formalism for quantum mechanics.

Recall that the real-valued Lagrange function L, appearing in the Euler-Lagrange equations of motion, was defined on the space $\mathbb{R}^n \times \mathbb{R}^n \times \mathbb{R}$.

The reader familiar with the elementary notions of differential geometry—see e.g. Subsections 4.2.a, b and c—will realize that the forthcoming presentation generalizes immediately to the cases where the configuration space $\mathbb{R}^n$, associated to the variable q, is replaced by a m-dimensional smooth manifold M, e.g. to account for certain constraints; the range $\mathbb{R}^n \times \mathbb{R}^n$ of $(q, \dot{q})$ is then to be replaced by the $2m$-dimensional smooth manifold TM, the tangent bundle of M. The didactic advantage to view L as a map from $TM \times \mathbb{R}$ to $\mathbb{R}$ is to emphasize the sense in which the space of Hamiltonian mechanics differs from the space of Lagrangean mechanics. Indeed, the space on which the real-valued Hamilton function H is to be defined is $T^*M \times \mathbb{R}$, where T^*M is the cotangent bundle of M. For what follows, the central feature is that to every $q \in M$ is uniquely associated a copy of $\mathbb{R}^m$ in TM and a copy of $\mathbb{R}^{*m}$ in T^*M, where $\mathbb{R}^{*m}$ is the vector space dual to $\mathbb{R}^m$; while $\mathbb{R}^{*m}$ can be identified with $\mathbb{R}^m$, such an identification would not be natural in the present context and will thus be ignored.

The canonical procedure, by which the Hamilton function H is associated to the Lagrange function L, is known under the name of Legendre transformation, and it can be described as follows. Since the variables q and t are unaffected by this transformation, we drop them temporarily from the notation, and we focus our attention on those smooth functions

$$(1) \qquad f : \dot{q} \in \mathbb{R}^m \mapsto f(\dot{q}) \in \mathbb{R}$$

that satisfy the condition that for each $p \in \mathbb{R}^{*m}$ the function

$$(2) \qquad F(p, \cdot) : \dot{q} \in \mathbb{R}^m \mapsto p\dot{q} - f(\dot{q})$$

admit a unique extremum.

Note that the uniqueness of such an extremum, if it exists, is ensured by the condition that f be convex, i.e. that the matrix

$$(3) \qquad (f_{jk}) \equiv \left(\frac{\partial^2 f}{\partial \dot{q}^j \partial \dot{q}^k} \right)$$

be positive definite for every $\dot{q} \in \mathbb{R}^m$. The value $\dot{q}(p) \in \mathbb{R}^m$ at which this extremum then occurs is determined by the equations

$$(4) \qquad p_k = \frac{\partial f}{\partial \dot{q}^k} \quad (k = 1, \ldots, m)$$

The condition that an extremum exist is therefore that the range of grad f coincide with $\mathbb{R}^{*m}$, i.e. that $\{(\partial f/\partial \dot{q}^1, \ldots, \partial f/\partial \dot{q}^m) | \dot{q} \in \mathbb{R}^m\} = \mathbb{R}^{*m}$.

The *Legendre transform* f^* of f is then defined as

$$(5) \qquad f^* : p \in \mathbb{R}^{*m} \mapsto F(p, \dot{q}(p))$$

where F is defined by (2) and $\dot{q}(p)$ is determined by (4).

The first step in the passage from Lagrangean mechanics to Hamiltonian mechanics is thus to check that, for each $(q,t) \in M \times \mathbb{R}$, the function

$$(6) \qquad f_{(q,t)} : \dot{q} \in \mathbb{R}^m \mapsto L(q, \dot{q}, t) \in \mathbb{R}$$

satisfies the conditions we just imposed, where L is the Lagrange function for the system considered. When this is the case, one defines the *Hamilton function* by:

$$(7) \qquad H : (q, p, t) \in T^*M \times \mathbb{R} \mapsto f^*_{(q,t)}(p) \in \mathbb{R}$$

For instance, in the simplest of all cases, where

$$(8a) \qquad L(x, \dot{x}, t) = \frac{1}{2} m \dot{x}^2 - V(x)$$

we have:

$$(8b) \qquad \frac{\partial^2 L}{\partial \dot{x}^j \partial \dot{x}^k} = m\, \delta_{jk} \quad \text{and} \quad p_k = m \dot{x}^k$$

and thus

$$(8c) \qquad H(x, p, t) = \frac{1}{2m} p^2 + V(x)$$

i.e. the Hamiltonian function coincides with the total energy of the system.

We next note that the Legendre transform is involutive; in particular, if f is convex, so is f^*, and we have $(\partial f^*/\partial p_k) = \dot{q}^k$ with $k = 1, \ldots, m$; we can, therefore, iterate the Legendre transformation, i.e. define $(f^*)^*$, and verify that $(f^*)^* = f$. Consequently, if the Hamilton function $H : (q, p, t) \in T^*M \times \mathbb{R} \mapsto H(q, p, t) \in \mathbb{R}$ is well-defined as the Legendre transform of the Lagrange function $L : (q, \dot{q}, t) \in TM \times \mathbb{R} \mapsto L(q, \dot{q}, t) \in \mathbb{R}$, then the passage from L to H does not involve any loss of information: we can recover L from H by taking the Legendre transform of H.

The last step in the passage from the Lagrangean formulation to the Hamiltonian formulation is to note that the differential of H can be written in two ways. Firstly, we have the tautology:

$$(9) \qquad dH = \frac{\partial H}{\partial q^k} dq^k + \frac{\partial H}{\partial p_k} dp_k + \frac{\partial H}{\partial t} dt \, .$$

Secondly, upon using the explicit definition of H, namely via (7), (5) and (2):

$$H(q,p,t) = p\dot{q}(p) - L(q,\dot{q}(p),t) \tag{10}$$

and the defining condition (4), i.e. the Poisson definition

$$p_k = \frac{\partial L}{\partial \dot{q}^k} \tag{11}$$

of the conjugate invariable p_k to q^k (see 1.2.25), we obtain:

$$dH = -\frac{\partial L}{\partial q^k} dq^k + \dot{q}^k \, dp_k - \frac{\partial L}{\partial t} dt \tag{12}$$

Hence, the comparison of (9) and (12) gives:

$$\frac{\partial H}{\partial q^k} = -\frac{\partial L}{\partial q^k} \; ; \; \frac{\partial H}{\partial p_k} = \dot{q}^k \; ; \; \frac{\partial H}{\partial t} = -\frac{\partial L}{\partial t} \tag{13}$$

Consequently, the Euler-Lagrange system of m second-order differential equations

$$\frac{d}{dt}\left(\frac{\partial L}{\partial \dot{q}^k}\right) - \frac{\partial L}{\partial q^k} = 0 \tag{14}$$

is equivalent, upon using again the definition (11), to the system of $2m$ first-order differential equations

$$\dot{q}^k = \frac{\partial H}{\partial p^k} \; ; \; \dot{p}_k = -\frac{\partial H}{\partial q^k} \tag{15}$$

These are the *Hamilton equations of motion.* Together with (9), these equations imply

$$\frac{d}{dt} H = \frac{\partial H}{\partial t} \tag{16}$$

and thus, upon using the third of the identities (13):

$$\frac{d}{dt} H = 0 \quad \text{if and only if} \quad \frac{\partial L}{\partial t} = 0 \tag{17}$$

which we already knew from (1.2.27). This remark closes what we need to say here on the equivalence between the Lagrangean and Hamiltonian formulations of mechanics.

We now want to pin down the mathematical structures that appear in the study of *autonomous Hamiltonian systems*. In the approach we followed so far, these systems are characterized by: (i) a state space, called the *phase space* of the system, which was identified as the cotangent bundle T^*M of a smooth m-dimensional manifold M; (ii) a smooth function $H : (q,p) \in T^*M \to H(p,q) \in \mathbb{R}$, referred to as the *Hamiltonian* of the system; and (iii) the *Hamiltonian equations of motion* (15).

We first remark that for every smooth function $f : (q,p) \in T^*M \to f(q,p) \in \mathbb{R}$, we can rewrite

$$\frac{d}{dt} f(q,p) = \sum_{k=1}^{m} \left(\frac{\partial f}{\partial q^k}(q,p)\,\dot{q}^k + \frac{\partial f}{\partial p_k}(q,p)\,\dot{p}_k \right) \tag{18}$$

in two different ways, upon taking into account the equations of motion (15). The first way is

$$\dot{f} = \varsigma_H\, f \tag{19a}$$

where the vector field ς_H is defined as

$$\varsigma_H = \sum_{k=1}^{m} \left(\frac{\partial H}{\partial p_k}\, \partial_{q^k} - \frac{\partial H}{\partial q^k}\, \partial_{p_k} \right) \tag{19b}$$

while the second way to rewrite (18) is

$$\dot{f} = \{f, H\} \tag{20a}$$

where the *Poisson bracket* $\{f,g\}$ between any two smooth functions f and g from T^*M to $\mathbb{R}$ is defined as

$$\{f,g\} = \sum_{k=1}^{m} \left(\frac{\partial f}{\partial q^k}\frac{\partial g}{\partial p_k} - \frac{\partial f}{\partial p_k}\frac{\partial g}{\partial q^k} \right) \tag{20b}$$

This Poisson bracket satisfies the following characteristic properties, where $f, g, h, \ldots$ are taken to run over $C^\infty(T^*M, \mathbb{R})$ and $\lambda, \mu, \ldots \in \mathbb{R}$. It is *smooth*, i.e.

$$\{.\,,.\} : C^\infty(T^*M,\mathbb{R}) \times C^\infty(T^*M,\mathbb{R}) \to C^\infty(T^*M,\mathbb{R}); \tag{21a}$$

it is *skew-symmetric*, i.e.

$$\{f,g\} + \{g,f\} = 0\,; \tag{21b}$$

it is *bilinear*, i.e.

$$\{\lambda f + \mu g, h\} = \lambda\{f,h\} + \mu\{g,h\} \tag{21c}$$

and similarly in its second argument, because of (21b); it is a *derivation in both of its arguments*, i.e.

(21d) $$\{f, gh\} = \{f, g\}h + g\{f, h\},$$

and similarly, again because of (21b), when the roles of the first and second arguments of the Poisson bracket are interchanged; finally, the Poisson bracket satisfies the *Jacobi identity*:

(21e) $$\{f, \{g, h\}\} + \{g, \{h, f\}\} + \{h, \{f, g\}\} = 0 .$$

A smooth manifold $\mathcal{M}$, equipped with an operation $\{.,.\}$ that satisfies the axioms (21 a–e), is said to be a *Poisson manifold*; this structure has been explored, in the context of a general theory of Hamiltonian mechanics, by Souriau (1970) and Lichnerowicz (1977); see also Marle (1982) and references mentioned therein. The cotangent bundle T^*M of a smooth manifold, when equipped with the operation (20b) is a Poisson manifold.

T^*M moreover possesses a more stringent structure: it is a *symplectic manifold*; this means that T^*M is a $2m$-dimensional smooth manifold that comes equipped with a *symplectic form* ω, i.e. a closed, non-degenerate 2-form, namely:

(22) $$\omega = \sum_{k=1}^{m} dp_k \wedge dq^k$$

The symplectic form ω on T^*M, which we just wrote in its Darboux diagonalization, is defined in a canonical (i.e. explicitly coordinate-free) manner as follows. Let $\pi : T^*M \to M$ be the natural projection from the fiber bundle T^*M onto its base manifold M, and θ be the 1-form, defined at every $x \in T^*M$ by:

(23) $$\theta : \varsigma \in T_x(T^*M) \mapsto x(\pi_*\varsigma) \in \mathbb{R} .$$

The symplectic form ω is then defined by:

(24) $$\omega \equiv d\theta$$

Note that in a coordinate system where ω can be written as in (22), θ is expressed as

(25) $$\theta = \sum_{k=1}^{m} p_k \, dq^k .$$

For a modern presentation of symplectic techniques, starting with their original applications to optics, see Guillemin & Sternberg (1984). In general, a symplectic form ω on a smooth $2m$-dimensional manifold $\mathcal{M}$ (e.g. T^*M)

associates, to every smooth function $f : \mathcal{M} \to \mathbb{R}$, a smooth vector field ς_f on $\mathcal{M}$, defined uniquely (recall that ω is non-degenerate, by definition) by:

(26) $$\varsigma_f \rfloor \omega \equiv -df$$

where $\rfloor$ denotes the usual contraction operation (see e.g. 4.2.100); note that in the case of T^*M, where ω is defined by (24), (26) gives a canonical definition of the vector field ς_H associated to a given Hamiltonian H; compare with (19). Similarly (20) is a reflection of the fact that a symplectic manifold $(\mathcal{M}, \omega)$ is a Poisson manifold, with Poisson bracket $\{.\,,\,.\}$ defined canonically by

(27) $$\{f, g\} = -\omega(\varsigma_f, \varsigma_g)\,.$$

Note that the underlying relation

(28) $$\varsigma_f\, g = \{g, f\}$$

can be taken as an alternative definition of ς_f, involving only the Poisson structure of the symplectic manifold $(\mathcal{M}, \omega)$.

In closing this section, we want to single out for comment one of the reasons why autonomous Hamiltonian systems are sometimes said to be *conservative.*

The $2m$-dimensional symplectic manifold $(\mathcal{M}, \omega)$ comes naturally equipped with a *volume-element*, the $2m$-form $\Omega \equiv \omega^m$. For any smooth vector field ς on $\mathcal{M}$, $d(\varsigma \rfloor \Omega)$ is again a $2m$-form; it can thus differ from Ω at most by a smooth, real-valued function on $\mathcal{M}$, called the *divergence* of ς, denoted divς, and defined by:

(29) $$d(\varsigma \rfloor \Omega) \equiv (\text{div}\,\varsigma)\,\Omega$$

Since $d\Omega = 0$, the LHS of (29) can be rewritten as

(30) $$d(\varsigma \rfloor \Omega) = L_\varsigma\, \Omega$$

where L_ς is the Lie derivative along the vector field ς. From (26) and (29), we have for every smooth $f : \mathcal{M} \to \mathbb{R}$

(31) $$\text{div}\,\varsigma_f = 0$$

Together with the interpretation (30) of (29), this implies that, under the flow μ^t generated on $\mathcal{M}$ by ς_H (i.e. under the flow obtained by integrating the Hamilton equations of motion), the volume $\Omega(D)$ of any domain $D \subset \mathcal{M}$ remains constant; this conservation law is known as the *Liouville theorem.*

One of the simplest applications of this result is as follows.

Let $V : q \in \mathbb{R}^2 \mapsto V(q) \in \mathbb{R}^+$ be a smooth function such that $V(q) \to \infty$ when $\| q \| \to \infty$; let μ^t be the flow generated on $\mathbb{R}^2 \times \mathbb{R}^{*2}$ by the Hamiltonian

(32) $$H : (q, p) \in \mathbb{R} \times \mathbb{R}^{*2} \mapsto \frac{1}{2}p^2 + V(q) \in \mathbb{R}^+$$

For any $E > 0$, let $D_E = \{(q,p) \in \mathbb{R}^2 \times \mathbb{R}^{*2} \,|\, H(q,p) < E\}$, and let Ω be the ordinary Euclidean volume element in $\mathbb{R}^2 \times \mathbb{R}^{*2}$. While the details of the motion in the finite volume domain D_E are not known, a general qualitative result can nevertheless be proven, as we shall presently see.

We first generalize the above special example and consider a smooth manifold $\mathcal{M}$, on which are given a volume element Ω and a flow $\{\mu^t \,|\, t \in \mathbb{R}\}$ that preserves Ω. We further suppose that there exists an open subset $D \subset \mathcal{M}$ such that $\mu^t(D) \subseteq D \;\; \forall t \in \mathbb{R}$, and its volume $\Omega(D)$ is finite. Let now $U \subset D$ be open, and choose $t > 0$. Since (i) U, $\mu^t(U), \ldots, \mu^{nt}(U), \ldots$ all have the same volume, (ii) all belong to D, and (iii) $\Omega(D) < \infty$, we conclude that there exist $j, k \in \mathbb{Z}^+$ with $j < k$, such that $\mu^{jt}(U) \cap \mu^{kt}(U) \neq \emptyset$; consequently, there exists $n \in \mathbb{Z}^+$, e.g. $n = k - j$, such that $\mu^{nt}(U) \cap U \neq \emptyset$; hence there exists $x \in U$ such that $\mu^{nt}(x) \in U$. This result, which properly belongs to ergodic theory, is known as the *Poincaré recurrence theorem.* Together with Liouville's theorem, it rules out the existence of asymptotically stable equilibrium points (or limit cycles) in the phase space of any autonomous Hamiltonian system.

CHAPTER 2. THERMODYNAMICS AND STATISTICAL MECHANICS

SYNOPSIS

In the view of a mathematical physicist, Gibbs (1889): "Clausius...first memoir on thermodynamics [published in 1850] ... marks an epoch in the history of physics ... the science of thermodynamics came into existence." The motivation of thermodynamics is to produce a description of processes during which heat is produced or absorbed. Concomitant to, but distinct from, this programme is the definition of thermodynamical equilibrium states, i.e. thermostatics. Once this distinction is recognized—and it took many years to reach even this point—one may say that, while a mathematically satisfactory formulation of a variational principle for thermostatics was achieved by Gibbs (1878), several conceptual clarifications, pertinent to the study of thermodynamics proper, were developed in the second half of this Century.

The first problem that the science of heat had to cope with was to realize that heat in its flow is not some sort of "caloric" fluid, but rather a form of energy. In achieving this, it simultaneously produced an absolute meaning for temperature, and a new fundamental concept, which Gibbs credits to Clausius, the entropy.

In his own first work on the subject, Gibbs (1873) starts out with a formula which we write:

$$dU = -pdV + TdS \tag{1a}$$

In the simplest case, this formula can be interpreted as follows. For a homogeneous fluid the internal energy U is a function of two variables, the volume V and the entropy S, such that

$$\frac{\partial U}{\partial V} = -p \quad \text{and} \quad \frac{\partial U}{\partial S} = T \tag{1b}$$

where p is the pressure and T the temperature.

This formula can be generalized in many directions. Firstly, we can rewrite (1a) in the form of a process, letting the variables be functions of time, and write:

$$\dot{U} = W + Q \tag{2}$$

where W is interpreted as the "working" done on the system, and Q as the "heating" to which this system is submitted. This in fact is in line with how the subject both had started out and is developed today; the problem then is to formulate axioms on the kind of processes which are to be allowed. Another way in which formula (1) can be generalized is to dream—see, for instance, the work of Duhem and Hadamard (1898–1900)—that one could treat in the same way inhomogeneous fluids by simply introducing a specific energy ϵ, a specific volume ν, a specific entropy η, the pressure p and the temperature T at each point of the fluid in such a manner that (1b) be replaced by:

$$\begin{gathered} \epsilon = \epsilon(\nu,\eta) \\ p = -\frac{\partial \epsilon}{\partial \nu} \quad \text{and} \quad T = \frac{\partial \epsilon}{\partial \eta} \end{gathered} \tag{3}$$

There, however, life is not that simple, and much work is still necessary to reach a concensus on the sense in which these definitions are useful far from equilibrium.

Statistical mechanics sets out with the task of relating the macroscopic theory which thermodynamics is—especially when it deals with continuous fluids or rigid bodies—to microscopic models of matter; hence the name of "molecular science" it still had at the turn of the century. Much of the ideas of statistical mechanics can be found in the so-called kinetic theory of gases. While D. Bernoulli (1738) proposed a microscopic model which was essentially mechanistic, it belonged to Maxwell (1860) and to Boltzmann (1871) to realize that statistical mechanics should be based on two premises: a mechanistic one (which is already in Bernoulli) and a probabilistic one (which is new with them). These led to a sharpening of the concepts of temperature and entropy. Temperature is interpreted as the average kinetic energy of the molecules; Maxwell indeed derived a simple formula for the equilibrium distribution of the velocities, namely:

$$f(\mathrm{v}) = N\left(\frac{m}{2\pi kT}\right)^{3/2} \exp\left(-\frac{1}{2}m|\mathrm{v}|^2/kT\right) \tag{4}$$

A probabilistic interpretation of the entropy could then be given in terms of Boltzmann's "eta-function", namely:

$$S(t) = -\int d^3v\, f(\mathrm{v},t)\, ln\, f(\mathrm{v},t) \tag{5}$$

a formula later proven, by Shannon (1948) and Khinchin (1957), to be the only reasonable measure of the information contained in the distribution f. This classical probabilistic interpretation of the entropy played a central role in Planck's derivation of the black body radiation formula, the very formula which led to quantum theory. This historical milestone strongly affected the choice of the materials to be presented in this chapter.

In spite of the successes of statistical mechanics, there were still physicists, at the turn of the Twentieth Century, who doubted the "reality" of the objects dealt with in the emerging microphysics. The investigations on Brownian motion, by Einstein (1905 BM), provided independent evidence that these concepts had more than a heuristic value; it also motivated a whole new development in mathematics, namely the study of stochastic processes which has made a surprising reentry in modern quantum theory.

Section 1. TEMPERATURE AND HEAT

The mathematician interested in making the intuitive notions of temperature and heat precise enough to allow for quantitative statements may want to examine first the attempts presented in the *Théorie analytique de la chaleur* which Fourier published in 1822 from the vantage point of investigations he had started some fifteen years earlier; the comments and explanatory notes with which Darboux enriched the 1887 edition—part of the collected *Oeuvres de Fourier*—indicate the seriousness in which this work was then held in the mathematical physics community. In addition to the original exposition of the harmonic analysis that now carries Fourier's name, the reader will find there its motivation, namely the study of the solutions of the parabolic equation

$$(1) \qquad (\Delta - \kappa\partial_t)\Psi(\mathbf{x}, t) = 0$$

known today as the *heat equation*. Fourier studied the solutions of this equation under a variety of boundary conditions, e.g.

$$(2) \qquad \partial_t\Psi = 0 \quad \text{or} \quad \mathbf{n} \cdot \text{grad}\ \Psi + \lambda\Psi = 0$$

imposed for $\mathbf{x} \in \partial\Omega$, the boundary of the region Ω where (1) is supposed to hold. Note that in the static case where $\partial_t\Psi = 0$ holds through Ω, (1) reduces to the Laplace equation: it is the restriction of this equation to two dimensions that provides the elementary textbooks on complex analysis with an abundance of "applications". Equation (1) can be used to describe several diffusion phenomena; Fourier proposed it as the mathematization of "the question of propagation of heat [which] consists in the determination of the temperature [Ψ] at each point [$\mathbf{x}$] of a body and each instant [t] ... " Cranked-up versions of this equation can be considered as well, e.g.

$$(3) \qquad [\nabla(\kappa^{-1}\nabla) - \partial_t]\Psi = 0$$

but our purpose here is to analyze what are the empirical meanings of the terms "temperature" and "heat", first as Fourier understood them.

We all have a crude sensory feeling for "cold" and "warm", or rather for "cooler" and "warmer"; we further learned from early everyday experience, e.g. with tea kettles and the like, that when a warmer body, say A, is brought into contact with a cooler body, say B, the warmer will cool down and the cooler will warm up ... until a point where no observable variation occurs anymore; we then say that A is in "thermal equilibrium" with B, and B with

A. We further know that if a body A is in thermal equilibrium with a body B, and B is in thermal equilibrium with a body C, then A is in thermal equilibrium with C. These reflexive and transitive properties are sometimes referred to by the glorified name of "zeroth law of thermodynamics"; the important point is that it leads to the idea of attributing to each body in the thermal equilibrium a *temperature* Θ, chosen from a totally ordered continuum, say $\mathbb{R}$.

This empirical notion of temperature has two major drawbacks. It is limited, firstly, to equilibrium situations: temperature is a parameter of the state of a body only if this body is in equilibrium; this begs for an extension to the situations that Fourier's equation (1) is supposed to describe: some concept of local temperature will have to be delineated. Secondly, any monotonically strictly increasing function T of Θ could serve as well as Θ in the above discussion. However, for (1) to hold with Ψ identified with Θ and with T, one needs T to be a linear function of Θ; in everyday parlance, Fourier's heat equation holds as well if one measures Θ in $°F$ or in $°C$, but one needs a linear temperature scale.

Fourier chose his temperature scale in accordance with the conventional wisdom of the time; first "one determines two fixed temperatures, e.g. the temperature of melting ice, which we call 0, and the temperature of boiling water, which we call 1; we suppose that the boiling of water is taking place at an atmospheric pressure of 760 mm of Hg, the mercury of the barometer being at temperature 0." Note that this empirical definition involves at least a crude knowledge of a class of phenomena, known today under the general name of phase transitions; some essential aspects were already appreciated by the end of the 17th century. For instance, it was known that, as long as there is ice in a bucket of liquid water, the temperature of the water remains the same; and as long as water is boiling into vapour, so does again its temperature remain constant, provided that the pressure is kept constant. The idea to use these phenomena to calibrate thermometers antedates Fourier by at least a century, as evidenced by the memoirs published in 1702 and 1724 by Amontons in France, and Fahrenheit in England.

Two points can thus be fixed on the temperature scale, but one still has to interpolate the points in between, and to extrapolate. For this purpose, Fourier proposed to use heat, however raising the point that this requires caution: "One can at best form hypotheses on the nature of heat, but the knowledge of the mathematical laws to which its effects are subject is independent of any such assumption." He then described how "one measures different quantities of heat by determining how many times they contain a quantity one has chosen as a unit." To choose this unit, Fourier appealed again to a knowledge of phase transitions, and defined it as "the quantity of heat one must provide to a standard mass of ice (1 kg) at zero temperature to melt it completely into liquid water." With this primitive, but quantitative notion of heat, Fourier proceeded: "if a body of given nature [i.e. chemical composition] and weight [i.e. mass] occupies a volume V at temperature 0, it will occupy a larger volume

$V + \Delta$ when it will have acquired the temperature 1 ... if instead of adding the quantity C_0 [of heat, to pass from $\Theta = 0$ to $\Theta = 1$] one adds a quantity $z\,C_0$, with z positive or negative, the new volume will be $V+\delta$ ... experiments made it known that ... in general the value of δ is $z\Delta$ ... the ratio of the two quantities of heat one has added, $z\,C_0$ and C_0, which is also the ratio of the increases in volume, δ and Δ, is what one calls the temperature." Fourier was aware that this empirical definition was subject to serious consistency problems as he notes: "this property [i.e. the linearity $\delta/\Delta = z\,C_0/C_0$] is exact only in the case where the bodies considered are at temperatures far from those at which one observes changes of their state [i.e. phase transitions]." It is, nevertheless, on the basis of these dubious choices of definitions and units that Fourier built his theory; some testimony to the fact that he was not too far off was found in the agreement he obtained between the conclusions he drew from equation (1) and the corresponding experiments.

It must, nonetheless, be recognized that Fourier's empirical definitions—of temperature and of heat—were fraught with all kinds of hidden assumptions on the constancy, over different values of Θ, of several material coefficients. Contemporary investigtors were, in fact, already at work to clear up this situation. The key was to find some ideal class of materials that could serve as universal references; once these were recognized, the basic concepts could be isolated, and the theory could move more freely away from these ideal(ized) materials and gain the desired generality, in a case—typical of much of the development of thermodynamics—of the process by which, as Gillispie (1960) puts it: "we learn about the real by contemplation of the impossible."

Historically, the study of gases provided the necessary focalization of thought. The first result was obtained when Boyle (and Mariotte, some sixteen years later) noticed in 1660 that gases under "ordinary conditions" (i.e. away from condensation) exhibit a remarkable regularity, namely that if we denote by V the volume of the vessel in which such a gas (in equilibrium) is contained, and by p the pressure the gas exerts on the walls of this container, then the product $p\,V$ is constant—provided the temperature of the gas is kept constant. We write this in the form of the *Boyle-Mariotte law*

$$pV = f(\Theta) \tag{4}$$

where Θ is any empirical measure of the temperature. The form of the function f depends evidently on the empirical temperature scale one chooses. With a temperature scale akin to that still used by Fourier, the form of f was determined by the investigations Gay-Lussac carried in 1802 on the isobaric (i.e. $p =$const.) expansion of a variety of gases: "oxygen, hydrogen, nitrogen, nitrous oxide, ammonia, muriatic acid, sulphurous acid, ..., carbonic acid, ..., ether, ... " (at the time Gay-Lussac was employed as a demonstrator of chemistry at the Ecole Polytechnique in Paris, and was "guided", as he writes, in his investigations by the advice of "Citizen Laplace" who, as we shall see below, put to good use the information he collected first-hand from

Gay-Lussac, then and in the subsequent years). Gay-Lussac proposed two conclusions; the first is that "all gases in general expand equally between the same [temperatures], provided that they are all brought under the same conditions"; the second of his conclusions was that "the increase in volume received by each of them between the temperature of melting ice and that of boiling water is equal to ... [the same fraction] of the original volume of the gas". Specifically, if we denote by p the pressure of the gas, by V the volume it occupies at equilibrium temperature Θ (which Gay-Lussac measured in $°C$), by m the mass of the gas (measured, say in grammes) and by m_0 its molecular mass (i.e. $32\,gr$ for diatomic oxygen, $2\,gr$ for diatomic hydrogen, etc.), then Gay-Lussac's claims amount to saying that there exist two universal constants R and T_0 such that, for all the gases he experimented with:

$$p(V - V_0) = \frac{m}{m_0} R(\Theta - \Theta_0) \tag{5}$$

$$\frac{V_1 - V_0}{V_0} = \frac{1}{T_0}(\Theta_1 - \Theta_0) \tag{6}$$

These two relations can be rewritten in the form of the *Gay-Lussac law*

$$pV = (\frac{m}{m_0})RT \tag{7}$$

where T is measured on a "universal" temperature scale, shared by all ideal gases; this scale has an absolute zero that is theoretically approached when the volume V approaches zero, with p and m fixed, but arbitrary. This is named the Kelvin temperature scale, in honour of Thomson who later recognized (see below) its truly universal meaning: it is logically (even if not historically) independent of the actual laboratory existence of ideal gases that satisfy (7) over the whole range of parameters p, V, and T. The Celcius scale can then be defined (see 5-7) by $\Theta = T + (\Theta_0 - T_0)$ with $\Theta = \Theta_0 = 0\,°C$ when $T = T_0 = 273.15\,°K$, a value that could in principle be determined experimentally from (6): $(V_1 - V_0)/V_0 = (T_1 - T_0)/T_0$.

The next question is: what is heat? It is easy to say today that heat is a form of energy. The theoretical insight that energy is a universally conserved quantity came about in the middle of the 19th century, and it is a textbook case for the study of simultaneous discoveries: Kuhn (1959) enumerates more than a dozen original contributors, all essentially contemporaries, and some definitely independent of the others.

To appreciate what is involved, the reader may first want to recall how the potential energy was introduced in mechanics: with the obvious changes of notation often required when one passes from one science to another, we remember that we started (see 1.2.3) with the definition of the kinetic energy K; we then realized that for Newtonian systems where forces derive from a potential Φ, the quantity $E = K + \Phi$ is conserved in the course of the motion (see 1.2.4-8); we thus called E the total energy of the system, and Φ its potential energy.

By the middle of the 19th century, it was known that heat can be absorbed or produced in chemical reactions; that some chemical reactions generate electricity (e.g. the Volta pile) and conversely that electricity could be used to chemically decompose water in its elements, hydrogen and oxygen; that electric and magnetic phenomena bear on one another (see Section 3.1) and can be used to produce mechanical energy (Joule in fact experimented on electromagnetic engines before doing his fundamental work on the mechanical equivalent of heat); it was also evidently known that mechanical energy could produce heat (by friction) and conversely that heat could be used to produce mechanical energy (the steam engine explicitly motivated the pioneering work of Carnot, which Clapeyron reformulated in a mathematically tighter form; it seems that it is largely through this reformulation that Clausius knew about Carnot's work). As Kuhn (1959) points out, however, being aware of the convertibility of energies is not yet quite the same as being able to articulate the bold postulate that energy is universally conserved. This further step was taken, in the late 1840's, by Mayer, Clausius and Helmholtz: very much in the same way as Lavoisier had rationalized his knowledge of chemical reactions by postulating that matter is conserved, they postulated that energy is conserved, in order to put on a firm footing their rationalization of the processes in which heat, now a form of energy, is exchanged. The principle of conservation of energy proposed by these pioneers, and known as the first law of thermodynamics is usually written in the form

$$(8) \qquad \Delta U = \Delta W + \Delta Q$$

This is to be interpreted as follows. At this level, a system is characterized by a quadruple $(M, P, A, \Delta W)$. M is called the space of all equilibrium states of the system. P is called the space of all processes by which any state $\mu_0 \in M$ can be transformed into any other state $\mu_1 \in M$; to emphasize that several processes may exist, that transform the same initial state μ_0 into the same final state μ_1, we denote a process by the symbol $\mu_0\pi\mu_1$; two processes $\mu_0\pi\mu_1$ and $\mu'_0\pi'_1\mu'_1$ can be composed if $\mu_1 = \mu'_0$, and we denote the composite process by the symbol $\mu_0(\pi\mu'_0\pi')\mu'_1$. A is a subset of P, closed under composition; the elements $\mu_0\alpha\mu_1$ of A are called adiabatic processes. It is assumed that A connects M in the following sense: given any μ and μ' in M, there exists at least one adiabatic process $\mu_0\alpha\mu_1$ with either $(\mu_0 = \mu, \mu_1 = \mu')$ or $(\mu_0 = \mu', \mu_1 = \mu)$, i.e.

$$(9) \qquad [\{\mu\pi\mu'\} \bigcup \{\mu'\pi'\mu\}] \bigcap A \neq \emptyset$$

Finally, ΔW is a function from P to $\mathbb{R}$; one refers to $\Delta W[\mu_0\pi\mu_1]$, as the *(mechanical) work* done on the system in the process $\mu_0\pi\mu_1$; it is assumed that whenever two processes $\mu_0\pi\mu_1$ and $\mu'_0\pi'\mu'_1$ can be composed (i.e. whenever

$\mu_1 = \mu_0')$

(10) $$\Delta W[\mu_0(\pi\,\mu_0'\,\pi')\mu_1'] = \Delta W[\mu_0\,\pi\mu_0'] + \Delta W[\mu_0'\pi'\mu_1'].$$

It is further assumed that if μ_0 and μ_1 are the initial and final states of two adiabatic processes $\mu_0\alpha\mu_1$ and $\mu_0\alpha'\mu_1$, one has

(11a) $$\Delta W[\mu_0\alpha\mu_1] = \Delta W[\mu_0\alpha'\mu_1];$$

and that if π_0 denotes the trivial process consisting in doing nothing at all, then for all $\mu \in M$

(11b) $$\mu\pi_0\mu \in A \quad \text{and} \quad \Delta W[\mu\pi_0\mu] = 0\,.$$

Consequently, one has for every cyclic adiabatic process $\mu\alpha\mu$

(12a) $$\Delta W[\mu\alpha\mu] = 0$$

and thus, whenever $\mu_0\alpha\mu_1 \in A$ and $\mu_1\alpha'\mu_0 \in A$

(12b) $$\Delta W(\mu_0\alpha\mu_1) = -\Delta W(\mu_1\alpha'\mu_0)$$

Since A connects M (in the sense of (9)), these relations allow one to define unambiguously for every pair $(\mu, \mu') \in M \times M$:

(13a) $$\Delta U(\mu, \mu') = \begin{cases} \Delta W[\mu\alpha\mu'] & \text{if} \quad \mu\alpha\mu' \in A \\ -\Delta W[\mu'\alpha'\mu] & \text{if} \quad \mu'\alpha'\mu \in A \end{cases}$$

The antisymmetric function

(13b) $$\Delta U : (\mu, \mu') \in M \times M \mapsto \Delta U(\mu, \mu') \in \mathbb{R}$$

is called the increment of *internal energy* acquired by the system in any adiabatic process $\mu\alpha\mu'$; note that it depends only on μ and μ'. Upon choosing an arbitrary reference state μ_0, one can define a function

(14) $$U_{\mu_0} : \mu \in M \mapsto \Delta U(\mu_0, \mu) \in \mathbb{R}$$

called the *internal energy* of the system with respect to μ_0. Clearly,

(15a) $$\Delta U(\mu, \mu') = U_{\mu_0}(\mu') - U_{\mu_0}(\mu) \quad \forall\, (\mu, \mu') \in M \times M$$

and, if another reference state, say μ_1, is chosen, one has:

(15b) $$U_{\mu_1}(\mu) = U_{\mu_0}(\mu) + U_{\mu_1}(\mu_0) \quad \forall\, \mu \in M$$

Hence, one can define, up to a constant, a unique real-valued function on M, which one can call the internal energy of the system. With these definitions, formula (8) can now be understood to read

$$\Delta Q[\mu_0 \pi \mu_1] \equiv \Delta U(\mu_0, \mu_1) - \Delta W(\mu_0 \pi \mu_1) \tag{16}$$

In this perspective, the *first law of thermodynamics* therefore appears as the definition of the function $\Delta Q : P \to \mathbb{R}$, and $\Delta Q[\mu_0 \pi \mu_1]$ is interpreted as the *heat* absorbed by the system in the process $\mu_0 \pi \mu_1$; we have thus

$$\Delta Q[\mu_0 \alpha \mu_1] = 0 \quad \forall \ \mu_0 \alpha \mu_1 \in A \tag{17}$$

which is the reason why the elements of A were called *adiabatic processes.*

The second law of thermodynamics requires more structure, and will thus be stated only after an illustration is provided; only then will a thermodynamical system be fully defined.

For our illustration, we come back to the historical example of the ideal gas, the equilibrium states of which are given by the Gay-Lussac law (7). In 1807, Gay-Lussac conducted an experiment which could have been used to determine the internal energy U of such a system ... had the concept been formulated then; as a point of fact, this experiment later played a central role in the very elaboration of the theory, and it was repeated with greater care in 1845 by Joule, and again a few years later by Joule and Thomson. We follow here the account given by Joule. Two copper vessels (recall that copper is a good conductor of heat), one containing a gas (Joule worked with air), the other being empty, were linked by a thin conduct that was initially closed by stopcocks. The assembly was embedded in a calorimeter, that is, a large tank of water equipped with a thermometer. "The water was first thoroughly stirred and its temperature taken ... the stopcocks were then opened ... and the air allowed to pass from the full to the empty [vessel] until equilibrium was established between the two. Lastly, the water was again stirred and the temperature taken ... no change of temperature occurs when the air is allowed to expand in this manner." From the observed equality of the temperatures before and after the expansion, Joule could conclude that the gas has neither produced nor absorbed heat: the free expansion of an ideal gas is an adiabatic process. Joule also noticed that in this process the gas does not "develop mechanical power", i.e. no mechanical work is performed either by or on the gas. From the conservation of energy, we conclude that the internal energy of the gas at equilibrium is the same at the end of the experiment as it was in the beginning. Since no gas is allowed to escape, the ratio $N \equiv m/m_0$ is fixed, and the space M of equilibrium states a priori available to an ideal gas is the two-dimensional manifold $M \subseteq \mathbb{R}^3$ described by the constitutive equation (7). Joule's free expansion experiment shows that U and T are the same before and after the expansion, while V, and thus p, have changed; consequently the function $U : M \to \mathbb{R}$ depends only on T.

We now introduce a restricted class of processes, for which we not only have that μ_0 and μ_1 belong to M, but for which we can describe the process by a differentiable curve $\mu : t \in I \mapsto \mu_t \in M$. These processes are called *quasi-static* in the physics literature, and we shall keep this nomenclature; note that the free expansion of a gas which we just described is not a quasi-static process.

From Chapter One, we infer that the mechanical work done on a (ideal) gas during a quasi-static process can be described by a 1-form on M, namely

$$\Omega = -p\,dV \tag{18}$$

which we refer to as the *working-form*. With U denoting the internal energy function on M, the differential expression of (16) now reads

$$\eta \equiv dU + p\,dV \tag{19}$$

and defines a 1-form on M, which we refer to as the *heating-form*. Since we learned from Joule's experiment that the free-energy of an ideal gas depends only on T, we can rewrite (19) as

$$\eta = N\,C_v\,dT + p\,dV \tag{20}$$

Upon using (7) we can rewrite (20) as

$$\eta = N\,C_p\,dT - V\,dp \tag{21}$$

with C_p and C_v linked by the Mayer relation

$$C_p = C_v + R \tag{22}$$

The quantities C_v (resp. C_p) are interpreted, from (20) and (21), as the specific heat at constant volume (resp. constant pressure); they are directly accessible to experiments, and were already measured between 1813 and 1822 by Delaroche & Bérard, Clément & Desormes, and Gay-Lussac & Welter. For an ideal gas, these are constants which are understood, today, on the basis of microscopic models; their values are

$$C_v = \frac{3}{2}R \quad \text{and} \quad C_p = \frac{5}{2}R \tag{23}$$

for monoatomic ideal gases, while

$$C_v = \frac{5}{2}R \quad \text{and} \quad C_p = \frac{7}{2}R \tag{24}$$

for diatomic ideal gases where the molecules are supposed to be rigid dumbbells (which they pretty much are for gases like O_2 and N_2 at room temperature).

The ratio

$$\gamma \equiv C_p/C_v \tag{25}$$

is of particular theoretical importance. Indeed, for an adiabatic quasi-static process α [i.e. a quasi-static process for which $\eta(\dot{\alpha}_t) = 0$], we obtain from (20) and (7)

$$pV^\gamma = \text{ constant on } \alpha \tag{26}$$

in contrast with the result obtained directly from (7) for an isothermic quasi-static process μ [i.e. $dT(\dot{\mu}_t) = 0$], namely

$$pV = \text{ constant on } \mu \tag{27}$$

The difference between (26) and (27) allowed Laplace and Poisson to solve, around 1822–1823, a challenge that had frustrated the efforts of an impressive array of mathematical physicists, from Newton to Euler and Lagrange. Sound was indeed recognized early to be a compression wave, solution of the equation

$$(\Delta - \frac{1}{u^2}\,\partial_t{}^2)\Psi(x,t) = 0 \tag{28}$$

where Ψ is the instantaneous displacement of the air along the direction of propagation of the sound; and u, the speed of propagation of the sound, is linked to the compressibility of air by the relation

$$u^2 = (\frac{\partial p}{\partial \rho})\ ; \tag{29}$$

in this formula $\rho = m/V$ is the density of the air and p its pressure. Newton computed (29) from (27), or (7) and he found

$$u_N = (\frac{RT}{m_0})^{\frac{1}{2}} \tag{30}$$

which is at variance with the measured velocity of sound in air by a factor of about 20%. Upon recognizing that air is a poor conductor of heat, Laplace and Poisson proposed, instead, to compute (29) from (26) and (7); in this manner one finds indeed

$$u_{LP} = (\frac{\gamma RT}{m_0})^{\frac{1}{2}} \tag{31}$$

which, together with (24), accounts for the discrepancy of 20% that had for so long marred Newton's formula (30).

We now want to mention four properties of the ideal gases, as they significantly stimulated the development of thermodynamics and, still today, can serve as remarkably accessible illustrations of the conceptual content of the general theory.

The first of these properties is very much in the engineering spirit that motivated Carnot's *Réflexions sur la puissance motrice du feu et sur les machines propres à développer cette puissance* (1824). Consider the cyclic process that an ideal gas undergoes when it passes successively through the following four quasi-static steps: (i) it is expanded isothermally (following 27) at a temperature T_1; (ii) it is then expanded adiabatically (following 26) to reach a temperature $T_0 < T_1$; (iii) it is compressed isothermally (following 27) at the temperature T_0, until it reaches a state from where (iv) it is compressed adiabatically (following 26) to return to its original state. A straightforward application of (26), (27), (18) and (19) shows that one has, with ΔQ_1 (resp. ΔQ_0) denoting the heat absorbed (resp. given up) by the system during step (i) [resp. step (ii)]:

$$\frac{\Delta Q_0}{\Delta Q_1} = \frac{T_0}{T_1} \tag{32}$$

If one is interested, as Carnot indeed was, in the efficiency of this cyclic process as a model of the steam engine, i.e. if one is interested in the ratio between the total work ΔW produced during a complete cycle, and the quantity of useable heat ΔQ_1 one has to provide to the system, (32) reads

$$\frac{\Delta W}{\Delta Q_1} = 1 - \frac{T_0}{T_1} \tag{33}$$

We shall indicate, later on, the use Thomson and Clausius made of these two formulas; the reader's attention should nevertheless be directed to examine the privileged role played now by the absolute temperature scale used to measure T.

The second interesting property evidenced by the ideal gases is that there exists a function of U, V, and N, namely

$$S(U,V,N) \equiv N\,\{C_v\,\ell n(\frac{U}{N}) + R\,\ell n(\frac{V}{N})\} \tag{34}$$

such that, with $N = (m/m_0)$ fixed i.e. with a fixed total mass of the gas,

$$S_N(U,V) \equiv S(U,V,N) \tag{35}$$

satisfies

$$T dS_N = dU + p dV \tag{36}$$

Upon comparing this with (19) one obtains

$$\eta = T\, dS_N \tag{37}$$

Hence $T^{-1}\eta$ is an exact 1-form on the manifold M of the equilibrium state of the ideal gas considered.

The third property of ideal gases, which we want to single out here, consists of three statements: (a) the function S, defined in (34) is homogeneous of degree one; this is an expression of the fact that S, U and V are extensive variables; (b) S is concave; (c) U, when expressed in terms of S, V and N, is a convex function; the last two of these statements can be interpreted as stability requirements. Notice further that, with N fixed—we thus drop N to lighten the notation, and simply write $S = S(U,V)$—we have

$$\frac{\partial S}{\partial U} = \frac{1}{T} \quad \text{and} \quad \frac{\partial S}{\partial V} = \frac{p}{T} \tag{38a}$$

or, upon writing $U = U(S,V)$:

$$\frac{\partial U}{\partial S} = T \quad \text{and} \quad \frac{\partial U}{\partial V} = -p; \tag{38b}$$

furthermore,

$$N C_v = \frac{T}{\partial^2 U/\partial S^2} \tag{38c}$$

Note that, since dU is an exact 1-form, the functions T and p of the variables S and V, that appear in (38b), satisfy

$$\frac{\partial T}{\partial V} = -\frac{\partial p}{\partial S} \tag{39a}$$

which is the first of the so-called four Maxwell thermodynamical relations. The other three are obtained as follows. Since U is convex, its Legendre transforms can be defined by the procedure described in the beginning of Section 1.3. With $T = \partial U/\partial S$ now interpreted as the conjugate variable to S, one thus obtains the *Helmholtz function* $F \equiv U - TS$ of the variables T and V; we have $dF = -SdT - p\,dV$, where the functions S and p of the variables T and V satisfy the second Maxwell relation

$$\frac{\partial S}{\partial V} = \frac{\partial p}{\partial T} \tag{39b}$$

Similarly, with $-p = \partial U/\partial V$ conjugate variable to V, one obtains the *enthalpy* $H \equiv U + pV$, a function of the variables S and p, with $dH = TdS + V\,dp$, where the functions T and V of the variables S and p thus satisfy the third Maxwell relation

$$\frac{\partial T}{\partial p} = \frac{\partial V}{\partial S} \tag{39c}$$

Finally, the Legendre transform of U with respect to both variables S and V is the *Gibbs function* $G \equiv U - TS + pV$ of the variables T and p; we have then $dG = -SdT + Vdp$ where the functions S and V of the variables T and p satisfy the fourth Maxwell relation

$$\frac{\partial S}{\partial p} = -\frac{\partial V}{\partial T} \,. \tag{39d}$$

Finally, the fourth property of the ideal gases, to be mentioned here for later use, is a straightforward comment on (26), or equivalently (34)–(37): given $\mu_0 \in M$, let

$$\begin{aligned} M_0 &\equiv \{\mu \in M | S(\mu) = S(\mu_0)\} \\ M_+ &\equiv \{\mu \in M | S(\mu) > S(\mu_0)\} \\ M_- &\equiv \{\mu \in M | S(\mu) < S(\mu_0)\} \end{aligned} \tag{40}$$

Then, for each $(\mu, \mu_+) \in M_0 \times M_+$ (resp. $(\mu, \mu_-) \in M_0 \times M_-$) there exists an adiabatic process $\mu\alpha\mu_+$ (resp. $\mu_-\alpha'\mu$), specifically the composition of a quasi-static adiabatic process and an adiabatic free expansion of the type described by Joule (see above); however, it is a fact of experience that no adiabatic process exists, which would be of the form $\mu_+\alpha''\mu$ or $\mu\alpha''\mu_-$.

The asymmetry just mentioned hints at a form of the irreversibility that is the essence of the somewhat Victorian prohibitions expressed in the 19th-century formulations of the second law of thermodynamics. The development of thermodynamics during this period can be understood, at least in part, as the growing recognition that one or the other of the four properties we singled out above extends in fact to systems that are much more general than the ideal gases.

Thomson (1848), reflecting upon the unsatisfactory theoretical status of a definition of absolute temperature that depends on the relatively precarious existence of ideal gases, suggested that there should be a "principle on which an absolute thermometric scale can be founded"; and he proposed that (33) could be interpreted as the lowest upper bound on the efficiency of *any* thermal engine that performs work while operating between two heat reservoirs at temperatures $T_1 > T_0$. In Thomson (1852), he succeeded in obtaining this conclusion from the following "axiom: It is impossible, by means of inanimate material agency, to derive mechanical effect from any portion of matter by cooling it below the temperature of the coldest of the surrounding objects." This formulation is not without semantic ambiguities; Buchdahl (1966) restates this principle in a form more attuned to modern requirements: "it is impossible to construct an engine which, operating in a cycle, will produce no effect other than the extraction of heat from a reservoir and the performance of an equivalent amount of work." Thomson himself realized that he had been anticipated by Clausius whom he credited with the following statement: "It is impossible for a self-acting machine, unaided by any external agency, to convey heat from one body to another at a higher temperature." We read indeed in Clausius (1850) that it is impossible "without any expenditure of force or any other change, to transfer ... heat ... from a cold to a hot body." Again in the modern formulation of Buchdahl (1966): "it is impossible to construct a device which, operating in a cycle, will produce no effect other than the transfer of heat from a cooler to a hotter body." Following his paraphrase of Clausius, Thomson added: "It is easily shown that, although this and the axiom I have used are different in form, either is a consequence of the other. The reasoning of each demonstration is strictly analogous to that which Carnot originally gave." Almost every physics text on thermodynamics—see e.g. Fermi (1936)—discusses this equivalence, which is essentially correct, modulo the usual language ambiguities due to "the subtlety with which mathematical notions and ostensibly physical notions are almost inextricably interwoven" (Buchdahl, 1966); for modern discussion of this, and related, points see Serrin (1978–79), Lu (1982), and references cited therein. We shall state below still another formulation—due to Carathéodory (1909)—of this *second law of thermodynamics*.

Before we do so, however, we should mention that the principal property of the quantity S defined by (34) for the ideal gases is (37) which Clausius (1850) introduced in the form

$$\int \frac{dQ}{T} = S - S_0 \tag{41}$$

(where Clausius' notation dQ should *not* be interpreted as the differential of a putative state function Q ... which actually does not exist since $\Delta Q[\mu_0\pi\mu_1]$ depends on the process $\mu_0\pi\mu_1$ and not only on μ_0 and μ_1, as we already learned, see eq. 16). While quantities similar to S had been introduced earlier by Rankine and Thomson, Clausius gave S its name and general significance: "I propose to name ... S the *entropy* of the body, for the Greek word $\eta\tau\rho o\pi\eta$, a transformation. I have intentionally formed the word entropy so as to be as similar as possible to the word energy, since both of these quantities ... are so nearly related to each other in their physical significance ... ". Clausius then enters into an explanation, the gist of which is that for any adiabatic process $\mu_0\alpha\mu_1 : S(\mu_1) \geq S(\mu_0)$; recall the fourth property of the ideal gases which we mentioned earlier. The point now is that Clausius postulated, for a general thermodynamical system, that there exist functions T (the absolute temperature) and S (the entropy) defined on the equilibrium state space M of the system, such that the heating 1-form η can be written as TdS, where S can only increase in an adiabatic process. To emphasize the universality of this postulate, Clausius even gave it cosmic proportions: "we can express the fundamental laws of the universe which correspond to the two fundamental laws of the mechanical theory of heat in the following simple form. 1. The energy of the universe is constant. 2. The entropy of the universe tends toward a maximum."

Back down to earth, Gibbs (1873–78) encapsulated the mathematical essence of the investigations of Clausius and Thomson by first listing the equilibrium properties of a homogeneous fluid, not just an ideal gas; one has to consider: its internal energy $U \in [U_0, \infty)$, its volume $V \in [0, \infty)$, its mass $N \in [0, \infty)$ measured in moles, its entropy $S \in \mathbb{R}$, its absolute temperature $T \in [0, \infty)$, its pressure $p \in [0, \infty)$, and two process-functions: the work (resp. the working) done on the fluid, and the heat (resp. the heating) received by the fluid in a process (resp. a quasi-static process) that links two equilibrium states of the fluid. One next assumes that the equilibrium states of the fluid to be characterized by (U, V, N); thus S, T and p are to be considered as (smooth) functions of the three variables U, V and N. The relations (18), (19) and (37) are assumed, and the following properties of S and U are postulated. S and U, like V and N, are supposed to be extensive variables, i.e. for every $\lambda \geq 0$:

$$S(\lambda U, \lambda V, \lambda N) = \lambda\, S(U, V, N) \tag{42}$$

Given (U, V, N), let us denote by $\phi(U, V, N)$ the set $\{(U_1, V_1, N_1; U_2, V_2, N_2) | U_1 + U_2 = U, V_1 + V_2 = V, N_1 + N_2 = N\}$; one then assumes the stability condition:

$$S(U, V, N) = \sup_{\phi(U,V,N)} \{S(U_1, V_1, N_1) + S(U_2, V_2, N_2)\} \tag{43}$$

Note that, (42) being assumed, the phenomenological condition (43) is equivalent to the following mathematical condition. With $0 \leq \alpha \leq 1$; $U = \alpha U_1 + (1-\alpha)U_2$; $V = \alpha V_1 + (1-\alpha)V_2$; $N = \alpha N_1 + (1-\alpha)N_2$:

$$S(U, V, N) \geq \alpha\, S(U_1, V_1, N_1) + (1-\alpha) S(U_2, V_2, N_2) \tag{44}$$

i.e. S is concave. Similarly, with U expressed as a function of S, V and N, U is assumed to be convex.

Upon inspection of the computations we presented in the particular case of the ideal gases, one verifies readily that one obtains, from the above postulates, the relations (36), (38) and (39), without ever needing the specific form of the constitutive equation

$$f(p, V, T) = 0 \tag{45}$$

generalizing the ideal gas law (7) of Gay-Lussac. At fixed N, the temperature T is defined on M as a function of both U and V; in the particular case where (7) holds, the consequence (38a) of the above postulates, implies already that

T is then a function of U alone; this is the result found experimentally by Joule from the free expansion of ideal gases. Back to the general case, Gibbs then proceeded to generalize this theory further so that he could consider the equilibrium properties of heterogeneous substances and discuss systems that undergo phase transitions. For a contemporary account of this approach and of its ramification, see Wightman's Introduction to Israel (1979); see also Callen (1960) and Tisza (1966).

The last question we want to address briefly in this section is one of economy of principles; in particular, does one need to postulate explicitly the existence of the functions T and S such that the heating 1-form can be written as $\eta = TdS$? In this respect, the mathematical reformulation of the *second law of thermodynamics* proposed by Carathéodory (1909)—see also Buchdahl (1966) and, from a somewhat different point of view, Jauch (1972)—should be mentioned. Let M, P, A, ΔW and U be defined as we did in our sketch of the first law of thermodynamics. Suppose further that M is a differentiable manifold, that U is a smooth function on M, and that a coordinate system $(U, V^1, \ldots, V^n)$ together with n smooth functions $p_1, \ldots, p_n$ are given on M such that

$$\Omega \equiv -p_k \; dV^k \tag{46}$$

(where the convention of summation over repeated indices is used) can be interpreted as the (mechanical) working 1-form. Carathéodory's postulate, which abstracts to this general situation the last property we listed for ideal gases, consists in assuming that for every $\mu_0 \in M$, and every open subset $\mathcal{O}$ of M, containing μ_0, there exists $\mu_1 \in \mathcal{O}$ such that there is no adiabatic process $\mu_0 \alpha \mu_1$. A fortiori, there is no quasi-static adiabatic process $\mu_0 \alpha \mu_1$, which is to say that μ_1 does not belong to a solution curve $\mu : t \in I \mapsto \mu(t) \in M$ of the equation

$$(dU - \Omega)(\dot{\mu}) = 0 \tag{47}$$

passing through μ_0. Carathéodory recognized that this amounts to an integrability condition for the 1-form

$$\eta \equiv dU - \Omega \tag{48}$$

and that there exist, therefore, two functions T and S on M such that

$$\eta = TdS\,. \tag{49}$$

This is the main insight provided by this approach, but one can go further. T and S are evidently not determined uniquely by this relation; this ambiguity can be removed, up to a universal scalar factor, by standard considerations on composite systems. Up to that point the Carathéodory postulate has been used only for quasi-static adiabatic processes; the full strength of the postulate, involving adiabatic processes that are not necessarily quasi-static, is used when it comes to choose the sign of T (say $T > 0$) or equivalently the sign of ΔS, and thus of dS.

The global equilibrium concepts of absolute temperature T, entropy S, and heating $\eta = TdS$—which we approached in this section from different 19th-century points of view—provide most of the thermodynamical background necessary for the materials to be presented in this book; in that sense, they will thus suffice as an introduction. The reader should nevertheless realize that they do not exhaust all there is to thermodynamics: among other things, the theory of chemical reactions, and of deformable bodies made out of dissipative materials, the delineation of the fundamental local quantities adapted to their description, and the field equations that govern them have been entirely left out; for a precise introduction to contemporary investigations along these lines, see Truesdell (1969). Still, we felt it worthwhile to present here a sketch of the historical developments that paved the way to the 20th-century thermodynamics.

Section 2. CLASSICAL STATISTICAL MECHANICS

The theory of temperature, heat and entropy that we presented in Section One—describing the embryo from which thermodynamics was to grow—takes no cognizance of the atomic constitution of matter. This is mostly a didactic device, since microscopic considerations were not absent from the writings of the 19th-century pioneers of thermodynamics where references are often made, more or less casually, to idealizations called atoms and molecules. The purpose of statistical mechanics is to make this link explicit, to interpret macroscopic properties of matter in bulk in terms of the microscopic properties of its constitutive elements, and to learn about the latter through the former.

For statistical mechanics, as for thermodynamics, the ideal gases played a central, stimulating role. A kinetic theory of gases, that could find a place in the conceptual mainstream of mathematical physics, resulted from investigations spread over a century and a half, starting with the speculations of Hermann in 1716 and Euler in 1727–1729, and culminating in the theories of Maxwell (1860) and Boltzmann (1871, 1877). To illustrate the difficulties with which these ideas met in the scientific community, we could mention the microscopic model proposed and analyzed by Daniel Bernoulli (1738); the insightful proposals made by Herapath between 1816 and 1821, and by Waterston between 1843 and 1846, both of which being deliberately suppressed by the contemporary British scientific officialdom; and the formalizations presented in the 1850's by Joule, Krönig and Clausius. All these investigations involved a subtle mixture of mechanistic and more or less implicit and ill-defined probabilistic arguments. To sort out these arguments, let us first play down the role of randomness by considering the following caricature of the sort of model discussed by Maxwell predecessors.

Consider a rigid cubic box Λ of volume $V = L^3$, containing n point-particles of mass m. These particles are supposed to be uniformly distributed in space and among six beams, each one of which impinges perpendicularly on one of the walls of the box. We further assume that all the particles move rectilinearly with constant speed v, except when they hit a wall, at which point they experience an elastic collision so that their momentum changes by an amount $2\,mv$. Let p be the pressure exerted by the particles on a given wall of the box; this is given by

$$(1) \qquad p = \frac{1}{L^2} F$$

where the force F is computed from

$$(2) \qquad F = \frac{1}{\Delta t} \Delta P$$

where ΔP is the total momentum change of the particles hitting the wall in any time interval Δt such that $v\Delta t < L$; ΔP is thus given by

$$(3) \qquad \Delta P = 2mv \cdot \Delta n$$

where Δn is the number of particles that hit the wall during the time interval Δt; we thus have

$$(4) \qquad \Delta n = \frac{1}{6} n \frac{\Delta V}{V} \quad \text{with} \quad \Delta V = L^2 \cdot v \cdot \Delta t$$

From (1–4) we obtain

$$(5a) \qquad p = \frac{1}{3} \rho\, v^2 \quad \text{with} \quad \rho = \frac{nm}{V}$$

which we can rewrite as

$$(5b) \qquad pV = n \cdot \frac{2}{3} \cdot \frac{1}{2} m v^2$$

Upon comparing this with the Gay-Lussac law (2.1.7)

$$(6) \qquad pV = NRT$$

we thus have

$$(7a) \qquad N \cdot \frac{3}{2} R \cdot T = n \cdot \frac{1}{2} m v^2$$

The RHS of this expression is the total kinetic energy U of the n point-particles of mass m, contained in the box V; we can, therefore, rewrite (7a) in the suggestive form

$$U = N C_v T \quad \text{with} \quad C_v \equiv \frac{3}{2} R \tag{7b}$$

This model, therefore, indicates that the internal energy of a monoatomic ideal gas can be interpreted as the total (translational) kinetic energy of its molecules. Furthermore, we notice that, while it is natural to measure the quantity of a macroscopic gas by the number N of moles it contains, the natural measure of this quantity in a microscopic model is the number n of molecules it contains; with this in mind, we rewrite (7a) in the form

$$3 \cdot \frac{1}{2} k T = \frac{1}{2} m v^2 \quad \text{with} \quad nk \equiv NR \tag{7c}$$

This model, therefore, also suggests to interpret the temperature in terms of the kinetic energy of an individual molecule; we shall see later that the factor 3 in (7c) is a faithful reflection of the fact that we modeled an assembly of structureless molecules moving in a 3-dimensional space. This model is evidently extremely naive: (i) it assumes the particles to be uniformly distributed in space; (ii) it models the molecules as point masses; (iii) it restricts their velocities to be perpendicular to the faces the cube V; (iv) it assumes that all molecules have the same speed v. While (i) is reasonable for a gas in equilibrium, neither (ii), (iii), nor (iv) should be assumed to hold for real molecules, which have spatial extension and do collide with one another.

Maxwell (1860) thus set out to "lay the foundation ... on strict mechanical principles" and "to demonstrate the laws of motion of an indefinite number of small, hard, and perfectly elastic spheres acting on one another only during impact. If the properties of such a system of bodies are found to correspond to those of gases, an important physical analogy will be established, which may lead to more accurate knowledge of the properties of matter." To this day, this ambitious programme has not been satisfactorily carried out. Maxwell succeeded in describing most succinctly the first important property of the canonical equilibrium states of such an assembly when it is enclosed in a cubic box Λ of volume V. Recognizing that the numerous collisions that occur in a gas call for a probabilistic language to account for the expected randomness, hc assumed first of all that when a gas is in thermal equilibrium, its constitutive particles are distributed uniformly in space. He thus concentrated his attention on the following problem: "To find the average number of particles whose velocities lie between given limits, after a great number of collisions among a great number of equal particles." He solved this problem by showing, through an elementary calculus-type of argument, that the one-particle distribution functions

$$f(\mathbf{v}) = n(\alpha^2 \pi)^{-\frac{3}{2}} \exp\{-v^2/\alpha^2\} \tag{8}$$

Upon using the symmetry of the collision term when integrated over v, and the fact that the streaming term identically vanishes, we obtain

$$\begin{aligned}\frac{dH}{dt}(t) = \frac{1}{4}na^2\int d^3v\int d^3v_1\int d^2u\ \mathbf{u}\cdot(\mathbf{v}-\mathbf{v}_1)\times \\ \{\varphi(\mathbf{w};t)\varphi(\mathbf{w}_1;t)-\varphi(\mathbf{v};t)\varphi(\mathbf{v}_1;t)\}\times \\ \{\ell n[\varphi(\mathbf{v};t)\varphi(\mathbf{v}_1;t)]-\ell n[\varphi(\mathbf{w};t)\varphi(\mathbf{w}_1;t)]\}\end{aligned} \tag{29}$$

Since the integrant in the RHS is negative, we have

$$\frac{dH}{dt}(t)\leq 0 \quad \forall\, t\geq 0 \tag{30}$$

with the equality holding if and only if

$$\varphi(\mathbf{v};t)\varphi(\mathbf{v}_1;t)=\varphi(\mathbf{w};t)\varphi(\mathbf{w}_1;t) \tag{31}$$

where $(\mathbf{v},\mathbf{v}_1;\mathbf{w},\mathbf{w}_1)$ must satisfy (24). Consequently φ is a spatially homogeneous equilibrium solution of the Boltzmann equation if and only if

$$\ell n\ \varphi(\mathbf{v})+\ell n\ \varphi(\mathbf{v}_1) \tag{32}$$

depends only on $(\mathbf{v}+\mathbf{v}_1)$ and $(\mathbf{v}^2+\mathbf{v}_1{}^2)$; this happens exactly when φ is of the form

$$\varphi(\mathbf{v})=(\alpha^2\pi)^{-\frac{3}{2}}\exp\{-(\mathbf{v}-\mathbf{v}_0)^2/\alpha^2\} \tag{33}$$

with $\alpha\in\mathbb{R}^+$ and $\mathbf{v}_0\in\mathbb{R}^3$ arbitrary. Hence the Maxwell distribution (8) [or, in the physicist's interpretation, (22)] is, up to normalization, the unique smooth, spatially homogeneous solution of the Boltzmann equation, satisfying

$$\int d^3v\ \varphi(\mathbf{v})\,\mathbf{v}=0 \tag{34}$$

The solutions of the form (33) are called *Maxwell-Boltzmann distributions*. Upon using the method of Lagrange multipliers, we see that (33) can also be viewed as the solution of the variational problem that consists in finding the extrema of

$$H\equiv\int d^3v\ \varphi(\mathbf{v})\,\ell n\ \varphi(\mathbf{v}) \tag{35}$$

subject to the constraints

$$\int d^3v\ \varphi(\mathbf{v})=1 \tag{36a}$$

$$\int d^3v\ \varphi(\mathbf{v})(\mathbf{v}-\mathbf{v}_0)^2=\frac{3}{2}\alpha^2 \tag{36b}$$

where

$$\mathbf{v}_0 \equiv \int d^3v\, \varphi(\mathbf{v})\, \mathbf{v}\,. \tag{36c}$$

The functional $H(t)$ introduced in (28), and its particular time-independent form (35) have been given two suggestive interpretations. The first of these interpretations is to view

$$I(t) = k\, H(t) \tag{37}$$

as the information content of the probability distribution $\varphi(\,\cdot\,;t)$, in the sense of the Shannon (1948)—Khinchin (1957) theory of information. The second interpretation is obtained by a change of normalization, defining

$$f(\mathbf{x},\mathbf{v}) \equiv \psi(\mathbf{v}) \equiv \frac{n}{V}\varphi(\mathbf{v}) \tag{38a}$$

so that

$$\int_\Lambda d^3x \int_{\mathbb{R}^3} d^3v\, f(\mathbf{x},\mathbf{v}) = n \tag{38b}$$

Upon taking for φ the Maxwell-Boltzmann distribution (33), with $<\mathbf{v}>= 0$ see (34), and $\alpha^2 = 2kT/m$ (see 21), and substituting ψ for φ in (35), one obtains

$$S \equiv -kVH = N\{C_v\, \ell n(\frac{U}{N}) + R\, \ell n(\frac{V}{N}) + \text{const.}\} \tag{39}$$

where we used the thermodynamical units N and R (see 7c), $C_v = 3R/2$ and $U = NC_vT$ (see 7b). This S happens to coincide with the thermodynamical entropy (1.1.34).

With these interpretations, (30) is called the Boltzmann H-theorem. It suggests a link between the Boltzmann equation and the irreversible tendency towards equilibrium observed in thermodynamical systems. It also indicates that a derivation of the Boltzmann equation must involve, in addition to purely mechanistic arguments, some other arguments the nature of which is to be understood, either from some probabilistic standpoint, or in terms of some limiting procedure, or both.

Before we discuss this point, we still want to mention that the symmetry of the collision term in (26), which we used to derive (30), can also be used to direct attention towards the kind of phenomena the Boltzmann equation can be expected to describe, and the level of the expected description.

Let A be any one-particle observable that is conserved in the course of a collision, i.e. A is a function defined on the one-particle phase space $\{(\mathbf{x},\mathbf{v})|\mathbf{x} \in$

$\Lambda, \mathbf{v} \in \mathbb{R}^3\}$ and satisfying

$$\text{(40a)} \qquad \begin{array}{l} A(\mathbf{x}, \mathbf{v}) + A(\mathbf{x}, \mathbf{v}_1) = A(\mathbf{x}, \mathbf{w}) + A(\mathbf{x}, \mathbf{w}_1) \\ \text{whenever } (\mathbf{v}, \mathbf{v}_1, \mathbf{w}, \mathbf{w}_1) \text{ obeys (24)} \end{array}$$

Together with the afore-mentioned symmetry of the collision term in (26), this implies

$$\text{(40b)} \qquad \int d^3v \left(\frac{\partial \varphi}{\partial t}\right)_c (\mathbf{x}, \mathbf{v}; t)\, A(\mathbf{x}, \mathbf{v}) = 0$$

for any smooth solution $\varphi(\cdot\,,\cdot\,;t)$ of the Boltzmann equation. Define now

$$\text{(41a)} \qquad \varphi(\mathbf{x}; t) \equiv \int d^3v\; \varphi(\mathbf{x}, \mathbf{v}; t)$$

$$\text{(41b)} \qquad \rho(\mathbf{x}; t) \equiv \frac{n}{V} m\; \varphi(\mathbf{x}; t)$$

$$\text{(41c)} \qquad < A > (\mathbf{x}; t) \equiv [\varphi(\mathbf{x}; t)]^{-1} \int d^3v\; \varphi(\mathbf{x}, \mathbf{v}; t) A(\mathbf{x}; \mathbf{v})$$

From the explicit form of the streaming term in (26), one obtains

$$\text{(42)} \qquad \frac{\partial}{\partial t}\rho < A > + \frac{\partial}{\partial x^i}\rho < v^i A > - \rho < v^i \frac{\partial A}{\partial x^i} > - \frac{\rho}{m} F^i < \frac{\partial A}{\partial v^i} > = 0$$

Three particular cases are of interest: $A = 1$, $A = v^i$ with $i = 1, 2, 3$ and $A = v^2$; from these, all spatially homogeneous observables satisfying (39) can be obtained by linearity. With $A = 1$, (42) reduces to the *continuity equation*:

$$\text{(43)} \qquad \frac{\partial}{\partial t}\rho + \mathrm{div}(\rho \mathbf{u}) = 0 \quad \text{with} \quad \mathbf{u} \equiv < \mathbf{v} >$$

With $A = v^i$ $(i = 1, 2, 3)$, (42) reduces then to the *Euler equation of motion* (compare with 1.1.25):

$$\text{(44)} \qquad \rho\, \dot{\mathbf{u}} = \frac{\rho}{m}\mathbf{F} + \mathrm{div}\, P$$

where $\mathbf{u}$ is defined in (43); $(^\bullet)$ denotes the *material derivative*

$$\text{(45)} \qquad (^\bullet) \equiv \left(\frac{\partial}{\partial t} + u^k \frac{\partial}{\partial x^k}\right);$$

and the stress-tensor P is linked, through the definition (41c), to the one-particle probability distribution φ by the relation

$$\text{(46)} \qquad P^{ij} \equiv -\rho < (v^i - u^i)(v^j - u^j) > .$$

Lastly, when $A = (v - u)^2$, (42) reduces then to the *generalized heat equation*

$$\rho \dot{T} = -\frac{2}{3}[\text{div } \Phi - tr(P\Lambda)] \tag{47}$$

where $(^\bullet)$ denotes again the material derivative (45); P is defined as in (46); and

$$\frac{3}{2}kT \equiv \langle \frac{1}{2}m(v-u)^2 \rangle \tag{48}$$

$$k\Phi^i \equiv \frac{1}{2}m\rho \langle (v^i - u^i)(v-u)^2 \rangle \tag{49}$$

$$k\Lambda_{ij} \equiv \frac{1}{2}m(\frac{\partial u_i}{\partial x^j} + \frac{\partial u_j}{\partial x^i}) \tag{50}$$

(with k the universal constant introduced in 7c). Note that (46) and (48) imply that

$$p = \frac{\rho}{m}kT \quad \text{where} \quad p \equiv -\frac{1}{3}tr\,P\,. \tag{51}$$

In particular, for the Maxwell-Boltzmann distribution (33), ρ, $\mathbf{u}$ and P, defined by (41b), (43) and (46) reduce to

$$\rho = \frac{nm}{V}\,, \quad \mathbf{u} = \mathbf{v}_0\,, \quad \text{and} \quad P^{ij} = -p\,\delta^{ij} \tag{52}$$

so that (51) reduces then to the Gay-Lussac law (2.1.7), thus suggesting to extend the interpretation of P, p and T for non-equilibrium distributions, and thus to refer to the values of these functions of $(\mathbf{x}, t)$ as describing the local instantaneous stress-tensor, pressure and temperature. It must, nevertheless, be realized that one is then, at best, on heuristic grounds due to the unresolved existence questions for spatially non-homogeneous solutions of the Boltzmann equation. With this caveat in mind, one can, nevertheless, try to gain some insight into the phenomenological content of the conservation equations just obtained. One way to proceed is to analyze small deviations from the Maxwell-Boltzmann distribution, assuming for instance that the one-particle probability distribution φ is of the form

$$\varphi(\mathbf{x}, \mathbf{v}; t) = \varphi_0(\mathbf{x}, \mathbf{v}; t) + \psi(\mathbf{x}, \mathbf{v}; t) \tag{53}$$

where: $\varphi_0(\mathbf{x}, \mathbf{v}; t)$ takes the form (33), with, however, $\mathbf{v}_0$ and α^2 now allowed to be "slowly varying" functions of $\mathbf{x}$ and t; and ψ is supposed to be a "small" correction, necessary to make φ a solution of the Boltzmann equation. It is then possible, see e.g. the physics texts by Huang (1963) or Wannier (1966), to make it plausible that $(\partial\varphi/\partial t)_c \simeq -\psi/\tau$, where τ is the mean time between two successive collisions, linked to the mean free path λ by the relation $\overline{v} = \lambda/\tau$, with $\bar{v}^2 = 2kT/m$ and $\pi a^2 \lambda = V/n$. At that level of heuristic approximations, (44) can be brought to the form of the *Navier-Stokes equation* of viscous hydrodynamics

$$\rho \dot{\mathrm{u}} = \frac{\rho}{m}\mathrm{F} - \nabla(p - \frac{1}{3}\eta \nabla \cdot \mathrm{u}) + \eta \Delta \mathrm{u} \,; \tag{54}$$

and (47) to the form of the *heat conduction equation*

$$(\dot{T} - \frac{K}{\rho C_v}\Delta T) = -\frac{1}{C_v}(\nabla \cdot \mathrm{u})T \,. \tag{55}$$

Note, in particular, that if $\mathrm{u} \equiv 0$, (55) reduces to the Fourier heat equation (2.1.1); note also that $\eta \neq 0$ in (54) describes a first order deviation from the Euler (non-viscous) hydrodynamic equation (1.1.27). In (54) and (55) η and K are thus interpreted as the viscosity and thermal conductivity coefficients. Still at the same heuristic level, the theory predicts

$$\frac{2}{5}K = \eta = \tau \cdot kT \cdot (\frac{n}{V}) \tag{56}$$

Hence, up to numerical factors, these coefficients are of the order of $a^{-2}\sqrt{mkT}$; consequently, for any given gas (i.e. given a and m) at fixed temperature T: η and K are independent of the pressure of the gas; this result was counterintuitive enough for Maxwell to have gone to the trouble of checking it experimentally for η, by observing the damping rate of a pendulum oscillating in a gas at different pressures; the experimental confirmation of this law was one of the early successes of the kinetic theory of gases.

The consequences of the Boltzmann equation (26), which we have reviewed can be summarized in the following three points. *Firstly*, as we saw explicitly for the spatially homogeneous case in the absence of external forces, solutions of this equation exhibit a *dissipative, irreversible behaviour*: the Boltzmann constant H, defined in (28), cannot increase—see (30)—and indeed decreases in the course of time, unless the solutions are stationary. *Secondly*, the spatially homogeneous stationary solutions of the Boltzmann equation are all of the form (33) which can be given an interpretation in term of *thermodynamical equilibrium* concepts. Indeed the term v_0 in (33) is only a reflection of the Galilean invariance (see Chapter Four) of the theory; it can be eliminated by a change of coordinate-system, to bring (33) in the form of the Maxwell distribution (17). This distribution is consistent with the relation $pV = nkT$—see (19) and (21)—characteristic of ideal gases in equilibrium; this allows to rewrite (17) in the interpretable form (22) which has been experimentally tested. Moreover, the information content kH of these stationary solutions of the Boltzmann equation can be interpreted in term of the thermodynamical entropy S of the ideal gas: see (39). *Thirdly*, the non-stationary solutions of the Boltzmann equation allow to give a *microscopic* interpretation of the *macroscopic* evolution equations (43), (44) [see also (54)] and (47) [see also (55)]; this interpretation is tested by the predictions it makes on the numerical values of the transport coefficients that appear in these equations. These equations are to be viewed, in this perspective, as a partial, or reduced, description of the total system; this description is adapted to the partial information resulting from the measurement of experimentally accessible, or macroscopic, observables: the local mass density ρ, the local velocity $\mathbf{u}$, the local temperature T, etc.; recall further that these macroscopic observables are defined—see (41b), (43), (48), etc.—in terms of averages of one-particle quantities that are conserved, in the sense of (24), in the course of interparticle collisions: the mass, the momentum, and the kinetic energy.

The above discussion of the consequences of the Boltzmann equation should, however, not mask the fact that serious questions remain as to the premises from which the Boltzmann equation itself can be understood. We now turn to these premises.

One starts with a mechanical system of n identical particles enclosed in a box Λ of volume V; the particles are modelled as hard spheres of diameter a, bouncing elastically off each other. The microscopic observables on this system are functions defined on the phase space

$$\Gamma^n = T\Lambda^n \setminus \Xi \tag{57a}$$

where a representative point $\xi \in \Gamma^n$ is written as

$$\xi = (\xi_1, \xi_2, \ldots, \xi_n) \text{ with } \xi_i = (\mathbf{x}_i, \mathbf{v}_i) \quad 1 \leq i \leq n \tag{57b}$$

subject to the hard-sphere exclusion

$$\| \mathbf{x}_i - \mathbf{x}_j \| > a \quad \text{whenever} \quad i \neq j\,. \tag{57c}$$

Since we are interested only in one-particle observables, i.e. in functions of the form

$$A : \xi \in \Gamma^n \mapsto \frac{1}{n} \sum_{i=1}^{n} A(\xi_i) \tag{58}$$

one can reduce the description to the one-particle phase space

$$M = T\Lambda \tag{59}$$

and rewrite the value of A on the state $\xi \in \Gamma^n$ as

$$A(\xi) = \int \mu_\xi(d^3x, d^3v)\, A(\mathbf{x}, \mathbf{v}) \tag{60a}$$

where μ_ξ is the discrete measure on M, corresponding to ξ through the definition

(60b) $$\mu_\xi \equiv \frac{1}{n}\sum_{i=1}^{n} \delta_{\xi_i}$$

where δ_{ξ_i} is the Dirac measure concentrated at the point $\xi_i \in M$.

One first remarks that the Boltzmann equation does not involve discrete measures of the type (60b), but rather measures of the form

(61a) $$\mu(d^3x, d^3v) = \varphi(\mathbf{x}, \mathbf{v})\, d^3x\, d^3v\ ;$$

more specifically, these should be probability measures on M, absolutely continuous with respect to Lebesgue measure, and such that the expectation value of the conserved quantities are finite. To accommodate both these measures and the measures of the tye (60b), one therefore considers the subset P of the space Prob(M) of the probability measures on M, defined by

(61b) $$P \equiv \{\mu\ \in\ \mathrm{Prob}(M) |\ \int \mu(dx^3, dv^3)\, v^2 < \infty\}.$$

Boltzmann argued that, for the systems he was considering, the discrete probability measures (60b) of ordinary mechanics can be approximated by continuous measures of the form (61a). To make this precise, one must then define a topology on P; one chooses the weak*-topology for measures, strengthened by the condition that the following map be continuous

(61c) $$\mu\ \in\ P \mapsto\ \int \mu(d^3x, d^3v)\, v^2\ \in\ \mathbb{R}$$

For instance, if

(62a) $$\{n(k)\ \in\ \mathbb{Z}^+ | k\ \in\ \mathbb{Z}^+\}\ \text{ and }\ \{\xi(k)\ \in\ \Gamma^{n(k)} | k\ \in\ \mathbb{Z}^+\}$$

with $n(k) \to \infty$ as $k \to \infty$, $\{\mu_{\xi(k)} | k\ \in\ \mathbb{Z}^+\}$ converges to a measure μ of the form (61a) in P if and only if the following two conditions are satisfied. Firstly, for each 6-dimensional parallelepiped $\Delta \subset M \simeq \Lambda \times \mathbb{R}^3$, the fraction of the particles that are in Δ converges to $\mu(\Delta)$, i.e.

(62b) $$\lim_{k\to\infty} \frac{1}{n(k)} \sum_{j=1}^{n(k)} \chi_\Delta(\xi(k)_j) = \int d^3x d^3v\ \varphi(\mathbf{x}, \mathbf{v})\, \chi_\Delta(\mathbf{x}, \mathbf{v})\,.$$

Secondly, the average kinetic energy per particle converges to the expected value of the kinetic energy for the one-particle distribution φ, i.e.

(62c) $$\lim_{k\to\infty} \frac{1}{n(k)} \sum_{j=1}^{n(k)} (v(k)_j)^2 = \int d^3x\, d^3v\, \varphi(\mathbf{x}, \mathbf{v})\, v^2$$

The first ingredient of the theory will thus be to consider a limiting process in which the number n of the particles tends to infinity, so that there is no conflict between the use of discrete measures, such as those occurring in the mechanics of particle systems, and the continuous distributions involved in the Boltzmann equation. This limit will also be instrumental in avoiding conflicts between the reversibility of the mechanical description one starts from, and the irreversibility of the thermodynamical description that corresponds to the Boltzmann equation.

The second ingredient of the theory is to let the diameter a of the particles become very small. One should indeed notice that a limiting process that would involve only $n \to \infty$ is hardly compatible with the Boltzmann equation itself, since n appears explicitly as a factor in (26c). Following a path opened by Grad (1958), one infers upon inspection of (26c) that the proper limiting process should involve a joint limit where

$$\{n \to \infty\,,\ a \to 0 \ \text{ with } \ na^2 \to B \in (0,\infty)\} \tag{63}$$

This is called the *Boltzmann-Grad limit.* Its physical meaning is that, at fixed volume V, the mean free path λ, defined by $\pi a^2 \lambda = V/n$ remains finite, namely $\lambda \to V(\pi B)^{-1}$. This defines, in this limiting process, a finite spatial scale for the phenomena to be described by the Boltzmann equation. Together with $a \to 0$ this implies $(a/\lambda) \to 0$, which formalizes the notion that a particle moves freely, except for some occasional collisions. Note also that $(a/\lambda) \to 0$ is equivalent to $(na^3/V) \to 0$ which, in turn, formalizes the notion that one works with a very dilute gas: collectively the particles only occupy effectively a very small fraction of the total volume V at their disposal—and this in spite of the fact that $n \to \infty$.

The third, and most controversial, ingredient of the theory is what is referred to either as the *Boltzmann Stosszahlansatz,* or as the hypothesis of persistent molecular chaos; this is a statement of statistical independence implicit in the interpretation of the collision term $(\partial\varphi/\partial t)_c$; see (26). Indeed, in this expression, the products

$$\varphi(\mathbf{x},\mathbf{v};t)\,\varphi(\mathbf{x},\mathbf{v}_1;t) \tag{64a}$$

appearing in the loss term—and similarly for

$$\varphi(\mathbf{x},\mathbf{w};t)\varphi(\mathbf{x},\mathbf{w}_1;t) \tag{64b}$$

in the gain term—are meant to be interpreted as the probability density of events in which two particles with velocities $\mathbf{v}$ and $\mathbf{v}_1$ enter into a collision in the place $\mathbf{x}$ at the time t. One assumes, therefore, that

$$\Phi(\mathbf{x},\mathbf{v},\mathbf{x},\mathbf{v}_1;t) = \varphi(\mathbf{x},\mathbf{v};t)\,\varphi(\mathbf{x},\mathbf{v}_1;t) \tag{65}$$

or, more generally, that the joint probability distribution for n particles factors in the form

$$\Phi(\mathbf{x}_1, \mathbf{v}_1, \ldots, \mathbf{x}_n, \mathbf{v}_n; t) = \prod_{k=1}^{n} \varphi(\mathbf{x}_k, \mathbf{v}_k; t) \tag{66}$$

To assume that these conditions hold at all times is, however, unwarranted: in a dynamical theory based on mechanics, as the Boltzmann equation purports to be, one may only make assumptions, statistical or otherwise, on the initial state of the system, say at $t = 0$; after that, the basic mechanistic laws should be allowed to fully take over and determine the state of the system at $t > 0$. It is then by no means clear that if statistical independence holds at time $t = 0$, it will persist at any future time when collisions will have occurred. The evidence, in non-stationary situations, is actually pointing to the contrary, as we shall argue later on.

Lanford (1975) succeeded in bringing together the three ingredients listed above by controlling a limiting process that establishes a mathematiclly meaningful connection between the Boltzmann equation and the laws of particle mechanics. To state Lanford's result, we need a few more definitions.

$\{\mu^k | k \in \mathbb{Z}^+\}$ is said to be an *approximating sequence* for a continuous probability distribution φ on M when the following three conditions are satisfied. Firstly, for each $k \in \mathbb{Z}^+$, $(n_k, a_k) \in \mathbb{Z}^+ \times \mathbb{R}^+$ is given, such that, as $k \to \infty : n_k \to \infty$, $a_k \to 0$ with $n_k {a_k}^2 \to na^2$. Secondly, for each $k \in \mathbb{Z}^+$, μ^k is a probability measure on Γ^{n_k}. Thirdly, as $k \to \infty$, and for every parallelepiped $\Delta \subset M$, the random variables

$$\begin{aligned} F^k_\Delta : \xi = (\mathbf{x}_1, \mathbf{v}_1, \ldots, \mathbf{x}_{n_k}, \mathbf{v}_{n_k}) &\in \Gamma^{n_k} \\ \mapsto \frac{1}{n_k} \sum_{j=1}^{n_k} \chi_\Delta(\mathbf{x}_j, \mathbf{v}_j) &\in \mathbb{R} \end{aligned} \tag{67a}$$

converge, in distribution with respect to μ^k, to the constant

$$\int d^x\mathbf{x} d^3\mathbf{v}\, \varphi(\mathbf{x}, \mathbf{v})\, \chi_\Delta(\mathbf{x}, \mathbf{v}) \tag{67b}$$

This definition thus expresses that $\{\mu^k\}$, viewed from M, concentrates on φ.

The problem is now to determine whether some condition can be found on the initial continuous probability distribution φ_0 on M, which would guarantee that some $t_0 > 0$ exists such that, if $\{\mu^k\}$ is an approximating sequence for φ_0, and if $T^k(\cdot)$ denotes the mechanistic evolution on Γ^{n_k}, then for every $t \in [0, t_0]$, $\{\mu^k \circ T^k(-t)\}$ is an approximating sequence for a probability distribution $\varphi(\cdot, \cdot; t)$ on M that solves the Boltzmann equation with initial value $\varphi(\cdot, \cdot; 0) = \varphi_0(\cdot, \cdot)$.

Lanford found such a sufficient condition, assuming that each μ^k is absolutely continuous with respect to Lebesgue measure and that the corresponding probability distribution Φ^k on Γ^{n_k} is invariant under every permutation

$\xi_j \equiv (\mathbf{x}_j, \mathbf{v}_j) \to \xi_{p(j)} = (\mathbf{x}_{p(j)}, \mathbf{v}_{p(j)})$. The condition is expressed in terms of two restrictions on the marginal distributions $\{\Phi_j^k | 1 \le j \le n_k\}$ defined by

$$\Phi_j^k(\xi_1, \ldots, \xi_j) \equiv \frac{(n_k)!}{(n_k - j)!} \frac{1}{(n_k)^j} \int d^6 \xi_{j+1} \ldots d^6 \xi_{n_k} \, \Phi^k(\xi) \tag{68}$$

Note that $\{\mu^k\}$ is then an approximating sequence for φ_0 if and only if, in the sense of weak *-convergence of measures,

$$w^* - \lim_{k\to\infty} \Phi_1^k(\xi_1)\, d\xi_1 = \varphi_0(\xi_1)\, d\xi_1 \tag{69a}$$

and

$$w^* - \lim_{k\to\infty} \Phi_2^k(\xi_1, \xi_2)\, d\xi_1 d\xi_2 = \varphi_0(\xi_1)\, \varphi_0(\xi_2)\, d\xi_1 d\xi_2 \tag{69b}$$

Notice in particular that (69) are sufficient to ensure that, for every $j \in \mathbb{Z}^+$:

$$w^* - \lim_{k\to\infty} \Phi_j^k(\xi_1, \ldots, \xi_j)\, d\xi_1 \ldots d\xi_j = \prod_{i=1}^{j} \varphi_0(\xi_j)\, d\xi_1 \ldots d\xi_j \tag{70}$$

The first of the two restrictions on $\{\mu^k\}$ is then

$$\lim_{k\to\infty} \Phi_j^k(\xi_1, \ldots, \xi_j) = \prod_{i=1}^{j} \varphi_0(\xi_j) \tag{71}$$

uniformly on compacta in $\Gamma_0^j \equiv T\Lambda^j \setminus \Xi_0$ where $\Xi_0 \equiv \{\xi \in T\Lambda^j | x_\ell \neq x_m \; \forall \, 1 \le \ell \neq m \le j\}$. Compare (71) with (66), and note that the present restriction is imposed only on the initial state. The second restriction on $\{\mu^k\}$ is that there exist constants c, ς and β, such that for all k, j, $\xi_1, \ldots \xi_j$

$$\Phi_j^k(\xi_1, \ldots, \xi_j) \le c\, \Psi_j^\beta(\xi_1, \ldots, \xi_j) \tag{72a}$$

with

$$\Psi_j^\beta(\xi_1, \ldots, \xi_j) \equiv \prod_{i=1}^{j} \psi^\beta(\xi_i) \tag{72b}$$

and ψ^β is the Maxwell distribution

$$\psi^\beta(\mathbf{x}, \mathbf{v}) = \varsigma\left(\frac{m\beta}{2\pi}\right)^{\frac{3}{2}} \exp\{-\frac{1}{2} m\beta\, v^2\}\,. \tag{72c}$$

Lanford (1975) showed that these conditions, which are satisfied by the distributions Boltzmann seemed to have in mind when he was talking of smoothing out discrete particle distributions over small cells in phase space, are sufficient to ensure the desired result that there exists some $t_0 > 0$ such that, for all $t \in [0, t_0]$, $\{\mu^k \circ T^k(-t)\}$ is an approximating sequence for a probability distribution that solves the Boltzmann equation with initial value φ_0. In this sense, the Boltzmann equation gives a macroscopic description that is mathematically consistent with the microscopic dynamics.

Concerning the assumptions under which this result has been obtained, we should emphasize that (71) and (72) were imposed only on the *initial* state. While (72) can still be satisfied at later times, possibly at the cost of readjusting the constants c, ς and β, (71) is a very strong restriction which can, at best, only be assumed to hold for the initial state; indeed, to assume that (71) would hold on $[0, t_0]$ would bring about contradictions with the irreversibility of the Boltzmann equation, as can immediately been seen by applying the Lanford theorem on the approximating sequence obtained from $\{\mu^k \circ T^k(-t_0/2)\}$ by reversing all velocities.

It should also be mentioned that the time interval $[0, t_0]$, for which the proof can be made, is very small; indeed

$$t_0 = \frac{1}{5}\left(\frac{m\beta}{3}\right)^{\frac{1}{2}}(\pi\, na^2\varsigma)^{-1} \tag{73}$$

which is to be interpreted as being of the order of a fraction of the mean time between two successive collisions. While this is insufficient for applications to fluid dynamics, where one is interested in solutions of the Boltzmann equation that extend over much larger time intervals, the irreversibility associated to the decrease of the Boltzmann function H is an indication that, even during such a small time interval as $[0, t_0]$, sufficiently many collisions do actually occur to make it impossible to keep track, at the macroscopic level, of the information contained in the initial state: the time that ellapses between two successive collisions is a stochastic variable with enormous dispersion.

Finally, it should be recalled that the limiting procedure considered not only involves a limit where the number n of the particles tends to infinity, but simultaneously requires that the diameter a of the particles tends to zero, in such a manner that na^3/V tends to zero; i.e. the theory is only valid in the limit of an infinite, infinitely dilute gas. The challenge remains unabatted to obtain from mechanics a kinetic theory that applies to gases at finite densities; see Cohen (1973).

Four guidelines emerge from the discussion carried out in this section.

Firstly, some stochastic elements are introduced when one deals with macroscopic observations of large systems, even when the many particles that constitute such systems are assumed to satisfy, at the microscopic level, the deterministic laws of mechanics. This aspect of the theory was to be emphasized most graphically in the investigations, boosted by the contributions made by Einstein (1905 BM), in the theory of Brownian motion where one studies the manifestly random effects many small particles (e.g. molecules of O_2 and N_2 in air) have on a test-particle (e.g. a grain of tobacco) upon which they act via mechanical collisions. As Einstein described it, his "major aim in this was to find facts which would guarantee as much as possible, the existence of atoms of definite finite size"; in the process he "developed the statistical mechanics and the molecular-kinetic theory of thermodynamics." For a mathematical outline of the theory of Brownian motion, see Nelson (1967) where the reader will also find a brief historical introduction, as well as a seminal discussion of a possible relation between this formalism and quantum mechanics.

Secondly, the use of continuous probability distributions, instead of the discrete distributions of particle mechanics, can be justified by a limiting process in which the number of particles tends to infinity. Such limiting processes, a variant of which is known as the thermodynamical limit where both the number of particles and the total volume tend to infinity with their ratio approaching a finite limit, have the further advantage of concentrating attention on the properties of matter in bulk.

Thirdly, the macroscopic equilibrium properties of matter can be recovered from microscopic models by a variational principle involving the information content—or entropy—of probability distributions subject to macroscopic contraints.

Fourthly, macroscopic transport equations that exhibit an irreversible behaviour can be consistent with a reversible microscopic dynamics, provided proper considerations are made for the scale at which macroscopic phenomena do occur.

These four guidelines form the backbone of the programme known under the name of classical statistical mechanics. Its implementation, consisting in building up concrete microscopic models that are both rich and simple enough to allow the replacement of general conjectures by precise theorems, is still in progress today, as we have just illustrated by Lanford's treatment of the Boltzmann equation. Moreover, since the times of Maxwell and Boltzmann, we have learned that at the microscopic level, matter obeys the laws of quantum rather than classical mechanics. The introduction of this last ingredient is the purpose of quantum statistical mechanics; see Section 10.1.

CHAPTER 3. ELECTROMAGNETISM

SYNOPSIS

For a classical physicist, electromagnetism is the theory which links together the following time-dependent objects defined over a configuration space (e.g. $\mathbb{R}^3$): the (scalar) "electric charge" density ρ; the (vector) "electric current" density $\mathbf{J}$; the (vector) "electric" field $\mathbf{E}$; and the (pseudo-vector) "magnetic" field $\mathbf{B}$. The heart of the theory thus lies in the following system of differential equations, known as the four Maxwell equations (in vacuum):

$$\text{(1)} \qquad \text{curl } \mathbf{B} - \dot{\mathbf{E}} = \mathbf{J} \ [\text{ also written } \nabla \times \mathbf{B} - \partial_t \mathbf{E} = \mathbf{J}]$$
$$\text{(2)} \qquad \text{div } \mathbf{E} = \rho \ [\text{ also written } \nabla \cdot \mathbf{E} = \rho]$$
$$\text{(3)} \qquad \text{curl } \mathbf{E} + \dot{\mathbf{B}} = 0 \ [\text{ also written } \nabla \times \mathbf{E} + \partial_t \mathbf{B} = 0]$$
$$\text{(4)} \qquad \text{div } \mathbf{B} = 0 \ [\text{ also written } \nabla \cdot \mathbf{B} = 0]$$

from which follows

$$\text{(5)} \qquad \text{div } \mathbf{J} + \dot{\rho} = 0 \quad [\text{ also written } \nabla \cdot \mathbf{J} + \partial_t \rho = 0]$$

which expresses the conservation of electric charge. Equation (1) is also known as the Biot-Savart-Ampère-Maxwell equation; for static fields, i.e. $\dot{\mathbf{E}} = 0$, this equation reduces to $\mathbf{J} =$ curl $\mathbf{B}$, while the term $\dot{\mathbf{E}}$ was introduced by Maxwell (1865). For a mathematician, this equation can be taken as a definition of $\mathbf{J}$ from the basic fields $\mathbf{E}$ and $\mathbf{B}$; phenomenologically, it can be taken as a prescription to produce $\mathbf{B}$. Equation (2) can be interpreted similarly, either as a definition of ρ or as a prescription to produce $\mathbf{E}$. Equation (3) is known as the Faraday-Maxwell equation; it incorporates the remnant of an empirical fact, called "induction": a change in $\mathbf{B}$ creates some $\mathbf{E}$. Equation (4) is expressed by saying that there exists (mathematically, one should add "locally") a magnetic potential $\mathbf{A}$ such that $\mathbf{B} =$ curl $\mathbf{A}$. Equation (3) can then be rewritten curl $(\mathbf{E} + \dot{\mathbf{A}}) = 0$, which is to say that there exits (again "locally") an electric potential Φ such that $\mathbf{E} + \dot{\mathbf{A}} = -\text{grad } \Phi$.

In the static case, and with the gauge div $\mathbf{A} = 0$ consistent with $\dot{\Phi} = 0$, equation (2) can be rewritten in the form of a Poisson equation, namely $\rho = -\Delta\Phi[\equiv -\text{ div grad } \Phi]$; similarly, equation (1) can be rewritten $\mathbf{J} = -\Delta\mathbf{A}$ [$\equiv$ curl curl $\mathbf{A}$]. One could pursue this line in the non-static case; even without going into this, the last two equations already seem to beg for the introduction of a quadrivector potential $\mathcal{A}$, a quadrivector density $\mathcal{J}$, and some object $\mathcal{F}$

unifying E and B, a remark which in fact lies at the foundation of the theory of relativity.

Still within the borders of the classical physicist's realm lie two important aspects of this theory of electromagnetism, and we will also discuss them briefly in this chapter. Firstly, an electromagnetic field can be made manifest by its action on a moving, electrically charged "test-particle": the resulting Lorentz force law is a very useful laboratory tool when one tries identifying which electrically charged elementary particles are produced in a high-energy experiment. Secondly, the Maxwell equations admit "electromagnetic wave" solutions which allow understanding Optics as a part of Electromagnetism, another step towards the unification of physical phenomena which the raw human senses would a priori feel to be very disparate; this synthesis was proposed by Maxwell (1865).

Section 1. PHENOMENOLOGICAL BACKGROUND

The first fact of experience, recognized by any theory of electromagnetism, is the existence of electric charges that can be perceived through the forces exerted on them by electric and magnetic fields.

We take, as a primary concept of the theory, the notion of *charge*, a property quantified by some $e \in \mathbb{R}$, that can be associated to material bodies.

As in our approach to the Newtonian theory of gravitation (see Chapter One), we first concentrate our attention on the idealization where the scale of the phenomena to be considered is such that the notion of *test-particle* makes sense. These are objects described by their trajectories $\mathbf{x} : t \in I \mapsto \mathbf{x}(t) \in \mathbb{R}^3$ (with $I \subseteq \mathbb{R}$), their inertial mass m, and their electric charge e. For these particles, the fact of experience referred to above is now quantifyable: these particles behave as if they were individually acted upon by a force, called the *Lorentz force*, of the form

$$\mathbf{F}(\mathbf{x}(t)) = e\,\{\mathbf{E}(\mathbf{x}(t),t) + \dot{\mathbf{x}}(t) \times \mathbf{B}(\mathbf{x}(t),t)\} \tag{1}$$

where $\mathbf{E}(\cdot,t)$ and $\mathbf{B}(\cdot,t)$ are vector fields on $\mathbb{R}^3$ that may depend explicitly on the time t. It is a part of the definition of a test particle that its presence disturbs the fields E and B so little that this disturbance can be ignored.

In any experimental situation, prescribed by non-vanishing electric and magnetic fields E and B, the Lorentz force (1) can be used to compare the charges of different test particles; this is done routinely—albeit at great costs—in both the preparation and the analysis of the final products of the scattering situations that constitute the heart of experimental elementary particle physics. Since enough experiments of this kind have been performed consistently, one can use (1) to probe E and B and verify that these fields satisfy the so-called Maxwell equations. In an emphatically not accidental manner, these equations are the simplest field equations one can write in a relativistic theory: E and

B are the components of a 2-form $\mathcal{F}$ on space time, that satisfy $d\mathcal{F} = 0$ and defines $J \equiv \operatorname{div} \mathcal{F}$; see Sections 5.1 and 6.2. We shall take here the more traditional description and write the Maxwell equations in the form

$$\operatorname{curl} \mathbf{B} - \dot{\mathbf{E}} = \mathbf{J} \tag{2a}$$
$$\operatorname{div} \mathbf{E} = \rho \tag{2b}$$
$$\operatorname{curl} \mathbf{E} + \dot{\mathbf{B}} = 0 \tag{2c}$$
$$\operatorname{div} \mathbf{B} = 0 \tag{2d}$$

One can view here (2a) and (2b) as definitions of **J** and ρ, and note that they imply

$$\operatorname{div} \mathbf{J} + \dot{\rho} = 0 \,. \tag{3}$$

While the above facts of experience provide a phenomenological basis on which a modern physicist could accept the Maxwell equations as describing, together with the Lorentz force (1), a law of nature, it is nonetheless instructive to study briefly how the above programme was actually implemented in the years that preceded the inception of the theory; the remainder of this section retraces some of these historical steps. To understand what went on in this process, we briefly take advantage of the hindsight provided by the knowledge of the final form of the theory, and remark that the Maxwell equations admit static solutions, in which **E** and **B** do not depend on time; in this case, the equations decouple to give on the one hand:

$$\operatorname{div} \mathbf{E} = \rho \tag{4a}$$
$$\operatorname{curl} \mathbf{E} = 0 \tag{4b}$$

the study of which is known as *electrostatics*; and on the other hand:

$$\operatorname{curl} \mathbf{B} = \mathbf{J} \tag{5a}$$
$$\operatorname{div} \mathbf{B} = 0 \tag{5b}$$

which similarly delineate the realm of *magnetostatics*. These two pairs of equations synthetize the two classes of experimental situations that were first studied, very much independently from one another, before they were realized to belong to a single unified theory.

The word "electricity" seems to have been first used in 1600, by Gilbert, to characterize a quality that many substances share with amber ($\equiv$ *Ηλεκτρος*): when rubbed, they attract small particles (our test-particles, which Gilbert visualized as chaff). Du Fay is credited for having noticed, in 1734, that there are two kinds of electricities: those of the same kind repel one another, while those of opposite kinds attract each other. Several authors, among them B. Franklin, emphasized that there should be some additive measure of the quantity of electricity.

The quantification of these rather vague notions came in 1784 when Coulomb determined by experiment (with a torsion balance) that "the repulsive forces between two small spheres charged with the same kind of electricity is in the inverse ratio of the square of the distance between the center of the two spheres ... [and] the law of inverse squares was found to hold also the case of attraction." This is the basis for what we know to this day as the *Coulomb law of electrostatics*:

$$F = (4\pi)^{-1}\, e_0\, e\, r^{-2} \tag{6}$$

Compare with (1.1.16), and note that $(4\pi)^{-1}$ stands for a universal constant that takes this particular value as a consequence of a choice of the units in which we elect to measure the electric charges. Notice also that (6) gives the magnitude of the force; this force is a central force: specifically the force, that e_0 exerts on e, is applied on e and directed along the line joining e_0 to e if e_0 and e are of the same "kind", and in the opposite direction if e and e_0 are of opposite kinds.

More than a decade before Coulomb published his results, Cavendish had arrived at the same conclusion, namely that the force between two point charges decreases as the inverse of the square of their distance. An argument about priorities would be out of place here; it is, nevertheless, of interest to note that the methods used by Coulomb and Cavendish to arrive at this conclusion were quite different. Coulomb's was a direct measurement of the force as a function of distance. Cavendish considered instead the force exerted by a large metallic (i.e. uniformly charged) sphere on a test-particle placed inside or outside this sphere. He then sized on the analogy between this situation and that considered by Newton in his treatment of gravitational forces: a test-particle placed anywhere within the sphere will experience no net force exactly when the two-body central force between point charges is exactly proportional to r^{-2}, while a net force toward (or away from) the center of the large sphere would be observed if $F(r) \sim r^{-2+\epsilon}$ with $\epsilon \neq 0$. This conditional statement is what Cavendish published; the collection of Cavendish's laboratory notes, edited by Maxwell (1879), shows however, that Cavendish had actually performed the experiment with great care, and concluded "that the electric attraction and repulsion is inversely as the square of the distance, or to speak more properly, that the theory will not agree with experiment on the supposition that it varies according to any other law". Cavendish even put an upper bound on ϵ : $|\epsilon| < 1/60$; by the end of the 19th century this was improved to 10^{-4}, and modern direct evidence from atomic physics reduced this estimate to better than 10^{-9}. Notice further that Cavendish also used Newton's argument to show that a test-particle, placed outside a uniformly charged sphere, is submitted to a force, equal to that due to an identical total charge concentrated at the center of the sphere, and that the force decreases indeed as the inverse of the square of the distance between the test-particle and the center of the sphere.

Cavendish's approach lends itself naturally to a modern reformulation that shows the logical circumstances under which Coulomb's law (6) is equivalent to the basic field equations (4) of electrostatics, as we shall now see.

The first ingredient in Cavendish's approach is the use of a test-particle to probe the effect of a charge e_0. This can be described by introducing the *electric vector field* E_0 defined by $\mathrm{F} = \mathrm{E}_0\, e$, where F is the force that the charge e_0, located at $\mathbf{x}_0$, exerts on a test particle of charge e, located at $\mathbf{x} \neq \mathbf{x}_0$; from (6) we obtain:

$$\mathrm{E}_0(\mathbf{x}) = \frac{1}{4\pi} \cdot \frac{e_0}{|\mathbf{x}-\mathbf{x}_0|^2} \cdot \frac{\mathbf{x}-\mathbf{x}_0}{|\mathbf{x}-\mathbf{x}_0|} \tag{7}$$

The second ingredient in Cavendish's approach is the principle of superposition, according to which the effect of a collection of n charges e_k, located at $\mathbf{x}_k$, is the sum of the effect of the individual charges, i.e.

$$\mathrm{E}(\mathbf{x}) = \frac{1}{4\pi} \sum_{k=1}^{n} \frac{e_k}{|\mathbf{x}-\mathbf{x}_k|^3} (\mathbf{x}-\mathbf{x}_k) \tag{8}$$

Finally, Cavendish used implicitly the continuum limit, where the discrete collection $\{e_k \,|\, k = 1, 2, \ldots, n\}$ is replaced by a charge density ρ, to give the following generalization of (8):

$$\mathrm{E}(\mathbf{x}) = \frac{1}{4\pi} \int d^3x_0 \frac{\rho(\mathbf{x}_0)}{|\mathbf{x}-\mathbf{x}_0|^3} (\mathbf{x}-\mathbf{x}_0) \tag{9}$$

As already noted, Cavendish derived from (9) the electric field created by a uniformly charged spherical shell, of vanishing thickness, and of radius a:

$$\mathrm{E}_0(\mathbf{x}) = \begin{cases} \frac{e_0}{4\pi} \cdot \frac{\mathbf{X}-\mathbf{X}_0}{|\mathbf{X}-\mathbf{X}_0|^3} & \text{for} \quad |\mathbf{x}-\mathbf{x}_0| > a \\ 0 & \text{for} \quad |\mathbf{x}-\mathbf{x}_0] < a \end{cases} \tag{10}$$

From this (or any similarly obtained) formula, the idealization involved in the characterization of a point charge can be comfortably inferred. Hence (9) contains Coulomb's law (6); with this understanding, we will refer to (9) as the *Cavendish-Coulomb law*, and show that it is equivalent to the field equations (4) of electrostatics.

Note indeed that (9) (or in a simplistic version, (7) itself) implies that for any (regular) open domain $V \subset \mathbb{R}^3$ and any (regular) open two-dimensional surface $S \subset \mathbb{R}^3$:

$$\int_{\partial V} d\mathbf{s} \cdot \mathrm{E} = \int_V dv\, \rho \quad \text{and} \quad \int_{\partial S} d\mathbf{l} \cdot \mathrm{E} = 0 \tag{11}$$

where E is the electric field created by that part of the charge distribution ρ which is contained in V.

The general Stokes theorem

$$\int_{\partial U} \omega = \int_U d\omega \tag{12}$$

in its special forms (known as Gauss and Stokes formulas, respectively)

$$\int_{\partial V} d\mathbf{s} \cdot \mathbf{E} = \int_V dv \ \mathrm{div}\ \mathbf{E} \quad \text{and} \quad \int_{\partial S} d\mathbf{l} \cdot \mathbf{E} = \int_S d\mathbf{s} \cdot \ \mathrm{curl}\ \mathbf{E} \tag{13}$$

allows to rewrite (11) as

$$\mathrm{div}\ \mathbf{E} = \rho \quad \text{and} \quad \mathrm{curl}\ \mathbf{E} = 0 \tag{14}$$

which are indeed (4a) and (4b), the fundamental differential equations of electrostatics, which we wanted to justify phenomenologically.

Note further that these equations imply that

$$\mathbf{E} = -\mathrm{grad}\ \Phi \quad \text{with} \quad \Delta\Phi = -\rho \tag{15}$$

where Δ is the *Laplace operator* (div grad) and the second of these equations is referred to as the *Poisson equation*. The physical interpretation of the real-valued function Φ follows from its definition (15) written in the form:

$$e\,[\Phi(\mathbf{x}_2) - \Phi(\mathbf{x}_1)] = -e \int_{\mathbf{x}_1}^{\mathbf{x}_2} d\mathbf{l} \cdot \mathbf{E} \tag{16}$$

which is the *work* to be performed in order to bring a test-particle of charge e, from the position $\mathbf{x}_1$ to the position $\mathbf{x}_2$, and is independent (see the second of the equations 11) of the path followed from $\mathbf{x}_1$ to $\mathbf{x}_2$. Hence Φ is referred to as the *electrostatic potential* for the field E; this function is evidently defined only up to an additive constant. With this remark, it is easy to verify, for instance, that the electrostatic potential created by a smooth charge distribution ρ, with compact support, can be chosen to vanish at infinity (and in fact to fall to zero at least as fast as $|\mathbf{x}|^{-1}$ when $|\mathbf{x}| \to \infty$); with this condition Φ then satisfies

$$\Phi(\mathbf{x}) = \frac{1}{4\pi} \int d^3x_0 \ \frac{\rho(\mathbf{x}_0)}{|\mathbf{x} - \mathbf{x}_0|} \tag{17}$$

from which the Cavendish-Coulomb law (9) is recovered.

We shall need, later on, the expression of the field created by an *electric dipole*, i.e. by a pair of charges, of magnitude e and of opposite sign, separated by a distance $\mathbf{d}$. At a point $\mathbf{x}$ far away from this arrangement, or more specifically in the limit where $d \to 0$, with $ed \equiv \mathbf{p}$ fixed, one obtains readily:

$$\Phi(\mathbf{x}) = -\frac{1}{4\pi}\mathbf{p} \cdot \ \mathrm{grad}\ \left(\frac{1}{|\mathbf{x}|}\right) = \frac{1}{4\pi|\mathbf{x}|^3}\ \mathbf{p} \cdot \mathbf{x} \tag{18}$$

In particular, at fixed $\mathbf{x}/|\mathbf{x}|$, the electric field thus falls as $|\mathbf{x}|^{-3}$.

Upon returning to the general case, and pursuing the interpretation of Φ one step further, one can associate to a collection $\{(e_k, \mathbf{x}_k) \mid k = 1, 2, \ldots, n\}$ of charges the total energy necessary to assemble this collection, namely

$$U \equiv \frac{1}{4\pi} \sum_{\text{pairs}} \frac{e_j\, e_k}{|\mathbf{x}_j - \mathbf{x}_k|} \tag{19}$$

or more generally, to a smooth charge distribution ρ with compact support, one associates

$$\begin{aligned} U &\equiv \frac{1}{2} \int d^3x \int d^3x_0 \cdot \frac{1}{4\pi} \cdot \frac{\rho(\mathbf{x})\rho(\mathbf{x}_0)}{|\mathbf{x} - \mathbf{x}_0|} \\ &= \frac{1}{2} \int d^3x\, \rho(\mathbf{x})\Phi(\mathbf{x}) \\ &= -\frac{1}{2} \int dv (\Delta\Phi)\Phi = \frac{1}{2} \int dv |\operatorname{grad} \Phi|^2 \end{aligned} \tag{20}$$

In this manner, one can, therefore, associate to an electric field $\mathbf{E}$ at $\mathbf{x}$ an *electrostatic energy density*

$$u(\mathbf{x}) = \frac{1}{2} |\mathbf{E}(\mathbf{x})|^2 \tag{21}$$

With this remark, we end our excursion into the fundamentals of electrostatics, and engage in a brief exploration of the basic facts of magnetostatics.

For centuries, the "effluvium" later known under the name of magnetic field was detected through its action on elongated iron or steel needles that had been "touched" with a piece of the naturally magnetized mineral, called magnetite —$Fe_3 0_4$—or lodestone, already mentioned by Thales of Miletus and Plato for the property it has to attract iron objects. While the origin of the idea to draw practical use from the fact that free magnetic needles "consistently" point to the North is a matter of national legends now blurred by time, the existence of mariner compasses has been documented in China, Western Europe, Arabia and Scandinavia from written sources dating variously from the 11th to the early 13th centuries.

Some attempts at a quantitative theory were made in the 17th and 18th centuries; the reading of these, however, is often confusing, due to a tentative analogy with electrostatics which is unfortunately faulted by nature's choice not to allow for magnetic mono-poles, the would-be equivalents of the electric charges. A theory, that was to be free of such mono-poles, was initiated in 1820 by the investigations published in that same year by Oersted, Biot & Savart, Ampère, and Arago.

Oersted, while still speaking of separate austral and boreal magnetic poles, correctly identified experimentally the source of the magnetic field, namely the electric current; news of his discovery spread immediately.

which his light particles initially travelled. This flaw was corrected by Fermat (1601–1665) in two notes, published posthumously in 1679. It was also in these papers that he enunciated the famous principle according to which "nature operates by means and ways that are the easiest: ... we do not consider paths that are shorter in space, but those that are shorter in time". According to this principle, the light rays are the solutions of the variational problem

$$\delta \int_{x_0}^{x_1} \frac{ds}{v} = 0 \tag{2}$$

where v is the instantaneous speed of light on its trajectory. In the particular case of a homogeneous medium, where v is constant, the distinction between the shortest paths in time and in space evidently vanishes, and the principle reduces to that used already by Hero of Alexandria (1st century A. D.) to rationalize the first two laws of geometrical optics; this historical precedent explains the caveat form in which Fermat enunciated his principle. In the case where we have two media separated by a sharp boundary, the Fermat principle gives

$$\frac{1}{v_i} \sin \alpha = \frac{1}{v_r} \sin \gamma \tag{3}$$

which, upon comparison with (1), leads to the interpretation of the refractive index n as a quantity proportional to v^{-1}; the universal constant of proportionality c, which has the dimension of a velocity, can be taken to be equal to 1 by a convenient choice of the units in which one measures space and time. With this interpretation of n, (2) can be rewritten as

$$\delta \int_{x_0}^{x_1} ds\, n = 0 \tag{4}$$

Fermat's principle appeared to be compatible with a corpuscular theory of light, so much so in fact that, in the variational form just stated, it was to exert a profound influence on the formulation of mechanics itself (See Section 1.2).

Still another aspect of light, namely its colour, was to be first interpreted in terms of such a corpuscular theory. This approach was expounded by Newton in his *New theory about light and colours* (1672), and in his *Opticks* (1704). Newton's basic experimental evidence was collected when he passed beams of light through sequences of glass prisms; he observed the by-now familiar rainbow patterns and was able to conclude that: (a) the primary colours that appear in the rainbow (or spectrum) are intrinsic properties of light, and not properties of its passage through prisms; in particular, a beam of any primary colour, when selected from the rainbow pattern emerging after a beam of sunlight had passed through a first prism, does not produce another rainbow upon its impinging on a second prism; but on the contrary, it emerges from the second prism with the same colour it had upon entering this prism; (b) a given sunlight ray is an aggregate of several rays of different colours; white light can be synthesized back from these colours; and (c) the refractive index is a function of the colour.

Ignoring altogether the phenomena of diffraction (see below) which however had been discovered by then, Newton argued: "Are not the rays of light very small bodies emitted from shining substances? For such bodies will pass through uniform mediums in right lines, without bending into the shadow ... they will also be capable of several properties, and be able to conserve these properties unchanged in passing through several mediums." For Newton, primary colours were such intrinsic properties of light; Newton also knew about a phenomenon we shall describe below, namely double refraction, and he attributed it to the same cause, interpreting what we call today polarization as yet another intrinsic property of the particles that constitute light.

Diffraction phenomena not withstanding, the theory of a corpuscular nature of light was not sterile. For instance, it was on the basis of this theory that Bradley (1729) discovered the phenomenon of stellar aberration; after having made careful observations of the directions in which a telescope must be pointed towards a distant star, at different seasons of the year, to compensate for the velocity u of the earth on its orbit, he was able to compute that the ratio of the speed of the earth to the speed of light is 1:10210 (for the relativistic treatment of this aberration, see Section 5.2); the interest of Bradley's computation is that it can be completely understood by an analogy with the streaks marked by drops of rain on the side windows of a car in motion; the result agree with 5.2.21 to first order in u, as the reader will want to verify.

The discovery of the phenomenon we know today under the name of *diffraction* has been credited to Grimaldi (1618–1663) and the experiments are described in a book published posthumously in 1665 under his name. He observed that if light passes through a pin-hole made in a screen, or if a wedge is placed in a beam of light, then a close inspection in the immediate vicinity of the expected geometric image reveals a complex system of alternating regions of shadow and light, of varying width and brightness. Hooke immediately understood the potential significance of this phenomenon and, in his *Micrographia* (1665), he advanced the idea that light might be a wave; however, in the mind of their British contemporaries at least, Hooke was not able to withstand the weight of Newton's arguments in favour of a corpuscular theory.

We find again the conjecture that light is a wave in Huygens' *Traité de la lumière* (1690), where he proposed a model that could "explain" Fermat's principle, and thus answer the question as to why should light "choose" to follow the shortest path. He postulated that light is a disturbance of the medium in which it travels; considering first the case of a homogeneous isotropic medium, he further assumed that any such disturbance acts as a center creating a spherical wave which expands at constant speed; he then completed his picture by the consistency requirement that each point of the medium, upon being reached by this wave, becomes in turn the center of a spherical wave that expands with the same speed, characteristic of the medium. The instantaneous wave front is thus the envelope of the spherical waves emanating from points that have been simultaneously excited by an earlier wave front. The three laws of geometrical optics follow readily from this picture, upon identifying the light rays with the curves normal to the successive wave fronts, e.g. of a typical plane wave. For reasons that will become clear when we will describe the phenomenon of double refraction, Huygens also considered the modifications to be brought to the above scheme in the case of an anisotropic, but homogeneous, medium where the spherical waves have to be replaced, as the building elements of the wave fronts, by ellipsoidal waves.

Huygens' conjecture on the wave nature of light was revived in the beginning of the 19th century, principally under the impetus of the investigations of Young who realized that the diffraction of light should be described as an interference phenomenon, i.e. as the result of a linear superposition of several coherent waves. Young's seminal contribution was to simplify the original diffraction situations considered by Grimaldi, and to study first the interference patterns produced by two coherent sources of light. From his *Course of lectures on natural philosophy and the mechanical arts* (1807) we read the original description of what came to be known as the Young two slits experiment and which was to play a central role in the discussions of the foundations of quantum mechanics: "In order that the effects of two portions of light may be thus combined, it is necessary that they be derived from the same origin, and that they arrive at the same point by different paths, in directions not much deviating from each other ... the simplest case appears to be when a beam of homogeneous light falls on a screen in which there are two very small holes or slits which may be considered as centers of divergence ... ". At this point the reader should recall Huygens' model for the propagation of light; returning to Young, we read further: "In this case, when the two newly formed beams are received on a surface placed as to intercept them, their light is divided by dark stripes into portions nearly equal ... so as to subtend very nearly equal angles from the apertures ... and wider also in the same proportion as the apertures are closer to each other." In modern notation, this amounts to the following formula, valid in the limit where the observation is made very far from the screen in which the apertures have been opened:

$$\text{(5)} \qquad \sin \varphi_m = \left(\frac{\lambda}{d}\right) m \quad \text{with} \quad m = 0, \pm 1, \pm 2, \ldots$$

where φ_m are the angles corresponding to the bright stripes, d is the distance between the apertures, and λ depends only on the colour of the light. Young correctly identified λ with the wave length—see definition below—and estimated that in the red $\lambda \simeq 1/36,000$ inch, and in the violet $\lambda \simeq 1/60,000$ inch.

The mathematical synthesis that came out of this and subsequent Young's investigations was that, in an isotropic dispersion-free medium, light can be described as a phenomenon that satisfy the linear wave equation

$$(6) \qquad (\Delta - \frac{1}{v^2}\frac{\partial^2}{\partial t^2})\,\Psi(\mathbf{x},t) = 0$$

where Δ is the Laplace operator $\sum_i \partial^2/(\partial x^i)^2$, v is the speed of light in the medium considered, and

$$(7) \qquad I(\mathbf{x},t) = |\Psi(\mathbf{x},t)|^2$$

is the intensity of the light at the point $\mathbf{x}$ and the time t. The sources give the boundary conditions under which (6) must be solved; a great variety of diffraction phenomena (by extended apertures and obstacles, wedges, etc.) can then be reformulated in terms of integral equations that provided 19th-century analysts with a wealth of computational challenges. The simplest of the solutions of (6) is the free plane wave

$$(8) \qquad \Psi(\mathbf{x},t) = a\cdot\exp\{i(\omega t - \mathbf{k}\cdot\mathbf{x})\} \quad \text{with} \quad \omega^2 = v^2|\mathbf{k}|^2$$

which one often describes in term of its *frequency* $\nu = \omega/2\pi$, its *wave length* $\lambda = 2\pi/|\mathbf{k}|$ or its *wave vector* $\mathbf{k}$.

Besides Young's two slits experiment, another experimental set up used to demonstrate the interference of light is to let a plane wave impinge with an incidence angle φ on a parallel plate of transparent material with relative refractive index n, and to observe, separately on each side of the plate, the interference patterns between the plane waves emerging after multiple reflections and refractions. A straightforward computation shows that a maximally positive interference between the plane waves emerging from the face of incidence (resp. the face of the plate opposite to the face of incidence) after an odd (resp. even) number of reflections, occurs whenever the angle of refraction γ satisfies

$$(9a) \qquad \cos\gamma = \frac{1}{2}(\frac{\lambda}{d})(m + \frac{1}{2}) \quad \text{with} \quad m = 0, \pm 1, \pm 2, \ldots$$

respectively

$$(9b) \qquad \cos\gamma = \frac{1}{2}(\frac{\lambda}{d})\,m \quad \text{with} \quad m = 0, \pm 1, \pm 2, \ldots$$

where λ is the wave length of the light inside the plate, and d is the thickness of the plate. This theoretical prediction is confirmed by experiment, thus giving us additional confidence in the model described by equations (6–8).

One could multiply the examples ... and miss the point, namely that equation (6) is not the end of the story: light is *not* a scalar wave. The first indication of this fact was obtained by Huygens' refinement of an observation due to Bartholinus who had discovered a phenomena he called *double refraction*. Bartholinus had reported, in 1669, that the anisotropic crystaline structure of a transparent mineral—Iceland spar—is such that a light ray, upon impinging normally on a face of the crystal, splits into two rays; the first, or "ordinary ray", pursues the straight normal path predicted by the usual law of refraction; there is, however, a second ray, the "extraordinary ray", that splits away from the ordinary ray at an angle, and emerges from the crystal at a distance from the ordinary ray, proportional to the thickness of the crystal; from there on, the two rays are parallel to one another. This phenomenon remains qualitatively the same if the initial ray is allowed to make an angle with the normal to the face of the crystal on which it impinges. When the crystal is rotated around an axis normal to this face, the "ordinary ray" conforms to the usual law of refraction, and is thus not affected by this rotation; the "extraordinary ray", on the contrary, follows the crystal in its rotation. Only a few years after Bartholinus' investigations, Huygens had the idea to perform the following crucial experiment. Two such birefringent crystals are placed, one behind the other, on the path of light. Huygens first verified that if the axes of the second crystal are parallel to the axes of the first crystal, then the rays emerging from the first crystal do not split again upon traversing the second crystal: the ordinary ray (resp. the extraordinary ray) makes a second ordinary (resp. extraordinary) refraction upon entering in the second crystal. Huygens further noticed that if he rotated the second crystal, around an axis normal to the face of incidence, by an angle $\alpha = \pi/2$, then again the two rays do not split a second time upon entering in the second crystal; however, the ordinary ray (resp. the extraordinary ray) emerging from the first crystal now makes an extraordinary (resp. ordinary) refraction upon entering in the second crystal. Finally, Huygens observed that for any intermediary angle $0 < \alpha < \pi/2$, the rays emerging from the first crystal both split again upon entering in the second crystal, resulting thus in four rays of various intensities that add up to the intensity of the initial ray. No such phenomenon had every been observed for water waves, nor for sound waves; Huygens conceded that this is "a wonderful phenomenon ... although I have not been able to find a reason for it, I will not refrain from pointing it out, so as to give an opportunity to others to investigate it." Since then, we have learned that Huygens' discovery is an essential ingredient in devising the following model. Light can be decomposed into two orthogonal polarizations, transverse to the direction of the light ray; the fact that, in such an orthogonal decomposition, the sum of the intensities is conserved indicates that the intensity of the light is a measure of the square of the amplitude of the corresponding vector components. Furthermore, the two orthogonal polarizations may have different refractive indices, and they do in some anisotropic substances such as the Iceland spar with which Bartholinus and Huygens experimented. This model of light, however, was only to take hold in the beginning of the 19th century, ultimately to lead to the identification of light as an electro-magnetic wave. The two further evidences that convinced the scientific community of the transverse character of the luminous waves are now briefly reported.

The first of these two experimental evidences was discovered, almost accidentally, by Malus, who published in 1809 a memoir from which we quote: " ... light, reflected by a surface of water at an angle of 52° 45′ with the vertical, has all the characteristics of one of the beams produced by double refraction; ... the angle at which light experiences this modification when it is reflected at the surface of transparent bodies is different for each of them. In general it is larger for bodies that refract light more. Above or below this angle a part of the ray is more or less modified in a way analogous to that which occurs when light passes through two crystals whose principal sections are neither parallel nor perpendicular." Hence polarization is a property that light may have independently of its passage through a birefringent substance. The curious angle $\alpha = 52^\circ 45'$ will be interpreted (see 27) from the Maxwell equations: when light is reflected from a surface separating two transparent media of relative refractive index $n = n_r/n_i$, the angle $0 < \alpha < \pi/2$ is given by the solution of the pair of equations

$$\sin\alpha = n \sin\gamma \quad \text{and} \quad \alpha + \gamma = \pi/2 \tag{10a}$$

known as *Brewster's law*, and amounting to

$$\alpha = \arctan n \tag{10b}$$

so that α is indeed an increasing function of n as Malus had noted.

In a series of sophisticated experiments Arago & Fresnel brought together the phenomenon of polarization and the wave character of light, by observing the interference pattern between beams of light having varied degrees of polarization. In their *Mémoire sur l'action que les rayons de lumière polarisée exercent les uns sur les autres* (1819) they conclude: "In the same situations in which two rays of ordinary [i.e. unpolarized] light seem to destroy each other ... two rays polarized at right angle ... exert on each other no appreciable action; ... the rays of light polarized in the same direction act on one another as ordinary rays do: so that ... [in this case] the phenomena of interference are absolutely the same [as they are when the rays of light are unpolarized]." These experiments were properly understood as the most compelling evidence to date in favour of a theory where light is described as a transverse wave.

Except for the discovery, made in 1846 by Faraday—see also the related investigations of Verdet—that a magnetic field can produce a rotation of the plane of polarization of light in matter, there was very little evidence of any direct relation between light and electromagnetism when Maxwell (1865) published his famous theoretical paper.

To approach Maxwell's contribution to the understanding of the nature of light, let us consider the equations (3.1.42 & 43) which we obtained for the electromagnetic potentials Φ and $\mathbf{A}$. In a region free of sources, i.e. where and when $\rho = 0 = \mathbf{J}$, these equations reduce to

(11a) $$(\Delta - \partial^2{}_t)\Phi = 0$$
(11b) $$(\Delta - \partial^2{}_t)\mathbf{A} = 0$$
(11c) $$\operatorname{div}\mathbf{A} + \partial_t\Phi = 0$$

For E and B defined by (3.1.38 & 39) these equations imply

(12a) $$(\Delta - \partial^2{}_t)\mathrm{E} = 0$$
(12b) $$(\Delta - \partial^2{}_t)\mathrm{B} = 0$$

which we could have derived directly from the Maxwell equations (3.1.2); note however that the two decoupled equations (12) contain less information that the Maxwell equations themselves.

Returning to the equations (11), we see that they admit the fundamental solutions

(13a) $$\Phi = \Phi_0 \sin(\omega t - \mathrm{k}\cdot\mathbf{x})$$
(13b) $$\mathbf{A} = \mathbf{A}_0 \sin(\omega t - \mathrm{k}\cdot\mathbf{x})$$

with

(13c) $$|\mathrm{k}|^2 - \omega^2 = 0 \quad \text{and} \quad \mathrm{k}\cdot\mathbf{A}_0 - \omega\Phi_0 = 0$$

from which we obtain:

(14a) $$\mathrm{E} = \mathrm{E}_0 \cos(\omega t - \mathrm{k}\cdot\mathbf{x}) \quad \text{with} \quad \mathrm{E}_0 = -\omega\mathbf{A}_0 + \mathrm{k}\Phi_0$$
(14b) $$\mathrm{B} = \mathrm{B}_0 \cos(\omega t - \mathrm{k}\cdot\mathbf{x}) \quad \text{with} \quad \mathrm{B}_0 = -\mathrm{k}\times\mathbf{A}_0$$

and thus

(15) $$\mathrm{k}\cdot\mathrm{E} = \mathrm{k}\cdot\mathrm{B} = \mathrm{E}\cdot\mathrm{B} = 0$$
(16) $$|\mathrm{E}_0|^2 = |\mathrm{B}_0|^2 = \omega^2(|\mathbf{A}_0|^2 - \Phi_0{}^2)$$
(17) $$\mathrm{S} \equiv \mathrm{E}\times\mathrm{B} = \frac{1}{2}(|\mathrm{E}|^2 + |\mathrm{B}|^2)\mathrm{u} \quad \text{with} \quad \mathrm{u} = \frac{\mathrm{k}}{|\mathrm{k}|}$$

Hence these fundamental solutions of the Maxwell equations in free space are electromagnetic plane waves $(\mathrm{E}(\mathbf{x},t))$, $(\mathrm{B}(\mathbf{x},t))$ that are *transverse* to the direction u in which they propagate, and such that the electric field $\mathrm{E}(\mathbf{x},t)$ and the magnetic field $\mathrm{B}(\mathbf{x},t)$ are mutually orthogonal. In the units we have chosen, the speed of propagation of these waves is equal to 1. When Maxwell found these solutions, he wrote: "this velocity is so nearly that of light, that it seems we have strong reasons to conclude that light itself ... is an electromagnetic disturbance in the form of waves propagated ... according to electromagnetic laws." This conjecture was proven to be correct, before the close of the century, when electromagnetic waves were produced in the laboratory by clearly electromagnetic means, and observed to obey the same laws of propagation as light including reflection, diffraction and interference of polarized waves analogous to those observed for light by Arago & Fresnel.

For the discussion of refraction, a modification has to be worked into the theory in order to account for the influences that a material medium exerts on the propagation of light: for one thing, the velocity of light must be allowed to differ from one medium to the other. The simplest phenomenological model one can build to this effect will describe an isotropic, non-absorbant dielectric medium which is homogeneous, at least over distances (and times) that are very large compared to the wave lengths (and frequencies) of the waves one wants to consider. Mathematically, this can be achieved by rewriting the basic equations (11) in the following modified form:

$$(\Delta - \epsilon\,\partial_t{}^2)\Phi = 0 \tag{18a}$$
$$(\Delta - \epsilon\,\partial_t{}^2)\mathbf{A} = 0 \tag{18b}$$
$$\text{div }\mathbf{A} + \epsilon\,\partial_t\Phi = 0 \tag{18c}$$

where the "dielectric constant" ϵ is assumed to satisfy $\epsilon \geq 1$. Upon defining again

$$\mathbf{E} \equiv -\partial_t\mathbf{A} - \text{grad }\Phi \tag{19a}$$
$$\mathbf{B} \equiv \text{curl }\mathbf{A} \tag{19b}$$

one obtains from (18) and (19) the modified Maxwell equations:

$$\text{curl }\mathbf{B} - \epsilon\dot{\mathbf{E}} = 0 \quad \text{i.e.} \quad \int_{\partial S} d\mathbf{l}\cdot\mathbf{B} - \partial_t\int_S d\mathbf{s}\cdot(\epsilon\,\mathbf{E}) = 0 \tag{20a}$$

$$\text{div }\mathbf{E} = 0 \quad \text{i.e.} \quad \int_{\partial V} d\mathbf{s}\cdot\mathbf{E} = 0 \tag{20b}$$

$$\text{curl }\mathbf{E} + \dot{\mathbf{B}} = 0 \quad \text{i.e.} \quad \int_{\partial S} d\mathbf{l}\cdot\mathbf{E} + \partial_t\int_S d\mathbf{s}\cdot\mathbf{B} = 0 \tag{20c}$$

$$\text{div }\mathbf{B} = 0 \quad \text{i.e.} \quad \int_{\partial V} d\mathbf{s}\cdot\mathbf{B} = 0 \tag{20d}$$

and thus the wave equations

$$(\Delta - \epsilon\,\partial_t{}^2)\mathbf{E} = 0 \tag{21a}$$
$$(\Delta - \epsilon\,\partial_t{}^2)\mathbf{B} = 0 \tag{21b}$$

The only modifications of the fundamental solutions (13)–(17) resulting from the replacement of (11) by (18) are that (13c), (16) and (17) now take the form

$$|\mathbf{k}|^2 - \epsilon\omega^2 = 0 \quad \text{and} \quad \mathbf{k}\cdot\mathbf{A}_0 - \epsilon\omega\Phi_0 = 0 \tag{22}$$

$$|\mathbf{E}_0|^2 = \frac{1}{\epsilon}|\mathbf{B}_0|^2 = \omega^2|\mathbf{A}_0|^2 - |\mathbf{k}|^2\Phi_0 \tag{23}$$

$$\mathbf{S} \equiv \mathbf{E}\times\mathbf{B} = \sqrt{\epsilon}|\mathbf{E}|^2\mathbf{u} \quad \text{with} \quad \mathbf{u} = \frac{\mathbf{k}}{|\mathbf{k}|} \tag{24}$$

The interpretative remarks we made after (17) thus remain valid, with now the additional flexibility that these electromagnetic waves propagate with a speed $v = 1/\sqrt{\epsilon}$; hence $n \equiv \sqrt{\epsilon}$ is to be interpreted as the refractive index of the medium.

This model can be extended to the situation where two homogeneous media of the type just described, but differing by the value of their respective ϵ, share a common smooth two-dimensional boundary, say a plane π: we assume the integral equations (20) to hold throughout the space occupied by both media, and verify that they imply that the components of $\mathbf{E}$ and $\mathbf{B}$ parallel to π are continuous across π. With these boundary conditions, the Fresnel formulas for the reflection and refraction of polarized light follow by a straightforward computation. In addition to the laws of geometrical optics enunciated at the beginning of this section, one also derives from this computation the intensities $I = |\mathbf{S}|$ of the reflected and refracted waves as a function of the intensity I_0 of the incident wave. In particular, for the reflected waves, one obtains: (a) if $\mathbf{E}$ is perpendicular to the plane of incidence

$$I_a(\alpha) = I_0\frac{\sin^2(\alpha-\gamma)}{\sin^2(\alpha+\gamma)} \tag{25a}$$

where α (resp. γ) is the angle of incidence (resp. refraction); and (b) if $\mathbf{E}$ is in the plane of incidence

$$I_b(\alpha) = I_0\frac{\tan^2(\alpha-\gamma)}{\tan^2(\alpha+\gamma)} \tag{25b}$$

For normal incidence, i.e. for $\alpha \to 0$, these two intensities coincide—as to be expected—and give

$$I_a(0) = I_b(0) = \left(\frac{n-1}{n+1}\right)^2 \tag{26}$$

where $n = n_r/n_i = (\epsilon_r/\epsilon_i)^{1/2}$ is the relative refractive index of the second medium with respect to the first. Back to (25), we notice that a singular situation occurs precisely when $\alpha + \gamma = \pi/2$; one obtains then

$$I_a = \left(\frac{n^2-1}{n^2+1}\right)^2 \quad \text{and} \quad I_b = 0 \tag{27}$$

which corresponds (see 10) to Malus' observation, and Brewster's law, on the polarization that occurs upon reflection off a dielectric material. The above formula contains moreover some additional information: first of all, it gives the intensity of the light reflected at the Brewster angle; more importantly, it provides an interpretation of the plane of polarization of light in term of the plane determined by k and E.

This remark will close this section on the nature of light. The reader should however notice that we have been singularly scant on the interaction of light with matter in bulk: a satisfactory treatment of that subject would require a microscopic knowledge of matter, and thus involve quantum mechanics ... which we do not have yet at this point.

CHAPTER 4. GEOMETRY

SYNOPSIS

Nine plane geometries appear in the Cayley-Klein classification; we single out three of them for study here: the Euclidean geometry, the doubly parabolic geometry and a pseudo-Euclidean geometry. From a mathematical point of view these are characterized as the three geometries for which the distance is an additive function along straight lines. From a physical point of view, the doubly parabolic geometry will be shown to correspond to Galilean relativity, whereas the pseudo-Euclidean geometry is the two-dimensional version of the Minkowski space-time of Einstein's special relativity; the Euclidean geometry is evidently used here for comparison purposes only. Section 1 thus serves as an introduction to Chapter Five, by way of simple examples. The same didactic purpose presided over the selection of the examples of curved space-times chosen in Section 2 to introduce the reader to the kind of geometries encountered in Chapter Six.

Section 1. GALILEAN AND MINKOWSKIAN GEOMETRIES

In a lecture before the Prussian Academy of Sciences (January 27, 1921) Einstein explained the sense in which "geometry is evidently a natural science" provided one carefully distinguishes two stages in its advances. These are: (i) "axiomatics" which in Einstein's words "purges mathematics of all extraneous elements", i.e. those elements which relate to experience and intuition; and (ii) the setting up of a correspondence between "real objects" and "the system of concepts of axiomatic geometry". This separation seems to have played a central methodological role in the elaboration of Einstein's ideas; he states indeed: "I attach special importance to the view of geometry which I have just set forth, because without it I should have been unable to formulate the theory of relativity". Einstein's characterization of the axiomatic approach to geometry calls to mind Hilbert's *Grundlagen der Geometrie* which went through ten editions, the first being published in 1899; the successive editions comprise numerous and substantial additions and the fifth edition (1922), which is essentially contemporary to Einstein's lecture, contains eight commented appendices. Einstein's ideas, while drawing most of their motivation from physics, were clearly touching upon a fiber of current mathematical preoccupations.

For our purpose in this chapter, we should actually go back further in time and recall that in the course of the Nineteenth Century serious doubts took shape concerning the ineluctability of the system of axioms—in particular the parallel postulate—defining Euclidean geometry; this flashback discounts, perhaps too harshly, the pioneering, but mistaken, work of Saccheri (1733). The first successful attempts to build alternatives to Euclidean geometries are due to Gauss (1816), Lobachevski (1826) and Bolyai (1831) who, independently, produced the hyperbolic geometry. A metric approach, on which we will have more to say in the next section, was then proposed by Riemann (1854) who also constructed an elliptic geometry, Beltrami (1866, 1868) and von Helmoltz (1868). The latter actually came close to a third point of view, associated with the names of Cayley (1859), Klein (1872, 1893) and Lie (1885, 1893); we see there geometry characterized as the study of invariant properties of figures, i.e. properties which are unchanged under a given group of "admissible" transformations. This is the point of view emphasized in this section, and we refer to it, for short, as the "Erlanger spirit" (see Klein, *loc. cit.*).

Nine plane geometries appear in the Cayley-Klein classification. Their relative positions are indicated in the following table.

ELLIPTIC	*Euclidean*	*HYPERBOLIC*
CO — Euclidean	*GALILEAN*	*CO — MINKOWSKIAN*
CO — HYPERBOLIC	*MINKOWSKIAN*	*DOUBLY — HYPERBOLIC*

Table 1

These geometries owe their names to the quadratic forms which one chooses to define, for each geometry, a measure of distance (the form is elliptic for the geometries in the first column, parabolic for those in the second, and hyperbolic for those in the third) and a measure of angle (again the form is elliptic for the geometries in the first line, parabolic for those in the second, and hyperbolic for those in the third). As we will concentrate (in this section) our attention on geometries in the second column, we choose to refer to them by names germane to the context of this work.

Models for all nine of these geometries obtain as follows. Firstly, for every pair $(t,x) \in \mathbb{R}^2$, define $z \equiv t + \eta x$,

$$\begin{aligned} z_1 + z_2 &= (t_1 + t_2) + \eta(x_1 + x_2) \\ z_1 \cdot z_2 &= (t_1 t_2 + \eta^2 x_1 x_2) + \eta(t_1 x_2 + t_2 x_1) \\ \overline{z} &= t - \eta x \end{aligned} \tag{1}$$

and η^2 to be either $-1, 0$ or $+1$. (The three lines will result from this choice of η). Note that in all three cases $z\overline{z} = t^2 - \eta x^2$. To distinguish between the columns, a notion of "distance" is chosen, namely:

$$\tan^2\{\frac{1}{2}d(z_1, z_2)\} = \frac{(z_1 - z_2)(\overline{z}_1 - \overline{z}_2)}{(1 + z_1\overline{z}_2)(1 + \overline{z}_1 z_2)} \tag{2a}$$

for the first column,

$$d(z_1, z_2)^2 = (z_1 - z_2)(\overline{z}_1 - \overline{z}_2) \tag{2b}$$

for geometries in the second column, and

$$\tanh^2\{\frac{1}{2}d(z_1, z_2)\} = \frac{(z_1 - z_2)(\overline{z}_1 - \overline{z}_2)}{(1 - z_1\overline{z}_2)(1 - \overline{z}_1 z_2)} \tag{2c}$$

for the third column.

Clearly, the geometries of the second column are then characterized among these nine geometries, by the requirement that the distance be "additive"; i.e. for colinear points z_1, z_2 and z_3, one has $d(z_1, z_2) + d(z_2, z_3) = d(z_1, z_3)$. This is at least a didactic reason for limiting, at first, one's attention to these geometries. We do so from now on, and to the end of this section. There is, however, an even better reason to do this. Indeed these three geometries are singled out by the requirement that their group of invariance transformations be given as follows: to each pair

$$\begin{gathered}\{p \equiv \alpha + \eta\beta \quad ; \quad q \equiv \sigma + \eta\varsigma\} \\ \text{with } \alpha, \beta, \sigma, \varsigma \in \mathbb{R} \text{ and } p \cdot \overline{p} = 1\end{gathered} \tag{3a}$$

corresponds an ("admissible") transformation defined by

$$z \mapsto \hat{z} = p \cdot z + q \tag{3b}$$

One verifies immediately that these transformations preserve the distance, see (2b) above, i.e. $d(\hat{z}_1, \hat{z}_2) = d(z_1, z_2)$, so that the distance is an invariant; hence, in the Erlanger spirit, the distance is a geometrical concept indeed.

We first notice that these transformations map (ordinary) straight lines into (ordinary) straight lines; hence a straight line is a geometric concept (and it had better be so, since our first reason to single out these geometries, namely the distance being additive, did involve the concept of colinearity: we do not want non-geometric concepts to sneak in).

We now review and compare these three geometries.

The first of these geometries obtains for $\eta^2 = -1$. We have thus in this case:

$$d(z_1, z_2)^2 = (t_1 - t_2)^2 + (x_1 - x_2)^2 \tag{4a}$$

which is the usual distance in two-dimensional Euclidean geometry. In the last two thousand years a lot has been said on this geometry, and we shall thus limit our comments to noticing that the transformation law (3) now reads

$$\begin{gathered}t \to \hat{t} = t \cdot \cos\alpha + x \cdot \sin\alpha + t_0 \\ x \to \hat{x} = -t \cdot \sin\alpha + x \cdot \cos\alpha + x_0 \\ \text{with } (t_0, x_0) \in \mathbb{R}^2 \text{ and } \alpha \in [0, 2\pi]\end{gathered} \tag{4b}$$

i.e. the group of transformations is a semidirect product of $\mathbb{R}^2$ with S^1, to which one may, or may not, want to add the reflections for good measure.

The second of the geometries we want to consider obtains when $\eta^2 = 0$. We have thus in this case:

$$d(z_1, z_2)^2 = (t_1 - t_2)^2 \tag{5a}$$

which is independent of $(x_1 - x_2)$; hence this notion of distance is degenerate: it produces only a semi-norm $|z| \equiv +\sqrt{d(z,0)^2}$, and we do *not* identify two points z_1 and z_2 for which $|z_1 - z_2| = 0$. The condition $p\bar{p} = 1$, with $p = \alpha + \eta v$, in the transformation law (3) now reads $\alpha^2 = 1$. For simplicity in the presentation, we shall ignore central reflections, i.e. $\alpha = -1$, and thus consider only that part of the group for which $\alpha = +1$. The transformation law (3) now reads:

$$\begin{aligned} & t \to \hat{t} = t + t_0 \\ & x \to \hat{x} = vt + x + x_0 \\ & \text{with } (t_0, x_0) \in \mathbb{R}^2 \text{ and } v \in \mathbb{R} \end{aligned} \tag{5b}$$

which thus corresponds to a translation, by (t_0, x_0), preceded by a "shear" along the x-direction; the latter is best illustrated (see Figure 1) by its effects on an arbitrary straight line through the origin.

We already noticed that, since (3) is linear, straight lines are transformed into straight lines; there is, however, a new fact: some straight lines (see Figure 2), called "special", are invariant under the homogeneous group of all shears; thus, in the Erlanger spirit, they must have an absolute geometrical meaning; this will also be given a physical interpretation later on; for the moment, let us only remark that under the full group (5b) of this geometry, special lines are mapped into special lines.

We now define, as in Euclidean geometry, a circle as the loci of all points at equal distance r of a fixed point (t_0, x_0). Such a circle is represented in Figure 2: $c = \ell_1 \cup \ell_2$.

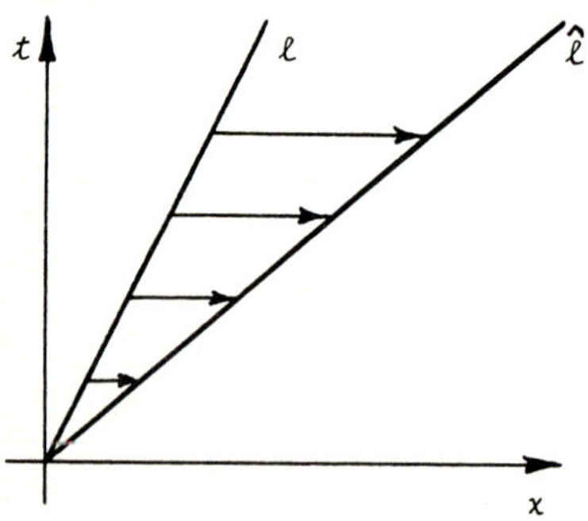

Fig.1

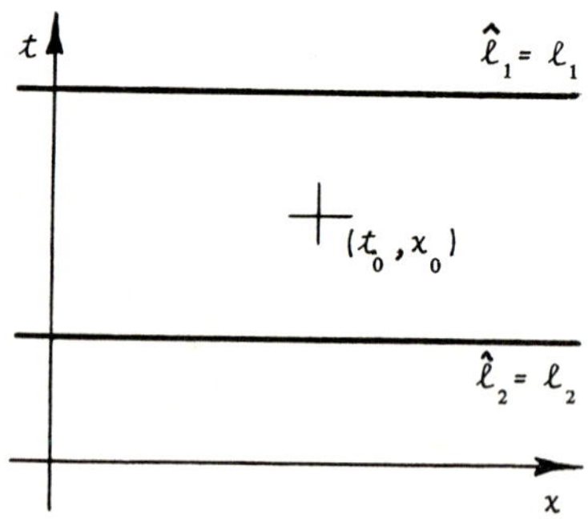

Fig.2

Again circles are mapped into circles, so that the notion of a circle is a geometrical concept indeed. It will be used now to define another geometric object: the angle between two lines intersecting at a point (t_0, x_0). These two lines, of equation

$$x_k : t \in \mathbb{R} \mapsto v_k(t - t_0) + x_0 \quad k = 1, 2 \tag{6}$$

intersect the unit circle of radius 1 centered at (t_0, x_0) at the points $(t_0 \pm 1, x_0 \pm v_k)$, intercepting segments of Euclidean length $(v_2 - v_1)$. Although the concept of Euclidean length does not belong to this geometry, $(v_2 - v_1)$, does: it is an invariant under the group of transformations (5b), i.e. $(\hat{v}_2 - \hat{v}_1) = (v_2 - v_1)$; this is called the "angle" between the two lines.

Another peculiarity of Galilean geometry is that, for each fixed $a \in \mathbb{R}$, the family $\Gamma(a)$ of curves:

$$x : t \in \mathbb{R} \mapsto \frac{1}{2}a(t - t_0)^2 + v_0(t - t_0) + x_0 \, , \tag{7}$$

parametrized by $(v_0, t_0, x_0) \in \mathbb{R}^3$, is invariant under the group of transformations (5b). Indeed, under any transformation of the form

$$\begin{aligned} & t \to \hat{t} = t + \tau \\ & x \to \hat{x} = \beta t + x + \xi \\ & \text{with } (\tau, \xi) \in \mathbb{R}^2 \text{ and } \beta \in \mathbb{R} \end{aligned} \tag{8}$$

the curves (7) become

$$\begin{aligned} & \hat{x} : \hat{t} \in \mathbb{R} \mapsto \frac{1}{2}a(\hat{t} - \hat{t}_0)^2 + \hat{v}_0(\hat{t} - \hat{t}_0) + \hat{x}_0 \\ & \text{with } : \hat{v}_0 = v_0 + \beta \, ; \, \hat{t}_0 = t_0 + \tau \, ; \, \hat{x}_0 = \beta t_0 + x_0 + \xi \end{aligned} \tag{9}$$

Note that each of these curves still belongs to $\Gamma(a)$, with the same a!

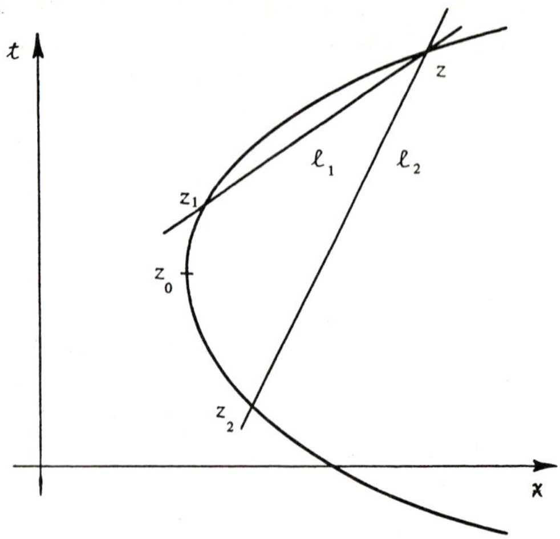

Fig.3

The geometric significance of these special curves results from the following computation (see Figure 3).

The slope of the straight line $\ell_k\,(k = 1,2)$ joining z_k and z is given by

$$v_k = \frac{x(t) - x_k}{t - t_k} = \frac{1}{2}a(t_k + t - 2t_0) + v_0 \tag{10}$$

so that what, in this geometry, we called the "angle" between the lines ℓ_1 and ℓ_2 is given by:

$$v_2 - v_1 = \frac{1}{2}a|z_2 - z_1| \tag{11}$$

which is independent of z. Hence the "special" lines (7) can be defined, in a geometrically invariant manner, as the loci of all the points z such that the angle between two lines ℓ_1 and ℓ_2, issued from two fixed points z_1 and z_2 is proportional to the distance $|z_2 - z_1|$ between these two points. This is reminiscent of an alternate definition of the circle in Euclidean geometry; accordingly, we shall call *cycles* the special curves (7); in the same way that the radius r of a circle in Euclidean geometry is a geometric invariant, so is the number a for a cycle in Galilean geometry.

Notice also that Figure 3 illustrates that the measure of angles in Galilean geometry is "parabolic" indeed.

So much for the axiomatic of Galilean geometry. Its physical relevance can be seen as follows. Interpret the point (t, x) as the mark that an event occurs, at time t, in position x. The distance (5a) between two points measures how far two events are separated in time; in particular, it is zero if and only if these events are simultaneous. *Time is a Galilean absolute*, in the sense that not a word of the preceding sentence is changed by a change of coordinates (5b); this change of coordinates corresponds to the kind of change you make when, instead of looking at whatever there is to be seen of the World when sitting in Lake Wobegon, you decide you would rather describe that same part of the World while driving straight by at constant speed v, *and* that you choose arbitrarily the origin of space and of time. There, as elsewhere, the curves (7) describe uniformly accelerated motions; while your Uncle, sitting in Lake Wobegon, and you passing by, will disagree on the value of the velocity v_0 to be introduced in (7), see formula (9), you will both agree, precisely, on the value of a: *acceleration is a Galilean invariant*, velocity is not—moreover, in contradistinction with what happens in Minkowskian geometry, there is no privileged velocity.

However natural the above physical intuition seems to be, it is wrong in its claim on "absolutes", and it is correct only as a first approximation of what the World looks like when seen from relatively slow moving platforms like ships on the sea (Galileo), cars on the highways (our Prairie Home Companion), or even the Earth on its orbit (where it took the Michelson-Morley experiment to observe a measurable departure from this approximation). A better geometry will turn out to be that of Minkowski, which we now first approach from the Erlanger point of view again.

Going back to equations (1), (2b) and (3), we saw that $\eta^2 = -1$ (resp. $\eta^2 = 0$) corresponds to Euclidean geometry (resp. to Galilean geometry). We now let $\eta^2 = +1$. The distance (2b) becomes

$$d(z_1, z_2)^2 = (t_2 - t_1)^2 - (x_2 - x_1)^2 \tag{12a}$$

with the *RHS* taking negative as well as positive values. The condition $p\bar{p} = 1$, with $p = \alpha + \eta\beta$ in the transformation law (3), now reads $\alpha^2 - \beta^2 = 1$ which we can write $\alpha = \cosh\lambda$ and $\beta = \sinh\lambda$, so that (3) becomes

$$\begin{aligned} t \to \hat{t} &= t\ \cosh\lambda + x\sinh\lambda + t_0 \\ x \to \hat{x} &= t\ \sinh\lambda + x\cosh\lambda + x_0 \\ &\text{with } (t_0, x_0) \in \mathbb{R}^2 \text{ and } \lambda \in \mathbb{R} \end{aligned} \tag{12b}$$

which makes easy a formal comparison with the Euclidean formula (4b). Upon writing $v = \tanh\lambda$, we can rewrite (12b) as

$$\begin{aligned} t \to \hat{t} &= t \cdot \frac{1}{\sqrt{1-v^2}} + x \cdot \frac{v}{\sqrt{1-v^2}} + t_0 \\ x \to \hat{x} &= t \cdot \frac{v}{\sqrt{1-v^2}} + x \cdot \frac{1}{\sqrt{1-v^2}} + x_0 \\ &\text{with } (t_0, x_0) \in \mathbb{R}^2 \text{ and } |v| < 1 \end{aligned} \tag{12c}$$

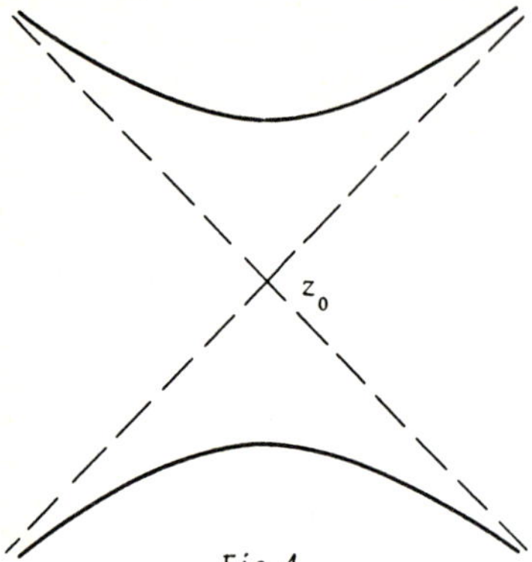

Fig.4$_+$

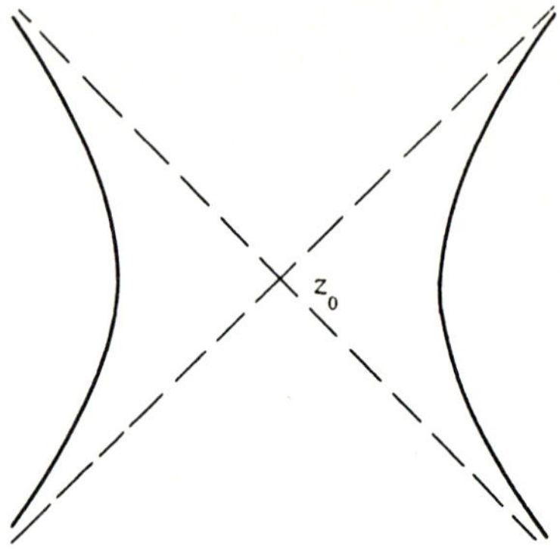

Fig.4$_-$

which makes tempting a comparison with the Galilean formula (5b), especially for $|v| \ll 1$. Here, however, we are in search of geometric absolutes, and we shall thus refrain from any such approximation.

We already noticed that, since the transformation (3) are linear, straight lines are mapped onto straight lines, so that the notion of straight line is a legitimate concept again for the Minkowski geometry. We further notice that, given any point z_0, the loci of all points z such that $d(z_1, z_0) = 0$ consists of two 45° straight lines through the point z_0. These are called the *null-lines* of the geometry (two such lines are pictured as dotted lines in Figure 4). More generally, for every $\rho \geq 0$ we define two kinds of circles, $c_+(\rho, z_0)$ and $c_-(\rho, z_0)$ centered at a point z_0, namely

$$c_\pm(\rho, z_0) = \{z \,|\, d(z_1, z_0)^2 = \pm\rho^2\} \tag{13}$$

pictured in Figure $4_\pm$ above; notice in particular that the null-lines appear as the circle $c_+(0, z_0) = c_-(0, z_0)$.

Since the distance is an invariant notion, the families

$$\Gamma_\pm(\rho) \equiv \{c_\pm(\rho, z_0) | z_0 \in \mathbb{R}^2\} \tag{14}$$

are geometrically meaningful notions: circles are mapped onto circles of the same kind.

We say that a straight line ℓ is *time-like* (resp.*space-like*) if it is not a null-line and if it intersects $c_+(1, z_0)$ (resp. $c_-(1, z_0)$) for any (and thus all) $z_0 \in \ell$. This again is a geometrically meaningful notion: time-like lines (resp. space-like lines) are mapped onto time-like lines (resp. space-like lines) by any transformation in (12c).

We claim next that the Euclidean area is a meaningful (i.e. invariant) notion in Minkowski geometry. To see this, we first notice that the two null-lines through the origin suggest the change of variables

$$\xi = \frac{1}{\sqrt{2}}(t+x) \quad ; \quad \eta = \frac{1}{\sqrt{2}}(t-x) \tag{15a}$$

and that under any homogeneous (i.e. $x_0 = 0 = t_0$) transformation in (12c) we have:

$$\hat{\xi}\hat{\eta} = \frac{1-v}{\sqrt{1-v^2}} \cdot \frac{1+v}{\sqrt{1-v^2}}\xi\eta = \xi\eta \tag{15b}$$

so that the Euclidean area is invariant under the homogeneous transformations. Since it is also obviously invariant under the translations, we conclude that the Euclidean area is invariant under the group of all transformations (12c).

We now use this geometric invariant to define the angle between two time-like lines (resp. space-like lines), intersecting at z_0, as twice the Euclidean area between these two lines and the circle $c_+(1, z_0)$ (resp. $c_-(1, z_0)$). This angle is in fact equal to the Minkowski length of the segment they intercept on the circle. By invariance, it is sufficient to verify this for the situation described in Figure 5 (the proof being conducted similarly for the angle between two space-like lines).

For the equations of the line ℓ and the circle $c_+(1,0)$ we have:

$$x : t \in \mathbb{R} \mapsto vt \in \mathbb{R} \text{ and } t^2 - x^2 = +1 \tag{16}$$

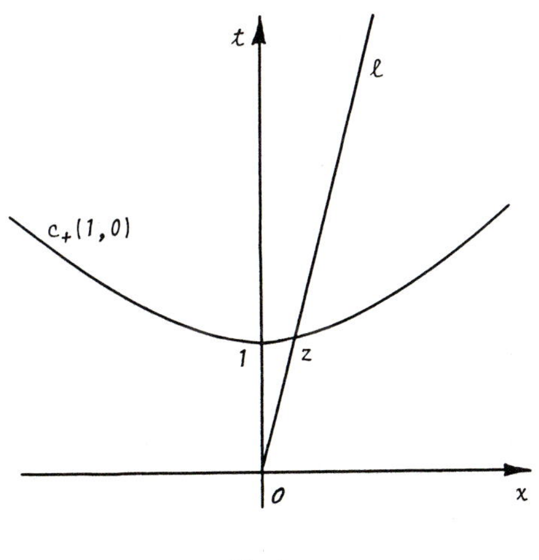

Fig.5

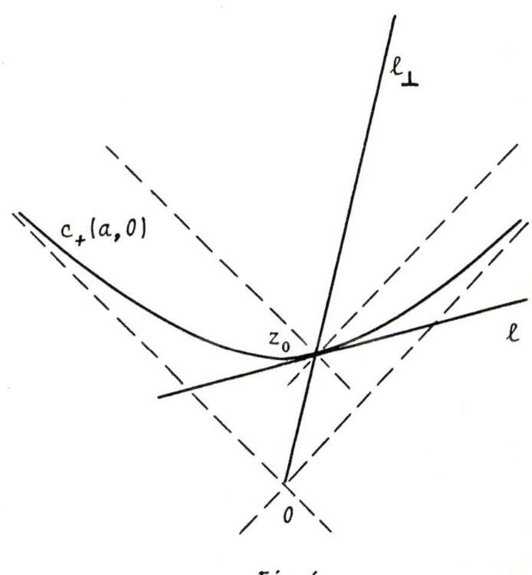

Fig.6

We thus have, for the Minkowski length of the segment $\widehat{1z}$ on the circle:

(17)
$$\begin{aligned}\varphi &= \int_1^z ds = \int_1^t dt\sqrt{\left(\frac{dx}{dt}\right)^2 - 1} \\ &= \int_1^t dt\{\left(\frac{t}{\sqrt{t^2-1}}\right)^2 - 1\}^{\frac{1}{2}} = \int_1^t dt\frac{1}{\sqrt{t^2-1}} \\ &= \ell n\,(t + \sqrt{t^2-1}) = \cosh^{-1} t\end{aligned}$$

so that

(18)
$$v = \tanh \varphi$$

For the Euclidean area of the curvilinear triangle $0z1$, we have:

(19)
$$\begin{aligned}A &= \frac{1}{2}\,tx - \int_1^t dt\; x(t) = \frac{1}{2}\,t\sqrt{t^2-1} - \int_1^t dt\sqrt{t^2-1} \\ &= \frac{1}{2}\int_1^t dt[\frac{d}{dt}(t\sqrt{t^2-1}) - 2\sqrt{t^2-1}] = \frac{1}{2}\int_1^t dt\frac{1}{\sqrt{t^2-1}}\end{aligned}$$

Upon comparing (17) and (19), we obtain indeed our assertion

(20)
$$\varphi = 2A\,.$$

We should remark that not only does this relation verify that the angle φ is a geometric invariant (since the RHS is one, as we have seen), but also that it establishes a parallel between this definition of the angle with that used in the Galilean case (see Figure 3, and the related discussion, establishing a parallel between that definition and the usual definition of angles in Euclidean geometry). This illustrates the statement that in Minkowski geometry the measure of angles is "hyperbolic". Moreover, since the Euclidean area is obviously an additive concept, the angle $(\varphi_2 - \varphi_1)$ between two lines such as the one drawn in Figure 5 is linked to their slopes by the formula (see 18):

(21)
$$\tanh(\varphi_2 - \varphi_1) = \frac{v_2 - v_1}{1 - v_1 v_2}$$

In particular, we should notice that

(22)
$$\lim_{v_2 \to \pm 1} \frac{v_2 - v_1}{1 - v_1 v_2} = \pm 1$$

a mathematically trivial formula which however translates the fundamental postulate of Einstein's special relativity, namely that the velocity of light is the same in all reference frames. Let us, however, not be distracted at this point by interpretative commentaries: we still have one purely geometrical concept which we want to bring out.

We still have no measure of angle between a time-like line and a space-like line. We actually do not really need one, except for a relation of orthogonality. We obtain the latter as follows (see Figure 6).

Let $c_+(a,0)$ be the Minkowski circle, of radius a, centered at 0; $z_0 = (t_0, x_0) \in c_+(a,0)$, and

$$tt_0 - xx_0 = a^2 \tag{23}$$

be the equation of the tangent to $c_+(a,0)$ at z_0. This is a space-like line ℓ. Let now $\ell_\perp$ be the straight line issued from 0 and joining the point $z \in \ell$ such that $d(z,0)$ be extremal. One verifies easily that $z = z_0$ and that

$$tx_0 - xt_0 = 0 \tag{24}$$

is the equation of the time-like line $\ell_\perp$. By analogy of this construction with the situation encountered in the Euclidean case, we call $\ell_\perp$ the *normal* to ℓ at z_0. This is clearly a geometrically invariant concept, independent in fact of the circle $c_+(a', z')$, tangent to ℓ at z_0, used to define it. Notice further that all normal pairs $(\ell, \ell_\perp)$ through z_0 are obtained from one another by some transformation in the group of all transformations in (12c) which leave invariant the point z_0, again a geometrically invariant statement; and that the null-lines through z_0 bisect the Euclidean angle between ℓ and $\ell_\perp$ (compare (23) and (24)).

This concludes our study of the elementary geometrical aspects of the two-dimensional Minkowski space. Its physical relevance is the central import of special relativity, the object of Chapter Five. Let it suffice here to say that, as in the Galilean case, the points in this space are used to track "events". The language, however, has to be refined since, contrary to the Galilean case, there is no geometrically invariant way to single out a time axis: time is not an absolute, and every time-like line can serve equally well; nor is there a geometrically invariant notion of simultaneity, only a notion of present which comprises all points linked to a fixed z_0 by space-like lines. Finally, if one measures, within a reference frame $(\ell, \ell_\perp)$ the velocities v_1 and v_2 of two time-like lines (interpreted as the trajectories of free particles), then their relative velocity is not $(v_2 - v_1)$ which is not a geometric invariant, but is given by the RHS of formula (21), which is a geometric invariant.

Section 2. CURVED SPACE-TIME GEOMETRIES

There are several ways to generalize the Minkowski geometry $M^{1,1}$ discussed in the previous section. The most immediate are to increase the number of dimensions, and define the (n,m)-Minkowski spaces $M^{n,m}$ as the linear spaces

$$\mathbb{R}^{n+m} \equiv \{z = (x^1, \ldots, x^n, t^1, \ldots, t^m) | x^j, t^k \in \mathbb{R}\}$$

i) A topological space $(S, \mathcal{O})$ is a set S together with a collection $\mathcal{O}$ of subsets of S, called *open sets*, such that: (a) $\emptyset \in \mathcal{O}$, $S \in \mathcal{O}$; (b) $U, V \in \mathcal{O}$ implies $U \cap V \in \mathcal{O}$; and (c) for any collection $\{U_i \in \mathcal{O} | i \in I\}$: $\bigcup_{i\in I} U_i \in \mathcal{O}$. A subset $B \subset S$ is said to be *closed* if its complement in S is open. The *closure* $\overline{X}$ of an arbitrary subset $X \subset S$ is the smallest closed subset containing X.

ii) With $(S, \mathcal{O})$ a topological space, a collection $\mathcal{B} \subseteq \mathcal{O}$ is said to be a *basis* if for every $U \in \mathcal{O}$ there exists $\{U_i \in \mathcal{B} | i \in I\}$ such that $U = \bigcup_{i\in I} U_i$. The topology is said to be *second countable* if it has countable basis.

iii) In the above definition of a differentiable manifold, the requirement that the topology be second countable is equivalent to the requirement that $(S, \mathcal{S})$ admit an atlas with a countable family of local charts.

iv) A topological space $(S, \mathcal{O})$ is said to be Hausdorff if and only if $x, y \in S$ with $x \neq y$ implies $\exists\, X, Y \in \mathcal{O}$ such that $x \in X$, $y \in Y$ and $X \cap Y = \emptyset$.

Remark 2. Some authors require explicitly in their definition of a differentiable manifold that S be a priori given a paracompact (Hausdorff) topological structure. This calls for some commentary and more definitions.

i) One feature of this requirement is that every paracompact space is *normal*, i.e. $\Xi, H \subset S$, Ξ, H closed, and $\Xi \cap H = \emptyset$ imply $\exists\, X, Y \in \mathcal{O}$ such that $\Xi \subset X$, $H \subset Y$, and $X \cap Y = \emptyset$. Hence this requirement appears as a strengthening of the condition that S be Hausdorff.

ii) Paracompactness (to be defined below) often comes naturally: every second countable, locally compact Hausdorff space is paracompact.

iii) With $(S, \mathcal{O})$ a topological space, a covering of S by open subsets is a family $\mathcal{F} = \{V_i \in \mathcal{O} | i \in I\}$ such that $\bigcup_{i\in I} V_i = S$. $(S, \mathcal{O})$ is said to be *compact* if in every covering $\mathcal{F}$ of S by open sets one can select a finite subcovering $\mathcal{F}_0$ of S (by open sets).

iv) For an arbitrary subset A of a topological space $(S, \mathcal{O})$ the *relative topology* of A is the topology the open subsets of which are $\mathcal{O}_A = \{U \cap A | U \in \mathcal{O}\}$.

v) A topological space $(S, \mathcal{O})$ is said to be *locally compact* if for every $x \in S$, $\exists\, U \in \mathcal{O}$ such that $x \in U$, $(\overline{U}, \mathcal{O}_{\overline{U}})$ is compact (where $\overline{U}$ is the *closure* of U, i.e. the smallest subset of S which contains U and the complement of which is open).

vi) With our definition of a differentiable manifold, $(S, \mathcal{O})$ is locally compact. Indeed, let $x \in S$, (U, φ) be a chart with $x \in U$. Since $\varphi(U) \subset \mathbb{R}^n$ $\exists\, W \subset \mathbb{R}^n$ such that $\varphi(x) \in W \subseteq \varphi(U)$, W open and $\overline{W}$ compact; then with $V = \varphi^{-1}(W)$ and $\psi = \varphi|V$: (V, ψ) is a chart. Since φ is continuous, V is open and $\overline{V}$ compact.

vii) For completeness, recall now what is meant by paracompactness. Let S be a set, $\{U_\alpha | \alpha \in A\}$ (resp. $\{V_i | i \in I\}$) be a covering of S. $\{U_\alpha | \alpha \in A\}$ is said to be a *refinement* of $\{V_i | i \in I\}$ if for every $\alpha \in A\ \exists\, i \in I$ such that $U_\alpha \subset V_i$. A covering $\{U_\alpha | \alpha \in A\}$ of a topological space $(S, \mathcal{O})$ is said to be *locally finite* if for every $x \in S$ $\exists\, U \in \mathcal{O}$ with $x \in U$ and $\#\{\alpha \in A | U \cap U_\alpha \neq \emptyset\} < \infty$. A topological space $(S, \mathcal{O})$ is said to be *paracompact* whenever $(S, \mathcal{O})$ is Hausdorff and every covering of S by open sets has a locally finite refinement by open sets.

viii) The requirement that a differentiable manifold be paracompact is important in that it guarantees the existence of partitions of unity, the later being essential for the definition of the integral.

ix) The reason, apparently, why some authors want to assume paracompactness is that they do not want to assume second countable (e.g. $\mathbb{R}^{n+1} = \mathbb{R} \times \mathbb{R}^n$ with $\mathcal{A} = \{U = x \times \mathbb{R}^n, \varphi(x,y) = y \,|\, x \in \mathbb{R}\}$ is a n-dimensional manifold which is not second countable; we shall not need to consider such perverse situations).

We now give five examples of differentiable manifolds; these will be used throughout this section to illustrate, or to motivate, the notions which we will introduce.

Example 1. $M^{n,1}$ with the differentiable structure generated by the trivial atlas generated by the identity map $\varphi : M^{n,1} \to \mathbb{R}^{n+1}$.

Example 2. $S^n \equiv \{x \in \mathbb{R}^{n+1} | (x^1)^2 + \ldots + (x^{n+1})^2 = \rho^2\}$ $(\rho > 0)$, with the differentiable structure generated by the atlas consisting of two charts obtained by stereographic projections relative to two different points $N \in S^n$. One of these charts is pictured in Figure 2; notice φ is defined on $S \setminus \{N\}$. Remark that the topology defined by this differentiable structure coincides with the relative topology which S^n inherits as a subset of $\mathbb{R}^{n+1}$.

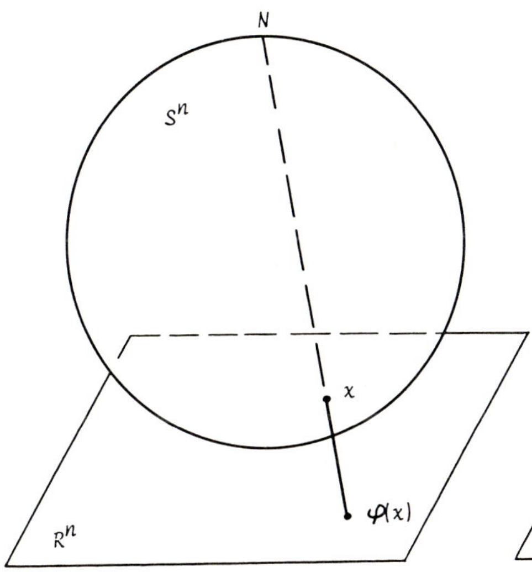

Fig.2

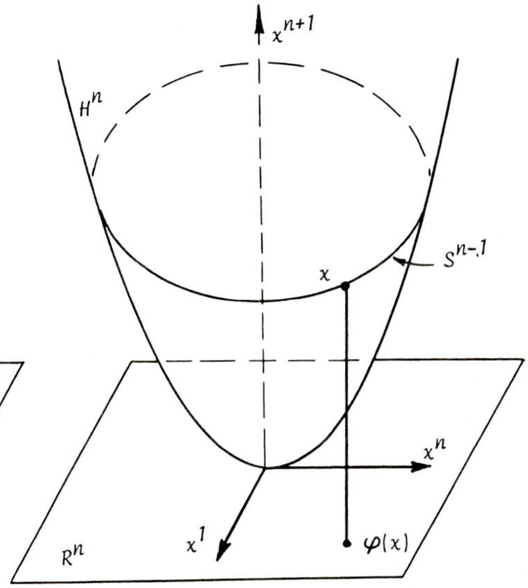

Fig.3

(i.e. $\varphi(x) = x^j e_j$); these are smooth, so that we can use (10) to define the n 1-forms dx^i on U. Let further, for every $x \in U, (\epsilon_x)_j$ be the equivalence class of the curve:

$$\gamma_j : t \in I \mapsto \varphi^{-1}(\varphi(x) + te_j) \tag{12}$$

Define then the action $(\partial_j \cdot)_x$ of $(\epsilon_x)_j$ on $\mathcal{F}(U)$ by:

$$(\partial_j \cdot)_x : f \in \mathcal{F}(U) \to (\partial_j f)_x \equiv df_x((\epsilon_x)_j) \in \mathbb{R} \tag{13}$$

One then verifies that:

$$(\partial_j f)_x = \partial_j(f \circ \varphi^{-1})|_{\varphi(x)} \tag{14}$$

where the *RHS* indicates the usual partial derivative with respect to the j-th variable in the function $f \circ \varphi^{-1} : \mathbb{R}^n \to \mathbb{R}$; moreover:

$$dx^k((\epsilon_x)_j) = \delta_j^k \quad \text{and} \quad df_x = (\partial_j f)_x dx^j \tag{15}$$

In such local coordinates, a metric g on $\mathcal{M}^n$, as defined previously, will take the form

$$g : x \in \mathcal{M}^n \mapsto g_x = (g_x)_{jk}\, dx^j \otimes dx^k \tag{16}$$

where $[(g.)_{jk}]$ is a smooth function from $\mathcal{M}^n$ to the algebra of all $n \times n$ symmetric and invertible matrices. We denote by $[(g_x)^{jk}]$ the inverse of the matrix $[(g_x)_{jk}]$.

We are now ready to have a closer look at the analogies we noticed from Table 1.

In line with formula (1a) and (1b) in the beginning of this section, an invariant metric on $M^{n,m}$ will be given by

$$\begin{aligned} &g_x = g_{jk}\, dx^j \otimes dx^k \text{ with} \\ &g_{jk} = \epsilon_j\, \delta_{jk} \text{ and } \epsilon_j = \begin{cases} +1 & 1 \le j \le n \\ -1 & n+1 \le k \le n+m \end{cases} \end{aligned} \tag{17}$$

Hence $\mathbb{R}^n \equiv M^{n,0}$ becomes a Riemann manifold, and $M^{n,1}$ becomes a Lorentz manifold. We now observe how these metrics restrict on the embedded manifolds $S^n, H^n, \mathcal{M}_+^{n,1}$ and $\mathcal{M}_-^{n,1}$.

S^n is embedded in $\mathbb{R}^{n+1}$. A change of local coordinates in $\mathbb{R}^{n+1}$ will help; consider:

$$\begin{aligned} x^1 &= \rho \cos\theta^1 \cos\theta^2 \cos\theta^3 \cdots \cos\theta^n \\ x^2 &= \rho \sin\theta^1 \cos\theta^2 \cos\theta^3 \cdots \cos\theta^n \\ x^3 &= \rho \sin\theta^2 \cos\theta^3 \cdots \cos\theta^n \\ &\vdots \\ x^{n+1} &= \rho \sin\theta^n \\ \text{with} \quad \rho^2 &= \sum_{j=1}^{n+1} (x^j)^2 \end{aligned} \tag{18}$$

in which the Riemann metric on $\mathbb{R}^{n+1}$ takes the form

(19) $$g_{R^{n+1}} = g_{S^n} + d\rho \otimes d\rho$$

with

(20) $$g_{S^n} = \rho^2 \sum_{k=1}^{n} [\prod_{j=k+1}^{n} \cos^2 \theta^j] d\theta^k \otimes d\theta^k$$

from which we conclude that S^n shares with $\mathbb{R}^n$ the property of being a Riemann manifold.

H^n is embedded in $M^{n,1}$. The local change of variables

(21) $$\begin{aligned} x^1 &= \rho \sinh \lambda \cos \theta^1 \cdots \cos \theta^{n-1} \\ &\vdots \\ x^n &= \rho \sinh \lambda \sin \theta^{n-1} \\ x^{n+1} &= \rho \cosh \lambda \\ \text{with } \sum_{j=1}^{n} (x^j)^2 - (x^{n+1})^2 &= -\rho^2 \end{aligned}$$

allows to rewrite the Lorentz metric on $M^{n,1}$ in the form:

(22) $$g_{M^{n,1}} = g_{H^n} - d\rho \otimes d\rho$$

with

(23) $$g_{H^n} = \rho^2 \{\sinh^2 \lambda \sum_{k=1}^{n-1} [\prod_{j=k+1}^{n-1} \cos^2 \theta^j] d\theta^k \otimes d\theta^k + d\lambda \otimes d\lambda\}$$

We conclude, therefore, that H^n also shares with S^n the property of being a Riemannian manifold, and this in spite of the fact that it is embedded in a Lorentzian manifold. Two incidental remarks may be in order. Firstly, a comparison of (23) and (20) shows that, as suggested in Figure 3, S^{n-1} with its Riemann structure is embedded in H^n. Secondly, $H^2_{\rho=1}$ is the hyperbolic geometry constructed by Gauss, Lobachevski and Bolyai, and is isomorphic as a Riemann manifold to the Poincaré-half plane, i.e. $H^2 \cong \{(x,y) \in \mathbb{R}^2 | y > 0\}$ with $g = y^{-2}\,(dx \otimes dx + dy \otimes dy)$.

$\mathcal{M}^{n,1}_+$ is embedded in $M^{n+1,1}$. The local change of variables

(24) $$\begin{aligned} x^0 &= \rho \cosh \tau \cdot \cos \theta^1 \cdots \cos \theta^n \\ &\vdots \\ x^n &= \rho \cosh \tau \sin \theta^n \\ x^{n+1} &= \rho \sinh \tau \\ \text{with } \sum_{j=0}^{n} (x^j)^2 - (x^{n+1})^2 &= \rho^2 \end{aligned}$$

Note: (32c-f) imply that D is $\mathbb{R}$-bilinear. (32f) implies that if ς and ς' coincide on a neighbourhood of x, then $D(\xi,\varsigma) = D(\xi,\varsigma')$ i.e. $D(\xi,\cdot)$ is a local operator. (32g) only expresses a smoothness property of D.

At fixed $\xi \in T\mathcal{M}$, the map

(33) $$\nabla_\xi : \varsigma \in \mathcal{V}(\mathcal{M}) \mapsto \nabla_\xi\varsigma \equiv D(\xi,\varsigma) \in T\mathcal{M}$$

is called the *covariant derivative* with respect to ξ; if ξ is a vector field, then the vector field $\nabla_\xi\varsigma$ is called the covariant derivative of ς along ξ.

These definitions are coordinate-free, i.e. independent of any choice of coordinate system; in the physics literature they are often replaced by coordinate dependent definitions where the fundamental objects are the Christoffel symbols Γ^ℓ_{jk} which emerge from the above definitions as follows. Making use of the identification (13) we write $\xi_x = \xi^i(x)\partial_i$ and introduce the notation:

(34) $$\Gamma^\ell_{jk} \equiv D(\partial_j,\partial_k)^\ell$$

Let us briefly check that, given the system of coordinates chosen to define them, these symbols are sufficient to compute the connection in the given chart. We have indeed (upon using the definition (32)):

(35) $$\begin{aligned} D(\xi,\varsigma) &= D(\xi^j\partial_j, \varsigma^k\partial_k)^\ell\partial_\ell \\ &= \xi^j(\partial_j\varsigma^k\cdot\delta^\ell_k + \varsigma^k\Gamma^\ell_{jk})\partial_\ell \\ &= \{\xi\cdot\varsigma^\ell + \xi^j\varsigma^k\Gamma^\ell_{jk}\}\partial_\ell \end{aligned}$$

An important aspect of a general connection is the following result. Given a differentiable manifold $\mathcal{M}$, a smooth curve $\gamma : I \to \mathcal{M}$, an open set $\mathcal{N}$ in $\mathcal{M}$ with $\gamma(I) \subset \mathcal{N}$, and a connection D on $\mathcal{N}$; let $\xi \equiv \dot\gamma$; t, t_0 and $t_0' \in I$; then the linear differential equation

(36) $$\begin{aligned} &D(\xi,\varsigma) = 0 \\ &\varsigma_{\gamma(t_0)} = \varsigma_0 \in T\mathcal{M}_{\gamma(t_0)} \end{aligned}$$

admits a unique solution $\varsigma_{\gamma(t)}$ which defines a linear isomorphism

(37) $$\tau_{t,t_0} : \varsigma_{\gamma(t_0)} \in T\mathcal{M}_{\gamma(t_0)} \mapsto \varsigma_{\gamma(t)} \in T\mathcal{M}_{\gamma(t)}$$

called the *parallel transport* associated to the connection D, along γ, from $\gamma(t_0)$ to $\gamma(t)$. Furthermore

(38) $$\tau_{t,t_0} \circ \tau_{t_0,t_0'} = \tau_{t,t_0'}$$

It is important to realize that, in general, the parallel transport depends on the curve along which it is defined, and not only on the end points. The difference between ζ_γ and $\zeta_{\gamma'}$ can even be used to measure the curvature of the connection considered. As an illustration, we compare in Figure 7 two parallel transports for a particular connection—the Levi-Cevita connection to be defined below—resulting from two different paths on S^2. For the benefit of the experimentally inclined mind, we should remark that for any smooth surface Σ^2 in $\mathbb{R}^3$, the parallel transport resulting from the Levi-Cevita connection is precisely that transport which is obtained by rolling, without twisting, Σ^2 along γ over $\mathbb{R}^2$ and lifting, all along, from $\mathbb{R}^2$ to Σ^2, the parallel tangent vectors in $\mathbb{R}^2$.

A geodesic is defined as a curve along which its tangent vector is parallel transported: analytically, this means that a curve $\gamma : I \to \mathcal{M}$ is a *geodesic* for the connection D whenever (i) D is defined on some open set $\mathcal{N} \subseteq \mathcal{M}$ with $\gamma(I) \subseteq \mathcal{N}$, and (ii) $D(\dot\gamma, \dot\gamma) = 0$. In local coordinates this equation reads (see (34) for the definition of the Christoffel symbols Γ^ℓ_{jk}):

$$\ddot\gamma^\ell + \Gamma^\ell_{jk}\dot\gamma^j\dot\gamma^k = 0 \tag{39}$$

In Euclidean geometry, where $\Gamma^\ell_{jk} \equiv 0$ as we shall soon see, this reduces to the equation of a straight line.

A differentiable manifold $\mathcal{M}$ equipped with a connection D is said to be *geodesically complete* whenever for every geodesic $\gamma : t \in I \mapsto \gamma(t) \in \mathcal{M}$ there exists a geodesic $\overline{\gamma} : t \in \mathbb{R} \to \overline{\gamma}(t) \in \mathcal{M}$ such that $\overline{\gamma}|I = \gamma$. While geodesic

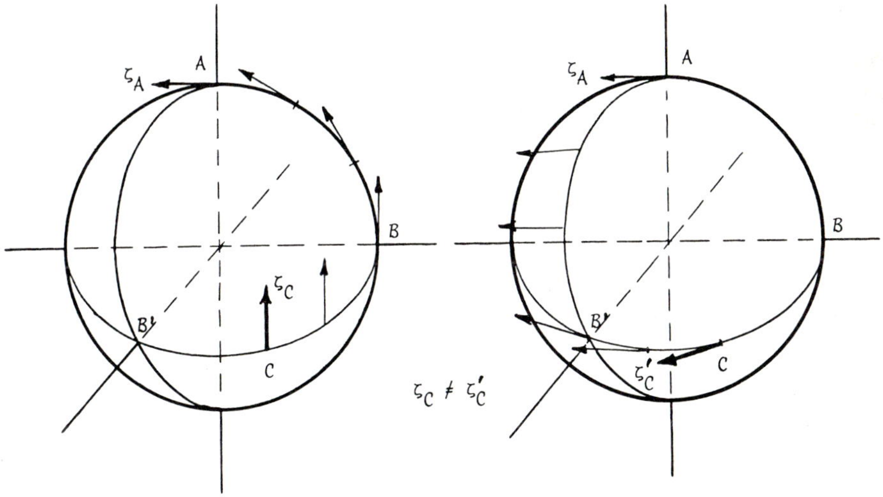

Fig.7

completeness is an attractive feature, we must point out that several space-time models of traditional cosmological interest are *not* geodesically complete; for instance, the Schwarzschild space-times are not geodesically complete (see Chapter Six). Whether geodesic incompleteness is a law of nature, or simply a misleading feature of current models and theories, is a question best left open. Anticipating on our next step, we should, nevertheless, mention here that all the examples considered so far in this section are geodesically complete when equipped with their Levi-Cevita connection.

Up to this point, the basic definitions of a connection and of the allied notion of geodesics were formulated on an arbitrary differentiable manifold $\mathcal{M}$, independently of whether or not $\mathcal{M}$ was endowed with a metric g. We now bring the two structures together.

A connection D on a Lorentzian (resp. Riemannian) manifold $(\mathcal{M}, g)$ is said to be *compatible* with g whenever the following two conditions are satisfied for all $\xi, \eta, \varsigma \in \mathcal{V}(\mathcal{M})$:

$$g(D(\xi,\eta),\varsigma)+g(\eta,D(\xi,\varsigma))=\xi\cdot g(\eta,\varsigma) \tag{40}$$

$$D(\xi,\eta)-D(\eta,\xi)-[\xi,\eta]=0 \tag{41}$$

where $[\xi, \eta]$, called the Lie bracket of ξ and η, is the vector field defined, through its action on $\mathcal{F}(\mathcal{M})$, by:

$$[\xi,\eta]\cdot f=\xi\cdot(\eta\cdot f)-\eta\cdot(\xi\cdot f) \tag{42}$$

This composition law satisfies the two properties

$$[\xi,\eta]=-[\eta,\xi] \tag{43}$$

$$[[\xi,\eta],\varsigma]+[[\eta,\varsigma],\xi]+[[\varsigma,\xi],\eta]=0 \tag{44}$$

(43) is referred to by saying that the Lie bracket is *antisymmetric*, and (44) by saying that it satifies the *Jacobi identity.* (41) is the condition that the connection D be *symmetric*; certain authors properly omit it in the definition of compatibility: throughout this work, however, it will systematically be assumed to hold.

Proposition. *Let $(\mathcal{M}, g)$ be a Lorentzian (resp. Riemannian) manifold. Then there exists exactly one connection D on $\mathcal{M}$ compatible with g.*

This unique connection is called the *Levi-Cevita connection* of g. The *proof* of this proposition is instructive and we thus reproduce it below.

Uniqueness: Since a connection is determined locally by its Christoffel symbols, we first compute these and show that they are uniquely determined by

g. Let $\{\partial_j | j = 1, \ldots, n\}$, where n is the dimension of $\mathcal{M}$, be a local coordinate system in $T\mathcal{M}|U$ where U is the domain of the local chart considered. We first notice that $[\partial_i, \partial_j] = 0$; indeed for all $f \in \mathcal{F}(U)$: $[\partial_i, \partial_j]f = \partial_i\partial_j f - \partial_j\partial_i f = 0$. Consequently, the definition of the Christoffel symbols and the assumed symmetry of D imply:

$$\Gamma^k_{ij} = \Gamma^k_{ji} \quad \forall\, i, j, k = 1, \ldots, n \tag{45}$$

From the compatibility condition (40) we have:

$$g(D(\partial_i, \partial_j), \partial_k) + g(\partial_j, D(\partial_i, \partial_k)) - \partial_i g(\partial_j, \partial_k) = 0 \tag{46}$$

i.e.

$$\Gamma^\ell_{ij}\, g_{\ell k} + \Gamma^\ell_{ik}\, g_{j\ell} - \partial_i g_{jk} = 0 \tag{47a}$$

and, by a cyclic permutation of the indices i, j, k:

$$-\Gamma^\ell_{jk}\, g_{\ell i} - \Gamma^\ell_{ji}\, g_{k\ell} + \partial_j g_{ki} = 0 \tag{47b}$$

$$\Gamma^\ell_{ki}\, g_{\ell j} + \Gamma^\ell_{kj}\, g_{i\ell} - \partial_k g_{ij} = 0 \tag{47c}$$

Upon adding (47a), (47b) and (47c) taking into account that g is a symmetric tensor field, and the symmetry (45) of the Christoffel symbols, we find:

$$2\Gamma^\ell_{ik}\, g_{j\ell} = \partial_i g_{jk} + \partial_k g_{ij} - \partial_j g_{ik} \tag{48}$$

and thus, upon defining $[g^{ik}]$ by:

$$g^{ik} g_{k\ell} \equiv \delta^i_\ell \tag{49}$$

we find:

$$\Gamma^\ell_{ik} = \frac{1}{2} g^{j\ell}(\partial_i g_{jk} + \partial_k g_{ij} - \partial_j g_{ik}) \tag{50}$$

which is the result announced in the first sentence of the proof. Uniqueness is thus established. Moreover, this explicit formula will be most useful in computations. The coordinate-free analog of this formula, obtained similarly by playing off symmetry against cyclic permutations is

$$\begin{aligned} & g(D(\xi, \eta), \varsigma) = \frac{1}{2}\Delta(\xi, \eta, \varsigma) \text{ with} \\ & \Delta(\xi, \eta, \varsigma) \equiv \xi g(\eta, \varsigma) + \eta g(\varsigma, \xi) - \varsigma g(\xi, \eta) \\ & \qquad - g(\xi, [\eta, \varsigma]) + g(\eta, [\varsigma, \xi]) + g(\varsigma, [\xi, \eta]) \end{aligned} \tag{51}$$

Since the expression of Δ does not involve D, and since g is non-degenerate, this formula indeed determines D uniquely from g.

Existence: Given g, we can write down Δ as in (51) and use the argument just given to define D. One then has to prove that D is a connection on $\mathcal{M}$. This is done by checking first the following properties of Δ which follow in a straightforward manner from the linearity (over $\mathbb{R}$) of g and the Lie bracket, and from the symmetry of g coupled with properties (43) and (44) of the Lie bracket

(52) Δ is additive in each of its variables

(53) $$\Delta(\xi,\eta,\varsigma)+\Delta(\xi,\varsigma,\eta)=2\xi\cdot g(\eta,\varsigma)$$

(54) $$\Delta(\xi,\eta,\varsigma)-\Delta(\eta,\xi,\varsigma)=2g([\xi,\eta],\varsigma)$$

(55) $$\Delta(\xi,\eta,f\varsigma)=f\Delta(\xi,\eta,\varsigma)$$

(56) $$\Delta(\xi,f\eta,\varsigma)=f\Delta(\xi,\eta,\varsigma)+2(\xi\cdot f)g(\eta,\varsigma)$$

(57) $$\Delta(f\xi,\eta,\varsigma)=f\Delta(\xi,\eta,\varsigma)$$

Now, from (52) and (55) we conclude that at fixed ξ and η:

(58) $$\Delta(\xi,\eta,\cdot):\varsigma\in\mathcal{V}(M)\mapsto\Delta(\xi,\eta,\varsigma)\in\mathcal{F}(\mathcal{M})$$

is a 1-form. Since g is non-degenerate, there exists then a vector field, depending on ξ and η, and denoted $D(\xi,\eta)$ such that

(59) $$g(D(\xi,\eta),\varsigma)=\frac{1}{2}\Delta(\xi,\eta,\varsigma)$$

We have, therefore, obtained a map

(60) $$D:(\xi,\eta)\in\mathcal{V}(\mathcal{M})\times\mathcal{V}(\mathcal{M})\mapsto D(\xi,\eta)\in\mathcal{V}(\mathcal{M})$$

satisfying all the conditions (32); specifically (32c) and (32d) follow from (52), (32e) from (57), (32f) from (56). Moreover, D satisfies (41) which follows from (54), and (40) which follows from (53). (Note that (55) was used in proving that (58) is a 1-form). □

Let γ be any curve in $\mathcal{M}$, η_0 and $\varsigma_0\in T\mathcal{M}_{\gamma(0)}$; define η_t (resp. ς_t) the parallel transport (see (36)), with respect to the Levi-Cevita connection, of η_0 (resp. ς_0) along γ; upon writing $\dot\gamma(t)$ for ξ in (40), one obtains $g_{\gamma(t)}(\eta_t,\varsigma_t)=g_{\gamma(0)}(\eta_0,\varsigma_0)$, i.e. *angles are preserved by Levi-Cevita parallel transport.* If γ is moreover a geodesic, then $g_{\gamma(t)}(\dot\gamma(t),\eta_t)$ is itself constant in t.

The metric of the manifolds $\mathbb{R}^n$ and $M^{n,1}$, when expressed in Cartesian coordinates (see (17)) is independent of $x\in\mathcal{M}$. Hence, see (50), the Christoffel symbols of the Levi-Cevita connection identically vanish, so that (see (39)) the geodesics are straight lines with constant $\dot\gamma$.

Upon using the explicit form of the metric for S^2 (resp. H^2, $\mathcal{M}_+^{1,1}$ and $\mathcal{M}_-^{1,1}$) given in (20) (resp. (23), (26), and (29)), and formula (50), the reader will verify by straightforward computations that the only non-vanishing Christoffel symbols of the Levi-Cevita connection are

$$\text{(61)} \qquad \text{for } S^2 : \Gamma^2_{11} = \sin\theta^2 \cdot \cos\theta^2 \qquad \Gamma^1_{12} = \Gamma^1_{21} = -\tan\theta^2$$

$$\text{(62)} \qquad \text{for } H^2 : \Gamma^\lambda_{\theta\theta} = -\sinh\lambda\cosh\lambda \qquad \Gamma^\theta_{\lambda\theta} = \Gamma^\theta_{\theta\lambda} = \coth\lambda$$

$$\text{(63)} \qquad \text{for } \mathcal{M}^{1,1}_{+} : \Gamma^\tau_{\theta\theta} = \sinh\tau \cdot \cosh\tau \qquad \Gamma^\theta_{\tau\theta} = \Gamma^\theta_{\theta\tau} = \tanh\tau$$

$$\text{(64)} \qquad \text{for } \mathcal{M}^{1,1}_{-} : \Gamma^\tau_{\lambda\lambda} = -\sin\tau \cdot \cos\tau \qquad \Gamma^\lambda_{\tau\lambda} = \Gamma^\lambda_{\lambda\tau} = -\tan\tau$$

From these expressions, one can solve the non-linear differential equations (39) to compute the explicit form of the geodesics: for each one of these four manifolds the geodesics turn out to be the plane curves obtained as the intersection of the manifold considered with an arbitrary plane passing through the origin of the embedding space. To see this without gruelling computation, notice first that since each of these manifolds $\mathcal{M}$ is homogeneous, it is sufficient to find all geodesics through a well-chosen point $x_0 \in \mathcal{M}$, and then use the fact that there exists an element Λ in the symmetry group of $\mathcal{M}$ (see Table 1) such that $\Lambda[x_0] = x$; it then follows from the compatibility condition of the Levi-Cevita connection with the metric that every geodesic through x is the image through Λ of a geodesic through x_0. For instance, by choosing x_0 to lie on the "equator" in case of S^2, at the bottom of the hyperboloid H^2, or on the "bottle-neck" in the cases of $\mathcal{M}^{1,1}_{\pm}$, one verifies easily (using the stabilizer of x_0) that all geodesics through x_0 satisfy indeed the announced property. This property is finally carried over to $x = \Lambda[x_0]$ since Λ, considered as a transformation of the embedding space, is linear and thus: (i) leave the origin invariant, and (ii) maps straight lines onto straight lines and hence planes onto planes. In particular, the reader will check that all null-geodesics of $\mathcal{M}^{1,1}_{\pm}$ are straight lines: they are the generators of these ruled surfaces.

Looking back at (61), (63) and (64) the reader will further notice that there is a point in the corresponding manifolds where all Christoffel symbols vanish. Since the Christoffel symbols depend on the coordinate system in which they are expressed, this fact should tell us something, not about the manifold, but rather about the coordinate system. This is indeed the case, as the following result shows.

Scholium. *Let $\mathcal{M}$ be a n-dimensional differentiable manifold; D be a connection on $\mathcal{M}$, and x be an arbitrary point in $\mathcal{M}$. Then there exists a local coordinate system around x such that the Christoffel symbols of D, when expressed in this coordinate system all vanish simultaneously at x.*

The essential steps of the proof can be sketched as follows. One first remarks that if $\gamma : t \in (-a, b) \to \gamma(t) \in \mathcal{M}$ is a geodesic with $\gamma(0) = x$, and if $\lambda \in \mathbb{R}$ with $\lambda \neq 0$, then $\gamma_\lambda : t \in (-a/\lambda, b/\lambda) \to \gamma(\lambda t) \in \mathcal{M}$ is again a geodesic with $\gamma_\lambda(0) = x$. One then verifies that there exists an open $V \subseteq TM_x$ with $O \in V$ such that for all $\varsigma \in V$ there exists an open interval $I \supseteq [0,1]$ with $\gamma_\varsigma : t \in I \to \gamma_\varsigma(t) \in \mathcal{M}$ a geodesic satisfying $\gamma_\varsigma(0) = x$ and $\dot{\gamma}_\varsigma(0) = \varsigma$. After defining $\exp_x : \varsigma \in V \mapsto \gamma_\varsigma(1) \in \mathcal{M}$, one verifies that for all $\lambda \in [0,1]$ and all $\varsigma \in V$, $\exp_x(\lambda\varsigma) = \gamma_\varsigma(\lambda)$ and in particular $\exp_x(0) = x$. One can moreover choose V small enough so that the geodesics issued from x do not intersect in $\exp_x(V) \equiv U$, i.e. so that $\exp_x$ is injective. It is then easy to check that $(U, \exp_x^{-1})$ is an admissible local chart. Let finally $\{e_i | i = 1, \ldots, n\}$ be a basis in TM_x and $\varsigma \in V$. One can then identify $y = \exp_x \varsigma \in U$ with $(\varsigma^1, \ldots, \varsigma^n) \in \mathbb{R}^n$ (where $\varsigma = \varsigma^i e_i$). The $\{\varsigma^i | i = 1, \ldots, n\}$ are called the geodesic coodinates of ς with respect to $\{e_i | i = 1, \ldots, n\}$. For the geodesic $\gamma_\varsigma(t) = \exp_x(t\varsigma)$, one has then $\gamma_\varsigma(t)^i = t\varsigma^i$ so that $\dot{\gamma}_\varsigma(0)^i = \varsigma^i$ and $\ddot{\gamma}_\varsigma(0)^i = 0$. Consequently, the geodesic equation (39) reduces, at $t = 0$ i.e. at x, to $\Gamma^\ell_{jk}\varsigma^j\varsigma^k = 0$. Since ς is arbitrary in V, this implies $\Gamma^\ell_{jk} = 0$ at x. □

For reasons which should be obvious from the above demonstration, a coordinate system satisfying the conclusion of the Scholium is said to be a *geodesic* system of local coordinates; it is also said to be *normal*. The reader will have noticed that the above scholium is valid for any connection, not just the Levi-Cevita connection: the above argument did not involve any metric notion; notice also that the scholium only asserts existence, not uniqueness (which would obviously be wrong).

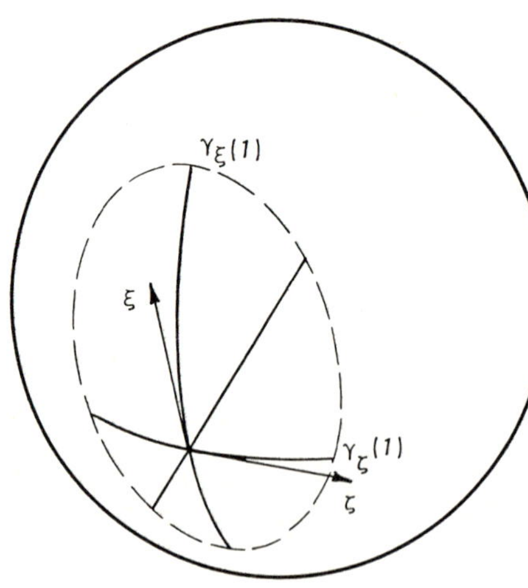

Fig. 8

e. Tensorial expressions of curvature

Recall that we introduced the notion of a connection right after we had asked the question of what distinguishes from one another the three Riemann manifolds $\mathbb{R}^n$, S^n or H^n, or the three Lorentz manifolds $M^{n,1}$, $\mathcal{M}_+^{n,1}$ and $\mathcal{M}_-^{n,1}$. We merely announced the answer: their "curvature" being zero, positive or negative. We are now ready to formalize this with the help of the Levi-Cevita connection D uniquely associated to the metric g of the manifold considered.

To define a tensor, starting with a connection $D : \mathcal{V}(\mathcal{M}) \times \mathcal{V}(\mathcal{M}) \to \mathcal{V}(\mathcal{M})$ we first have to overcome the fact that D is not linear over $\mathcal{F}(\mathcal{M})$ in its second variable: see (32 f). The most immediate way to achieve this is to define the *torsion*:

$$T_D(\xi,\eta) \equiv D(\xi,\eta) - D(\eta,\xi) - [\xi,\eta] \tag{65}$$

This, however, is not interesting for us, since we decided to consider only torsion-free (i.e. symmetric) connections: see (41). Our next step in this direction is to define the *curvature operator* of a (symmetric) connection D as a map which assigns to each ordered pair $(\eta,\varsigma) \in \mathcal{V}(\mathcal{M}) \times \mathcal{V}(\mathcal{M})$ an operator $R_{\eta,\varsigma}$ defined by:

$$R_{\eta,\varsigma} : \xi \in \mathcal{V}(\mathcal{M}) \mapsto (\nabla_\eta\nabla_\varsigma - \nabla_\varsigma\nabla_\eta - \nabla_{[\eta,\varsigma]})\xi \in \mathcal{V}(\mathcal{M}) \tag{66}$$

Consider now the map

$$R : (\xi,\eta,\varsigma) \in \mathcal{V}(\mathcal{M}) \times \mathcal{V}(\mathcal{M}) \times \mathcal{V}(\mathcal{M}) \mapsto R_{\eta,\varsigma}\xi \in \mathcal{V}(\mathcal{M}) \tag{67}$$

From its definition, one checks easily that:

$$R(\xi,\eta,\varsigma) = -R(\xi,\varsigma,\eta) \tag{68}$$

(69) R is additive in each of its variables

A straightforward computation further shows that

(70) R is linear over $\mathcal{F}(\mathcal{M})$ in each of its variables, i.e.
$$fR(\xi,\eta,\varsigma) = R(f\xi,\eta,\varsigma) = R(\xi,f\eta,\varsigma) = R(\xi,\eta,f\varsigma)$$

With α a one-form on $\mathcal{M}$, and $\xi,\eta,\varsigma \in \mathcal{V}(\mathcal{M})$ one verifies that $\alpha(R(\xi,\eta,\varsigma))$ evaluated at $x \in \mathcal{M}$ depends only on the value of $\alpha,\xi,\eta,\varsigma$ at x, and that

$$R : x \in \mathcal{M} \mapsto \alpha_x(R(\xi_x,\eta_x,\varsigma_x)) \in \mathbb{R} \tag{71a}$$

is smooth. From (69) and (70) we conclude then that (71a) defines a (1,3)-tensor field on $\mathcal{M}$. This tensor field, denoted by R in the sequel, is called the *Riemann curvature tensor* of the connection D. When a local system of coordinates is introduced, we have:

(71b)
$$R(\alpha_\ell dx^\ell, \xi^i\partial_i, \eta^j\partial_j, \varsigma^k\partial_k) = R^\ell_{ijk}\alpha_\ell\xi^i\eta^j\varsigma^k$$
$$\text{with } R^\ell_{ijk} \equiv R(dx^\ell, \partial_i, \partial_j, \partial_k)$$

In terms of the Christoffel symbols of the connection D, this reads

(71c)
$$R^\ell_{ijk} = (\partial_j\Gamma^\ell_{ki} - \partial_k\Gamma^\ell_{ji}) + (\Gamma^m_{ki}\,\Gamma^\ell_{jm} - \Gamma^m_{ji}\,\Gamma^\ell_{km})$$

From the antisymmetry (68) one obtains immediately

(72a)
$$R(\alpha, \xi, \eta, \varsigma) = -R(\alpha, \xi, \varsigma, \eta)$$

i.e.

(72b)
$$R^\ell_{ijk} = -R^\ell_{ikj}$$

The reader will notice that the antisymmetry (72b) is manifest in (71c). To review the definitions, let us perform explicitly at least one straightforward computation:

$$\begin{aligned}
&R(\xi,\eta,\varsigma) + R(\eta,\varsigma,\xi) + R(\varsigma,\xi,\eta) = \\
&D(\eta, D(\varsigma,\xi)) - D(\varsigma, D(\eta,\xi)) - D([\eta,\varsigma],\xi) + \\
&D(\varsigma, D(\xi,\eta)) - D(\xi, D(\varsigma,\eta)) - D([\varsigma,\xi],\eta) + \\
&D(\xi, D(\eta,\varsigma)) - D(\eta, D(\xi,\varsigma)) - D([\xi,\eta],\varsigma) = \\
&D(\eta, D(\varsigma,\xi) - D(\xi,\varsigma)) - D([\varsigma,\xi],\eta) + \\
&D(\varsigma, D(\xi,\eta) - D(\eta,\xi)) - D([\xi,\eta],\varsigma) + \\
&D(\xi, D(\eta,\varsigma) - D(\varsigma,\eta)) - D([\eta,\varsigma],\xi) = \\
&D(\eta,[\varsigma,\xi]) - D([\varsigma,\xi],\eta) + \\
&D(\varsigma,[\xi,\eta]) - D([\xi,\eta],\varsigma) + \\
&D(\xi,[\eta,\varsigma]) - D([\eta,\varsigma],\xi) = \\
&[\eta,[\varsigma,\xi]] + [\varsigma,[\xi,\eta]] + [\xi,[\eta,\varsigma]]
\end{aligned}$$

and thus, upon using the Jacobi identity (44):

(73a)
$$R(\xi,\eta,\varsigma) + R(\eta,\varsigma,\xi) + R(\varsigma,\xi,\eta) = 0$$

i.e.

(73b)
$$\underset{i,j,k}{\circlearrowleft} R^\ell_{ijk} = 0$$

(where $\underset{i,j,k}{\circlearrowleft}$ means: summation over all cyclic permutations of the indices (i, j, k)). (73), in either form (a) or (b), is called the *first Bianchi identity*. The reader

CHAPTER 5. THE PRINCIPLES OF SPECIAL RELATIVITY

SYNOPSIS

In Newtonian physics, one readily assumes that a change of the origin of time

(1a) $$t \to t + \tau$$

or of the origin of space

(1b) $$x^i \to x^i + \xi^i \quad i = 1, 2, 3$$

does not affect the form of the laws of physics, nor does another choice of an orthogonal reference frame in $\mathbb{R}^3$ (leaving thus t untouched):

(2) $$x^i \to \Lambda^i_j x^j \quad i, j = 1, 2, 3; \; \Lambda^T \Lambda = I;$$

the summation over repeated indices is understood throughout. As Galileo demonstrated in his *Dialogue Concerning the Two Chief World Systems* (1632) a touch of common sense is enough "to show that all experiments practicable upon the Earth are insufficient measures for proving its mobility, since they are indifferently adaptable to an earth in motion or at rest". We now refer to this, with Poincaré (1904), as "the principle of relativity according to which the laws of physical phenomena should be the same whether for an observer at rest, or for an observer carried along in a uniform movement"; this transformation of coordinates (which again leaves t untouched) reads:

(3) $$x^i \to x^i + v^i t \quad i = 1, 2, 3$$

In line with the views of Cayley, Klein (the Erlanger programme dates from 1872) and Lie, accordinng to whom "geometry is the study of invariant properties of figures, i.e. properties which are unchanged under a given group of admissible transformations", one would thus see space-time as that geometric structure which is invariant under the Galilean group, i.e. the group generated by the transformations (1), (2) and (3) above.

At this point, however, we should inject into the discussion the seminal remark of Lorentz and Poincaré that the Maxwell equations of electromagnetism are invariant under a different group, now referred to as the inhomogeneous Lorentz group. Einstein, followed by Minkowski, took in 1905 the daring conceptual step to take this new group as the group of invariance of space-time, i.e. to view it as the group of isometries of the vector space $\mathbb{R}^4$ equipped with the pseudo-metric

$$g = \sum_{i=1}^{3} dx^i \otimes dx^i - c^2 dt \otimes dt \tag{4}$$

where c is the velocity of light.

As $c \to \infty$, the inhomogeneous Lorentz group "contracts" to the Galilean group, a mathematical fact the physical meaning of which is that Galilean relativity appears as an approximation of Einstein relativity corresponding to those experimental situations where the velocities involved are negligible compared to the velocity of light ($c \cong 3 \cdot 10^8 m/sec.$; we shall henceforth chose units of time and length such that $c = 1$; in these units the velocity of the earth on its orbit is $\cong 10^{-4}$).

Einstein's postulate brought with it some drastic revisions of the classical views on space and time. Taken separately, space and time have no absolute meaning, i.e. no meaning independent of a particular observer. Einstein realized that the correct language of discourse should be rooted instead in the absolute concept of space-time. At a more immediate level, this new postulate has numerous consequences which, however simple they may now appear, must have been somewhat baffling to the classical physicist—or mathematician, for that matter! Among these consequences we shall discuss in the present Chapter: the Lorentz-FitzGerald contraction of space-length, the slowing-down of clocks, the Michelson-Morley experiment, the Doppler shift, the aberration of light, and the Fresnel drag.

Section 1. FORMULATION OF THE PRINCIPLES

The physical basis of special relativity, as proposed in the introduction of Einstein's paper, *Zur Elektrodynamik bewegter Körper* (1905), reads: "...the same laws of electrodynamics and optics will be valid for all frames of reference for which the equations of mechanics hold good. We will raise this conjecture (the purport of which will hereafter be called the "Principle of Relativity") to the status of postulate, and also introduce another postulate, which is only apparently irreconcilable with the former, namely, that light is always propagated in empty space with a definite velocity c which is independent of the state of motion of the emitting body. These two postulates suffice for the attainment of a simple and consistent theory of the electrodynamics of moving bodies based on Maxwell's theory of stationary bodies."

This is what we have to understand.

Mathematically, the content of the theory can be divided along a line which is different from that separating the above two principles; this dividing line is reflected in Einstein's own presentation of his work under two main headings: I. Kinematical Part, and II. Electrodyamical Part.

Conceptually, Part I of Einstein's paper is mostly geometrical and, according to Einstein's own view (reported in the beginning of our Section 4.1) we should distinguish there the axiomatic stage and the interpretative stage.

The axiomatic stage of Part I of Einstein's paper amounts to establishing the geometry of what we now call the Minkowski space-time $M^{3,1}$. We defined this space-time in the beginning of section 4.2; the reader will easily infer its most important geometric properties from those of the space $M^{1,1}$ studied in detail in section 4.1; in addition, we saw in section 4.2 that $M^{3,1}$ is a flat Lorentz manifold so that its geodesics coincide with the Euclidean straight lines.

We will, therefore, concentrate our attention, in the present section, on the interpretation of $M^{3,1}$ as a substitute for the Galilean space time $\mathbb{R}^4 = \mathbb{R}^3 \times \mathbb{R}$ where $\mathbb{R}^3$ is "space" and $\mathbb{R}$ is "time". To present the interpretative rules of Einstein's special relativity, linking "geometric notions" with the "objects of physics" we proceed by contrasting them with the principal tenets of what can properly be called Galilean relativity, to be surveyed now.

The delineation of a relativity principle occupies much of Galileo's efforts in his *Dialogues*...(1632) as he proposes there a battery of experimental tests (in fact, thought-experiments, or experiences) to prove the inconclusive character of the arguments invoked by the "Aristotelean philosophers" against the motion of the earth. Amongst the most telling of the evidences marshalled by Galileo are those in which he describes how the conclusions, reached by passengers on a sea voyage, on events occuring within the ship, cannot depend on whether the ship is moving uniformly along a straight line or standing still with respect to the shore. Such events include, for instance, the point of impact of a stone falling from the top of the high mast, or even more picturesquely, the easiness with which "butterflies and flies will continue their flights indifferently toward every side, nor will it ever happen that they are concentrated toward the stern, as if tired out from keeping up with the course of the ship, from which they have been separated during long intervals by keeping themselves in the air. And if smoke is made by burning some incense, it will be seen going up in the form of a little cloud, remaining still and moving no more toward one side than the other. The cause of all these correspondences of effects is the fact that the ship's motion is common to all the things contained in it ... you have never felt within yourself your participation in that motion, except when the boat has been stopped by running aground ... when you ... taken by surprise, have stumbled perilously."

The principle of relativity we thus owe to Galileo is that the laws of nature are the same for all systems of reference which differ from one another by a uniform rectilinear motion, or by a choice of the origin (x_0, t_0) of space ($x_0 \in \mathbb{R}^3$) and time ($t_0 \in \mathbb{R}$).

Mathematically, the primary object of Galilean relativity is, therefore, the vector space $\mathbb{R}^4 = \{z = (x,t) | x \in \mathbb{R}^3, t \in \mathbb{R}\}$, the points (x,t) of which are interpreted as the "tag" marking that an event has occured, at time $t \in \mathbb{R}$, in position $x \in \mathbb{R}^3$. The principle of Galilean relativity is then the requirement that the geometry of this space must be invariant under the transformations

$$\text{(1a)} \qquad \begin{aligned} t \to \hat{t} &= t + t_0 \\ x \to \hat{x} &= A[x] + vt + x_0 \end{aligned}$$

with

$$\text{(1b)} \qquad \begin{aligned} (x_0, t_0) \in \mathbb{R}^4 &\equiv \text{ group of translations} \\ A \in O^3 &\equiv \text{ group of rigid rotations in } \mathbb{R}^3 \\ v \in \mathbb{R}^3 &\equiv \text{ group of velocities (or Galilean boosts)} \end{aligned}$$

The reader will recognize here the extension to $\mathbb{R}^4$ of the transformation laws (4.1.5b) equipping $\mathbb{R}^2$ with the structure of a Galilean geometry; $\mathbb{R}^4$ comes now equipped with the Galilean distance

$$\text{(1c)} \qquad d(z_1, z_2) = |t_1 - t_2|$$

interpreted as the *time elapsed* between two events z_1 and z_2. The absolutes of this geometry are interpreted as follows. The "special hyperplanes" $\{(x,t)|x \in \mathbb{R}^3\}$ are interpreted as *instants*: they are the locus of *simultaneous* events. A curve $\gamma : s \in I \mapsto \gamma(s) \in \mathbb{R}^4$, the tangent to which is never contained in a "special hyperplane" is called a *motion*. The straight lines, not contained in a "special hyperplane", are interpreted as *uniform motions*. The Galilean angle $(v_2 - v_1)$ between two straight lines ℓ_2 and ℓ_1, with ℓ_k $(k = 1,2)$ given by $(x_k - x_0) = v_k(t - t_0)$, is interpreted as the *relative velocity* of the uniform motion ℓ_2 with respect to the uniform motion ℓ_1. The Galilean cycles $(x - x_0) = a(t-t_0)^2 + v(t-t_0)$ are interpreted as motions at *constant acceleration* a, i.e. as *uniformly accelerated motions.*

To complete the interpretation of the theory, one must formalize the notion of an observer, i.e. of a reference frame in the Galilean space-time $(\mathbb{R}^4, d)$. An *instantaneous observer* Z at $z \in \mathbb{R}^4$ is identified with a straight line, called the *time-axis* of the observer, $Z : s \in \mathbb{R} \mapsto Z(s) \in \mathbb{R}^4$ with $Z(0) = z$ and $d(Z(s), z) = s$, completed—when necessary—by a local chart (U, φ) of the Euclidean hyperplane $\Sigma \equiv \{z' \in \mathbb{R}^4 | d(z', z) = 0\}$; in many cases, it is convenient—but not necessary—to take for φ the identity map and to equip Σ with a Cartesian coordinate system. Clearly, an instantaneous observer equips $\mathbb{R}^4$ with a local chart, or local reference frame. More generally, an *observer* is a curve $Z : s \in I \to Z(s) \in \mathbb{R}^4$ such that its tangent at every point is an instantaneous observer at this point.

Note again, that any two instantaneous observers Z_1 and Z_2 will ascribe the same acceleration to a uniformly accelerated motion, and that they will agree on the relative velocity of a uniform motion ℓ_2 with respect to another uniform motion ℓ_1, however much they may disagree on the velocity of a single of these uniform motions: no velocity is an absolute in Galilean relativity.

It is against this background, now stated in a mathematically precise manner, that we want to formulate the basic assumptions of Einstein's special theory of relativity.

While Einstein kept, from the Galilean world picture, the notion that events are tagged by points z in $\mathbb{R}^4$, he proposed that the geometry be now given by the "distance":

$$\text{(2a)} \qquad \begin{aligned} &d(z_1, z_2)^2 = g_{\mu\nu}(z_1 - z_2)^\mu (z_1 - z_2)^\nu \\ &\text{with } g_{\mu\nu} = \epsilon_\mu \delta_{\mu\nu} \text{ and } \epsilon_\mu = \begin{cases} +1 & 1 \leq \mu \leq 3 \\ -1 & \mu = 4 \end{cases} \end{aligned}$$

invariant under the group of linear transformations

$$\text{(2b)} \qquad \begin{aligned} &z \to \hat{z} = \Lambda[z] + z_0 \\ &\text{with } z_0 \in \mathbb{R}^4 \text{ and } g_{\mu\nu}\Lambda^\mu_\rho \Lambda^\nu_\sigma = g_{\rho\sigma} \end{aligned}$$

Recall from Chapter Four that straight lines are geometrically meaningful notions for this geometry, i.e. straight lines are mapped onto straight lines by the admissible transformations (2b); and, in fact, the straight lines are the geodesics of (the Levi-Cevita connection associated to) the metric g issued from (2a). Recall also that since $M^{3,1}$ is flat, the distinction between $M^{3,1}$ and $TM^{3,1}_x$, however real, is somewhat pedantic; consequently, unless we absolutely need it, we will omit this distinction in this section in order to keep the notation to a minimum. For a discussion of the meaning of the assumption that Λ be linear, see Borchers & Hegerfeldt (1972) and references quoted therein.

Clearly the Cartesian decomposition $z \in \mathbb{R}^4 \mapsto (\mathbf{z}, z^4) \in \mathbb{R}^3 \times \mathbb{R}$ is not an invariant under the group of transformations (2b); this decomposition is therefore *not* a geometrically meaningful concept, in the sense that it is not a geometric absolute. Consequently, one should not attempt to give it a universal meaning, e.g. trying to find absolute meanings for space and time.

Quite to the contrary, Einstein recognized that space and time are relative notions, namely relative to an observer, the latter being defined as follows (compare to the corresponding notion defined above in the case of Galilean relativity). An *instantaneous observer* Z at $z \in M^{3,1}$ is a straight line $Z : s \in \mathbb{R} \mapsto Z(s) \in M^{3,1}$ with $Z(0) = z$ and

$$\text{(3a)} \qquad g_{\mu\nu}(Z(s) - z)^\mu (Z(s) - z)^\nu = -s^2$$

a condition expressed intrinsically (i.e. in a coordinate-free manner) using the metric g, namely:

$$\text{(3b)} \qquad g(\dot{Z}, \dot{Z}) = -1$$

As in the Galilean case, this definition is completed by giving a local chart (U, φ) around z in the Euclidean hyperplane Σ spanned here by the straight lines $X : s \in \mathbb{R} \mapsto X(s) \in M^{3,1}$ specified by the conditions

$$\text{(4a)}\qquad \begin{aligned} &X(0) = z \\ &g_{\mu\nu}(Z(s) - z)^{\mu}(X(s) - z)^{\nu} = 0 \\ &g_{\mu\nu}(X(s) - z)^{\mu}(X(s) - z)^{\nu} = +s^2 \end{aligned}$$

i.e. in an intrinsic manner:

$$\text{(4b)}\qquad \begin{aligned} &X(0) = z \\ &g(\dot{Z}, \dot{X}) = 0 \\ &g(\dot{X}, \dot{X}) = +1 \end{aligned}$$

Notice that Σ can be described alternately as:

$$\text{(5)}\qquad \Sigma \equiv \{\xi \in TM_z^{3,1} | g(\xi, \dot{Z}) = 0\}$$

and that the restriction $g|\Sigma$ of the metric g of $M^{3,1}$ to the hyperplane Σ indeed equips Σ with the structure of the Euclidean space $\mathbb{R}^3$. In his original paper, Einstein was not considering a general coordinate system (U, φ) in Σ, and was satisfied with a geodesic, i.e. straight line, orthogonal system; we shall follow him in this—it is not necessary, but it is convenient here.

It is important to take note of the fact that the notion of instantaneous observer is a geometrically meaningful concept, namely under the transformations (2b) instantaneous observers are mapped into instantaneous observers. The same holds true for the general notion of an *observer*, defined here again as a curve in the manifold such that its tangent at every point is an instantaneous observer at this point.

The question of the physical justification for this new choice of axioms for space-time and observers is best answered in the perspective emphasized by Einstein in the title of his memoir, namely electrodynamics, and more specifically Part II of that paper. From this, we now abstract, in a modern language which will immediately translate to the more general framework of Chapter Six, an observer-free formulation of electromagnetism; this formulation will, in turn, reduce for any instantaneous observer Z to the Maxwell equations where the objects entering these equations will depend on the observer Z, in a precise way which we shall define.

An *electromagnetic field* is now simply defined as a closed two-form $\mathcal{F}$ on the Lorentz manifold $\mathcal{M} = M^{3,1}$.

In anticipation to Chapter Six, we should mention in passing that this definition is, in fact, so intrinsic that it will carry over verbatim to the framework of general relativity where $\mathcal{M}$ is a general Lorentz manifold.

Here, however, we must worry about the opposite question, namely whether this definition is sufficient to establish proper contact with the objects called electromagnetic fields in Chapter Three. To answer this, let us "peel" our intrinsic definition and extract from it, with the help of an arbitrary instantaneous observer, the Maxwell equations relative to this observer.

Recall that an instantaneous observer at $z_0 \in \mathcal{M}$ is a vector $Z \in T\mathcal{M}_{z_0}$ with $g_{z_0}(Z, Z) = -1$, *and* a local chart (U, φ) in $\Sigma \equiv \{X \in T\mathcal{M}_{z_0} | g_{z_0}(X, Z) = 0\}$. We can choose $U = \Sigma, \varphi = id$, and $\{X_k | k = 1, 2, 3\} \subset \Sigma$ with $g_{z_0}(X_j, X_k) = +\delta_{jk}$. This defines a system of coordinates around z_0. Let $\xi_k (k = 1, 2, 3)$ and $\xi_4 \equiv \varsigma$ be the corresponding vector fields around z_0. Let now Ω be the four-form defined around z_0 by:

$$\Omega_z((\xi_1)_z, (\xi_2)_z, (\xi_3)_z, \varsigma_z) = +1 \tag{6}$$

We start by noticing that the map

$$\eta_z \in T\mathcal{M} \mapsto \mathcal{F}_z(\eta_z, \varsigma_z) \in \mathbb{R} \tag{7}$$

defines a one form, namely $-\varsigma \rfloor \mathcal{F}$; since g is non-degenerate there exists a unique vector field $\mathbf{E}$ on $\mathcal{M}$, such that:

$$g(\eta, \mathbf{E}) = \mathcal{F}(\eta, \varsigma) \;\; \forall \, \eta \in \mathcal{V}(\mathcal{M}) \tag{8}$$

Moreover, the antisymmetry of $\mathcal{F}$ (it is a two-form!) gives immediately

$$g(\varsigma, \mathbf{E}) = 0 \tag{9}$$

We define similarly the vector field $\mathbf{B}$ on $\mathcal{M}$ by

$$\begin{array}{l} \Omega_z(\xi_z, \eta_z, \mathbf{B}_z, \varsigma_z) = \mathcal{F}_z(\xi_z, \eta_z) \\ \forall \, z \in \mathcal{M} \;\text{ and }\; \forall \, \xi_z, \eta_z \in T\mathcal{M}_z \\ \text{satisfying } g(\xi, \varsigma) = 0 = g(\eta, \varsigma) \end{array} \tag{10a}$$

supplemented by the condition:

$$g(\varsigma, \mathbf{B}) = 0 \tag{10b}$$

The reader should notice that the vector field $\mathbf{E}$ depends only on the vector field ς, and thus only on the vector Z; and that $\mathbf{B}$ depends similarly on Z, but in addition $\mathbf{B}$ depends also on the orientation Ω chosen by the observer. They are thus, beyond that, independent of the orthogonal basis $\{X_k | k = 1, 2, 3\}$ in Σ chosen by the observer. Furthermore, (9) and (10b) imply, since $g(\varsigma, \varsigma) = -1$

$$\begin{array}{l} g_z(\mathbf{E}_z, \mathbf{E}_z) > 0 \;\text{ whenever }\; \mathbf{E}_z \neq 0 \\ g_z(\mathbf{B}_z, \mathbf{B}_z) > 0 \;\text{ whenever }\; \mathbf{B}_z \neq 0 \end{array} \tag{11}$$

Using now the convenience of the Cartesian coordinate system $\{(\xi_1)_z, (\xi_2)_z, (\xi_3)_z, (\xi_4)_z \equiv \varsigma_z\}$ we introduce the functions:

$$\begin{array}{l} \mathcal{F}_{\mu\nu} : z \in \mathcal{M} \mapsto \mathcal{F}_z((\xi_\mu)_z, (\xi_\nu)_z) \\ E^\mu : z \in \mathcal{M} \mapsto g_z((\xi_\mu)_z, \mathbf{E}_z) \\ B^\mu : z \in \mathcal{M} \mapsto g_z((\xi_\mu)_z, \mathbf{B}_z) \end{array} \tag{12}$$

This formula allows our two observers to compare notes. They clearly do not agree on what they each call *the* electric field, and *the* magnetic field. Having followed the procedure by which they arrived at these objects, we are not surprised: although the electromagnetic field $\mathcal{F}$ was defined in an intrinsic manner, i.e. is an absolute in our geometry, E and B are not absolutes—they depend on the observer who defines them. At this point, our observers may want to reject the role we make them play, faulting the script. Suppose, however, they stay with us, and try to resolve their conflict within the theory. To induce them to do so, we will first point out to them that they were treated on equal footing; this is reflected in formula (31) which is symmetric in the exchange of $(\mathrm{E}, \mathrm{B}$ and $\mathrm{v})$ and $(\mathrm{E}', \mathrm{B}'$ and $-\mathrm{v})$: no observer is privileged over the other. Secondly, we will invoke classical electrodynamics and remind each observer that he knows from this theory that a test particle of electric charge e moving with velocity v^* in an electric field E^* and a magnetic field B^* experiences an "electromotive" force

(32a) $$\mathrm{F}^* = e(\mathrm{E}^* + \mathrm{v}^* \times \mathrm{B}^*)$$

i.e. a force

(32b) $$\mathrm{F}^{*\parallel} = e\mathrm{E}^{*\parallel}$$

along the direction of its motion, and a force

(32c) $$\mathrm{F}^{*\perp} = e(\mathrm{E}^* + \overline{v}^* \times \mathrm{B}^*)^{\perp}$$

in the plane perpendicular to the direction of its motion. Pondering now over the meaning of (31) in the light of (32) our two observers will thus arrive at the conclusion that they move with respect to one another with a *relative velocity* v, allowing us to give a physical interpretation to the geometric transformation (27). Since the other transformations in the group (2b) defining the geometry of Minkowski space are readily interpreted in terms of ordinary rotations, and of translations of the origin of our space-time $M^{3,1}$, we now have gained a physical understanding of this geometry.

This is the central conclusion of this section.

Lest the reader may think that we have extensively belabored the conceptual process involved in reaching this conclusion, we should mention that it will carry almost verbatim to the case of the general theory of relativity (see Chapter Six). Moreover, it seems that our exposition, except for a modernization of the mathematical language, is faithful to Einstein's own ideas. In his original paper of 1905, he wrote, indeed, at the end of the section in which he derived formula (31): the "electromagnetic force plays in the developed theory merely the part of an auxiliary concept, which owes its introduction to the circumstance that electric and magnetic forces do not exist independently of the state of motion of the system of coordinates". As for his original motivation, he reminisced almost half a century later (see Einstein (1952)): "What led me directly to the Special Theory of Relativity was the conviction that the electromotive force acting on a body in motion in a magnetic field was nothing else but an electric field."

We feel the case is made. However, our two observers, still pondering over formula (31), may still argue over the presence of the factor $1/\sqrt{1-v^2}$. Two arguments will temporarily pacify them. The first is irreproachable: this factor is necessary to make the formula symmetric under an exchange of the observers. The second looks a bit of a cop-out: in the units we choose, the velocity of light c is taken to be equal to 1; this means, for instance, that the velocity of the Earth on its orbit, when expressed in these units, is of the order of 10^{-4}; hence our neglecting the factor $1/\sqrt{1-v^2}$ would only produce an error of the order of 10^{-8}, well beyond the precisions of the experiments which established the validity of classical electrodynamics. Nevertheless, since the realm of Twentieth Century experimental physics does well encompass situations in which this factor is relevant, we will come back to this point in the next section.

Section 2. IMMEDIATE CONSEQUENCES OF THE PRINCIPLES

The first important consequence of the principles enounced in the previous section is this: whenever an electromagnetic field $\mathcal{F}$ is such that a single instantaneous observer Z_0 sees it as a plane-wave solution of his Maxwell equations, so will every other instantaneouos observer Z see this electromagnetic field as a plane-wave solution of his own Maxwell equations. Since all instantaneous observers agree about the form of the Maxwell equations, *all observers will agree on the value of the speed of light, namely* $c = 1$.

Accordingly, we will call *light ray* any geodesic (i.e. straight line) with tangent satisfying

$$g(\dot{\gamma}, \dot{\gamma}) = 0 \tag{1a}$$

i.e. in any instantaneous observer's coordinate system

$$g_{\mu\nu}\dot{\gamma}(s)^{\mu}\dot{\gamma}(s)^{\nu} = 0 \tag{1b}$$

These lines are clearly geometric invariant, i.e. any transformation (5.1.2b) maps light rays onto light rays.

The light rays split the tangent space $T\mathcal{M}$ of $\mathcal{M} = M^{3,1}$ into the following three regions

$$\begin{aligned} \mathcal{L} &\equiv \{\varsigma_z \in T\mathcal{M} | g(\varsigma_z, \varsigma_z) = 0, \varsigma_z \neq 0\} \\ \mathcal{T} &\equiv \{\varsigma_z \in T\mathcal{M} | g(\varsigma_z, \varsigma_z) < 0\} \\ \mathcal{S} &\equiv \{\varsigma_z \in T\mathcal{M} | g(\varsigma_z, \varsigma_z) > 0\} \cup \{0\} \end{aligned} \tag{2}$$

For the reason just indicated, the vectors in $\mathcal{L}$ are said to be *light-like*. Notice that for every instantaneous observer, the vector spanning his time-axis lies in $\mathcal{T}$; for this reason, the vectors ς in $\mathcal{T}$ are said to be *time-like*. Similarly, for every instantaneous observer, the vectors spanning the hyperplane this observer calls space are all in S; the vectors ξ in S are said to be *space-like*.

We further notice that $\mathcal{T}$ is an open submanifold of TM and has two connected components. Each one of these is invariant under the component of the identity in the group (5.1.2b) of symmetries of $M^{3,1}$. We arbitrarily denote by $\mathcal{T}_+$ one of the components of $\mathcal{T}$ and by $\mathcal{T}_-$ the other. There are now two disjoint classes of observers: those whose tangent vectors belong to $\mathcal{T}_+$, and those for which they belong to $\mathcal{T}_-$. The observers belonging to the first of these two classes are said to be *orthochronous*. Mathematically (as will be explained in Chapter Six when we need the general concept involved in this) what we have just done is to prove that $M^{3,1}$ is "time-orientable" and we have chosen a "time-orientation". Physically, we have eliminated those "conservative" observers whose clocks run backward in time; as just explained, this choice is arbitrary, and by another choice of words we might just as well have kept this class of observers and have eliminated the other. Which choice we make is irrelevant; what is relevant is that we can make the choice, i.e. that $M^{3,1}$ is time-orientable.

With that choice now made, we call *future* (resp. *past*) of the point $z_0 \in M^{3,1}$, the collection of points z_1 in $M^{3,1}$ such that there exists a curve $\gamma : s \in \mathbb{R} \mapsto \gamma(s) \in M^{3,1}$ with $\gamma(0) = z_0$, $\gamma(1) = z_1$, and $\dot\gamma(s) \in \mathcal{T}_+$ (resp. $\mathcal{T}_-$) $\forall s \in [0,1]$. We can obviously choose this line to be a straight line generated by a time-like vector $\varsigma \in \mathcal{T}_\pm M_{z_0} \equiv TM_z \cap \mathcal{T}_\pm$.

Similarly, we call *present* of z_0 the collection of points $z_1 \in M^{3,1}$ such that there is a curve $\gamma : s \in \mathbb{R} \mapsto \gamma(s) \in M^{3,1}$ with $\gamma(0) = z_0$, $\gamma(1) = z_1$, and $\dot\gamma(s) \in S$ $\forall s \in [0,1]$. We can obviously (again) choose this line to be a straight line generated by a (now) space-like vector $\xi \in S_{z_0} \equiv TM_z \cap S$.

It is a fact of life (or "physical experience") that we do not know how to transmit information along a curve γ which is space-like, i.e. for which $g(\dot\gamma, \dot\gamma) > 0$. That some may want to call this a "law of nature" is a matter of philosophical debate. What we must realize, in any case, is that this "fact of life" is a geometric invariant: the symmetries of $M^{3,1}$ map space-like curves onto space-like curves.

It is also important to realize that the concept of "present", just defined in Einstein's special relativity, is in sharp contrast with what one calls the present in Galilean relativity. Figure 1 illustrates the situation, with an arbitrary instantaneous observer indicated in dotted lines.

Indeed, if you ask a Galilean observer what it means for two events to be simultaneous, his answer will be that they occur at the same time. This means that $z_1 = (x_1, t_1)$ and $z_2 = (x_2, t_2)$ are simultaneous iff $t_1 = t_2$, i.e. iff $d(z_1, z_2) = 0$, i.e. iff z_1 and z_2 are contained in the same special hyperplane $\Sigma = \{(x,t) | x \in \mathbb{R}^3, t = t_1 = t_2\}$. This is a geometrically meaningful (i.e. invariant) statement: all Galilean observers will agree on whether two events are simultaneous or not. This neat notion of simultaneity has to be given up

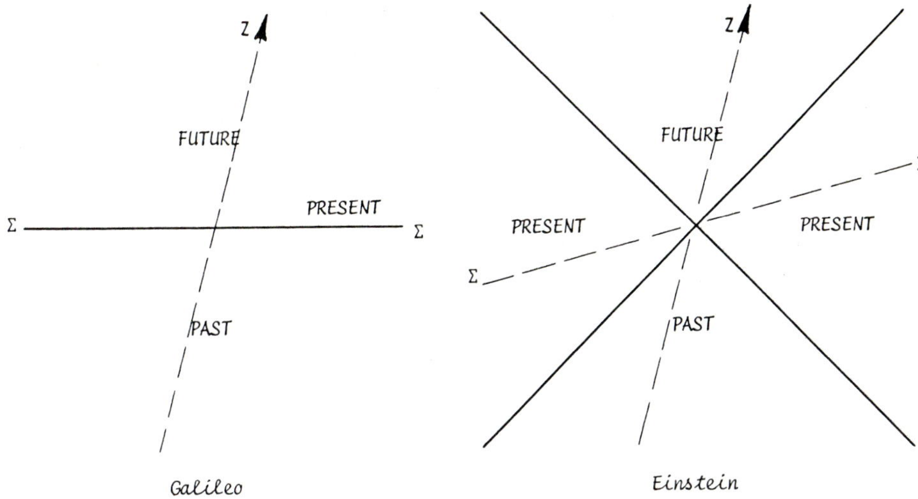

Fig. 1

in special relativity. Indeed, an instantaneous observer Z may want to tell you that two events are simultaneous if both are contained in a hyperplane Σ orthogonal to his own time axis. This, however, is not a geometric invariant: two different instantaneous observers may have non-parallel time-axes, and thus non-parallel hyperplanes Σ orthogonal to their respective time-axes. Such observers may, therefore, disagree on whether two events are simultaneous or not. *Simultaneity is not an absolute.* The next best thing is the notion of the present, which in Einstein relativity is a four-dimensional, connected submanifold of $M^{3,1}$ whereas in Galilean relativity it was simply a three-dimensional special hyperplane $\Sigma \subset \mathbb{R}^4$. This is a contributing factor to the difficulties one encounters when one tries to establish a relativistically covariant Hamiltonian formalism describing several particles in interaction.

Let us, however, remain here within the realm of the simplest consequences of the postulates of special relativity.

Suppose we are sitting with an instantaneous observer Z observing a rigid rod XY moving with a uniform velocity v. For this observer, the space coordinates of the end points of the rod are

$$\left.\begin{aligned} x^i(t) &= x^i(0) + v^i t \\ y^i(t) &= y^i(0) + v^i t \end{aligned}\right\} \quad i = 1, 2, 3 \tag{3}$$

Without loss of generality, we can assume $\mathrm{v} = (v, 0, 0)$. Let us now jump to an observer comoving with the rod, and consider what this observer $\hat{Z}$ sees when he passes by our original observer. His coordinate system is related to the coordinate system of our original observer by the transformation:

$$\Lambda_v = \frac{1}{\sqrt{1-v^2}} \begin{pmatrix} 1 & 0 & 0 & -v \\ 0 & \sqrt{1-v^2} & 0 & 0 \\ 0 & 0 & \sqrt{1-v^2} & 0 \\ -v & 0 & 0 & 1 \end{pmatrix} \tag{4}$$

Hence for the second instantaneous observer, the space coordinates of the end points of the rod are

$$\begin{aligned} \hat{x}^1 &= \frac{1}{\sqrt{1-v^2}}[x^1(t) - vt] = \frac{1}{\sqrt{1-v^2}} x^1(0) \\ \hat{x}^2 &= x^2(t) = x^2(0) \\ \hat{x}^3 &= x^3(t) = x^3(0) \end{aligned} \tag{5}$$

and similarly for $\hat{y}^i$. The rod is indeed at rest for $\hat{Z}$. Let now:

$$\begin{aligned} d^i &\equiv y^i(t) - x^i(t) \\ \hat{d}^i &\equiv \hat{y}^i - \hat{x}^i \end{aligned} \tag{6}$$

From (3) and (5), these are related by:

$$\hat{d}^1 = \frac{1}{\sqrt{1-v^2}} d^1 \quad ; \quad \hat{d}^2 = d^2 \quad ; \quad \hat{d}^3 = d^3 \tag{7}$$

Hence, if the rod is perpendicular to the direction of its motion, the two observers attribute the same length to it; this result conforms to Galilean experience. If, however, the rod is parallel to the direction of its motion, the observers disagree on its length: the first observer attributes to it a length d which is *shorter* than the length $\hat{d}$ found by the observer comoving with the rod, specifically:

$$d = \sqrt{1-v^2}\ \hat{d} \tag{8}$$

Hence, using a language fraught with Galilean residues, one could say that to an observer "at rest" rigid objects seem to "contract" in the direction of their motion, i.e. "are" shorter than they "would be" if they were "at rest". Formula (8) goes under the name of *Lorentz-FitzGerald contraction*: H. A. Lorentz (1892) and G. F. FitzGerald (1892) had indeed postulated something like this as a rather ad hoc hypothesis apparently disconnected from any fundamental principle (see S. G. Brush (1967) and G. Holton (1973)). In his 1905 paper, where he derived (8), Einstein offers no comment on these previous proposals. In contrast with these, Einstein's analysis is rooted in the Maxwell equations themselves, and on that basis he provides a general perspective—known today as the special theory of relativity—from which this phenomenon appears just as another way—a particularly picturesque one at that—to say that space and time, taken separately, are not absolutes.

Upon replacing the rod in the previous discussion by a clock, we can compare the times seen by our two observers. We have for the trajectory of the clock, in the first observer's coordinate system:

$$x^i(t) = x^i(0) + v^i t \tag{9}$$

and for the change of coordinates (thus completing (5)):

$$\hat{t} = \frac{1}{\sqrt{1-v^2}}(-v\, x^1(t) + t) \tag{10}$$

Hence, to a time interval $\Delta\hat{t}$ for the comoving observer corresponds a time interval Δt for the first observer, given by:

$$\Delta t = \frac{1}{\sqrt{1-v^2}}\ \Delta\hat{t} \tag{11}$$

This formula, also derived in Einstein's 1905 paper, is known as the *"slow-down"* of moving clocks. For instance, if an unstable particle with life-time $\Delta\hat{t}$ travels at a slightly subluminal speed (say $v = 1 - 1/5000$) with respect to an observer, then this observer will measure a larger life time (here $\Delta t \cong 50\Delta\hat{t}$); there is plenty of experimental confirmation for this phenomena, which clearly goes hand-in-hand with the Lorentz-FitzGerald contraction.

The attentive reader will notice that the factor $\sqrt{1-v^2}$ occurs at opposite places in (8) and (11). This is nothing but another expression of the fact (already noticed for the case of $M^{1,1}$ in formula (4.1.16)) that the Euclidean volume in $\mathbb{R}^4$ is preserved by transformations in (5.1.2b). We have indeed, since there is no Lorentz-FitzGerald contraction in the spatial directions perpendicular to the direction of motion:

$$\begin{aligned} &\Delta\Omega \equiv \Delta V \cdot \Delta t \equiv \Delta x^1 \cdot \Delta x^2 \cdot \Delta x^3 \cdot \Delta t = \\ &(\sqrt{1-v^2}\Delta\hat{x}^1) \cdot \Delta\hat{x}^2 \cdot \Delta\hat{x}^3 \cdot \left(\frac{\cdot 1}{\sqrt{1-v^2}}\Delta\hat{t}\right) \equiv \Delta\hat{V} \cdot \Delta\hat{t} \equiv \Delta\hat{\Omega} \end{aligned} \tag{12}$$

We now have to say a few words about the *Michelson-Morley experiment.* It was first performed in 1887 and refined further over the subsequent years. For an intellectual history of this experiment, documenting its meaning and impact on the scientific community, see G. Holton (1973), pp.261-352. In a nutshell, let it suffice here to say that Holton provides convincing scholarly support to two contentions. The first is that, as one could have conjectured from the organization and very title of Einstein's 1905 paper, the Michelson-Morley experiment played at most a minor, implicit role in the genesis of

Einstein's special relativity. The second important point of Holton's analysis is that the Michelson-Morley experiment, however, happened to be instrumental in the acceptance of Einstein's theory by the scientific community, so much so that in his textbook *Das Relativitätsprinzip* von Laue (1911) could already write: "... the experiment became, as it were, the fundamental experiment for the relativity theory."

Let us place ourselves back in the historical context in which the experiment was first conducted. In analogy with the way sound waves are carried by (i.e. are vibrations of) a medium such as the air, Nineteenth-Century physicists postulated that light waves are carried by a medium through which they travel with characteristic velocity c ; they called this medium the *aether*, or the "luminiferous ether". Note that the existence of this aether, pervading the whole universe, could then be used to give a meaning to absolute space, so that the postulate could be interpreted as a departure from Galilean invariance. The Michelson-Morley experiment was, in part, designed to test whether this could be evidenced by measuring the velocity v of the Earth through the aether.

Schematically, the experimental set up is illustrated in Figure 2, where v represents the velocity of the Earth through the aether.

A beam of light enters at A, falls on a semi-transparent mirror B which splits the beam in two beams: one proceeding to C, the other to D, where these secondary beams are reflected back to B. There B again splits the incoming secondary beams so that two coherent beams arrive at E: one having traveled the path $ABCBE$, and the other the path $ABDBE$. At E the position of the interference pattern (see Chapter Three for the basic notions of interferometry) produced by the superposition of these two coherent beams allows one to measure the difference between the travel times of the two beams.

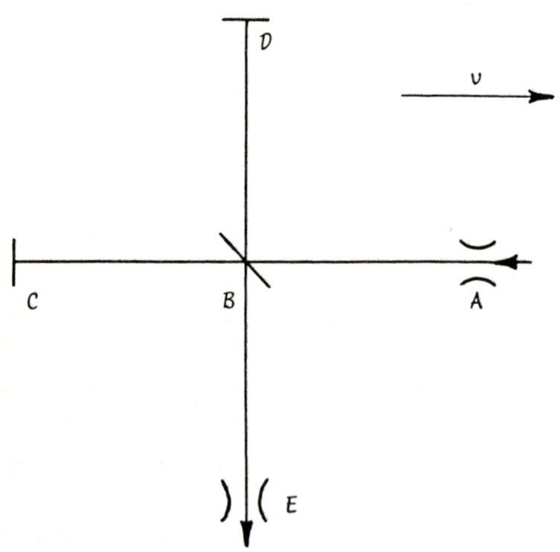

Fig.2

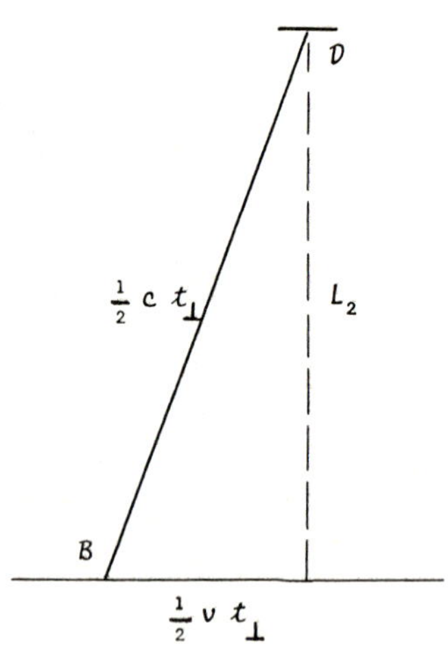

Fig.3

(*Stricto sensu,* this difference of travel times is experimentally determined only modulo the period T of the light wave; since, however, the analysis will turn out to involve only comparisons of travel time differences which differ among themselves by less than this period, this mod(T) is irrelevant and will, therefore, be ignored in our presentation).

Denote by L_1 (resp. L_2) the length of the arm BC (resp. BD). The time the light takes to travel the path BCB (resp. BDB) is

$$t_{\|} = \frac{L_1}{c-v} + \frac{L_1}{c+v} = 2L_1 \frac{c}{c^2 - v^2} \tag{13}$$

respectively (see Figure 3)

$$t_{\perp} = 2L_2 \frac{1}{\sqrt{c^2 - v^2}} \tag{14}$$

Choose now L_1 and L_2 in such a manner that $t_{\|} - t_{\perp} = 0$; we have then

$$L_1 \cdot \frac{c}{c^2 - v^2} = L_2 \cdot \frac{1}{\sqrt{c^2 - v^2}} \equiv \frac{L}{c} \tag{15}$$

The astucious experimental step now is to turn rigidly this apparatus around by 90°, thus exchanging effectively the roles of L_1 and L_2. One obtains in this manner

$$\begin{aligned} t'_{\|} &= 2L_2 \cdot \frac{c}{c^2 - v^2} = \frac{2L}{c} \cdot \frac{c}{\sqrt{c^2 + v^2}} \\ t'_{\perp} &= 2L_1 \cdot \frac{1}{\sqrt{c^2 - v^2}} = \frac{2L}{c} \cdot \frac{\sqrt{c^2 + v^2}}{c} \end{aligned} \tag{16}$$

Taking into account that $v/c \cong 10^{-4}$, one obtains

$$t'_{\perp} - t'_{\|} \cong \frac{2L}{c} \left(\frac{v}{c}\right)^2 \tag{17}$$

Playing further on multiple reflections, Michelson and Morley were able to get for L an effective length of about 10 meters. With these figures in hand, the reader will verify that, for ordinary light, this results in a displacement of the interference pattern by about 0.4 of the distance between the fringes. The beauty of the experimental set-up is that it allowed Michelson and Morley to conclude that "if there is any displacement due to the relative motion of the earth and the luminiferous ether, this cannot be much greater than 0.01 of the distance between the fringes". This is clearly a major discrepancy: the classical theory predicted an effect at least 40 times (!) larger than what could possibly be observed. Subsequent experiments, carried out with increasing sensitivity, confirmed this null result.

By contrast, in the frame of special relativity, where the speed of light is the same for every observer, the analysis of this experiment gives *immediately*

$$t'_{\perp} - t'_{\parallel} = 0 \tag{18}$$

in sharp contrast with (17), and in agreement with experiment. No wonder then that the Michelson-Morley experiment "became...the fundamental experiment for the relativity theory" (see M. von Laue, loc.cit.), and that, later on, many enquired whether it could not have motivated Einstein to formulate his theory. Marvelous as this may now sound, Einstein's primary concern was rather with a proper formulation of "the electrodynamics of moving bodies": a theoretical physicist's dream!

The above comment should not be construed as implying that Einstein was not interested in working out definite, immediately measurable, consequences of his general postulates. Quite to the contrary. For instance, here are two particularly elegant results which he already offered, among others, in his 1905 paper.

Suppose an observer moves with respect to a star S so that he sees, in his coordinates systems, the star moving with velocity **u**. Let him place his x^1-axis parallel to **u**, and his x^2-axis so that the plane (x^1, x^2) contain S. We have thus the situation described in Figure 4.

We now want to compare this situation with what would happen if the velocity **u** were zero, i.e. if our observer were at rest with respect to the star. This would correspond to an observer with coordinate system $(x_0{}^1, x_0{}^2, x_0{}^3, t_0)$;

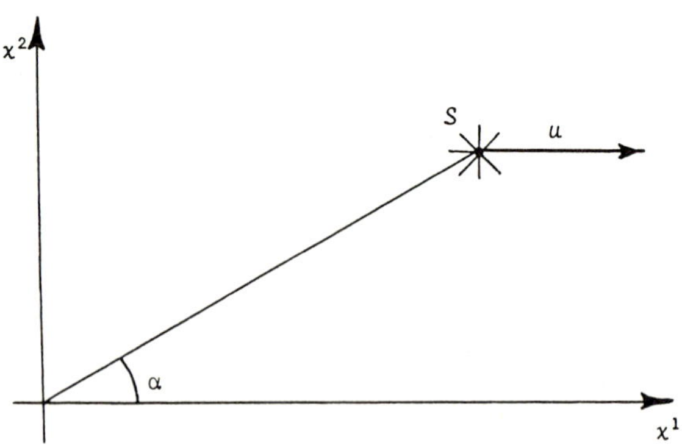

Fig.4

these two coordinate systems can be arranged in such a manner that:

$$
(19) \qquad
\begin{aligned}
x^1 &= \frac{1}{\sqrt{1-u^2}}({x_0}^1 + ut_0) \\
x^2 &= {x_0}^2 \\
x^3 &= {x_0}^3 \\
t &= \frac{1}{\sqrt{1-u^2}}(u{x_0}^1 + t_0)
\end{aligned}
$$

Since both observers see a light ray coming from the star as a plane-wave solution of their respective Maxwell equations, we must have for the phase of that wave:

$$
(20) \qquad \nu(t - x^1 \cos\alpha - x^2 \sin\alpha) = \nu_0(t_0 - {x_0}^1 \cos\alpha_0 - {x_0}^2 \sin\alpha_0)
$$

where ν (resp. ν_0) denotes the frequency of the light for the observer moving (resp. at rest) with respect to the star.

Upon substituting (19) in (20), and using the fact that ${x_0}^1$, ${x_0}^2$ and t_0 are independent variables, one obtains

$$
(21) \qquad \tan\frac{\alpha}{2} = \sqrt{\frac{1+u}{1-u}} \cdot \tan\frac{\alpha_0}{2}
$$

and

$$
(22) \qquad \nu = \nu_0 \cdot \frac{1 - u\cos\alpha}{\sqrt{1-u^2}}
$$

Formula (21) gives thus an *aberration*, which is a first order effect in u. For instance, with the velocity of the Earth on its orbit being of the order of 10^{-4} (in units were the speed of light $c = 1$), one concludes from this formula that, in the course of a year, the "fixed stars" should describe in the sky small ellipses with axes of the order of one minute of arc; which they do.

Formula (22) gives a shift in frequency; such a thing was known in classical theory under the name of *Doppler shift*. In the classical theory there are, in fact, two Doppler shifts; the first (resp. the second) obtains when the observer (resp. the source) is in motion relatively to the medium supporting the waves, here the hypothetical aether, and the source (resp. the observer) is at rest in that medium. These are given by:

$$
(23) \qquad \nu' = \nu_0(1 - u\cos\alpha)
$$

when the observer is in motion, with the source at rest; and

$$
(24) \qquad \nu'' = \nu_0 \cdot \frac{1}{1 + u\cos\alpha}
$$

when the observer is at rest, and the source is in motion.

One should first notice that (22), (23) and (24) all agree on a qualitative point, namely as $u \cos \alpha > 0$, i.e. with the source and the observer moving apart from one another, they all predict that the observed frequency is smaller than ν_0, i.e. visible light is shifted towards the red; hence the connection between a general astronomical *red shift* and a *universe in expansion*. Moreover, in first order in u, the three formulas give the same quantitative results. To observe differences between Einstein's formula (22) and the classical predictions (23) and (24) one needs, therefore, to be able to detect deviations of the order of u^2, which means speeds much greater than that of the Earth on its orbit. Such corrections are necessary when observing far away galaxies; then $\cos \alpha \cong 1$, and the differences between (22), (23) and (24) are illustrated in Figure 5.

At the qualitative level, one should notice that at $\cos \alpha = 0$, the classical theory predicts

$$\nu' = \nu_0 = \nu'' \tag{25}$$

whereas Einstein's formula gives:

$$\nu = \nu_0 \frac{1}{\sqrt{1 - u^2}} \tag{26}$$

This *tangential Doppler effect*, totally absent from the classical theory (see 25), has been measured in the laboratory, for $u \cong 10^{-2}$, by Ives and Stilwell (1941).

Finally, one should remark that after the negative result of the Michelson-Morley experiment, one might be inclined to think that the aether is a very

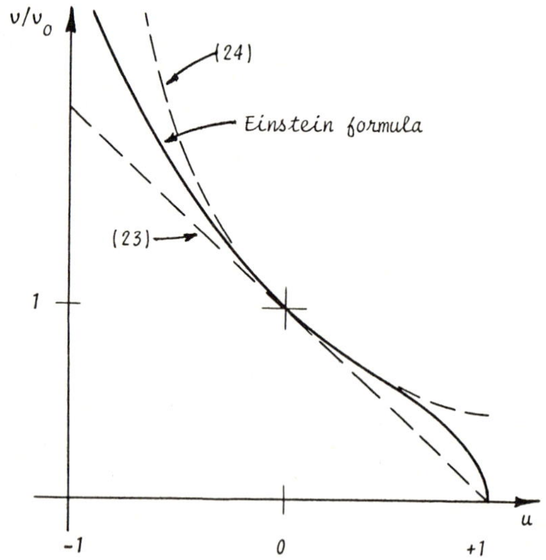

Fig. 5

elusive medium indeed and that one would be all the better without it; there is, in fact, no such thing in Einstein's relativity. In the absence of a medium, there is then no way to decide which one of (23) or (24) to use, since only relative motions make sense. Accordingly, Einstein has a single formula, namely (22).

Let us now examine another of the immediate consequences of the principles of special relativity. Suppose that an (instantaneous) observer Z sees two other (instantaneous) observers Z_1 and Z_2 moving in the same direction, with instantaneous velocities v_1 and v_2. If these were Galilean observers, Z_1 would be saying that Z_2 moves with an instantaneous velocity $(v_2 - v_1)$. Recall that, in special relativity, two observers decide in principle (see Section 1) about their relative velocity by comparing their respective electric and magnetic fields: see (5.1.31 and 5.1.32). Looking back on the geometric relations between these observers' coordinate systems (5.1.27 and 5.1.30), see (4), one finds here that Z_k $(k = 1, 2)$ is obtained from Z by the transformation

$$\Lambda_{v_k} = \frac{1}{\sqrt{1-v_k^2}}\begin{pmatrix} 1 & 0 & 0 & -v_k \\ 0 & \sqrt{1-v_k^2} & 0 & 0 \\ 0 & 0 & \sqrt{1-v_k^2} & 0 \\ -v_k & 0 & 0 & 1 \end{pmatrix} \tag{27}$$

so that Z_2 is obtained by Z_1 by the transformation $\Lambda_v = \Lambda_{v_2}\Lambda_{v_1}^{-1}$ given by:

$$\Lambda_v = \frac{1}{\sqrt{1-v^2}}\begin{pmatrix} 1 & 0 & 0 & -v \\ 0 & \sqrt{1-v^2} & 0 & 0 \\ 0 & 0 & \sqrt{1-v^2} & 0 \\ -v & 0 & 0 & 1 \end{pmatrix} \tag{28}$$

with

$$v = \frac{v_2 - v_1}{1 - v_1 v_2} \tag{29}$$

Mathematically, one should notice that (29) is precisely the geometric invariant (4.1.21), i.e. the Minkowski angle between the lines $x_k = v_k t$: hence v depends only on Z_1 and Z_2, but not on Z.

Physically, (29) says that the statement $-1 < v_1 < 1$ is an absolute (compare with Fig. 1b), since $-1 < v_1, v_2 < 1$ implies $-1 < v < 1$. Note also that, in special relativity, velocities do not add linearly. Einstein derived formula (29) in his 1905 paper; he did not, however, make there the following simple remark. Suppose that in (29) $v_1/v_2 \to 0$; one can then approximate (29) as

$$v \cong v_2 - v_1(1 - (v_2)^2) \tag{30}$$

and we will need to formalize our concept of matter to change this hint into a precise statement (see Section Three below). In the present section, we only take this remark as a stimulant for a further examination into the geometric and kinematic properties of curved space-times.

Following Sachs and Wu (1977) we now want to give a kinematic interpretation for one of the most fundamental objects in the study of curved space-times, namely the Ricci tensor associated to the metric.

A few preliminary remarks and some notation will help. Let Z be an instantaneous observer at $z \in \mathcal{M}$. Notice first that one can show that there exists a geodesic vector field ς, defined in some neighbourhood U of z, and a unique integral curve $\gamma : s \in I \mapsto \gamma(s) \in U$ satisfying $\gamma(0) = z$ and $\dot{\gamma}(0) = Z = \varsigma_z$, with γ inextendible in U. Clearly, since γ is by definition a geodesic,

$$(12) \qquad g_{\gamma(s)}(\dot{\gamma}(s), \dot{\gamma}(s)) = -1 \quad \forall\, s \in I$$

so that γ is an observer through z. Moreover, for every $z' \in U$ there exists similarly a unique integral curve γ' of ς such that $\gamma'(0) = z'$, $\dot{\gamma}'(0) = \varsigma_{z'}$ and γ' inextensible in U. Define then $\mu_t(z') = \gamma'(t)$. Clearly, for s and t small enough $\mu_t \circ \mu_s = \mu_{s+t}$. This μ is called the *flow* generated by the vector field ς. At fixed t, μ_t maps smooth curves (in its domain) to smooth curves in $\mathcal{M}$; let us denote by $(\mu_t)_*$ the corresponding map of tangent vectors to $\mathcal{M}$ to tangent vectors.

Let us go back to our observer through the point z we chose in the beginning, and define for every $s \in I$:

$$(13) \qquad \Sigma_s = \{\xi \in T\mathcal{M}_{\gamma(s)} | g_{\gamma(s)}(\dot{\gamma}(s), \xi) = 0\}$$

Notice that, because of (12), we have for every $\xi \in \Sigma_s$

$$(14) \qquad g_{\gamma(s)}(\xi, \xi) \geq 0$$

with equality holding if and only if $\xi = 0$; hence $(\Sigma_s, g|\Sigma_s)$ is isometrically isomorphic to the ordinary Euclidean space $\mathbb{R}^3$.

Let now $\xi : s \in I \mapsto \xi_s \in T\mathcal{M}$ be smooth and such that $\pi \circ \xi = \gamma$, where $\pi : \eta_z \in T\mathcal{M} \mapsto z \in \mathcal{M}$ is the canonical projection of $T\mathcal{M}$ onto $\mathcal{M}$. This object is called a *vector field over* γ.

We now say that a vector field ξ over γ is a *neighbour* of γ if the following two conditions are satisfied

$$(15) \qquad \xi_s \in \Sigma_s \quad \forall\, s \in I$$

and

$$(16) \qquad (\mu_t)_*(\xi_s) = \xi_{s+t} \quad \forall\, s, t \in I \text{ with } s + t \in I$$

Note that we can show then, without loss of generality, that there exists a vector field $\overline{\xi}$ defined on a neighbourhood V of z, with $V \subset U$, such that on V:

$$[\varsigma, \overline{\xi}] = 0 \tag{17}$$

For this reason, (16) is referred to by saying that ξ is Lie-transported along γ. As a last preliminary remark, we should mention that the notion of neighbour can be extended from the case of a geodesic observer to arbitrary observers (for this and the related concept of a general Fermi-Walker connection, see e.g. R. K. Sachs and H. Wu (1977)); we do not need this generality here.

With these preliminaries out of the way, let now γ be our geodesic observer through z, and ξ be a neighbour of γ. Recall: (i) the symmetry of the Levi-Cevita connection (see 4.2.41), namely

$$\nabla_\varsigma \overline{\xi} - \nabla_{\overline{\xi}} \varsigma - [\varsigma, \overline{\xi}] = 0 \; ; \tag{18}$$

(ii) the definition of the curvature tensor R (see 4.2.66), implying in particular:

$$R(\varsigma, \varsigma, \overline{\xi}) = (\nabla_\varsigma \nabla_{\overline{\xi}} - \nabla_{\overline{\xi}} \nabla_\varsigma - \nabla_{[\varsigma, \overline{\xi}]}) \varsigma \; ; \tag{19}$$

and (iii) the fact that ς is geodesic, i.e.

$$\nabla_\varsigma \varsigma = 0 \; . \tag{20}$$

Then from condition (16) on ξ being a neighbour of γ (see 17), we obtain:

$$\nabla_{\dot{\gamma}}{}^2 \xi = R_z(\dot{\gamma}, \dot{\gamma}, \xi) \tag{21}$$

which is to say that ξ is a Jacobi field along γ.

Similarly, from the compatibility condition of the Levi-Cevita connection (see 4.2.40), implying in particular:

$$g(\nabla_\varsigma \overline{\xi}, \varsigma) + g(\overline{\xi}, \nabla_\varsigma \varsigma) = \varsigma \; g(\overline{\xi}, \varsigma) \tag{22}$$

and the fact that ς is geodesic (see (20) above), one obtains from condition (15) on ξ being a neighbour of γ:

$$g(\nabla_{\dot{\gamma}} \xi, \dot{\gamma}) = 0 \;\text{ i.e. }\; (\nabla_{\dot{\gamma}} \xi)_s \in \Sigma_s \tag{23}$$

We can, therefore, repeat this argument to obtain similarly

$$g(\nabla^2_{\dot{\gamma}} \xi, \dot{\gamma}) = 0 \text{ i.e. } (\nabla^2_{\dot{\gamma}} \xi)_s \in \Sigma_s \tag{24}$$

The three-vector $(\nabla_{\dot{\gamma}} \xi)_s \in \Sigma_s$ is interpreted as the *relative spatial velocity* at $\gamma(s)$, with respect to our geodesic observer γ, of the neighbour ξ of γ. Similarly, $(\nabla^2_{\dot{\gamma}} \xi)_s \in \Sigma_s$ is interpreted as its *relative spatial acceleration*.

It is this acceleration, through its average over all spatial directions which is linked to the Ricci tensor. To see this, write Σ_z for Σ_0, and consider the maps associated to the instantaneous observer $Z = \dot{\gamma}(0)$, by:

$$\psi_Z : \xi \in \Sigma_z \mapsto R_z(Z, Z, \xi) \in T\mathcal{M}_z \tag{25}$$

Notice first that ψ_Z takes in fact its values in Σ_z. We have indeed:

$$g(\psi_Z \xi, Z) = g_z(Z, R_z(Z, Z, \xi)) = \hat{R}_z(Z, Z, Z, \xi) \tag{26}$$

and thus by the antisymmetry of $\hat{R}$, the *RHS* vanishes. This operator from Σ_z to Σ_z is, in fact, the map which attributes to any neighbour ξ of γ its relative spatial acceleration at z, which is why we are interested in it.

Notice, also, as a result of the symmetry properties of $\hat{R}$ that ψ_Z is a self-adjoint operator, i.e.

$$g(\psi_Z \xi, \eta) = g(\xi, \psi_Z \eta) \quad \forall\, \xi, \eta \in \Sigma_z \tag{27}$$

To extract the Ricci tensor from this operator, let $\{X_\mu | \mu = 1, 2, 3, 4\}$ be an orthonormal basis in $T\mathcal{M}_z$ with $X_4 = Z$, so that $X_k \in \Sigma_z$ for $k = 1, 2, 3$. Let $\{\omega^\mu | \mu = 1, 2, 3, 4\}$ be the dual basis in $T^*\mathcal{M}_z$. Compute

$$\begin{aligned} tr\ \psi_Z &\equiv \sum_{k=1}^{3} \omega^k(\psi_Z X_k) \\ &= \sum_{k=1}^{3} R_z(\omega^k, Z, Z, X_k) \\ &= \sum_{\mu=1}^{4} R_z(\omega^\mu, Z, Z, X_\mu) \\ &= -\sum_{\mu=1}^{4} R_z(\omega^\mu, Z, X_\mu, Z) \end{aligned}$$

i.e.

$$tr\ \psi_Z = -Ric_z(Z, Z) \tag{28}$$

To complete our programme, we now have to show that the *LHS* of (28) can be interpreted as the average over all spatial directions of the relative spatial acceleration of the neighbours of our observer.

To see this, let

$$\begin{aligned} O^3 \equiv \{\varphi : T\mathcal{M}_z \to T\mathcal{M}_z | \varphi(Z) = Z \text{ and} \\ g(\varphi[\xi], \varphi[\eta]) = g(\xi, \eta) \quad \forall\, \xi, \eta \in T\mathcal{M}_z\} \end{aligned} \tag{29}$$

i.e. O^3 is isomorphic to the group of isometries of $\mathbb{R}^3 \cong \Sigma_z$ with the origin as fixed point; let O^3_+ be the connected component of the identity in O^3, i.e. O^3_+ is the group of rotations. This is a compact Lie group and it, therefore, has a finite invariant measure μ.

For any self-adjoint operator ψ on Σ_z we define the average of ψ over O^3_+ as:

$$\overline{\psi} \equiv \frac{1}{\mu(O^3_+)} \int_{O^3_+} d\mu(\varphi)\varphi^{-1}\psi\,\varphi \tag{30}$$

One has then, by the invariance property of the trace:

$$tr\ \overline{\psi} = tr\ \psi \tag{31}$$

Moreover, from (30) one has

$$\varphi^{-1}\overline{\psi}\,\varphi = \overline{\psi}\ \forall\ \varphi \in O^3_+ \tag{32}$$

so that $\overline{\psi}$ is a multiple of the identity, and thus

$$\overline{\psi} = \frac{1}{3}(tr\ \overline{\psi})\ I \tag{33}$$

Hence from (31) and (33):

$$\overline{\psi} = \frac{1}{3}(tr\ \psi)\ I \tag{34}$$

Upon bringing together (34) and (28), we obtain the result we were looking for: a kinematic interpretation of the Ricci tensor.

Proposition 2. *Let Z be an instantaneous observer at $z \in \mathcal{M}$; ς be a geodesic vector field, defined on a neighbourhood of z, with $\varsigma_z = Z$. Let ψ_Z be the map which associates to every neighbour ξ of Z in ς its relative spatial acceleration. Then the average over space-directions of the relative spatial acceleration*

$$\overline{\psi}_Z = \frac{1}{\mu(O^3_+)} \int_{O^3_+} d\mu(\varphi)\varphi^{-1}\psi_Z\,\varphi \tag{35}$$

is given by:

$$\overline{\psi}_Z = -\frac{1}{3}Ric_z(Z,Z)\cdot I \tag{36}$$

As an illustration, this shows immediately the kinematic difference between three Lorentz manifolds which we studied in Section 4.2. For $M^{3,1}$, $\overline{\psi}_Z \equiv 0$.

For $\mathcal{M}_{+}^{3,1}$ (resp. $\mathcal{M}_{-}^{3,1}$), $\overline{\psi}_Z = +\rho^{-2}I$ (resp. $-\rho^{-2}I$) so that neighbours, on the average, accelerate away from (resp. towards) any (freely falling) observer. As we shall see in Section Three below, the cosmologies corresponding to these three manifolds are all matter-empty, so that these differences will only reflect the effect of the addition of a term Λg to the definition of the Einstein tensor G; Λ is called the Einstein cosmological constant and its values are $\Lambda = 0$ for $M^{3,1}$, $\Lambda > 0$ for $\mathcal{M}_{+}^{3,1}$, and $\Lambda < 0$ for $\mathcal{M}_{-}^{3,1}$.

Notice further that if the average spatial acceleration is the same for all observers, i.e. $\overline{\psi}_Z = \frac{1}{3}\Lambda \cdot I$ with $\Lambda \in \mathbb{R}$ independent of Z, then $Ric = +\Lambda g$. This follows from two facts, complementing the above proposition. The first fact is, as we already pointed out, that we can assume without loss of generality that, given $z \in \mathcal{M}$ and Z an instantaneous observer at z, there exists a geodesic vector field ς, defined on a neighbourhood of z, with $\varsigma_z = Z$. The second fact is spelled out in the following scholium; note that both Ric and g are symmetric $(0,2)$-tensor fields.

Scholium 3. *The only symmetric $(0,2)$-tensor field E, such that $E_z(Z,Z) = 0$ for all $z \in \mathcal{M}$ and all $Z \in T_+\mathcal{M}_z$ with $g_z(Z,Z) = -1$, is $E \equiv 0$.*

Proof: For arbitrary $X \in T_+\mathcal{M}_z$, $Z \equiv |g(X,X)|^{-1/2}X$ satisfies $g_z(Z,Z) = -1$; from the linearity of E in each of its variables it follows that $E(X,X) = 0$. Let now Y be an arbitrary element in $T\mathcal{M}_z$, and $X \in T_+\mathcal{M}_z$. Since $T_+\mathcal{M}_z$ is open, $\exists\, a > 0$ such that $Z(x) \equiv X + xY \in T_+\mathcal{M}_z \;\forall\; x \in [0,a)$. We have then (using the linearity and symmetry of E) for all $x \in [0,a)$:

$$\begin{aligned} O &= E_z(Z(x), Z(x)) \\ &= E_z(X,X) + 2xE_z(X,Y) + x^2E(Y,Y) \end{aligned}$$

which implies:

$$E(Y,Y) = 0 \quad \forall\; Y \in T\mathcal{M} \tag{37}$$

Finally, for arbitrary $X, Y \in T\mathcal{M}_z$, we obtain by the usual polarization method, upon using again the symmetry of E : $2E_z(X,Y) = E(X+Y, X+Y) - E(X,X) - E(Y,Y)$ which, by (37) vanishes. □

We now understand the kinematic meaning of the condition $Ric = \Lambda g$. The Lorentz manifolds which satisfy this condition are called *Einstein spaces*. The context in which they appear is clarified in the following statement.

Proposition 4. *Let $(\mathcal{M}, g)$ be a connected n-dimensional (with $n \geq 3$) Lorentz or Riemann manifold. Then:*

(a) The following conditions are equivalent

$$G + \Lambda g = 0 \text{ with } \Lambda \in \mathbb{R} \tag{38}$$

$$G + \frac{n-2}{2n} Sg = 0 \tag{39}$$

$$Ric = \frac{2}{n-2}\Lambda g \text{ with } \Lambda \in \mathbb{R} \tag{40}$$

$$Ric = \frac{1}{n} Sg \tag{41}$$

(b) the conditions of (a) are satisfied if there exists a function $K : z \in \mathcal{M} \mapsto K_z \in \mathbb{R}$ such that $\forall\, \xi_z, \eta_z \in T\mathcal{M}_z$ and $\forall\, z \in \mathcal{M}$:

$$\hat{R}_z(\xi,\eta,\xi,\eta) = K_z[g_z(\xi,\xi)g_z(\eta,\eta) - g_z(\xi,\eta)^2] \tag{42}$$

(c) Any of the above conditions implies that S is constant over $\mathcal{M}$

(d) In (b) one can assume, without loss of generality that K is constant over $\mathcal{M}$.

Proof: (38) implies (39). Indeed (38) reads $Ric - \frac{1}{2}Sg + \Lambda g = 0$ so that

$$S \equiv (Ric)^{\mu}_{\mu} \equiv g^{\mu\nu}(Ric)_{\nu\mu} = [\frac{1}{2}S - \Lambda]g^{\mu\nu}g_{\nu\mu}$$

i.e.

$$S = n[\frac{1}{2}S - \Lambda] \text{ or } \Lambda = \frac{n-2}{2n}S\,.$$

and thus (39) follows from (38) by substitution.

(39) implies (41). Indeed (39) reads $Ric - \frac{1}{2}Sg + \frac{n-2}{2n}Sg = 0$ so that (41) follows immediately.

(41) implies that S constant over $\mathcal{M}$. Indeed (41) reads

$$(Ric)^{\mu}_{\nu} = \frac{1}{n}S\delta^{\mu}_{\nu}$$

and thus

$$(Ric)^{\mu}_{\nu;\mu} = \frac{1}{n}\partial_{\mu}S \cdot \delta^{\mu}_{\nu}$$

But we know (4.2.91) that

$$(Ric)^{\mu}_{\nu;\mu} = \frac{1}{2}\partial_{\nu}S \tag{43}$$

and thus, upon comparing these two equations, and noticing $n \geq 3$, we obtain indeed $\partial_{\nu}S = 0$.

(41) implies (40). Indeed we just learned that S is constant over $\mathcal{M}$; so Ric is indeed a constant multiple of g, which is what (40) says.

(40) implies (38). Indeed (40) implies:

$$S \equiv (Ric)^{\mu}_{\mu} = \frac{2}{n-2}\Lambda g^{\mu}_{\mu} = \frac{2n}{n-2}\Lambda$$

which, when inserted in the definition of G gives indeed (38). Incidentally, the proof shows that it is the same real number Λ which appears in (38) and in (40). Part (a) of the Proposition is proven. □

To prove (b), we shall first show that

$$\hat{R}_{\mu\nu\rho\sigma} - K(g_{\mu\rho}g_{\nu\sigma} - g_{\mu\sigma}g_{\nu\rho}) = 0 \tag{44}$$

For this purpose, let $S_{\mu\nu\rho\sigma}$ denote the *LHS* of (44). From the corresponding properties of $\hat{R}$ one verifies immediately

$$S_{\mu\nu\rho\sigma} = S_{\rho\sigma\mu\nu} \tag{45}$$
$$S_{\mu\nu\rho\sigma} + S_{\mu\nu\sigma\rho} = 0 \tag{46}$$
$$\underset{\nu,\rho,\sigma}{\circlearrowleft} S_{\mu\nu\rho\sigma} = 0 \tag{47}$$

Define further

$$S(\xi,\eta) \equiv S_{\mu\nu\rho\sigma}\xi^\mu\eta^\nu\xi^\rho\eta^\sigma \tag{48}$$

and notice that condition (42) is precisely $S(\xi,\eta) = 0$. Upon playing with the linear independence of ξ and η, one obtains successively from this:

(49) $S_{\mu\nu\mu\nu} = 0$
(50a) $S_{\mu\nu\rho\nu} + S_{\rho\nu\mu\nu} = 0$
and thus by (45)
(50b) $S_{\mu\nu\rho\nu} = 0$
and, with (46)
(50c) $S_{\mu\nu\mu\sigma} = 0$
(51a) $S_{\mu\nu\rho\sigma} + S_{\mu\sigma\rho\nu} + S_{\rho\nu\mu\sigma} + S_{\rho\sigma\mu\nu} = 0$
and thus
(51b) $S_{\mu\nu\rho\sigma} + S_{\mu\sigma\rho\nu} = 0$

From these relations, one obtains the system of simultaneous conditions

$$\begin{aligned} S_{\mu\nu\rho\sigma} + S_{\mu\nu\sigma\rho} &= 0 \\ 2S_{\mu\sigma\nu\rho} + 2S_{\mu\sigma\rho\nu} &= 0 \\ -S_{\mu\rho\nu\sigma} - S_{\mu\nu\sigma\rho} - S_{\mu\sigma\rho\nu} &= 0 \\ S_{\mu\sigma\nu\rho} + S_{\mu\rho\nu\sigma} &= 0 \\ -S_{\mu\sigma\rho\nu} - S_{\mu\nu\rho\sigma} &= 0 \end{aligned} \tag{52}$$

Upon adding the five equations (52) one obtains $3S_{\mu\sigma\nu\rho} = 0$ which proves (44).

(44) now gives in succession:

$$R^\mu_{\nu\rho\sigma} = K(\delta^\mu_\rho g_{\nu\sigma} - \delta^\mu_\sigma g_{\nu\rho}) \tag{53}$$

$$(Ric)_{\nu\rho} = (n-1)Kg_{\nu\rho}, \text{ i.e. } \quad Ric = (n-1)Kg \tag{54}$$

$$S \equiv (Ric)^\rho_\rho = n(n-1)K \tag{55}$$

(54) and (55) give (41), thus proving the second part of the proposition.

We already proved that (41) implies that S is constant over $\mathcal{M}$, which is part (c) of the proposition. Part (d) finally follows from (b), (c) and (55). □

This proposition calls for a few comments.

Firstly, it is now manifest from part (a) of the proposition that any one of the four conditions (38)–(41) can be used to characterize Einstein spaces.

Secondly, condition (42) requires only that at each point $z \in \mathcal{M}$, taken separately, the manifold $\mathcal{M}$ has constant *sectional curvature* K_z, and that K_z is a smooth function of z. It is, therefore, quite remarkable that this local condition implies—see part (d)—a global result, namely that K_z is a constant over the whole manifold. For Riemann manifolds this statement is known as Schur's theorem and is colloquially paraphrased by saying that isotropy implies homogeneity.

Thirdly, condition (42) is indeed a very strong condition (see e.g. J. A. Wolf (1974)). It is known that every geodesically complete, connected and simply connected, four-dimensional Lorentz manifold $\mathcal{M}$ of constant sectional curvature K is isometrically isomorphic to: $M^{3,1}$ if $K = 0$, $\mathcal{M}^{3,1}_{+}$ with $\rho = K^{1/2}$ if $K > 0$, or $\mathcal{M}^{3,1}_{-}$ with $\rho = (-K)^{-1/2}$ if $K < 0$. As we saw in Section 4.2, these are homogeneous manifolds; let us denote any of these three manifolds by $\overline{\mathcal{M}}_K$ (with $K = 0$, > 0 or < 0) and by G_K the symmetry group of $\overline{\mathcal{M}}_K$ (see Table 4.2.1). Notice next that one does not gain very much if one drops the requirement that the manifold $\mathcal{M}$ is simply connected; let indeed $\mathcal{M}_K$ be such a manifold, indexed by its curvature K. One has then that there exists a subgroup $\Gamma \subset G_K$ acting freely and properly discontinuously on $\overline{\mathcal{M}}_K$ such that $\mathcal{M}_K$ is the left quotient $\Gamma \setminus \overline{\mathcal{M}}_K$ of $\overline{\mathcal{M}}_K$ by Γ. Consequently, if one wants to get seriously away from $\overline{\mathcal{M}}_K$ (and thus obtain manifolds which can serve as space-times of non-empty cosmologies) one must drop either geodesic completeness, or constant sectional curvature, or both.

The question comes naturally as to whether there exist Einstein spaces which are not of constant sectional curvature. That this is indeed the case is shown by the following example, due to Schwarzschild (1916).

We proceed as conservatively as possible, and consider first the Minkowski space $M^{3,1}$ of special relativity, with metric

$$g^{(0)} = \sum_{k=1}^{3} dx^k \otimes dx^k - dx^4 \otimes dx^4 \tag{56a}$$

which we rewrite in polar coordinates on $\mathbb{R}^3$,

$$\begin{aligned} g^{(0)} &= -dt \otimes dt + dr \otimes dr + g_S(r) \text{ with} \\ g_S(r) &= r^2(\sin^2\theta\, d\varphi \otimes d\varphi + d\theta \otimes d\theta) \end{aligned} \tag{56b}$$

We now want to modify this metric, but keep three of its essential properties, namely that it be a stationary Lorentz metric, that the corresponding Einstein tensor satisfy $G \equiv 0$, and that it be invariant under the group of rotations of

$\mathbb{R}^3$. The simplest modification one can think of is to write:

$$g = -e^{A(r)} dt \otimes dt + e^{B(r)} dr \otimes dr + g_S(r) \tag{57}$$

This is not only the simplest, but, in fact, the only reasonable modification of $g^{(0)}$ satisfying the above three requirements. We shall not concern ourselves with that aspect here, but instead forewarn the reader that (r, θ, φ, t) should be thought of as local coordinates only, for a manifold the extent of which is still to be determined.

On the one hand, it is clear that g is indeed a Lorentz metric which is invariant under the group of rotations of $\mathbb{R}^3$.

On the other hand, part (a) of our proposition shows that the condition $G \equiv 0$ is equivalent to the condition $Ric \equiv 0$. A straightforward computation shows that the non-identically zero components of the Ricci tensor are:

$$\begin{aligned}
(Ric)_{rr} &= \frac{1}{2}A'' - \frac{1}{4}A'B' + \frac{1}{4}A'^2 - \frac{1}{r}B' \\
(Ric)_{tt} &= -[\frac{1}{2}A'' - \frac{1}{4}A'B' + \frac{1}{4}A'^2 + \frac{1}{r}A']\, e^{(A-B)} \\
(Ric)_{\theta\theta} &= [1 + \frac{1}{2}r(A' - B')]\, e^{-B} - 1 \\
(Ric)_{\varphi\varphi} &= \sin^2\theta \cdot R_{\theta\theta}
\end{aligned} \tag{58}$$

From $(Ric)_{rr} = 0 = (Ric)_{tt}$ one obtains in particular

$$A' + B' = 0 \text{ i.e. } A + B = k \in \mathbb{R} \tag{59}$$

The integration constant k can be absorbed by the trivial change of variable $t \to e^{-k/2}t$. We will, therefore, put $k = 0$, so that

$$A(r) + B(r) = 0 \tag{60}$$

Together with the condition $(Ric)_{\theta\theta} = 0$, this gives

$$(1 + rA')\, e^{A} = 1 \tag{61}$$

and thus

$$\begin{aligned}
e^{A(r)} &= (1 - \frac{2m}{r}) \\
e^{B(r)} &= (1 - \frac{2m}{r})^{-1}
\end{aligned} \tag{62}$$

Note that these, in spite of the cancellation involved in getting (59), do indeed insure $Ric \equiv 0$. In (62) m is only an integration constant. Upon considering the Newtonian limit (11), one finds that $m \geq 0$ would give for Φ the gravitational potential of a point mass m at $r = 0$; taking up this hint, we concentrate on the case where indeed m is a positive real constant.

On the strength of these preliminary remarks, we define the Lorentz manifold $(\mathcal{M}_{ext}, g_{ext})$ as:

$$\begin{aligned} &\mathcal{M}_{ext} \equiv \mathbb{R} \times (2m, \infty) \times S^2 \\ (63) \qquad &g_{ext} \equiv -(1 - \frac{2m}{r})\, dt \otimes dt + (1 - \frac{2m}{r})^{-1} dr \otimes dr + g_S(r) \\ &\text{with } g_S(r) = r^2(\sin^2\theta\, d\varphi \otimes d\varphi + d\theta \otimes d\theta) \end{aligned}$$

Notice incidentally that for $m = 0$, this metric coincides with the Minkowski and that $(\mathcal{M}_{ext}, g_{ext})$ can be smoothly extended to $(M^{3,1}, g^{(0)})$. Note also that for all $m \geq 0\,(\mathcal{M}_{ext}, g_{ext})$ is orientable and time-orientable. It can, therefore, be made into an acceptable space-time.

More to the point here, let thus $m > 0$. A straightforward computation then shows that

$$(64) \qquad R_{trtr} = -\frac{2m}{r^3}$$

Hence we have constructed an Einstein manifold which is Ricci-flat, but not flat. If it were of constant sectional curvature, (54) would imply $K = 0$, and thus (53) would imply $R \equiv 0$ which contradicts (64). We have thus produced, in answer to our question, an Einstein space which is not of constant sectional curvature. Condition (42) is, therefore, strictly stronger than the conditions (38)–(41) in Proposition 4.

The Lorentz manifold (63), when a choice of orientation and time orientation has been made, is called the *exterior Schwarzschild solution* (of the equation $G \equiv 0$). We shall see in Section Three below that it can be used to describe the motion of a planet around the Sun. The reader might worry at this point about the consequences of having restricted the range of the variable r to $(2m, \infty)$. Upon filling in the mass of the Sun and the value of the universal gravitation constant, he will find that the lower limit of this range of r is about 3km, which is well inside the Sun itself (the radius of the Sun is indeed about 7.10^5 km!) A similar computation for the motion of the Moon around the Earth would give about 1cm (compared to the radius of the Earth which is about 6.10^3km). Hence, for such applications, the manifold $(\mathcal{M}_{ext}, g_{ext})$ is perfectly adequate. One might, nevertheless, wonder at what would possibly happen if one were to have a star so dense that all its matter would be concentrated well inside its *Schwarzschild radius* $2m$. In other words, could the manifold $(\mathcal{M}_{ext}, g_{ext})$ be extended inwards across $r = 2m$ or could it be, to the contrary, that the Schwarzschild radius is a natural boundary? The following few pages are devoted to the mathematical exploration of this question.

Let us first notice that

$$\begin{aligned} &\mathcal{M}_{int} \equiv \mathbb{R} \times (0, 2m) \times S^2 \\ (65) \qquad &g_{int} \equiv -(1 - \frac{2m}{r}) dt \otimes dt + (1 - \frac{2m}{r})^{-1} dr \otimes dr + g_s(r) \\ &\text{with } g_s(r) = r^2(\sin^2\theta\, d\varphi \otimes d\varphi + d\theta \otimes d\theta) \end{aligned}$$

shares with $(\mathcal{M}_{ext}, g_{ext})$ all the following properties: it is a connected, four-dimensional, orientable and time orientable Lorentz manifold (so that it can be made into an acceptable space-time); it is Ricci flat, i.e. $Ric \equiv 0$ and thus $G \equiv 0$, but it is not flat, i.e. $R \not\equiv 0$ (see 64 again) and it is invariant under the group O^3 of rotations of $\mathbb{R}^3$. This manifold is called the *interior Schwarzschild* solution.

The question we want to address is whether $(\mathcal{M}_{ext}, g_{ext})$ and $(\mathcal{M}_{int}, g_{int})$ can be made into an extension of one another. At first glance, one may think of using $\mathcal{M} = \mathbb{R} \times \{(0, 2m) \cup (2m, \infty)\} \times S^2$ with the metric given in (63) and (65); this, however, would not do, since this manifold is not connected. What then about $\mathcal{M} = \mathbb{R} \times (0, \infty) \times S^2$? One has there to face another difficulty, namely that there is an apparent singularity of the metric at $r = 2m$; the question now is whether this singularity is real, or is just a misleading feature of a poor choice of coordinates (and/or of differential structure). To answer that question, we first look again at the manifolds $(\mathcal{M}_{ext}, g_{ext})$ and $(\mathcal{M}_{int}, g_{int})$ separately.

On $(\mathcal{M}_{ext}, g_{ext})$ we introduce the change of variables $(t, r) \in \mathbb{R} \times (2m, \infty) \mapsto (v, r) \in \mathbb{R} \times (2m, \infty)$ defined by:

(66a) $$v \equiv t + r + 2m\, \ell n(r - 2m)$$

i.e.

(66b) $$dt = dv - (1 - \frac{2m}{r})^{-1} dr$$

Upon substituting in (63) one obtains

(67) $$\begin{aligned} &\mathcal{M}_{ext} = \mathbb{R} \times (2m, \infty) \times S^2 \\ &g_{ext} = -(1 - \frac{2m}{r}) dv \otimes dv + 2dv \otimes dr + g_s(r) \\ &\text{with } g_s(r) \text{ as in (63)} \end{aligned}$$

On $(\mathcal{M}_{int}, g_{int})$ we introduce a different change of variables $(t, r) \in \mathbb{R} \times (0, 2m) \mapsto (v, r) \in \mathbb{R} \times (0, 2m)$ defined by:

(68a) $$v \equiv t + r + 2m\, \ell n(2m - r)$$

i.e.

(68b) $$dt = dv - (1 - \frac{2m}{r})^{-1} dr$$

Upon substituting in (65), one obtains:

(69) $$\begin{aligned} &\mathcal{M}_{int} = \mathbb{R} \times (0, 2m) \times S^2 \\ &g_{int} = -(1 - \frac{2m}{r}) dv \otimes dv + 2dv \otimes dr + g_s(r) \\ &\text{with } g_s(r) \text{ as in (65)} \end{aligned}$$

Three remarks have to be made about the change of variables (66) and (68).

Firstly, these changes of variables are not as incongruous as they may seem to be: their interpretation follows directly from the following computation. Consider, for instance, the curves defined in $(\mathcal{M}_{ext}, g_{ext})$ by:

$$(70)\qquad \begin{array}{l}\gamma(s) \equiv (t(s), r(s), \theta(s), \varphi(s)) \text{ with} \\ t(s) = s\,;\ \theta(s) = \theta_0\,;\ \varphi(s) = \varphi_0\,; \text{ and} \\ r(s) + 2m\,\ell n[r(s) - 2m] = -s + v_0 \\ \text{ where } v_0 \in \mathbb{R} \text{ is fixed.}\end{array}$$

It can be checked that these curves are geodesics and satisfy $g(\dot\gamma, \dot\gamma) = 0$. Hence the level curves $v =$ constant of $(\mathcal{M}_{ext}, g_{ext})$ when this manifold is coordinated as prescribed by (66) can be interpreted as incoming (since $dr/ds < 0$) light rays (since $g(\dot\gamma, \dot\gamma) = 0$ and γ have geodesics trajectories).

Secondly, the changes of coordinates (66) and (68) are diffeomorphisms. We have, therefore, done nothing to the differential structures of our two manifolds.

Thirdly, these changes of coordinates allow to embed smoothly both the exterior and the interior Schwarzschild solutions in a single Lorentz manifold, namely

$$(71)\qquad \begin{array}{l}\mathcal{M}' \equiv \mathbb{R} \times (0,\infty) \times S^2 \cong \mathbb{R}^4 \setminus \mathbb{R} \\ g' \equiv -(1 - \dfrac{2m}{r})dv \otimes dv + 2dv \otimes dr + g_s(r) \\ \text{with } g_s(r) \text{ as in (63) and (65)}\end{array}$$

The remarkable feature of this is now that g' shows no singularity whatsoever across the three-dimensional manifold separating $(\mathcal{M}_{ext}, g_{ext})$ and $(\mathcal{M}_{int}, g_{int})$, namely across

$$(72)\qquad \mathcal{M}'_{2m} \equiv \{(v, r, \theta, \varphi) \in \mathcal{M}' | r = 2m\}$$

Since the change of variables $(t, r) \to (v, r)$ is not smooth across this manifold, the Lorentz manifold $(\mathcal{M}', g')$ is not the same as the object we would have obtained by naively glueing together the exterior and interior Schwarzschild solutions as given in their original coordinates.

The smooth extension of $(\mathcal{M}_{ext}, g_{ext})$ to $(\mathcal{M}_{int}, g_{int})$ has a complicated history; see e.g. Rindler (1969, 1977), note 7, p.150 who cites: E. Kasner (1921), A. S. Eddington (1924), Lemaitre (1933), D. Finkelstein (1958), C. Fronsdal (1959), M. D. Kruskal (1960) and G. Szekeres (1960). Let it suffice to mention here that the form (67), (69) and (71) of the Schwarzschild metric is often referred to as the *Eddington-Finkelstein metric*; and that the extension $(\mathcal{M}', g')$ is not unique. The second point is worth some closer examination as it shows that $(\mathcal{M}', g')$ itself can be extended, and that space times which are not geodesically complete do occur in the theory.

Indeed, instead of the change of variables corresponding to an emphasis on incoming light rays, see (66), (68) and (70), namely

$$\varphi' : (t,r) \in \mathbb{R} \times \{(0,2m) \cup (2m,\infty)\} \mapsto (v,r) \in \mathbb{R} \times \{(0,2m) \cup (2m,\infty)\}$$

defined by

$$v = t + r + 2m \begin{Bmatrix} \ell n(r-2m) & r>0 \\ \ell n(2m-r) & r<0 \end{Bmatrix} \tag{73}$$

consider now the change of coordinates

$$\varphi'' : (t,r) \in \mathbb{R} \times \{(0,2m) \cup (2m,\infty)\} \mapsto (w,r) \in \mathbb{R} \times \{(0,2m) \cup (2m,\infty)\}$$

defined by

$$w = t - r - 2m \begin{Bmatrix} \ell n(r-2m) & r>0 \\ \ell n(2m-r) & r<0 \end{Bmatrix} \tag{74a}$$

i.e.

$$dt = dw + (1 - \frac{2m}{r})^{-1} dr \tag{74b}$$

As with $(\mathcal{M}', g')$ one verifies that this new change of variables allows to embed smoothly both the exterior and the interior Schwarzschild solutions in a single Lorentz manifold, namely

$$\begin{aligned} &\mathcal{M}'' \equiv \mathbb{R} \times (0,\infty) \times S^2 \cong \mathbb{R}^4 \setminus \mathbb{R} \\ &g'' \equiv -(1-\frac{2m}{r}) dw \otimes dw - 2dw \otimes dr + g_s(r) \\ &\text{with } g_s(r) \text{ as in (63) and (65)} \end{aligned} \tag{75}$$

Again this metric shows no singularity whatsoever across the three-dimensional manifold separating $(\mathcal{M}_{ext}, g_{ext})$ and $(\mathcal{M}_{int}, g_{int})$, namely across

$$\mathcal{M}''_{2m} \equiv \{(w,r,\theta,\varphi) \in \mathcal{M}'' | r = 2m\} \tag{76}$$

Notice, however, that the map

$$\varphi'' \circ (\varphi')^{-1} : (v,r) \in \mathbb{R} \times \{(0,2m) \cup (2m,\infty)\} \mapsto (w,r) \in \mathbb{R}\{(0,2m) \cup (2m,\infty)\}$$

given by

$$w = v - 2[r + 2m\, \ell n|r-2m|] \tag{77}$$

does not extend smoothly across $r = 2m$, so that $(\mathcal{M}'', g'')$ and $(\mathcal{M}', g')$ have different differential structures, and this although they both extend the exterior and the interior Schwarzschild solutions.

It is quite a remarkable fact, which we know from the work of Kruskal (1960), that there exists a Lorentz manifold in which both $(\mathcal{M}', g')$ and $(\mathcal{M}'', g'')$ can be embedded smoothly and isometrically. This can be seen as follows.

Consider the successive change of coordinates

$$\begin{aligned}(t,r) \in \mathbb{R} \times (2m,\infty) &\mapsto (v,w) \in \mathbb{R}^2 \setminus \{0\} \\ &\mapsto (\hat{v},\hat{w}) \in (0,\infty) \times (-\infty,0) \\ &\mapsto (\tau,\rho) \in \{(\tau,\rho) \in \mathbb{R}^2 | \rho > 0; \tau^2 - \rho^2 < 0\}\end{aligned} \tag{78a}$$

defined by:

$$\begin{aligned} v &= t + r + 2m\,\ell n(r - 2m) \\ w &= t - r - 2m\,\ell n(r - 2m)\end{aligned} \tag{78b}$$

$$\begin{aligned} \hat{v} &= exp(v/4m) \\ \hat{w} &= -exp(-w/4m)\end{aligned} \tag{78c}$$

$$\begin{aligned} \tau &= \frac{1}{2}(\hat{v} + \hat{w}) \\ \rho &= \frac{1}{2}(\hat{v} - \hat{w})\end{aligned} \tag{78d}$$

Upon following the metric g_{ext} in (63) through these successive changes of variables, one can identify the exterior Schwarzschild solution as:

$$\begin{aligned} &\mathcal{M}_{ext} = \{(\tau,\rho,\theta,\varphi) \in \mathbb{R}^2 \times S^2 | \rho > 0, \tau^2 - \rho^2 < 0\} \\ &g_{ext} = \frac{16m^2}{r} exp(-r/2m)\{-d\tau \otimes d\tau + d\rho \otimes d\rho\} + g_S(r) \\ &\text{with } g_S(r) \text{ as in (63), and } r \text{ the unique solution of} \\ &\tau^2 - \rho^2 = -(r - 2m)\, exp(r/2m) \text{ with } r \in (2m,\infty)\end{aligned} \tag{79}$$

Once it is written in this form, the exterior Schwarzschild solution begs for an extension to

$$\begin{aligned} &\mathcal{M} \equiv \{(\tau,\rho,\theta,\varphi) \in \mathbb{R}^2 \times S^2 | \tau^2 - \rho^2 < 2m\} \\ &g \text{ as in (79) with range of } r \text{ extended to } (0,\infty)\end{aligned} \tag{80}$$

We will distinguish six submanifolds in this $\mathcal{M}$, namely:

(81,I) $\mathcal{M}_I \equiv \{z \in \mathcal{M} | \tau^2 - \rho^2 < 0; \rho > 0\}$

(81,II) $\mathcal{M}_{II} \equiv \{z \in \mathcal{M} | \tau^2 - \rho^2 > 0; \tau > 0\}$

(81,III) $\mathcal{M}_{III} \equiv \{z \in \mathcal{M} | \tau^2 - \rho^2 < 0; \rho < 0\}$

(81,IV) $\mathcal{M}_{IV} \equiv \{z \in \mathcal{M} | \tau^2 - \rho^2 > 0; \tau < 0\}$

(81,V) $\mathcal{M}_V \equiv \{z \in \mathcal{M} | \tau + \rho > 0\}$

(81,VI) $\mathcal{M}_{VI} \equiv \{z \in \mathcal{M} | \rho - \tau > 0\}$

Let us further denote by g_A the restriction of g to $\mathcal{M}_A$ (with $A = I, II, \ldots, VI$).

We already identified $(\mathcal{M}_I, g_I)$ with $(\mathcal{M}_{ext}, g_{ext})$. Next we notice that the map

$$\{(\tau,\rho) \in \mathbb{R}^2 | \tau^2 - \rho^2 < 2m; \tau + \rho > 0\} \to \{(v,r) \in \mathbb{R} \times (0,\infty)\} \tag{82a}$$

given by:

$$\begin{aligned} &v = 4m\, \ell n(\tau + \rho) \\ &r \text{ solution of } \tau^2 - \rho^2 = (2m - r)\exp(r/2m) \end{aligned} \tag{82b}$$

is smooth and extends to an isometric diffeomorphism of $(\mathcal{M}_V, g_V)$ onto $(\mathcal{M}', g')$; moreover, the submanifold $\{z \in \mathcal{M}_V | \rho = \tau\}$ corresponds then to the submanifold $\mathcal{M}'_{2m} = \{z \in \mathcal{M}' | r = 2m\}$. Hence, $(\mathcal{M}_{II}, g_{II})$ is identified with the interior Schwarzschild solution $(\mathcal{M}_{int}, g_{int})$.

Similarly, the map

$$\{(\tau,\rho) \in \mathbb{R}^2 | \tau^2 - \rho^2 < 2m; \rho - \tau > 0\} \to \{(w,r) \in \mathbb{R} \times (0,\infty)\} \tag{83a}$$

given by:

$$\begin{aligned} &w = -4m\, \ell n(\rho - \tau) \\ &r \text{ solution of } \tau^2 - \rho^2 = (2m - r)exp(r/2m) \end{aligned} \tag{83b}$$

is smooth and extends to an isometric diffeomorphism of $(\mathcal{M}_{VI}, g_{VI})$ onto $(\mathcal{M}'', g'')$; moreover, the submanifold $\{z \in \mathcal{M}_{VI} | \rho = -\tau\}$ corresponds then to the submanifold $\mathcal{M}''_{2m} = \{z \in \mathcal{M}'' | r = 2m\}$. Hence $(\mathcal{M}_{IV}, g_{IV})$ is another copy in $(\mathcal{M}, g)$ of the interior Schwarzschild solution $(\mathcal{M}_{int}, g_{int})$.

A similar argument allows to identify a second copy of the exterior Schwarzschild solution in $(\mathcal{M}, g)$, namely $(\mathcal{M}_{III}, g_{III})$.

From there the full picture follows readily (see Fig. 1). $(\mathcal{M}, g)$ is a connected, four-dimensional, orientable and time-orientable Lorentz manifold. It is Ricci-flat, but not flat. $(\mathcal{M}, g)$ contains two disjoint copies of the exterior Schwarzschild solution, namely, $(\mathcal{M}_I, g_I)$ and $(\mathcal{M}_{III}, g_{III})$; two disjoint copies of the interior Schwarzschild solution, namely $(\mathcal{M}_{II}, g_{II})$ and $(\mathcal{M}_{IV}, g_{IV})$; two disjoint copies of $(\mathcal{M}', g')$, namely $(\mathcal{M}_V, g_V)$ and its image through $(\tau, \rho) \to (-\tau, -\rho)$; and two copies of $(\mathcal{M}'', g'')$, namely $(\mathcal{M}_{VI}, g_{VI})$ and its image through $(\tau, \rho) \to (-\tau, -\rho)$.

Notice that within each copy of either $(\mathcal{M}', g')$ or $(\mathcal{M}'', g'')$ everything, including the metric, is smooth across $r = 2m$. So nothing can possibly happen when crossing through the imaginary boundaries $\mathcal{M}'_{2m}$ or $\mathcal{M}''_{2m}$: the singularity of the original form of the metric (see (63) and (65)) is, therefore, a delusion due to a poor choice of the coordinate system in which it was expressed.

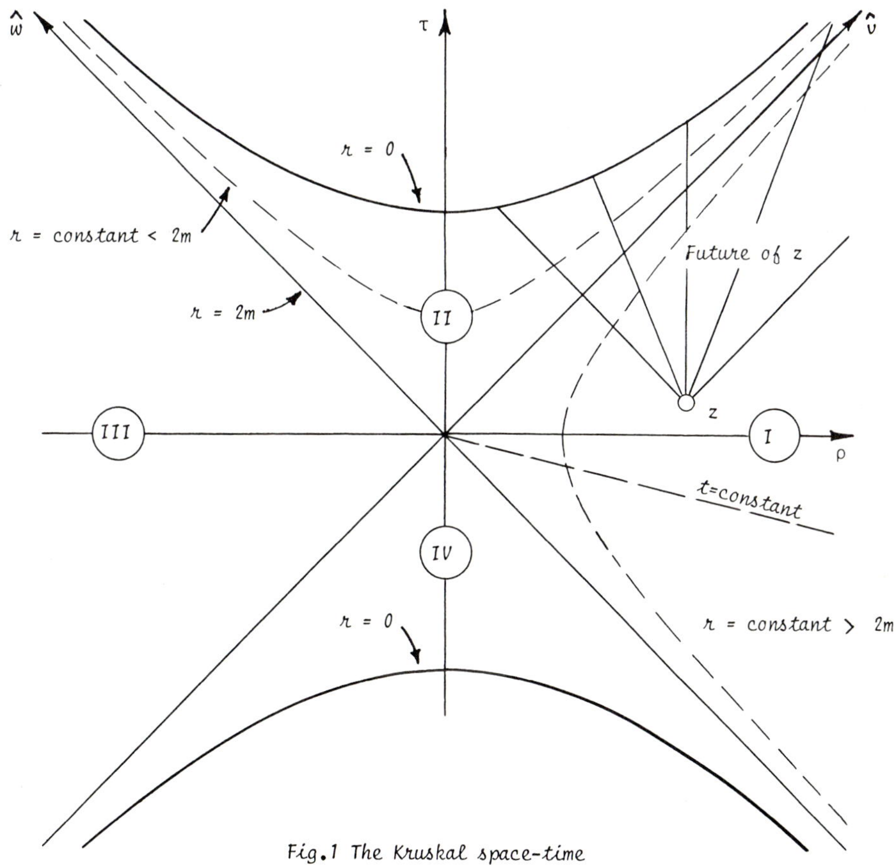

Fig. 1 The Kruskal space-time

Although nothing happens locally at $r = 2m$, Figure 1 suggests that something does happen globally which is worth mentioning. Let us time-orient $(\mathcal{M}, g)$ as indicated on Figure 1, and look at the future of any point z in the copy $(\mathcal{M}_I, g_I)$ of the Schwarzschild solution. Any causal, future-pointing curve (e.g. observer, particle or light-ray) through z will remain in $\mathcal{M}_V = \mathcal{M}_I \cup \mathcal{M}'_{2m} \cup \mathcal{M}_{II}$. As it crosses $\mathcal{M}'_{2m}$ nothing happens; however, once inside $\mathcal{M}_{II}$ it cannot escape this region—ever! $\mathcal{M}_{II}$, therefore, acts as a perfect trap: you do not feel anything as you enter, but once inside, you cannot escape, nor can you even communicate with the outside world. This situation (of a star so dense that all its matter is concentrated well inside its Schwarzschild radius) is dramatized in the popular literature by saying that it is a black hole...without much regard to the question of whether black holes do really exist in nature, or even could be seen if they existed. Still there is some evidence that such a strange creature is a partner in the binary system Cygnus $X - 1$. The exploration of the possible properties of black holes is the object of many serious theoretical speculations, and more is getting known on what to look for at the observational level.

Back to the mathematical objects just defined, the reader will verify that the space-time $(\mathcal{M}, g)$ is not geodesically complete (not even causally geodesically complete), nor is any of the manifolds $(\mathcal{M}_A, g_A)$ with $A = I, II, \ldots, VI$. Also, there is no way to incorporate $r = 0$ into a smooth extension of $(\mathcal{M}, g)$: this manifold is not extendable any further.

Up to this point, all the examples we discussed were of Einstein spaces, i.e. space-times which are characterized mathematically by the condition that the Einstein tensor be proportional to the metric tensor (see Proposition 4), and physically by either the fact that the average spatial acceleration is the same for all observers (see Proposition 2) or equivalently (see Section Three below) by the absence of active matter.

The simplest example of a space-time not belonging to this class is obtained from the following result.

SCHOLIUM 5. *Let* $\mathcal{M} \equiv \mathbb{R}^3 \times (0, \infty)$ *be equipped with the differential structure it inherits as a submanifold of the standard* $\mathbb{R}^4$*, and with the metric*

$$g_{(x,t)} = R(t)^2 \sum_{k=1}^{3} dx^k \otimes dx^k - dt \otimes dt \tag{84}$$

with $R : (0, \infty) \to \mathbb{R}^+$.

Then the following conditions are equivalent on $(\mathcal{M}, g)$.

(85) (i) $G_{kk} = 0 \quad k = 1,2,3$

(86) (ii) $R(t) = (\alpha t + c)^{2/3}$ *for some* $(\alpha, c) \in \mathbb{R}^+ \times \mathbb{R}^+$

(87) (iii) *With* $Ric^{\sim}$ *the* $(1,1)$ — *tensor equivalent to* Ric, *there exists* $f : (0,\infty) \to \mathbb{R}$ *such that*

$$Ric^{\sim} = f(t)[\sum_{k=1}^{3} dx^k \otimes \partial_k - dt \otimes \partial_t]$$

(iv) *same as (iii) with* $f(t) = \frac{1}{2} S(x,t)$ *where* S *is the Ricci scalar of the metric* g.

Proof: A straightforward computation gives that the only non-identically zero components of the Ricci tensor for a metric of the form (84) are

$$(Ric)_{kk} = RR'' + 2(R')^2 \quad k = 1,2,3 \tag{88a}$$

$$(Ric)_{44} = -\,3R''R^{-1} \tag{88b}$$

so that

$$S = 6[R''R^{-1} + (R'R^{-1})^2] \tag{89}$$

and thus

$$G_{kk} = -[2RR'' + (R')^2] \tag{90a}$$

$$G_{44} = 3(R'R^{-1})^2 \tag{90b}$$

The equivalence of (i) and (ii) follows then directly from (90a). Upon recalling that $(Ric^{\sim})^{\mu}_{\nu} \equiv g^{\mu\rho}(Ric)_{\rho\nu}$, we see that condition (iii) reads

$$R^{-2}(Ric)_{kk} - (Ric)_{44} = 0 \tag{91}$$

i.e. upon using (88a and b)

$$2[2R''R^{-1} + (R')^2] = 0 \tag{92}$$

Hence, upon comparing (92) and (90a), we see that (i) and (iii) are equivalent. The equivalence of (iii) and (iv) follows directly from the definition of $S \equiv (Ric^{\sim})^{\mu}_{\mu}$. □

Let $(\mathcal{M}, g)$ be as in the scholium. Upon inserting (86) into (89) and (90) one obtains

$$G = S(t)\, dt \otimes dt \tag{93}$$

Consequently, $(\mathcal{M}, g)$ is an Einstein space, i.e. G proportional to its metric g, if and only if $S \equiv 0$ and thus R is a constant, which we can assume without loss of generality (by performing a scale change in $\mathbb{R}^3$) to be 1. Hence $(\mathcal{M}, g)$ is an Einstein space if and only if it is a submanifold of the flat Minkowski space-time $M^{3,1}$.

To obtain something new, i.e. a space-time which is not an Einstein space, it is, therefore, sufficient to assume $\alpha > 0$ in (86). Again by adjusting the origin of time and the scale in $\mathbb{R}^3$, we can assume without loss of generality $\alpha = 1$ and $c = 0$. We have then

$$g = t^{4/3} \sum_{k=1}^{3} dx^k \otimes dx^k - dt \otimes dt \tag{94}$$

Although Robertson (1933) already pointed out in his review of relativistic cosmologies the precedents of Lemaitre (1925), Robertson (1928, 1929) and Heckmann (1931), the metric (94) is usually called the *Einstein-de Sitter metric* after Einstein & de Sitter (1932).

For our purpose in this section, we note that $\mathbb{R}^3 \times (0, \infty) \equiv \mathcal{M}$ equipped with the metric (94) is connected, four-dimensional, Lorentzian, orientable (e.g. by $\Omega = dx^1 \wedge dx^2 \wedge dx^3 \wedge dt$) and time-orientable (e.g. by the vector field $\xi = \partial_t$), so that it can be made into an acceptable space-time; this is not an Einstein space (see 93) since

$$S(t) = \frac{4}{3} t^{-2} \tag{95}$$

Notice, incidentally, that this $(\mathcal{M}, g)$ cannot be extended through $t = 0$ since the Ricci scalar (see 95) diverges as $t \to +0$. Verify also that if ς is a vector field in $\mathcal{T}_+(\mathcal{M})$ with $g(\varsigma, \varsigma) = -1$ and $G(\varsigma, \cdot) = f g(\varsigma, \cdot)$ for some $f \in \mathcal{F}(\mathcal{M})$, then $\varsigma = \partial_t$; hence ∂_t is canonically distinguished, i.e. is distinguished in an intrinsic manner since we can characterize it using only the metric and the time-orientation of $(\mathcal{M}, g)$ without appealing to any special coordinate-system. This simple fact is more than just a mathematical curiosity, and is indeed linked to our next remark.

Since the Einstein-de Sitter space just described is not an Einstein space, one might suspect that it contains some active matter which would cause, by gravitation, a deviation from the uniformity of the average space-acceleration (see Proposition 2 and the discussion following it). As we shall see in Section Three below, this is indeed the case: the Einstein tensor (93) corresponds to a form of matter one calls "dust", i.e. a gas with zero pressure.

The Einstein-de Sitter space-time can be generalized to a class of space-times usually referred to as *Robertson-Walker* spaces and characterized by the fact that the metric is of the form

$$g = R(t)^2 g_\Sigma - dt \otimes dt \tag{96}$$

where g_Σ is the metric of a three-dimensional Riemann manifold Σ of constant sectional curvature K_Σ; the Einstein-de Sitter case thus obtains with $\Sigma = \mathbb{R}^3$, and thus $K_\Sigma = 0$, supplemented by the conditions of the above Scholium.

For Robertson-Walker spaces, one can prove a natural generalization of the remark made after (93), namely:

Scholium 6. *A Robertson-Walker space is an Einstein space if and only if it has constant sectional curvature.*

Proof: Since a Robertson-Walker by definition is a space-time, it is a connected four-dimensional Lorentz manifold; we, therefore, already know (see Proposition 4) that if it has constant sectional curvature, it is an Einstein space. We now prove the converse. To simplify the notation we use $i, j, k, \ell, \ldots$ for indices taking the values 1, 2, 3; and $\mu, \nu, \rho, \sigma, \ldots$ when the value ranges extend to 1, 2, 3, 4. A straightforward computation shows that the only non-identically vanishing components of the Riemann tensor of a Robertson-Walker space are, up to the usual symmetries:

$$\hat{R}_{ijk\ell} = R^{-2}[K_\Sigma + (R')^2](g_{ik}g_{j\ell} - g_{i\ell}g_{jk}) \tag{97a}$$

$$\hat{R}_{4i4j} = R^{-1}R''(g_{44}g_{ij} - g_{4j}g_{i4}) \tag{97b}$$

so that the only non-identically zero components of the Ricci tensor are:

$$(Ric)_{ij} = R^{-2}\{RR'' + 2[K_\Sigma + (R')^2]\}g_{ij} \tag{98a}$$

$$(Ric)_{44} = 3R^{-1}R''g_{44} \tag{98b}$$

The condition that we have an Einstein space (namely $Ric = \Lambda g$, see Proposition 4, and recall that here $n = 4$) reduces to:

$$K_\Sigma + (R')^2 = \frac{1}{3}\Lambda R^2 \tag{99a}$$

$$R^{-1}R'' = \frac{1}{3}\Lambda \tag{99b}$$

which, when inserted in (97a and b) give

$$\hat{R}_{\mu\nu\rho\sigma} = \frac{1}{3}\Lambda(g_{\mu\rho}g_{\nu\sigma} - g_{\mu\sigma}g_{\nu\rho}) \tag{100}$$

From this we read that the space considered has constant sectional curvature namely $K = \Lambda/3$. □

Combining this with what we already knew, we conclude that the space-times $M^{3,1}$, $\mathcal{M}_+^{3,1}$ and $\mathcal{M}_-^{3,1}$ are the only Robertson-Walker spaces which are geodesically complete, connected and simply connected Einstein spaces.

Notice, however, that for a general Robertson-Walker space which is an Einstein space, the cosmological constant $\Lambda \equiv 3K$ (where K is the total sectional curvature) does not determine the spatial sectional curvature K_Σ. Indeed, the following three Robertson-Walker metrics

$$g = \rho^2 \cosh^2(t/\rho)\, g_\Sigma - dt \otimes dt \tag{101a}$$

$$g = exp(2t/\rho)\, g_\Sigma - dt \otimes dt \tag{101b}$$

$$g = \rho^2 \sinh^2(t/\rho)\, g_\Sigma - dt \otimes dt \tag{101c}$$

all have cosmological constant $\Lambda = 3\rho^{-2}$, while their spatial sectional curvature is $K_\Sigma = +1$ (for 101a), 0 (for 101b) and -1 (for 101c).

Finally, and most importantly, there are Robertson-Walker spaces which are not Einstein spaces; a counterexample has indeed been presented, namely the Einstein-de Sitter space-time.

Section 2. ELECTROMAGNETISM REVISITED

The definition of an electromagnetic field given in the special theory of relativity (see Section 5.1) extends immediately to the general theory, namely an *electromagnetic field* on a space-time $(\mathcal{M}, g)$ is a closed two-form $\mathcal{F}$ on the differentiable manifold $\mathcal{M}$. The orientation Ω and the time-orientation $\mathcal{T}_+$ of the space-time (and thus its Lorentz metric g) allow to define again an *observer* as a curve $\gamma : s \in I \mapsto \gamma(s) \in \mathcal{M}$ such that $g_{\gamma(s)}(\dot\gamma(s), \dot\gamma(s)) = -1$ and $\dot\gamma(s) \in \mathcal{T}_+ \ \forall\, s \in I$. For every $s_0 \in I$ the vector $Z = \dot\gamma(s_0)$ is said to be an *instantaneous observer* at $z_0 = \gamma(s_0)$. A *reference frame* around z_0 is a vector field ς, defined in a neighbourhood U of z_0, such that all its integral curves are observers. As in Section 5.1, we define the *electric field* E_Z associated to $\mathcal{F}$ by an instantaneous observer Z at $z \in U$ as the vector field E_Z in U satisfying

$$g_z(\eta_z, \mathrm{E}_Z) = \mathcal{F}_z(\eta_z, Z) \quad \forall\, \eta_z \in T\mathcal{M}_z \tag{1}$$

Notice that this vector field is unique, since g is non-degenerate, and that E_Z is orthogonal (i.e. space-like) to Z since $\mathcal{F}$ is antisymmetric. Again here, the *magnetic field* B_Z associated to $\mathcal{F}$ by Z is the unique vector field B_Z satisfying

$$\begin{aligned} &g_z(\mathrm{B}_Z, Z) = 0 \quad \text{and} \\ &\Omega_z(\xi_z, \eta_z, \mathrm{B}_Z, Z) = \mathcal{F}_z(\xi_z, \eta_z) \\ &\forall\, z \in U\,;\ \forall\, \xi_z, \eta_z \in T\mathcal{M}_z \text{ with } g_z(\xi_z, Z) = 0 = g_z(\eta_z, Z) \end{aligned} \tag{2}$$

Let us now complete our reference frame $\varsigma \equiv \{\varsigma_z \equiv Z \equiv (\xi_4)_z | z \in U\}$ by introducing three vector fields ξ_1, ξ_2 and ξ_3 on U such that

$$\begin{aligned} &g_z((\xi_j)_z, (\xi_k)_z) = \delta_{jk} \\ &g_z((\xi_j)_z, Z) = 0 \\ &\Omega_z((\xi_1)_z, (\xi_2)_z, (\xi_3)_z, Z) = +1 \end{aligned} \tag{3}$$

Let further $\{\epsilon^\mu | \mu = 1,2,3,4\}$ be the basis of one-forms dual to $\{\xi_\mu | \mu = 1,2,3,4\}$. This allows to extend to our present context, the definition of the source J of $\mathcal{F}$, namely

$$(4)\qquad \begin{aligned} J \equiv \operatorname{div} \mathcal{F} : \omega_z \in T^*U \\ \mapsto \nabla_{(\xi_\mu)_z} \hat{\mathcal{F}}_z(\omega_z, (\epsilon^\mu)_z) \in \mathbb{R} \end{aligned}$$

where $\hat{\mathcal{F}}$ is the $(2,0)$ antisymmetric tensor associated to $\mathcal{F}$ by the following routine extension of what we did in Section 5.1.

$$(4a)\qquad \begin{aligned} \hat{\mathcal{F}} : (\omega_z, \chi_z) \in \bigcup_{z \in \mathcal{M}} T^*\mathcal{M}_z \times T^*\mathcal{M}_z \\ \mapsto \mathcal{F}_z((\xi_\omega)_z, (\xi_\chi)_z) \in \mathbb{R} \end{aligned}$$

with

$$(4b)\qquad \begin{aligned} &g(\xi_\omega, \eta) \equiv \omega(\eta) \quad \forall\, \eta \in T\mathcal{M}; \\ &\text{and similarly for } \xi_\chi \end{aligned}$$

Finally, we define (again in complete analogy with Section 5.1)

$$(5)\qquad \begin{aligned} \rho_Z &= -g_z(J_z, Z) \\ \mathbf{J}_Z &= J_z - \rho_Z Z \end{aligned}$$

The reader will verify that J is independent of the choice of $\{\xi_k | k = 1,2,3\}$ satisfying (3), and, in fact, that J (and thus $\mathbf{J}$ and ρ, as well as $\mathbf{E}$ are independent of the orientation Ω whereas $\mathbf{B}$ changes sign if the opposite orientation Ω is chosen. Moreover, $\{\mathbf{E}, \mathbf{B}, \mathbf{J}, \rho\}$ satisfy the Maxwell equations (5.1.15) and (5.1.25).

With these notations, we now prove the following central result of this section (where the contraction $\rfloor$ is defined in (4.2.100)).

Proposition 1. *Let $\mathcal{F}$ be an electromagnetic field on a space-time $(\mathcal{M}, g)$. Then there exists exactly one symmetric $(2,0)$-tensor field $\hat{E}$ on $(\mathcal{M}, g)$ such that for every instantaneous observer Z*

$$(6)\qquad E(Z,Z) = \frac{1}{2}(|\mathbf{E}_Z|^2 + |\mathbf{B}_Z|^2)$$

where $\mathbf{E}_Z$ (resp. $\mathbf{B}_Z$) is the electric (resp. magnetic) field associated to $\mathcal{F}$ by Z. Specifically

$$(7)\qquad \hat{E} = \hat{D} - \frac{1}{4}(tr\, \hat{D})\hat{g}$$

with $\hat{D}$ defined on $(\mathcal{M}, g)$ as the symmetric $(2,0)$-tensor field given by:

$$(8)\qquad \begin{aligned} \hat{D} : (\omega_z, \chi_z) \in \bigcup_{z \in \mathcal{M}} T^*\mathcal{M}_z \times T^*\mathcal{M}_z \\ \mapsto \hat{g}_z((\xi_\omega \rfloor \mathcal{F})_z, (\xi_\chi \rfloor \mathcal{F})_z) \in \mathbb{R} \end{aligned}$$

Proof: Equation (8) clearly defines a symmetric (2, 0)-tensor field on $\mathcal{M}$, since $\hat{g}$ is. Let $\tilde{D}$ be the (1, 1)-tensor field defined by:

$$\begin{aligned} \tilde{D} : (\chi_z, \xi_z) \in & \bigcup_{z \in \mathcal{M}} T^*\mathcal{M}_z \times T\mathcal{M}_z \\ & \mapsto \hat{D}_z(\chi_z, (\omega_\xi)_z) \in \mathbb{R} \end{aligned} \tag{9}$$

where ω_ξ is the one-form defined by

$$\omega_\xi : \eta_z \in T\mathcal{M} \mapsto g_z(\xi_z, \eta_z) \in \mathbb{R} \tag{10}$$

Define now

$$tr\, \hat{D} : z \in \mathcal{M} \mapsto \tilde{D}((\epsilon^\mu)_z, (\xi_\mu)_z) \tag{11}$$

where $\{\xi_\mu | \mu = 1, 2, 3, 4\}$ is an arbitrary orthonormal basis of vector fields in a neighbourhood of z, and $\{\epsilon_\mu | \mu = 1, 2, 3, 4\}$ is the dual basis of one-forms. Note that $tr\, \hat{D}$ is independent of the orthonormal basis $\{\xi_\mu\}$ used to define it. Hence (7) defines a symmetric (2, 0)-tensor field on $\mathcal{M}$.

We now show that this tensor field satisfies condition (6). From (8) and (1), we obtain upon using (10) repeatedly:

$$\begin{aligned} D(Z, Z) &= \hat{D}(\omega_Z, \omega_Z) = \hat{g}(Z \rfloor \mathcal{F}, Z \rfloor \mathcal{F}) \\ &= \hat{g}(-\omega_{\mathbf{E}_Z} - \omega_{\mathbf{E}_Z}) = g(\mathbf{E}_Z, \mathbf{E}_Z) = |\mathbf{E}_Z|^2 \end{aligned} \tag{12}$$

Upon using (2) in (11), as well as (8), (1) and (10), one obtains with some more work:

$$(tr\, \hat{D})_z = 2(-|\mathbf{E}_Z|^2 + |\mathbf{B}_Z|^2) \tag{13}$$

which is invariant, as it should be, under all homogeneous isometries of $T\mathcal{M}_z$. To derive (13), it is instructive and expedient to use local coordinates, i.e. to use $\{\xi_\mu\}$ as above, with $\xi_4 = Z$. Writing then $\mathcal{F}_{\mu\nu} \equiv \mathcal{F}(\xi_\mu, \xi_\nu)$, one obtains sucessively

$$\hat{D}^{\mu\nu} = \hat{\mathcal{F}}^{\mu\sigma} \tilde{\mathcal{F}}_\sigma^{\ \nu} \tag{14}$$

with

$$\hat{\mathcal{F}}^{\mu\sigma} = \hat{g}^{\mu\alpha} \hat{g}^{\sigma\beta} \mathcal{F}_{\alpha\beta} \tag{15}$$

$$\tilde{\mathcal{F}}_\sigma^{\ \nu} = \hat{g}^{\nu\alpha} \mathcal{F}_{\alpha\sigma} \tag{16}$$

and thus

$$\tilde{D}_\nu^{\ \mu} = \hat{\mathcal{F}}^{\mu\sigma} \mathcal{F}_{\nu\sigma} \tag{17}$$

$$tr\, \hat{D} = \hat{\mathcal{F}}^{\mu\nu} \mathcal{F}_{\mu\nu} \tag{18}$$

Notice, in particular, that this gives

$$\hat{E}^{\mu\nu} = \hat{\mathcal{F}}^{\mu\sigma} \tilde{\mathcal{F}}_\sigma^{\ \nu} - \frac{1}{4} \hat{\mathcal{F}}^{\rho\sigma} \mathcal{F}_{\rho\sigma} g^{\mu\nu} \tag{19}$$

The reader will notice that up to here we did not use more than the fact that $\{\xi_\mu\}$ is a basis of vector fields around z. Using now $\xi_4 = Z$, and for simplicity, the orthonormality of our basis, one obtains readily from (1), (2) and (18) the announced formula (13). From (12), (13), and (7) one obtains finally (6). The uniqueness asserted in the Proposition follows directly from Scholium 6.1.3. □

The symmetric (0, 2)-tensor field E defined uniquely by (6), is called the *stress-energy tensor* of the electromagnetic field $\mathcal{F}$. The "energy" part of its name comes precisely from (6), the *RHS* of which is interpreted by each observer Z as the total energy (density) of the electric and magnetic fields, $\mathbf{E}_Z$ and $\mathbf{B}_Z$, which he associates with $\mathcal{F}$. The interpretation of the other components of E is best explained from the following result.

COROLLARY 2. *Let $\mathcal{F}$ be an electromagnetic field, E be its stress-energy tensor, and $J \equiv \operatorname{div}\mathcal{F}$ be its source. Then*

$$\operatorname{div}\hat{E} = -\tilde{\mathcal{F}}J \tag{20}$$

Proof: Here again it is both expedient and instructive to conduct the proof using local coordinates. We have, using (19),

$$(div\hat{E})^\mu = \hat{E}^{\mu\nu}{}_{;\nu} = -K^\mu + N^\mu \tag{21}$$

with

$$-K^\mu \equiv \mathcal{F}^{\mu\sigma}\mathcal{F}^\nu_{\sigma;\nu} \tag{22}$$

$$N^\mu \equiv \mathcal{F}^\nu_\sigma \mathcal{F}^{\mu\sigma}{}_{;\nu} - \frac{1}{4}(\mathcal{F}^{\rho\sigma}\mathcal{F}_{\rho\sigma})_{;\nu}g^{\mu\nu} \tag{23}$$

where $(;\nu)$ denotes the covariant derivative in the direction ξ_ν. We also drop from the proof the symbols $(^\wedge)$ and $(^\sim)$ since the position of the indices make them superfluous. We first compute K^μ

$$K^\mu = -\mathcal{F}^{\mu\sigma}g_{\sigma\rho}\mathcal{F}^{\nu\rho}{}_{;\nu} = \mathcal{F}^\mu_\rho\mathcal{F}^{\rho\nu}{}_{;\nu} \tag{24}$$

i.e. upon recalling that by definition

$$J^\rho \equiv (div\hat{\mathcal{F}})^\rho = \hat{\mathcal{F}}^{\rho\nu}{}_{;\nu} \tag{25}$$

we obtain

$$K^\mu = \mathcal{F}^\mu_\rho J^\rho = (\tilde{\mathcal{F}}J)^\mu \tag{26}$$

Hence the proof will consist in showing that N vanishes identically. The first term in N^μ is:

$$\begin{aligned} &\mathcal{F}^\nu_\sigma\mathcal{F}^{\mu\sigma}{}_{;\nu} = g^{\mu\rho}\mathcal{F}^{\nu\sigma}\mathcal{F}_{\rho\sigma;\nu} = \\ &\frac{1}{2}g^{\mu\rho}\{\mathcal{F}^{\nu\sigma}\mathcal{F}_{\rho\sigma;\nu} + \mathcal{F}^{\sigma\nu}\mathcal{F}_{\rho\nu;\sigma}\} = \\ &\frac{1}{2}g^{\mu\rho}\mathcal{F}^{\nu\sigma}\{\mathcal{F}_{\rho\sigma;\nu} + \mathcal{F}_{\nu\rho;\sigma}\} \end{aligned} \tag{27}$$

The second term in N^μ is:

(28)
$$-\frac{1}{4}(\mathcal{F}^{\rho\sigma}\mathcal{F}_{\rho\sigma})_{;\nu}g^{\mu\nu} = -\frac{1}{4}\{\mathcal{F}^{\rho\sigma}{}_{;\nu}\mathcal{F}_{\rho\sigma} + \mathcal{F}^{\rho\sigma}\mathcal{F}_{\rho\sigma;\nu}\}g^{\mu\nu} = -\frac{1}{2}g^{\mu\nu}\mathcal{F}^{\rho\sigma}\mathcal{F}_{\rho\sigma;\nu} = +\frac{1}{2}g^{\mu\rho}\mathcal{F}^{\nu\sigma}\mathcal{F}_{\sigma\nu;\rho}$$

Upon adding up (27) and (28) we find:

(29)
$$N^\mu = \frac{1}{2}g^{\nu\rho}\mathcal{F}^{\nu\sigma}\{\mathcal{F}_{\rho\sigma;\nu} + \mathcal{F}_{\nu\rho;\sigma} + \mathcal{F}_{\sigma\nu;\rho}\}$$

where the term $\{\ldots\}$ vanishes identically since $d\mathcal{F} = 0$. □

We now look at the components of (20), with $\xi_4 = Z$, to find first that:

(30a)
$$(\operatorname{div} E)^4 = -(\mathcal{F}J)^4$$

reads

(30b)
$$\operatorname{div}\mathbf{S}_Z + \partial_t U_Z = -\mathbf{E}_Z \cdot \mathbf{J}_Z$$

with

(30c)
$$\mathbf{S}_Z \equiv (\hat{E}_z^{41}, \hat{E}_z^{42}, \hat{E}_z^{43}) = \mathbf{E}_Z \times \mathbf{B}_Z$$

and

(30d)
$$U_Z \equiv \frac{1}{2}(|\mathbf{E}_Z|^2 + |\mathbf{B}_Z|^2)$$

Recall that U_Z is what the instantaneous observer Z calls the total *energy* (*density*) of the electric and magnetic fields, $\mathbf{E}_Z$ and $\mathbf{B}_Z$, which he associates to the electromagnetic field $\mathcal{F}$. The *"Poynting vector"* $\mathbf{S}_Z$ is then to be interpreted as an *energy-flux* (*density*); this gives the physical interpretation of three more components, namely $\hat{E}^{4k}$ with $k = 1, 2, 3$, of our stress-energy tensor $\hat{E}$. Another way to look at $\mathbf{S}_Z$ is to think of it as the spatial part of the quadrivector $(E^{\mu 4}) = (\mathbf{S}_Z, U_Z)$; for this reason it is also called the *momentum* (*density*) of the electromagnetic field as seen by Z. To complete the interpretation of (30a or b), $\mathbf{E}_Z \cdot \mathbf{J}_Z$ is the *power* (*density*) of the electromagnetic field, relative to our observer Z at z.

Turning now to the spatial component of (20), we find that

(31a)
$$(\operatorname{div}\hat{E})^k = -(\tilde{\mathcal{F}}J)^k \quad k = 1, 2, 3$$

reads

(31b)
$$\operatorname{div}\overline{\mathbf{T}}_Z + \partial_t\mathbf{S}_Z = -(\rho\mathbf{E}_Z + \mathbf{J}_Z \times \mathbf{B}_Z)$$

where

$$(\overline{\mathrm{T}}_Z)^{ij} \equiv E_z^{ij} \quad i,j = 1,2,3; \tag{31c}$$

the other symbols entering (31b) have already been defined and interpreted. The conservation law (31b), together with the interpretation of S as a momentum (density) lead to the interpretation of the spatial tensor $\overline{\mathrm{T}}_Z$ as a stress-tensor, known in the literature as the *Maxwell stress-tensor*, while the *RHS* of (31b) is to be interpreted as a *force (density)*, known as the *Lorentz force* in the physics literature.

This completes the interpretation of our stress-energy tensor E.

As a small historical note, it may be mentioned that it was Minkowski (1908) who recognized, in the framework of the special theory of relativity, that (31b) was the spatial component of an equation, namely (20), the time-component of which is the "Poynting theorem" given by equation (30b). He did so in the limiting case where (J, ρ) is considered as a point charge-current density.

In closing this section, we should mention two mathematically evident, but physically important (see Section Three), properties of the electromagnetic stress-energy tensor, namely

$$tr\ \hat{E} = 0 \tag{32}$$

$$E(\varsigma, \varsigma) \geq 0 \quad \forall \varsigma \in T\mathcal{M} \text{ with } g(\varsigma, \varsigma) \leq 0 \tag{33}$$

(32) indeed immediately follows from (7), while (33) follows from (6) by linearity and continuity.

Section 3. MODELS FOR MATTER

Let $(\mathcal{M}, g)$ be a space-time, $G = Ric - \frac{1}{2}Sg$ be its Einstein tensor, and chose a "cosmological constant" Λ; let further $\mathcal{F}$ be an electromagnetic field on $(\mathcal{M}, g)$, and E be its stress-energy tensor. A *model for matter* will be a symmetric $(0,2)$-tensor field T on $(\mathcal{M}, g)$ such that

$$G + \Lambda g = E + T \tag{1}$$

and

$$T_z(Z, Z) \geq 0 \text{ for all instantaneous observers } Z \tag{2}$$

(1) is called the *Einstein field equation*; in his original paper, Einstein (1916) did not have the term Λg, and he lumped together $(E + T)$ as "the" stress-energy tensor. The term $\Lambda \neq 0$ was introduced for cosmological reasons by Einstein (1917) who, however, later deplored the fact that "the introduction

of this second term constitutes a complication of the theory, which seriously reduces its logical simplicity" (Einstein, 1956); more recently, arguments have been put forward to the effect that this term should be in the *RHS* rather than the *LHS* of the equation, meaning by this that it should not be interpreted as part of the geometry, but rather as belonging to the sources, arising purportedly from quantum fluctuations in the vacuum. Whatever the case may be, observational evidence shows that Λ must be a very small constant, probably $|\Lambda| \tilde{<} 10^{-56} - 10^{-54}\,\mathrm{cm}^{-2}$ (Sandage, 1961).

With or without Λ, the reader must realize that equation (1) is a new postulate. Einstein (1916) has this to say when he introduces it: "It must be admitted that this introduction of the energy-tensor of matter [either T or $(E+T)$ in eq. (1)] is not justified by the relativity postulate alone...but the strongest reason for the choice of this equation lies in its consequences..." The purpose of much of the material to be presented in this section is precisely to show what some of these consequences are.

Without having to specify the matter model any further, we obtain the following immediate consequence of the definition.

Proposition 1. *For every matter model T on a space-time $(\mathcal{M}, g)$ on which an electromagnetic field $\mathcal{F}$ with source J is defined, we have:*

(3) (i) $\operatorname{div} \hat{T} = \tilde{\mathcal{F}} J$

(4) (ii) $tr\, \hat{T} = -S + 4\Lambda$

(iii) *For every observer Z*

(5) $$\overline{\psi}_Z - \frac{1}{3}\Lambda \leq -\frac{1}{6}[tr\; T_z + 2T_z(Z,Z)]$$

where $\overline{\psi}_Z$ is the average spatial acceleration relative to Z.

(iv) *The inequality in (5) becomes an equality for all Z if and only if $\mathcal{F} \equiv 0$.*

Proof: From Corollary (6.2.2) we know that $\tilde{\mathcal{F}} J = -\operatorname{div} \hat{E}$. By definition of the covariant derivative, the compatibility condition of the Levi-Cevita connection with the metric implies $\operatorname{div} g = 0$ (see 4.2.89b). Consequently (1) implies that (3) is equivalent to $\operatorname{div} G = 0$, which is always satisfied by the very way we constructed G (see 4.2.91). This proves (i). Upon taking the trace of (1) written in the form

(6) $$\hat{Ric} - \frac{1}{2}\hat{g}S + \hat{g}\Lambda = \hat{E} + \hat{T}$$

and recalling that $S \equiv tr\, \hat{Ric}$, and $tr\, \hat{E} = 0$ (see 6.2.32) we obtain (4), thus proving (ii). From (6) we obtain

(7) $$Ric(Z,Z) + \frac{1}{2}S - \Lambda = E(Z,Z) + T(Z,Z)$$

Upon substituting (4) into (7) and recalling that $\overline{\psi}_Z = -\frac{1}{3}Ric(Z,Z)$ (see 6.1.36), we obtain

$$\overline{\psi}_Z - \frac{1}{3}\Lambda = -\frac{1}{6}[tr\, T + 2T(Z,Z)] - \frac{1}{3}E(Z,Z) \tag{8}$$

(5) then follows from (8) and (6.2.33) thus proving (iii). Finally (iv) follows from (8) and Scholium (6.1.3) together with (6.2.6) □

Notice in particular, that in an empty universe, where $T = 0 = E$, parts (iii) and (iv) of the above proposition imply $\overline{\psi}_Z = \frac{1}{3}\Lambda$ which we already know from Section 6.1. We are now going to explore what becomes of this when more specific models of matter are considered.

Proposition 2. *Let $(\mathcal{M}, g)$ be a space-time with cosmological constant Λ; and $\mathcal{F}$ be an electromagnetic field on $(\mathcal{M}, g)$, with source J. Let further ς be a vector field on $(\mathcal{M}, g)$, and suppose there exist $m \in \mathbb{R}^+$, $\epsilon \in \mathbb{R}$ and $\eta : \mathcal{M} \to \mathbb{R}^+$ smooth, such that:*

(9) $\varsigma_z \in \mathcal{T}_+ \; \forall\, z \in \mathcal{M}, \quad g(\varsigma,\varsigma) = -m^2\,, \; \nabla_\varsigma \varsigma = \epsilon \tilde{\mathcal{F}} \varsigma$

(10) $J = \eta\, \epsilon\, \varsigma$

Then:

(i) *there exists exactly one symmetric (2,0)-tensor field $\hat{T}$ on $(\mathcal{M}, g)$ such that*

(11) $T_z(Z,Z) = \eta_z\, g_z(Z, \varsigma_z)^2$

namely

(12) $\hat{T} = \eta\, \varsigma \otimes \varsigma,$

so that T is consistent with condition (2);

(13) (ii) $\operatorname{div} \hat{T} = \tilde{\mathcal{F}} J$

so that T is consistent with the consequence (3) of condition (1);

(iii) *under this condition, we have, moreover*

(14) $\overline{\psi}_Z - \frac{1}{3}\Lambda \leq -\frac{1}{6}(\eta_z\, m^2 + |\mathrm{E}_Z|^2 + |\mathrm{B}_Z|^2).$

Proof: With $\hat{T}$ as in (12), we have $T(Z,Z) = \hat{T}(\omega_Z, \omega_Z) = \eta_z\, \omega_Z(\varsigma_z)^2 = \eta_z\, g_z(Z,\varsigma_z)^2$, which proves the existence claimed in (i); uniqueness follows from Scholium (6.1.3). To prove (ii), compute:

$$(\operatorname{div} \hat{T})^\mu = T^{\mu\nu}{}_{;\nu} = (\eta\, \varsigma^\nu \varsigma^\mu)_{;\nu} = \eta\, \varsigma^\nu (\varsigma^\mu)_{;\nu} + (\eta\, \varsigma^\nu)_{;\nu}\, \varsigma^\mu$$

i.e.

$$\operatorname{div} \hat{T} = \eta\, \nabla_\varsigma \varsigma + \operatorname{div}(\eta\, \varsigma) \cdot \varsigma \tag{15}$$

Now from (10) $\operatorname{div} J = 0$ iff $\operatorname{div}(\eta\varsigma) = 0$ (and by 15) iff $\operatorname{div}\hat{T} = \eta\, \nabla_\varsigma \varsigma$ (so that by 9) iff $\operatorname{div}\hat{T} = \tilde{\mathcal{F}}(\epsilon\eta\varsigma)$ (and thus by (10) again), iff $\operatorname{div}\hat{T} = \tilde{\mathcal{F}} J$. Since $\operatorname{div} J = 0$ follows from the definition $J = \operatorname{div}\mathcal{F}$, we have proved (ii). From (11) we see that

$$tr\, \hat{T}_z + 2T_z(Z,Z) = \eta_z[-m^2 + 2g_z(Z\,,\varsigma_z)^2] \tag{16}$$

Since both Z and ς_z are time-like, we have

$$g_z(Z,\varsigma_z)^2 \geq g_z(Z,Z)\, g_z(\varsigma_z,\varsigma_z) = m^2 \tag{17}$$

so that (14) follows from (8), (16), (17) and (6.2.6). □

In analogy with classical hydrodynamics, matter models which are obtained from this proposition, together with the additional requirement that condition (1) be satisfied, are referred to as *particle flows.* Condition (9) indeed tells us that the integral curves of our vector field ς are test particles of mass m and electric charge ϵ for the electromagnetic field $\mathcal{F}$, while condition (10) tells us that this particle flow acts as the source of the electromagnetic field $\mathcal{F}$.

Note that the Einstein-de Sitter space of Section 6.1 provides an example of neutral particle flow.

Upon inserting (12) into (4), we see that the effect of matter on the curvature of space-time is to increase the curvature, as measured by the Ricci scalar, by a term proportional to the matter density. More sharply, let Z be an instantaneous observer at rest with the flow, i.e. such that $Z = \varsigma_z$; for this observer, the inequality (14) becomes an equality so that the effect of matter at z is to decrease the average spatial acceleration by a term proportional to the matter density at z. Hence, matter can be interpreted as the source of a gravitational field; note, in particular, the signs occurring in (14): gravity is an attactive force.

The link between the above and the Newton theory of gravitation can be seen as follows. Consider a particle flow with $\mathcal{F} = 0$ and $\Lambda = 0$ to simplify the argument. We have then $G = \eta\, \varsigma \otimes \varsigma$. For an instantaneous observer at rest with this flow, take a coordinate system around z such that, at z, $\partial_4 = Z$, and work in the Newtonian approximation discussed in the beginning of Section 6.1 (see especially, eqns. 6.1.7—6.1.11). From $G = \rho\, Z \otimes Z$ and $Ric = G - \frac{1}{2} tr\, G$ (which follows directly from the definition of G) we obtain, on the one hand, the exact expression:

$$(Ric)_{44} = \frac{1}{2}\rho \tag{18}$$

On the other hand, we have in the Newtonian approximation

$$\begin{aligned} (Ric)_{44} &\cong \partial_i \Gamma^i_{44} \\ \text{with } \Gamma^i_{44} &\cong -\frac{1}{2}\partial^i g_{44} \end{aligned} \tag{19}$$

so that

$$(Ric)_{44} = -\frac{1}{2}\Delta g_{44} \tag{20}$$

where $\Delta \equiv \partial_i \partial^i$ is the spatial Laplace operator. With (6.1.11), (18) and (20) lead to the following equation for the gravitational potential Φ:

$$\Delta\Phi = \frac{1}{2}\rho \tag{21}$$

This is the classical Poisson equation, which is solved by

$$\begin{aligned} \Phi(\mathbf{x}) &= \int d^3\mathbf{y}\, \rho(\mathbf{y})\, \varphi(\mathbf{x}-\mathbf{y}) \\ \text{with } \varphi(\mathbf{x}-\mathbf{y}) &= -\frac{1}{8\pi} \cdot \frac{1}{|\mathbf{x}-\mathbf{y}|} \end{aligned} \tag{22}$$

This is the law governing gravity in Newtonian physics: the gravitational force at $\mathbf{x}$ due to a point-mass at $\mathbf{y}$ is given by:

$$\begin{aligned} \mathbf{f}(\mathbf{x},\mathbf{y}) &\equiv -\operatorname{grad}_{\mathbf{x}-\mathbf{y}} \varphi(\mathbf{x}-\mathbf{y}) \\ &= -\frac{1}{8\pi} \cdot \frac{1}{|\mathbf{x}-\mathbf{y}|^2} \end{aligned} \tag{23}$$

i.e. gravitation is indeed attractive, and gravitational forces decrease as the inverse of the square of the distance. In this sense *Einstein's field equation* (1) *is a gravitational theory*: it generalizes the classical Newtonian theory of gravitation, which it contains as a controlable approximation. We shall see later in this section that it is more than just an elegant mathematical generalization: it predicts observable phenomena that the classical theory does not explain.

At this point of the exposition of the theory, we would like to pursue a little further our exploration of matter models and try to describe a relativistic gas.

Here again, we take stock from classical hydrodynamics, and take as our starting point the Bernoulli (1738) model of a gas. Suppose, indeed, that at z we have six particle flows which differ only in the spatial direction of ς, specifically we take:

$$\begin{aligned} \varsigma_1 &= m(\alpha,0,0,\beta) & \varsigma_4 &= m(-\alpha,0,0,\beta) \\ \varsigma_2 &= m(0,\alpha,0,\beta) & \varsigma_5 &= m(0,-\alpha,0,\beta) \\ \varsigma_3 &= m(0,0,\alpha,\beta) & \varsigma_6 &= m(0,0,-\alpha,\beta) \end{aligned} \tag{24}$$

with

$$g_z(\varsigma_k,\varsigma_k) = -m^2 \quad k = 1,\ldots,6 \quad \text{i.e. } \alpha^2-\beta^2 = -1 \tag{25}$$

We now construct, from (12), with $\eta > 0$:

$$\hat{T} = \sum_{k=1}^{6} \eta\, \varsigma_k \otimes \varsigma_k \tag{26}$$

which is diagonal in the basis in which we chose to express the ς_k's; specifically

$$\hat{T} = 2\eta m^2 \begin{pmatrix} \alpha^2 & & & \\ & \alpha^2 & & \\ & & \alpha^2 & \\ & & & 3\beta^2 \end{pmatrix} \tag{27}$$

We can, therefore, rewrite $\hat{T}$ in the form:

$$\hat{T} = \rho\, Z \otimes Z + p\,(\hat{g} + Z \otimes Z) \tag{28}$$

with

$$\begin{aligned} Z &= (0,0,0,1) \\ \rho &= 6\eta m^2 \beta^2 > 0 \\ p &= 2\eta m^2 \alpha^2 > 0 \end{aligned} \tag{29}$$

Notice:

$$\begin{aligned} &tr\, \hat{T} = -\rho + 3p < 0 \\ &T(X,X) \geq \rho > 0 \quad \forall\, X \text{ such that } g(X,X) = -1 \end{aligned} \tag{30}$$

The general structure, this particular model hints to, is elicited by the following result

PROPOSITION 3. *Let T be a symmetric $(0,2)$-tensor field on a space-time $(\mathcal{M}, g)$ and suppose that there exists a vector field ς on $(\mathcal{M}, g)$ such that for every $z \in \mathcal{M} : \varsigma_z \in \mathcal{T}_+$, $g_z(\varsigma_z, \varsigma_z) = -1$ and T_z is invariant under the group O^3 of spatial rotations of the space $\Sigma_z = \{\xi_z \in T\mathcal{M}_z | g(\xi_z, \varsigma_z) = 0\}$ equipped with the metric $g_\Sigma = g_z|\Sigma_z$. Then there exists two smooth functions $\rho, p : z \in \mathcal{M} \to \mathbb{R}$ such that*

(31) (i) $\hat{T} = \rho\, \varsigma \otimes \varsigma + p\, \{\hat{g} + \varsigma \otimes \varsigma\}$

(32) (ii) $tr\, \hat{T} \leq 0$ *iff* $\rho - 3p \geq 0$

(33) (iii) $T_z(Z,Z) \geq 0$ *for every instantaneous observer Z iff* $\rho \geq 0$ *and* $\rho + p \geq 0$

(34) (iv) $-T_z(Z,Z) \leq tr\, \hat{T}_z$ *for every instantaneous observer Z iff* $p \geq 0$ *and* $\rho + p \geq 0$.

Proof: For every $\Xi \in \Sigma_z \ \exists \ \psi \in 0^3$ such that $\psi[\Xi] = -\Xi$. Hence the spatial isotropy of T_z implies $T_z(\varsigma_z, \Xi) = T_z(\psi[\varsigma_z], \psi[\Xi]) = T_z(\varsigma_z, -\Xi) = -T_z(\varsigma_z, \Xi)$, i.e.

$$T_z(\varsigma_z, \Xi) = 0 \quad \forall \, \Xi \in \Sigma_z \tag{35}$$

For every $\xi \in T\mathcal{M}_z$, let $\Xi = \xi - x\varsigma_z$, with $x = -g_z(\xi, \varsigma_z)$; clearly $\Xi \in \Sigma_z$. Define similarly H and y for $\eta \in T\mathcal{M}_z$. Taking (35) into account, we obtain

$$T_z(\xi, \eta) = T_z(\varsigma_z, \varsigma_z) \ \varsigma_z \otimes \varsigma_z(\omega_\xi, \omega_\eta) + T_z(\Xi, H) \tag{36}$$

Upon using again the invariance of T_z under O^3, we conclude that there exists $p : z \in \mathcal{M} \to \mathbb{R}$ such that

$$\begin{aligned} T_z(\Xi, H) &= p_z \ g_z(\Xi, H) \\ &= p_z \ \{g_z(\xi, \eta) + \varsigma_z \otimes \varsigma_z(\omega_\xi, \omega_\eta)\} \end{aligned} \tag{37}$$

With

$$\rho : z \in \mathcal{M} \mapsto T_z(\varsigma_z, \varsigma_z) \in \mathbb{R} \tag{38}$$

(36) together with (37) gives (31) thus proving (i). From (31) follows immediately

$$tr \ T = -\rho + 3p \tag{39}$$

which makes (ii) clear. To prove (iii) and (iv), we compute from (31)

$$\begin{aligned} T_z(Z, Z) = \rho_z \ &\omega_{\varsigma_z} \otimes \omega_{\varsigma_z}(Z, Z) \\ &+ p_z \ \{g_z(Z, Z) + \omega_{\varsigma_z} \otimes \omega_{\varsigma_z}(Z, Z)\} \end{aligned}$$

i.e. for $Z \in T_+\mathcal{M}_z$:

$$T_z(Z, Z) = g_z(\varsigma_z, Z)^2 \ (\rho_z + p_z) - p_z \tag{40}$$

Since $g_z(\varsigma_z, Z)^2$ ranges over $[+1, \infty)$, (40) implies (iii). The same argument, in conjunction with (39) and (40) is used to prove (iv). □

Before we go into the physical interpretation of this proposition, let us still make another preliminary mathematical remark.

Corollary 4. *Let T be a symmetric $(0,2)$-tensor field on a space-time $(\mathcal{M}, g)$, and assume $tr \ T \leq 0$. Then at those points $z \in \mathcal{M}$ where $T_z \neq 0$ there exists at most one $\varsigma \in T\mathcal{M}_z$, such that $g_z(\varsigma, \varsigma) = -1$, $\varsigma \in T_+\mathcal{M}_z$, and T_z spatially isotropic with respect to ς.*

Proof: The condition that T_z be spatially isotropic with respect to ς, i.e. be invariant under the group O^3 of rotations of the space $\Sigma_z = \{\xi \in TM_z | g_z(\xi, \varsigma) = 0\}$ equipped with $g_z|\Sigma_z$, implies (31), namely

$$\hat{T}_z = \rho_z \, \varsigma \otimes \varsigma + p_z\{\hat{g}_z + \varsigma \otimes \varsigma\}$$

as we just learned. Moreover, from this a straightforward computation shows that

$$\tilde{T}_z \varsigma = -\rho_z \varsigma \tag{41}$$

Suppose now that T_z is also spatially isotropic with respect to another time-like, future oriented vector ς' with $g_z(\varsigma', \varsigma') = -1$. Notice then

$$g_z(\varsigma, \varsigma')^2 > 1 \quad \text{if } \varsigma' \neq \varsigma \tag{42}$$

and, with the help of (41):

$$\begin{aligned} -\rho_z \, g_z(\varsigma, \varsigma') &= g_z(-\rho_z \varsigma, \varsigma') = g_z(\tilde{T}_z \varsigma, \varsigma') = \\ T_z(\varsigma, \varsigma') &= T_z(\varsigma', \varsigma) = \\ g_z(\tilde{T}_z \varsigma', \varsigma) &= g_z(-\rho'_z \varsigma', \varsigma) = -\rho'_z g(\varsigma', \varsigma) \end{aligned} \tag{43}$$

From (42) and (43), we thus obtain $\rho_z = \rho'_z$. Let now

$$\begin{aligned} \alpha &\equiv g_z(\varsigma, \varsigma') + \sqrt{g_z(\varsigma, \varsigma')^2 - 1} \\ \xi &\equiv \varsigma + \alpha \varsigma' \end{aligned} \tag{44}$$

We have then, since (41) holds for ς and ς' with the same eigenvalue $-\rho_z$, and since $g_z(\xi, \xi) = 0$:

$$g_z(\tilde{T}_z \xi, \tilde{T}_z \xi) = 0 \tag{45}$$

However, a straightforward computation from (31), using $g(\xi, \xi) = 0$ shows:

$$g_z(\tilde{T}_z \xi, \tilde{T}_z \xi) = (-\rho_z^2 + p_z^2) \, g(\xi, \varsigma)^2 \tag{46}$$

Since $g(\xi, \varsigma)^2 \neq 0$ (see 42 and 44), we obtain from (45) and (46):

$$-\rho_z^2 + p_z^2 = 0 \tag{47}$$

Now, from $tr \, T \leq 0$ we know

$$\rho_z - 3p_z \geq 0 \tag{48}$$

(47) and (48) together imply $\rho_z = p_z = 0$ which contradicts $T_z \neq 0$. □

Concerning assumptions (ii), (iii) and (iv) of Proposition 3, namely: $-T(Z,Z) \leq tr\ T \leq 0$, we should notice that they hold for every particle flow, and, therefore, for any superposition $T = \Sigma_k T_k$ of particle flows $T_k = \eta_k\ \varsigma_k \otimes \varsigma_k$.

To illustrate the meaning of ρ and p in (31), suppose for simplicity that the tensor T of the Proposition is a finite superposition of particle flows:

$$T = \Sigma_k \eta_k\ \varsigma_k \otimes \varsigma_k \tag{49}$$

In this case, we have for the unique (see Corollary 4) instantaneous observer Z at z for which T_z is supposed to be spatially isotropic: $\rho_z = T_z(Z,Z)$ which, together with (49), gives

$$\rho_z = \Sigma_k \eta_k(z) \cdot e_k(z)^2 \tag{50}$$

where

$$e_k(z) \equiv -g_z((\dot{\gamma}_k)_z, Z) \equiv m_k(z) \tag{51}$$

is the energy (or effective mass) which the instantaneous observer Z associates to a test particle γ_k of the k-th flow as it passes through z. Recalling the meaning of $\eta_k(z)$ we conclude that ρ_z is to be interpreted as the (average, effective) *mass density* at z, with respect to the distinguished instantaneous observer Z.

Similarly, we obtain from (39), using (49):

$$p_z = \frac{1}{3}\Sigma_k\ \eta_k\ (z)\ |\mathbf{p}_k(z)|^2 \tag{52}$$

where

$$\mathbf{p}_k(z) \equiv (\dot{\gamma}_k)_z - e_k(z)Z \equiv m_k(z)\mathbf{v}_k(z) \tag{53}$$

is the momentum which the instantaneous observer Z associates to a test particle γ_k of the k-th flow as it passes through z. (52), therefore, suggests interpreting p_z as the *pressure* at z, with respect to the distinguished instantaneous observer Z. We can, indeed rewrite (52) in the form

$$p_z = \Sigma_k[\eta_k(z)\ m_k(z)]\ \frac{2}{3}\ [\frac{1}{2}m_k(z)\ |\mathbf{v}_k(z)|^2] \tag{54}$$

which calls to mind the classical equation for a perfect gas

$$p = (\frac{N}{V}) \cdot k \cdot T \tag{55}$$

first derived in the context of the nascent classical statistical mechanics by D. Bernoulli (1734) and then, in the proper generality by Maxwell (1860): see

Chapter Two. The fully relativistic equivalent of the classical perfect fluid, namely the matter tensor (31) was briefly mentioned in Einstein (1916) paper, discussed in some detail by Schwarzschild (1916), reworked and simplified by Weyl (1918); see illustration below. In special relativity (31) appeared in the papers of Herglotz (1911), Ignatowski (1911) and Lamla (1911). For a discussion of the physical thinking which went into these early relativistic applications, see e.g. Pauli (1921, 1958). We formalize the situation with the following mathematical definition.

A *perfect fluid* on a space-time $(\mathcal{M}, g)$ where an electromagnetic field $\mathcal{F}$ is defined, is an aggregate $(\varsigma, \rho, p, \sigma)$ consisting of a vector field ς on $\mathcal{M}$, and three real valued functions ρ, p and σ on $\mathcal{M}$, such that: (i) $\varsigma \in \mathcal{T}_+$, $g(\varsigma, \varsigma) = -1$, (ii) $\rho \geq 3p \geq 0$, (iii) $\operatorname{div} \mathcal{F} \equiv J = \sigma\varsigma$, and (iv) $G = T + E$ with $\hat{T} = \rho\, \varsigma \otimes \varsigma + p(\hat{g} + \varsigma \otimes \varsigma)$, G the Einstein tensor of $(\mathcal{M}, g)$ and E the stress-energy tensor of $\mathcal{F}$. For every $z \in \mathcal{M}$ where $T_z \neq 0$, $Z \equiv \varsigma_z$ is referred to as the instantaneous observer at z which is at rest with respect to the fluid. We assume $\Lambda = 0$.

Scholium 5. *Let $(\varsigma, \rho, p, \sigma)$ be a perfect fluid for $(\mathcal{M}, g)$ and $\mathcal{F}$. Then*

(56) (i) $\operatorname{div} \rho\varsigma = -p \operatorname{div} \varsigma$

(57) (ii) $(\rho + p)\nabla_\varsigma \varsigma + \varsigma(p)\varsigma + \operatorname{grad} p = \sigma \mathrm{E}_\varsigma$;
where E_ς is the electric field associated to $\mathcal{F}$ by the instantaneous observer at rest with the gas,
(iii) *for any instantaneous observer Z, the mean spatial acceleration $\overline{\psi}_Z$ satisfies*

(58) $$\overline{\psi}_Z \leq -\frac{1}{6}(\rho + 3p) - \frac{1}{3}E(Z, Z) \leq 0.$$

Proof: A straightforward computation gives, from the definition of $\hat{T}$:

(59a) $$\operatorname{div} \hat{T} = (\operatorname{div} \hat{T})_\parallel + (\operatorname{div} \hat{T})_\perp$$

with

(59b) $$(\operatorname{div} \hat{T})_\parallel \equiv (\operatorname{div} \rho\varsigma + p \operatorname{div} \varsigma)\varsigma$$

(59c) $$(\operatorname{div} \hat{T})_\perp \equiv (\rho + p)\nabla_\varsigma \varsigma + \varsigma(p)\varsigma + \operatorname{grad} p$$

and

(60) $$(\operatorname{div} \hat{T}_z)_\perp \in \Sigma_z \equiv \{\xi \in T\mathcal{M}_z | g_z(\varsigma_z, \xi) = 0\}$$

From (6.2.20) and the condition $J = \sigma\varsigma$, we have $\operatorname{div} \hat{E} = -\tilde{\mathcal{F}} J = -\sigma \tilde{\mathcal{F}} \varsigma$, and thus from the definition (6.2.1) of E_ς:

(61) $$\operatorname{div} \mathrm{E}_z = -\sigma \mathrm{E}_{\varsigma_z} \in \Sigma_z$$

From $G = E + T$ and $\operatorname{div} G = 0$, we see that (56) and (57) follow directly from (59), (60) and (61), thus proving (i) and (ii). Finally (58) follows immediately from (8), (39) and (40), and the fact that all the summands are separately negative. □

The inclusion of a cosmological constant only modifies the conclusions of the scholium to the extent that one should then add a term $(-\frac{1}{3}\Lambda)$ to the *LHS* of (58). For $p_z = 0$, our perfect gas reduces to a particle flow, i.e. a gas without pressure, commonly referred to as 'dust'. In this case (58) reduces to (14) which we already discussed. The reader will notice that the presence of a positive pressure $p_z > 0$ makes $\overline{\psi}_\varsigma$ even more negative; this is consistent with our interpretation of pressure as a mean kinetic energy: as far as the mean spatial acceleration $\overline{\psi}_\varsigma$ is concerned, increasing the energy, through p, works in the same direction as increasing the mass, through ρ. It should also be noted that in applications (56) and (57) are not enough to solve initial value problems: an additional equation of state, linking ρ and p is necessary. The particular case where p is simply a function of ρ—i.e. when we have an isentropic fluid—is of special interest as it allows an especially nice Lagrangean formulation of the theory (see e.g. Hawking and Ellis, 1973).

As an illustration of the kind of non-linear differential equations one encounters when one solves the Einstein field equation (1) for even very simple matter models, we now look at an incompressible, electrically neutral, perfect fluid in equilibrium. We, therefore, set out to find a space-time $(\mathcal{M}, g)$ such that G (assuming $\Lambda = 0$) is given by the *RHS* of (31), namely

$$\hat{G} = (\rho + p)\, \varsigma \otimes \varsigma + p\hat{g} \tag{62}$$

The symmetry and stationarity of the problem suggests to try

$$\mathcal{M} \subseteq S^3 \times \mathbb{R} \tag{63}$$

equipped with the metric

$$g = g^\Sigma - e^{2A}\, dt \otimes dt \tag{64}$$

where g^Σ is the metric of the sphere S^3 of radius a (to be determined), i.e.

$$g^\Sigma = a^2\{\cos^2\theta_2\, \cos^2\theta_3\, d\theta_1 \otimes d\theta_1 + \cos^2\theta_3\, d\theta_2 \otimes d\theta_2 + d\theta_3 \otimes d\theta_3\} \tag{65}$$

and

$$A = A(\theta_3) \tag{66}$$

Recall that in polar coordinates $r = a\cos\theta_3$, so that (66) anticipates the spherical symmetry: we expect the pressure p to depend on r only, while ρ should be a constant (the fluid is supposed to be incompressible).

We now have to compute the Ricci tensor for the metric (64—66); this computation is simplified by recalling that for the spatial S^3 : $(Ric^\Sigma)_{ij} = 2g_{ij}/a^2$, with $i, j = 1, 2, 3$. One then finds that the only non-vanishing components of the Ricci tensor are:

$$(Ric)_{11} = \frac{1}{a^2}\{2 + A' \tan\theta_3\}\, g_{11} \tag{67a}$$

$$(Ric)_{22} = \frac{1}{a^2}\{2 + A' \tan\theta_3\}\, g_{22} \tag{67b}$$

$$(Ric)_{33} = \frac{1}{a^2}\{2 - [(A')^2 + A'']\}\, g_{33} \tag{67c}$$

$$(Ric)_{44} = -\frac{1}{a^2}\{[(A')^2 + A''] - 2A' \tan\theta_3\}\, g_{44} \tag{68}$$

We now impose condition (62), substituting $g(\varsigma, \varsigma) = -1$ and $\varsigma^j = 0$ for $j = 1, 2, 3$ (by stationarity), to get that the only non-vanishing components of the Einstein tensor are:

$$G_{ii} = p\, g_{ii} \quad i = 1, 2, 3 \tag{69}$$

$$G_{44} = -\rho\, g_{44} \tag{70}$$

so that the only non-vanishing components of the Ricci tensor are

$$(Ric)_{ii} = \frac{1}{2}(\rho - p)\, g_{ii} \quad i = 1, 2, 3 \tag{71}$$

$$(Ric)_{44} = -\frac{1}{2}(\rho + 3p)\, g_{44} \tag{72}$$

The spatial isotropy (71) when inserted in (67) gives:

$$-A' \cdot \tan\theta_3 = (A')^2 + A'' \tag{73}$$

Upon comparing now (67) with (71), and (68) with (72) we obtain, with the help of (73):

$$\frac{1}{2}(\rho - p) = \frac{1}{a^2}\{2 + A' \cdot \tan\theta_3\} \tag{74}$$

$$-\frac{1}{2}(\rho + 3p) = \frac{3}{a^2}\, A' \cdot \tan\theta_3 \tag{75}$$

and thus

$$\rho = \frac{3}{a^2} \tag{76}$$

$$p = -\frac{1}{a^2}(1 + 2A' \cdot \tan\theta_3) \tag{77}$$

Note that (76) conforms with our condition that the fluid be incompressible: ρ is a constant. This equation shows that the density ρ of the fluid determines the spatial curvature, and thus the previously open parameter a in (65). Notice further that (77) gives indeed that the pressure p depends only on the radial coordinate $r = a\cos\theta_3$. Moreover, the third, and only interesting, component of (57) reads:

$$(\rho + p)\,\ell n\sqrt{-g_{44}} + \partial_3 p = 0 \tag{78}$$

which, when one substitutes in it (76) and (77), gives:

$$\frac{1}{a^2}\tan\theta_3\{-A'\cdot\tan\theta_3 - (A')^2 - A''\} = 0 \tag{79}$$

which is indeed satisfied for all θ_3 if and only if the non-linear equation (73) is satisfied. This equation admits the solution

$$\begin{aligned} &A = \ell n\left(\frac{3\Theta - \Theta_0}{2\Theta\Theta_0}\right) \quad \text{with} \\ &\Theta \equiv \frac{1}{\sin\theta_3}\,,\ \text{i.e.}\ \frac{1}{\Theta^2} = 1 - \frac{1}{a^2}r^2 \end{aligned} \tag{80}$$

which, together with (76) and (77) gives the equation of state

$$p = \rho\frac{\Theta_0 - \Theta}{3\Theta - \Theta_0} \tag{81}$$

The conditions $\rho \geq 3p \geq 0$ are satisfied when $0 \leq r \leq r_0 \leq a = \sqrt{3/\rho}$ where the radius r_0 of our spherical fluid—determined by the condition $p(r_0) = 0$ — is larger than the Schwarzschild radius associated to the total mass $m = \frac{4\pi}{3}r_0^3\rho$ of the fluid: there are then no singularity of the metric within the fluid.

In summary, we have seen that the Einstein field equation (1) does satisfy three important criteria that should be satisfied by a scientific theory.

Firstly, it incorporates as controlable approximations the validated aspects of pre-existing theories; here the Newton theory of gravitation, and special relativity, i.e. together with the latter, classical mechanics and classical electrodynamics.

Secondly, its structure is legible enough to allow for a precise mathematical formulation, the elegance of which derives from a greater economy of principles; here the general principles imply, without further ado, the Poisson equation (21) of Newton's gravitation theory, with its consequence that gravitational forces decrease with the square of the distance (see 22); more generally, the presence of matter translates into gravitational effects measurable by an attractive contribution to the mean spatial acceleration (see 58).

Thirdly, the theory is flexible enough to allow for the description of a wide variety of phenomena.

The mathematician will further note that the theory has provided motivating impetus to the development of interesting mathematical structures: the differential geometry of Lorentz manifolds, and non-linear differential field equations.

Of course, the physicist will, moreover, require that still a fourth criterion be satisfied for a scientific theory to catch his interest: the theory should make experimentally verifyable predictions which go beyond those of preexistent theories. The special theory of relativity did that very well (see Section 5.2 and Section 10.2). The Einstein field equation was shown very early to satisfy this criterion also, and this in a rather spectacular manner which we should at least briefly mention before we close this chapter.

In Newtonian mechanics, a solution of the gravitational two-body problem gives that the orbit of a planet, say Mercury, around the Sun is planar and is an ellipse with the center of mass of the Sun-planet system sitting at one of the foci of the ellipse. Under the influence of other planets, however, this ellipse is perturbed, resulting in particular in a precession of the major axis of the ellipse, measured as an advance of its perihelion—i.e. the position of closest proximity of the planet to the Sun. In addition, the kinematics of the motion of the Earth itself contributes to the observed value. Both effects can be computed theoretically by a rather horrendous perturbation technique, carried over by Leverrier as early as 1845 and published in 1859 in the case of interest here, namely the perihelion of the planet Mercury. Accurate observations of Mercury are going back to 1765, and Leverrier noticed a discrepancy of 35 seconds of arc per century between observation and theory. The computations involved have been repeatedly checked, in particular by Newcomb (1882) and Clemence (1943).

Today, the accepted observed value is $5600.73 \pm 0.41''$ per century while the accepted value of all classically accountable perturbation is $5557.62 \pm 0.20''$ per century, thus leaving a discrepancy

$$\Delta\varphi = 43.11 \text{ seconds of arc per century} \tag{82}$$

Several classical conjectures were advanced over the years to explain this minute, but systematic discrepancy. Leverrier (1859) postulated the existence of some small unknown planet(s) between Mercury and the Sun, while Newcomb (1882) and Seelinger (1906) proposed to attribute this discrepancy to the presence of more tenuous matter in that region; however, both the hypothetical planets and this tenuous matter failed to materialize under close observational scrutiny. Another proposal was a deviation from the inverse square law of gravitation; finally Dicke and Goldenberg (1967)—pursuing an idea already entertained at the time of Pauli's review article (1921), to the effect that a slight oblateness of the Sun could produce such a deviation—measured the oblateness of the Sun and found it falling short, by at least one order of magnitude, from explaining the observed discrepancy in the advance of the perihelion of Mercury. Moreover, this oblateness would add a term in

r^{-3} to the Newtonion gravitation potential in r^{-2}, resulting in an effect varying as r^{-2} on the perihelion; recent observations of the advance of the perihelia of other planets (Earth, Venus, Icarus) confirmed, however, a r^{-1}-dependence. For a full discussion of this problem, see Weinberg (1972) and references quoted therein. In a nutshell, classical theory has been unable to produce a clear-cut explanation for the 43″ per century advance of the perihelion of Mercury.

Let us now examine, in contrast, what Einstein's general theory of relativity, and especially equation (1), has to say about this problem.

We already discussed the metric generated by a spherical mass; this is the exterior Schwarzschild solution

(83a) $$g = -e^{A(r)}dt \otimes dt + e^{-A(r)}dr \otimes dr + g_S(r)$$

with

(83b) $$g_S(r) = r^2(\sin^2\theta \, d\varphi \otimes d\varphi + d\theta \otimes d\theta)$$

and

(83c) $$e^{A(r)} = (1 - \frac{2m}{r})$$

where m is now the mass of the Sun. The geodesic equation

(84) $$\ddot{\gamma}^\mu + \Gamma^\mu_{\nu\rho}\dot{\gamma}^\nu\dot{\gamma}^\rho = 0$$

for the metric (83) reads explicitly:

(84a) $$\ddot{r} - \frac{1}{2}A'\,\dot{r}^2 - e^A\,r\,\dot{\theta}^2 - e^A\,r\sin^2\theta\,\dot{\varphi}^2 + \frac{1}{2}e^{2A}A'\,\dot{t}^2 = 0$$

(84b) $$\ddot{\theta} + \frac{2}{r}\dot{\theta}\,\dot{r} - \sin\theta\cos\theta\,\dot{\varphi}^2 = 0$$

(84c) $$\ddot{\varphi} + \frac{2}{r}\,\dot{\varphi}\dot{r} + 2\cot\theta\,\dot{\varphi}\,\dot{\theta} = 0$$

(84d) $$\ddot{t} + A'\,\dot{t}\,\dot{r} = 0$$

As usual, when one tries to solve a system of coupled non-linear differential equations, the best is to look first for the constants of the motion.

We first notice that

(85) $$\theta(s) = \frac{\pi}{2} \quad \forall\, s \in \mathbb{R}$$

solves these equations. By a change in the origin of the coordinate θ, this solves the general case, and we do not need to worry about (84b) in the sequel. Classically (85) corresponds to the fact that the motion of a planet is planar.

Equation (84c) gives us then our first constant of the motion, namely

(86) $$J \equiv \dot{\varphi}\,r^2$$

Classically, this is Kepler's second law of planetary motion.

Equation (84d) gives us our second constant of the motion, namely

$$T \equiv \dot{t}\, e^{A} \tag{87}$$

By a rescaling of t we can assume without loss of generality $T = 1$. This corresponds classically, since $(2m/r) \ll 1$, to $\dot{t} = 1$, i.e. $t = s$, so our rescaling corresponds to the fact that in the classical approximation, the proper time coincides with the usual time with respect to which the motion is parametrized.

Finally, upon inserting (85), (86) and (87) with $T = 1$ into (84a), we receive

$$\ddot{r} - \frac{1}{2}A'\,\dot{r}^2 - e^{A}\,\frac{J^2}{r^3} + \frac{1}{2}A' = 0 \tag{88}$$

With A as in (82c) and in the approximation where $r \gg 2m$ (recall that $2m$ is the Schwarzschild radius which lies deep inside the Sun), and $\dot{r} \ll 1$ (non-relativistic velocities) (88) reduces to

$$\ddot{r} - \frac{J^2}{r^3} + \frac{m}{r^2} \cong 0 \tag{89}$$

which is as it should; indeed $LHS = 0$ is the Newtonian equation of planetary motion in the gravitational potential $\Phi = -m/r$.

From the exact equation (88) we obtain our third constant of the motion, namely

$$E \equiv e^{-A}(1 - \dot{r}^2) - \frac{J^2}{r^2} \tag{90}$$

Upon comparing (90) with (83), taking into account (85), (86) and (87) with $T = 1$, we see that our geodesic describes a particle moving slower than the velocity of light if and only if $E > 0$, whereas the geodesic is light-like exactly when $E = 0$. For our present problem, we should, therefore, consider the case $E > 0$ (more on $E = 0$ later). Consider, furthermore, the cases where (90) with $\dot{r} = 0$ admits two solutions r_+ and r_-, defining respectively the aphelion and the perihelion of our planet. Notice then, that r_+ and r_- are constants of the motion and can be used to express E and J.

We can now compute the advance of the perihelion in one revolution; this is defined as

$$\Delta\varphi = 2\{|\varphi(r_+) - \varphi(r_-)| - \pi\} \tag{91}$$

For this computation, we combine (90) and (86) to obtain

$$\frac{d\varphi}{dr} = \frac{1}{r^2}\left[\frac{1}{J^2} - e^{A(r)}\left(\frac{E}{J^2} + \frac{1}{r^2}\right)\right]^{-\frac{1}{2}} \tag{92}$$

which can be integrated to give

$$\Delta\varphi = 3\pi \cdot \frac{2m}{L} \text{ radians per revolution} \tag{93}$$
$$\text{with } \frac{1}{L} \equiv \frac{1}{2}\left(\frac{1}{r_+} + \frac{1}{r_-}\right)$$

Upon feeding in this formula the mass of the Sun $m = m_0 G = 1.475$ km (recall G is the universal gravitational constant which we put equal to 1 by a choice of units we now have undone: $2m$ is the Schwarzschild radius for the Sun); the value of L for Mercury, namely $L = 55.3 \cdot 10^6$ km; and the duration of the year of Mercury, namely 88 days, we obtain:

$$\Delta\varphi = 43 \text{ seconds of arc per century} \tag{94}$$

in spectacular agreement with the observed value (see 82). This agreement is further confirmed, although with less (experimental !) accuracy for the other planets for which data are available. This provides a rather solid pilar for the general theory of relativity, a standard against which competitive theories will have to be measured; recall, in particular, that no reasonable classical explanation had come even close to this agreement.

Another application of the above computation is to look at a light ray ($E = 0$) coming from a star and grazing the Sun on its way to the Earth. In this case $r_+ = +\infty$; and r_-, called now the *impact parameter*, is in the best possible conditions the radius r_0 of the Sun. For this problem, an analysis quite similar to the one we just went through in detail shows that the deflection of light is given by:

$$\delta\varphi = \frac{4m}{r_0} \text{ radians} \tag{95}$$

Filling in the numerical values, namely again the mass $m = 1.475$ km of the Sun, and the radius, $r_0 = 6.95.10^5$ km, of the Sun, one obtains

$$\delta\varphi = 1.75 \text{ seconds of arc} \tag{96}$$

This is a rather delicate experiment to conduct, involving comparing the apparent position of a star, once during an eclipse (so that one can see the star at the fringe of the solar disk), and then again at a six month interval (when the star lies at the opposite side of the Earth from the Sun, so that the light coming from the star does not pass by the Sun before reaching us). This experiment was first performed in 1919 and repeatedly several times thereafter; all observations agree, within experimental errors, with the value given in (96); so appear to do more recent radio-astronomical data. The 1919 solar eclipse expeditions were accompanied by much fanfare and brought general relativity to the attention of the public: on the very day - Nov. 6, 1919 - the results were made official by a meeting of the Royal Society, the Times reported that a scientific revolution had happened, that ruined Newton's ideas. A scientific revolution had occurred indeed, but it had been several years in the making and it did not ruin Newton's ideas: it incorporated them in a wider and richer conceptual framework. Truer to scientific spirit, Eddington, who was to lead one of the two 1919 expeditions, had characterized what was happening in the following words: "Whether the theory ultimately proves to be correct or not, it claims attention as being one of the most beautiful examples of the power of general mathematical reasoning." To this day, the theory, and its conceptual foundations, have not been falsified.

CHAPTER 7. THE "OLD" QUANTUM THEORY

SYNOPSIS

While the theory of relativity presented a severe criticism of the classical Newtonian notions on the separate essences of space and time, the new theory still won a rather rapid and widespread acceptance, perhaps for the reason that it provided a unification of previously disparate concepts. Not so with quantum theory, where a quarter of a century had to pass before some unifying picture would emerge. In view of the contradictions the "old" quantum theory generated in the course of its development, a chapter of mostly historical (if somewhat "whiggish") character seems to be called for, retracing the experimental motivations for the radical departure from classical paradigms.

To pin down the birthdate of a scientific theory is almost always a rather contrived exercise; in the case of quantum theory, however, the date (Dec. 14, 1900) of Planck's famous communication on the "black body radiation", made before the German Physical Society, certainly signaled the outbreak of a new era. Still, we know from Planck's own pen that he "had been wrestling for six years with the problem of equilibrium between radiation and matter", and that he was, as yet, not satisfied with the result. His contribution was two-fold. Firstly, he produced a formula

$$u_\nu = \frac{8\pi}{c^3} h\nu^3[\exp(h\nu/kT) - 1]^{-1}$$

which fitted experimental data and which was significantly different from the predictions of classical theories. Secondly, he proposed a derivation of his formula, for which he resorted, "as an act of desperation", to the ideas of Boltzmann on "the relationship between entropy and probability" complemented by an *ad hoc* postulate, now understood to be that light is absorbed and emitted by matter in discrete energy quanta.

It was only in the work of Einstein (1905) on the photoelectric effect (for which he was awarded the Nobel prize in 1922) that this "quantum hypothesis" was taken seriously, and this heralded the rebirth of the corpuscular theory of light. In 1909 Einstein showed in fact - in the paper containing the so-called Einstein formula for the energy fluctuations in black body radiation - that light has a dual nature: both particle and wave. Even so, Planck himself, wrote as late as in 1913 (while commenting on the proposal that Einstein be received to membership in the Prussian Academy of Sciences): "That he may sometimes have missed the target of his speculations, as for example in his hypothesis of light quanta, cannot really be held against him."

To compound the discomfort, Bohr (1913) proposed his theory of the discontinuous structure of the atom, and in 1917, Einstein, reconsidering again the problem of the interaction between matter and radiation, gave a new derivation of Planck's black body radiation formula, confirming the quantum hypothesis.

Up to the mid-1920's, the "quantum" view of the world registered remarkable successes, among which we mention a few: Einstein's quantitative predictions concerning the photoelectric effect; the Einstein-Debye theory for the temperature dependence of the specific heat of solids, marking the birth of quantum statistical mechanics; the formulation of the Bose-Einstein statistics governing the collective behaviour of particles like photons (the light quanta) and phonons (the elementary vibrations of crystals); the Fermi-Dirac statistics governing the collective behaviour of particles like electrons; the Pauli exclusion principle, soon to bring some rationale into the Mendelev table of chemical elements; the existence of matter waves, predicted by L. de Broglie and discovered soon afterwards by Davisson & Germer and by Thomson in their experiments on the diffraction of electrons; the Bohr-Sommerfeld construction of models for Hydrogen-like atoms, accounting for a wealth of spectroscopic data, including relativistic effects.

This saga will be retold in this chapter, while we will, nevertheless, try to bring into focus the radical departures from classical theory which the "quantum rules" implied as they were grafted on a recalcitrant classical substratum.

Among the puzzles left unsolved by the "old" quantum theory we should mention already here: the Bohr-Sommerfeld use of classical concepts in the elaboration of quantization rules which would ultimately violate in essential ways their classical premises; the Einstein particle-wave duality; and the appearance of classical theory as an asymptotic approximation manifest in such diverse contexts as the Rayleigh-Jeans radiation law, the Dulong-Petit specific heat law, and even Bohr's own correspondence principle.

Moreover, the blessed simplicity of the "old" quantum theory started, by the mid-1920's, to wither away as more complex problems were attacked, each needing in turn what one would call—depending on one's inclinations or prejudices—"new physical insights" or "fresh *ad hoc* assumptions". It was becoming clear that any global proposal to resolve these conceptual difficulties would require a more sharply defined mathematical formalism. In the next chapters, we will present successive attempts in this direction, and the axiomatically inclined reader may want to jump directly to Chapter Nine. We, however, thought it necessary to inform these attempts with an introduction to the basic empirical facts. To the latter, the present chapter is thus devoted.

Section 1. BLACK-BODY RADIATION

We shall be concerned in this section with the spectral properties of electromagnetic radiation—e.g. light; see Chapter Three—in thermal equilibrium with matter. That such an equilibrium can be reached evidently presupposes that there is some interaction between radiation and matter, but it does not require us to know the details of the mechanisms governing this interaction; we can, therefore, define our terms as we go.

In work done jointly in the late 1850's G. Kirchhoff and R. Bunsen established what they rightly saw to be a new and powerful tool of chemical analysis, namely *spectroscopy*; for a description of this potentiality in their method, see G. Kirchhoff and R. Bunsen (1860). From the point of view which interests us here, the seminal discovery was reported by Kirchhoff (1859, I and II): "Fraunhofer noticed that in the spectrum of a candle flame, two bright lines occur which coincide with the two dark lines D of the solar spectrum. We obtain the same bright lines in greater intensity from a flame in which common salt is introduced ... If we introduce lithium chloride into the flame ... its spectrum shows a very bright, sharply defined line which lies between the Fraunhofer lines B and C [of the solar spectrum] ... the dark D lines in the solar spectrum permit us to conclude that sodium is present in the sun's atmosphere ... [and since] there is no dark line in the solar spectrum coinciding with the red line of lithium, it seems probable that lithium either is not present in the sun's atmosphere or is there in relatively small quantity ... ". They then proceeded to identify these characteristic *spectral lines* for various elements, including the investigation of "the spectrum of the electric spark between iron electrodes and we found in it a great many bright lines which seem to coincide with dark lines of the solar spectrum .... When the spark is formed between other metals as, for example, between copper electrodes, these bright lines do not appear ... " from this, they draw "the conclusion that iron is present [in] the glowing atmosphere of the sun." This conclusion was further supported by the fact that they had succeeded in reproducing in the laboratory the dark lines from sodium and lithium in the solar spectrum by illuminating with a "strong-intensity source with continuous spectrum" an alcohol flame (low temperature!!) containing sodium- and lithium-chlorides, thus establishing one of the essential experimental facts at the foundation of the theory, namely the exact coincidence of emission—their "bright lines"—and absorption—their "dark lines"—spectra of each element. Kirchhoff actually went further: he discussed the relative intensity of the emission and absorption spectra and found " ... a general law, which seems ... to be of importance in several aspects ... it expresses a property of all bodies which is connected with the emission and absorption of heat and light". This law states that the ratio $(\epsilon_\nu/\alpha_\nu)$ of the emission and absorption power is a universal function of the frequency of the radiation and the temperature T of the substance; universal here means that it is independent of the substance considered. In the notation used later in the honour of Kirchhoff, this law is:

$$\frac{\epsilon_\nu}{\alpha_\nu} = K_\nu(T) \tag{1}$$

In modern language, this law is interpreted as governing the equilibrium between radiation and matter. The question which then remains is to determine theoretically the form of the universal function $K_\nu(T)$. *This is what Planck succeeded in doing.* Before we review Planck's contribution, we find it instructive to reproduce here the very simple heuristic argument Kirchhoff (1859) offered to support his contention that such a function $K_\nu(T)$ must exist.

Consider two parallel, infinite and perfectly reflecting plates, r and R, facing each other and coated on the inside with some materials c and C (see Figure 1), the whole assemblage being at constant temperature T.

Suppose further that c is capable of emitting and absorbing light at all possible wave-lengths λ, while C only emits and absorbs light of wave-length Λ. The temperature being constant, each of the two bodies c and C must retain the same temperature and thus take up by absorption as much heat as it loses by emission. Here Kirchhoff tacitly assumes that absorption (and emission) of radiation in the form of light is equivalent to absorption (and emission) of heat, which turns out to be correct.

Two cases are now considered in this imaginary set-up. First, consider a light ray of wave-length $\lambda \neq \Lambda$, emitted by c. In this case, C can be ignored and this light ray, bouncing off between the two mirrors r and R will ultimately be completely absorbed by c. Since this holds for all $\lambda \neq \Lambda$, Kirchhoff argues from the constancy of T (and the above tacit postulate) that c must absorb exactly as much of the rays at wave length Λ as it emits. For this wave-length

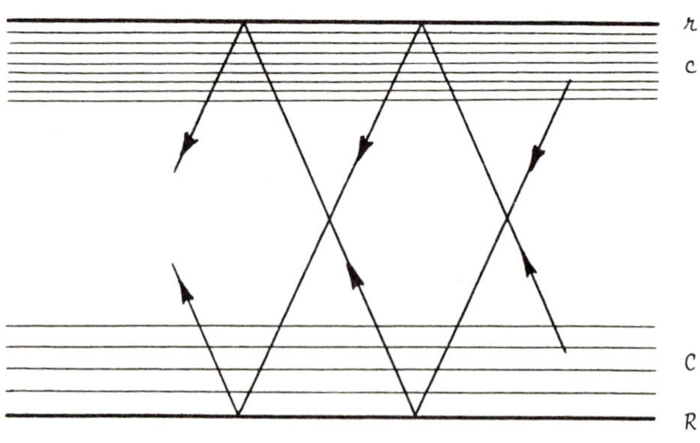

Fig. 1

Λ, let e (resp. E) and a (resp. A) be the emission and absorption power of c (resp. C). Consider now the case of light at wave-length Λ bouncing off the two mirrors. Of the total light emitted by both c and C, c must absorb the following fraction (computed from what happens at successive reflections):

$$\begin{aligned} & Ea \sum_{k=0}^{\infty} [(1-a)(1-A)]^k \\ & + e(1-A)\, a \sum_{k=0}^{\infty} [(1-a)(1-A)]^k \\ & = \frac{Ea + e(1-A)a}{1-(1-a)(1-A)} \end{aligned} \tag{2}$$

But this must be equal (see first step) to what it emits, i.e. to e; upon equating thus e with the RHS of (2), we receive:

$$\frac{e}{a} = \frac{E}{A} \tag{3}$$

For most physicists, this would establish (1); but Kirchhoff was proposing a new "general law"; to make sure his point was indeed made, he pursued the argument as follows: "We arrive at the same equation if we develop the condition that the temperature of C remains constant. If we suppose that the body C is replaced by another one at the same temperature, we find ... the same value for the ratio of the emissive power to the absorptive power of this body for rays of the same wave-length Λ. Now the wave-length Λ and the temperature are arbitrary. The law, therefore, follows that for waves of the same wave-length at the same temperature the ratio of the emissive power to the absorptive power is the same for all bodies ... The ratio common to all bodies ... is a function of the wave-length and the temperature." This is precisely (1).

This heuristic derivation, originally proposed by Kirchhoff, obviously leaves plenty of room for improvements; these were needed on two fronts: (i) to justify the various assumptions, explicit and implicit (or tacit) made in this original derivation; and (ii) to allow for more general boundary conditions. Kirchhoff (1860) provided some of these improvements; in that paper he also introduced the notion of what he called "a perfectly black, or more briefly a *black-body*", i.e. a body which absorbs all the radiation falling on it, i.e. for which $\alpha_\nu = 1$ and thus $\epsilon_\nu = K_\nu(T)$.

More sophisticated derivations were obtained by Pringsheim (1901, 1903), Evans (1910) and D. Hilbert (1912, 1913, 1914). By the time Hilbert's papers appeared, not only was Kirchhoff's law universally accepted, but even the form of the function $K_\nu(T)$ was set beyond reasonable doubts. Still questions lingered about the generality of the law and, more importantly, the nature of the hypotheses essential to its theoretical derivation. These factors did certainly play a role in Hilbert's motivation: "Physics is much too hard for physicists" he once remarked. Even the modern reader cannot fail to notice Hilbert's private satisfaction at presenting Kirchhoff's law as an application of the theory of linear integral equations which he had developed (see Hilbert, 1912), starting in 1901–1902 with a series of lectures and seminars—as was so often Hilbert's way to enter a new field of research. We may, therefore, be forgiven if we open here a small parenthesis to sketch the idea of Hilbert's derivation of Kirchhoff's law.

Hilbert considers the medium in thermal equilibrium into which radiation takes place as one with variable refractive index: he thus writes the velocity of light as a function $q(x, y, z)$; although the derivation is given with q continuous, he mentions towards the end of the paper that reasonable discontinuities would not invalidate his treatment of the problem. He then defines systematically his terms. For instance, a light ray between (x_1, y_1, z_1) and (x_2, y_2, z_2) is defined as that minimal curve solving the variation problem

$$\delta \int_{(x_1,y_1,z_1)}^{(x_2,y_2,z_2)} \frac{ds}{q} = 0 \quad \text{with } ds = \sqrt{dx^2 + dy^2 + dz^2} \tag{4}$$

The emission coefficient $\eta(x, y, z)$ is defined by stating that the energy radiating from a volume element $dv = dxdydz$ in a time interval dt, into a solid angle $d\chi$ is

$$\frac{1}{4\pi}\eta \; dt \; dv \; d\chi \tag{5}$$

The absorption coefficient $\alpha(x, y, z)$ is defined through the attenuation of the energy transported along a light ray:

$$dE = \alpha \; E \; dx \tag{6}$$

He then studies the (mechanical and thermal) equilibrium conditions and succeeds in expressing them as the integral equation

$$\eta(x, y, z) - \frac{\alpha(x, y, z)}{4\pi q(x, y, z)^2} \iiint dx_1 dx_2 dx_3 K(x, y, z; x_1, y_1, z_1)\eta(x_1, y_1, z_1) = 0 \tag{7}$$

with kernel

$$K(x, y, z; x_1, y_1, z_1) = \frac{e^{-A(x,y,z;x_1,y_1,z_1)}}{S(x, y, z; x_1, y_1, z_1)} \tag{8}$$

where

$$A(x, y, z; x_1, y_1, z_1) = \left| \int_{(x_1,y_1,z_1)}^{(x,y,z)} ds \; \alpha \right| \tag{9}$$

and

$$S(x,y,z;x_1,y_1,z_1) = q(x,y,z)^{-2}\frac{d\sigma}{d\chi} \tag{10}$$

where $d\sigma$ is the geometric element depicted in Figure 2.

Notice that in a medium with constant refractive index we can assume without loss of generality

$$\begin{aligned} &q(x,y,z) \equiv 1 \\ &S(x,y,z;x_1,y_1,z_1) = \sqrt{(x-x_1)^2+(y-y_1)^2+(z-z_1)^2} \end{aligned} \tag{11}$$

Hilbert does not need this particularization, but only the fact that in (10) S is symmetric, i.e.

$$S(x,y,z;x_1,y_1,z_1) = S(x_1,y_1,z_1;x,y,z) \tag{12}$$

For boundary conditions, he takes that the equation (7) holds in a region completely enclosed by walls, the parts of which are either perfectly black or perfectly reflecting; even that can be weakened. Now the problem is well posed, and Hilbert proves that given the functions α and q, equation (7) admits a unique solution for the function η, namely:

$$\eta = \frac{\alpha}{q^2}\varphi \tag{13}$$

where φ depends only on the given equilibrium temperature and on the frequency of the radiation but is independent of the position (x,y,z) and of the material considered. For $q \equiv 1$ this reduces exactly to the Kirchhoff's law (1); the generalization, therefore, consists in allowing for a variable refractive index which shows up in (13) through the q^{-2} dependence.

Through its derivation, this result (13) is the basis for Hilbert's interpretation of Kirchhoff's law as being of geometrical nature. The above sketch only

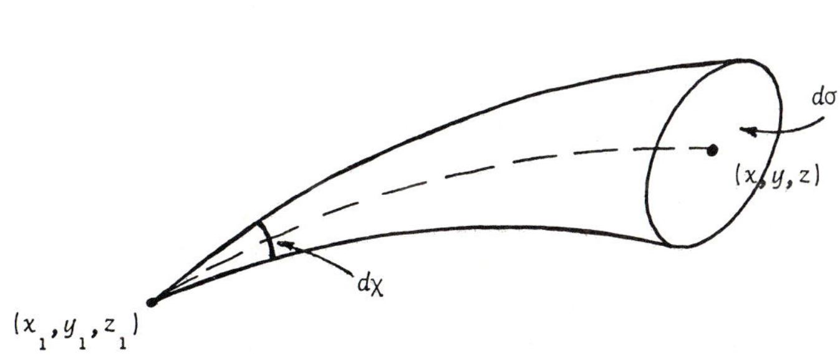

Fig. 2

covers the material of the first of Hilbert's three papers on the subject. In the second paper, he discusses briefly the axioms underlying the theory, while the third paper represents Hilbert's final version of the "Foundations of Radiation Theory". In characteristic Hilbert's fashion, the sections of this third paper are entitled: 1. The axioms of radiation theory; 2. Proof of Kirchhoff's theorem; 3. Radiation theory and elementary optics; 4. The self-consistency of the axioms. It is recommended reading for anyone who wants to know what Hilbert (1900) must have meant with the sixth of his famous problems, namely: "To axiomatize those physical sciences in which mathematics plays an important role".

At this point, it must be pointed out that Hilbert's derivation, just like Kirchhoff's original derivation, does not give the form of the function $K_\nu(T)$ in (1). By the time Planck (1900 a & b) proposed his famous formula for this function, other attempts had already been made, which, however, had failed to match the whole range of the experimental data by then available and this in ways which are best understood from Planck's own formula. We, therefore, postpone the description of these early attempts and turn first to the derivation of Planck's formula.

Although Planck himself probably did not think—and certainly did not write—in terms of separate and formal lemmata or theorems, it is convenient here to analyze his approach with some attention to a separation between the mathematics—which is elementary—and the physics—which is deep.

We, therefore, distinguish the following three steps in Planck's derivation.

The first step belongs to *thermodynamics*. Recall from Chapter Two that in a thermodynamical system in equilibrium, the entropy S, the internal energy U and the temperature T are linked by the relation $\partial S/\partial U = T^{-1}$.

Scholium 1. *Suppose the entropy S and the internal energy U of a thermodynamical system in equilibrium at temperature T are linked by the following relations*

$$S = \frac{k}{N}\,\ell n\left[\frac{(N+P)^{N+P}}{N^N P^P}\right] \tag{14}$$

$$U = \epsilon \cdot \frac{P}{N} \quad \text{with} \quad k, \epsilon, P \text{ and } N \in \mathbb{R}^+ . \tag{15}$$

Then :

$$U = \epsilon\left[e^{\epsilon/kT} - 1\right]^{-1} \tag{16}$$

Proof: Upon inserting (15) in (14), the latter can be rewritten as:

$$S = k\{(1+\frac{U}{\epsilon})\,\ell n(1+\frac{U}{\epsilon}) - \frac{U}{\epsilon}\,\ell n\frac{U}{\epsilon}\} \tag{17}$$

from which one reads

$$\frac{\partial S}{\partial U} = \frac{k}{\epsilon}\,\ell n(\frac{\epsilon}{U}+1) \tag{18}$$

The thermodynamical identity $T^{-1} = \partial S/\partial U$ inserted in (18) gives the desired equation (16). □

The second step belongs to *statistical mechanics*. Here Planck uses his own version of Boltzmann's formula $S = k\ \ell n W$ relating the entropy S with the number W of all possible "complexions" of the physical system he will want to consider; k is the Boltzmann constant (see Chapter Two).

Scholium 2. *Suppose we have P indistinguishable objects, each carrying the energy ϵ, distributed amongst N sites, for a total energy E. Then for N and P very large, the average entropy S per site (resp. the average energy U per site) satisfies asymptotically the assumption (14) (resp. 15) of the previous scholium.*

Proof: The number W of ways to distribute P objects amongst N sites is given by

$$W = \frac{(N+P-1)!}{(N-1)!P!} \tag{19}$$

The asymptotic formula (14) for the average entropy per site, namely $S \equiv N^{-1}k\ \ell n W$, then follows from Stirling's formula. As for the average energy per site, (15) follows trivially from $U \equiv N^{-1}E$ and $E = \epsilon P$. □

The third step is to relate the objects appearing in the above two scholia to those which are present in the physical situation at hand, namely radiation in thermal equilibrium with matter. Planck models the latter, at fixed temperature T and radiation frequency ν, as a collection of N_ν electric dipoles oscillating at resonance frequency ν. He then borrows from *classical electromagnetism* the fact that the spatial spectral energy density u of the radiation at frequency ν—the quantity which is experimentally observed—is linked, through the thermal equilibrium condition, to the average energy U of the resonators at frequency ν by the formula

$$u = \frac{8\pi}{c^3}\nu^2 U \tag{20}$$

Planck had worked on that formula for the last five years; it does not seem proper here to recount the details of that part of the argument beyond the mention that it involves the most classical views of a vibrating dipole in an electromagnetic field; for a systematic exposition of Planck's own view see Planck (1900,A) where he summarizes his earlier contributions. Then enters *Planck's "most essential point", namely that the total energy E of the collection of resonators at frequency ν is not "a continuously divisible quantity" but rather is "composed of a well-defined number of equal parts".* He introduces here the universal "constant of nature" h which now bears his name, and which he evaluates at $6.55\,10^{-27}$ergs sec—the modern value is $6.6256\,10^{-27}$ergs sec. "This constant multiplied by the common frequency ν of the resonators gives us the energy element ϵ... and dividing E by ϵ we get the number P of energy elements which must be divided over the N resonators" at frequency ν. We thus have the situation of our second scholium with

$$\epsilon = h\nu \tag{21}$$

Since this reasoning holds separately for all frequencies, inserting the expression (21) into (16) gives for the observable quantity u (see 20) as a function of the frequency ν and the temperature T:

$$u_\nu = \frac{8\pi}{c^3} h\nu^3 [\exp(h\nu/kT) - 1]^{-1} \tag{22}$$

This is *Planck's radiation law*, giving an explicit form to Kirchhoff's function $K_\nu(T)$.

A few comments are in order. First of all, this law fits experiment very well, and it was from this fit with the experimental data made available to him, that Planck calculated the numerical value of the constant h which he introduced. This fit was not accidental, and it appears from Planck's first paper (1900 a) that he actually worked backwards trying to fit experimental data to a simple curve and thus arrived at an ansatz amounting to the assumptions of our first scholium, specifically:

$$\frac{\partial^2 S}{\partial U^2} = \frac{\alpha}{U(\beta + U)} \tag{23}$$

where α and β are constants (compare with 18). Planck's satisfaction with this formula was that "it is by far the simplest of all expressions which lead to S as a logarithmic function of U—which is suggested from probability considerations", a statement he made more explicit two months later in his second paper (1900b): these are the assumptions of our second scholium. Still in his first papers, Planck seems to have been further encouraged in his ansatz (23) by the behaviour of his theory "for small values of U". This we can understand better from the finished product, namely (22). Indeed, when the frequency ν and the temperature T are such that $\epsilon \equiv h\nu \gg kT$ the exact formula (22) takes the asymptotic form:

$$u_\nu \cong \frac{8\pi}{c^3} h\nu^3 e^{-h\nu/kT} \tag{24}$$

a law Wien (1896) had proposed, however, as an exact one; this law Wien had obtained as a refinement of what he termed a "fortunate idea" by Michelson (1887); and he wrote it in the form

$$u_\nu = \alpha\nu^3 e^{-\beta\nu/T} \tag{25}$$

with α and β theoretically unspecified constants to be determined experimentally. Back to Planck's argument, notice that as $h\nu \gg kT$ (16) reduces indeed to

$$U \cong \epsilon \, e^{-\epsilon/kT} \tag{26}$$

so that indeed $U \ll \epsilon$, and thus (18) reduces to

$$\frac{\partial S}{\partial U} \cong \frac{k}{\epsilon} \, \ell n \frac{\epsilon}{U} \tag{27}$$

i.e. upon putting together (26) and (27)

$$\frac{\partial S}{\partial U} \cong \frac{1}{T} \tag{28}$$

a conspicuously thermodynamical-looking formula which, when taken as an exact one, Planck knew could be used to derive Wien's law (25). It is the observed discrepancy from Wien's law that Planck had set out to explain, and he thus viewed his initial ansatz (23) as a replacement for (27) which now becomes only an asymptotic formula as $h\nu \gg kT$.

Another interesting approximation of Planck's radiation formula (22) occurs in the opposite range, namely when $\epsilon \equiv h\nu \ll kT$. In this case (22) takes the asymptotic form

$$u_\nu \cong \frac{8\pi}{c^3} \nu^2 kT \tag{29}$$

a law which had been obtained by Rayleigh (1900) as an application of the equipartition theorem of statistical mechanics to electromagnetic waves in a cavity; in fact, Rayleigh's derivation contained a simple error which was corrected by Jeans (1905). Planck knew, through Rubens (see G. Hettner (1922) quoted by Kangro (1972)), that "for the longest waves measured ... the law obtained recently by Lord Rayleigh ... is valid." This is known today as the *classical approximation*: it is formally obtained from (22) by the—a priori unphysical—limit $h \to 0$, the true meaning of this limit being evidently in the present context that it gives an asymptotic behaviour as $h\nu/kT \to 0$. The fact that classical results are obtained in the limit of high temperature was to play an important role in the subsequent development of quantum statistical mechanics (for an early example, see Section 2 below). From the conceptual point of view, it should have been clear from the beginning that something definitely wrong was happening with the classical formula (29); u_ν grows as ν^2 which is incompatible with its being interpreted as a spectral energy density: this would lead to a divergent result for the total energy density u obtained from u_ν by integration in ν over $\mathbb{R}^+$. This unacceptable consequence of the classical theory, referred to since Ehrenfest (1911) as the *ultraviolet catastrophe*, is corrected by Planck radiation law (22); in this respect Wien's law (24) is acceptable, but it did not fit experimental data at low frequencies, which Planck's law does (see 29).

Finally—and with the kind of arrogance that only hindsight can give—it might be pointed out already here that although Planck's formula (22) seems to describe so well the black-body radiation, Planck's theory is rather vague on the reality of the underlying model for matter—not to speak of the fundamental mechanism governing the interaction of radiation and matter. Planck seems to have shyed away from—or perhaps did not feel the necessity of—interpreting his resonators as anything like our modern atoms; the physical reality of the atoms was not quite yet universally accepted amongst preeminent physicists of the time, although Planck himself, in one of the footnotes to his second paper, does recognize the usefulness, in another context, of what he still calls the "atomistic hypothesis". It is, in fact, one of the remarkable features of Planck's paper that he was able to reach his conclusion (22) without requiring more knowledge of the interaction between radiation and matter than the mere fact that (20) is satisfied. Similarly, Planck did not attach to $\epsilon = h\nu$ the meaning of an intrinsic property of radiation as well; never does he even mention in these two papers the words energy quantum which seem to have been first introduced by Einstein, as we shall see in the next section.

Section 2. ELECTRONS, PHOTONS AND PHONONS.

This section should open with three warnings. Our first warning is that it is often tempting, when one looks at the mathematical structure of a physical theory, to identify the theory with its fundamental equations. This can indeed be a useful approach—as when we identified an electromagnetic field via the Maxwell equations—but it can also be dangerous: the first formulas of this section belong to Einstein's paper (1905, PE); they look the same as those of Planck's papers (1900) which we discussed in the previous section; however, the interpretation Einstein gave of these apparently so similar formulas is completely different from Planck's, and it is this interpretation which opened vistas so daring that Planck recoiled in disbelief.

Our second warning concerns the question of the extent to which physicists at the turn of the Century believed in the reality of atoms—molecules, ions or elementary particles. Without doubt, atoms, ions and molecules were universally recognized as useful—but perhaps not indispensible—coding devices in Chemistry; their existence as theoretical constructs had also proved useful in the Kinetic Theory of Gases and in the attendant discussion of irreversible phenomena. Indirect evidence, based on these theories, provided some estimates on the size these objects would show up in, if they really existed. For instance, the order of magnitude of the mass of the hydrogen atom was known. Using "indirect" evidence to compute such a number is, however, quite a different thing from establishing the existence of atoms and molecules by a "direct" experiment; it is precisely such an experiment that Einstein (1905, BM) proposes in his paper on Brownian motion (see Chapter Two). Better still would be to isolate experimentally the object to be considered. Of all the elementary particles, this was done first for the *electron*. Free electrons were first seen—but not yet identified as such—in electric discharges made to occur, since the days of Faraday, in more and more rarefied gases. These "cathode rays" were seen to have some electric properties of their own when, in the late 1850's, it was observed that they are deflected by a magnetic field. With the development of better vacuum techniques, an electric field could be established in the immediate vicinity of the cathode rays, and its effect on them could be measured. On the basis of such measurements Perrin (1895) established that cathode rays are constituted of a stream of identical, negatively charged particles; this conclusion was confirmed by Thomson (1897) who measured the velocity of these particles, and the ratio of their charge over their mass. Two years later, Thomson succeeded in determining separately the charge and the mass of the electron. For our purpose, it is important to note that the mass of the electron was found to be only a minute part ($\sim$ 1/2000) of the mass of the hydrogen atom. The fact that an electron is an elementary grain—an atom in the $\alpha\tau o\mu o\varsigma$ sense of Leucippus and Democritus—of electricity was established beyond reasonable doubts by Millikan (1910) in the very method he used to determine the charge of the electron, within 1% of the value currently accepted (and supposed to be correct to within 3 parts per million, while the ratio e/m is known today to within 6 parts per million). The advent, more than half-a-century afterwards, of the quarks with their fractional charges need not concern us here. While the electron was being recognized as a fundamental particle, the atom itself was rapidly losing its identity as an indivisible object; the hunt, still going on today, for its elementary particle constituents, in fact started, at about the same time, with the discovery of radioactivity by Becquerel (1896). One more word of introduction concerning the electron: it was presented here as a particle; nevertheless, it should be mentioned that Hertz (1892) had thought he could establish that cathode rays were waves; in a strange way Hertz could not possibly have foreseen, there is a sense in which he too was right, as we shall see later on in this section.

The third, and last, warning we announced at the beginning of this section is to emphasize that, at the turn of the Century, the corpuscular theory of light, associated to the names of Descartes (1637) and Newton (1704) had been dead for the best part of the Nineteenth Century. Light was known to be a wave, and more specifically a periodic solution of the Maxwell equations which Einstein (1905, SR) had placed at the foundations of his theory of relativity. And now, Einstein (1905, PE) himself was writing: "The wave theory of light which operates with continuous functions in space has been excellently justified for the representation of purely optical phenomena ... however ... notwithstanding the complete experimental verification of the theory of diffraction, reflexion, refraction, dispersion, etc., it is quite conceivable that a theory of light involving continuous functions over space will lead to contradiction with experience when it is applied to the phenomena of the creation and conversion of light ... In fact ... the observations on black-body radiation, photoluminescence, production of cathode rays ... and other phenomena involving the emission or conversion of light can be better understood on the assumption that the energy of light is distributed discontinuously in space ... [and] consists of a finite number of energy quanta, localized in space, which move without being divided and which can be absorbed or emitted only as a whole."

This clearly is a different language and vision than that of Planck. And still Einstein takes off exactly where Planck stops, namely with Planck's radiation law

$$u_\nu = \frac{8\pi}{c^3} h\nu^3 [\exp(h\nu/kT) - 1]^{-1} \tag{1}$$

Einstein, however, takes (1) as a fundamental law of nature, irrespectively of any theoretical derivation, including Planck's which he finds contradictory; he notices indeed that Planck's resonators are assumed to behave discontinuously in the black-body radiation paper while they had been assumed to behave continuously in the derivation of (7.1.20). Looking at Planck's radiation law (1) Einstein sees in it only a property of the radiation in thermal equilibrium: in agreement with the principles of statistical mechanics, no reference need to be made to what the radiation might be in equilibrium with. As did Planck before him, Einstein notices that for $h\nu/kT \gg 1$ (1) reduces to Wien's law

$$u_\nu = \alpha\nu^3 e^{-\beta\nu/T} \tag{2a}$$

$$\text{with } \alpha \cong \frac{8\pi}{c^3} h \text{ and } \beta = \frac{h}{k} \tag{2b}$$

an approximation Einstein keeps through his paper; equivalently (2a) can be rewritten

$$\frac{1}{T} = -\frac{1}{\beta\nu}\, \ell n\left(\frac{u_\nu}{\alpha\nu^3}\right) \tag{2c}$$

Transposing now *to the radiation itself* the argument previously used by Planck (and Wien) for the ponderable matter, Einstein identifies the temperature T occurring in (2) with the integrating factor in $Td\sigma = du$ where σ is now interpreted as the spatial entropy density *of the radiation.* One thus obtains from (2c):

$$\frac{\partial\sigma}{\partial u} = -\frac{1}{\beta\nu}\, \ell n\left(\frac{u}{\alpha\nu^3}\right) \tag{3a}$$

and thus, by integration:

$$\sigma = -\frac{u}{\beta\nu}\left[\ell n\left(\frac{u}{\alpha\nu^3}\right) - 1\right] \tag{3b}$$

Assuming now that the radiation (at frequency ν) is spatially uniform in the volume V which it occupies, one can define the total entropy S and the total energy $\mathcal{E}$ of the radiation (at frequency ν) by:

(4) $$S \equiv V \cdot \sigma \text{ and } \mathcal{E} \equiv V \cdot u$$

so that (3b) reads:

(5) $$S = -\frac{\mathcal{E}}{\beta\nu}\left[\ell n\left(\frac{\mathcal{E}}{V\alpha\nu^3}\right) - 1\right]$$

Upon comparing the entropy S with that of a radiation of the same frequency ν and the same total energy $\mathcal{E}$, but enclosed in a volume V_0, we find from (5):

(6) $$S - S_0 = \frac{\mathcal{E}}{\beta\nu}\, \ell n(\frac{V}{V_0})$$

i.e. upon recalling (2b):

(7) $$S - S_0 = k \cdot \ell n\left(\frac{V}{V_0}\right)^{\mathcal{E}/h\nu}$$

Einstein has now the remarkable idea to compare (7) with the formula one obtains for the entropy of a dilute gas (see Chapter Two); for such an ideal gas, constituted of N modecules, in thermal equilibrium at temperature T, and occupying a volume V, one has:

(8) $$\frac{1}{N} S = C_V\, \ell n T + k\, \ell n V$$

from which follows

(9) $$S - S_0 = k\, \ell n\left(\frac{V}{V_0}\right)^N$$

Einstein, incidentally, derives this formula more directly, taking the following interpretation as his guiding line of argument:

(10) $$S - S_0 = k[\ell n W - \ell n W_0]$$

where W/W_0 is the probability that all particles of the gas are found in the subvolume V when the gas is enclosed in the total volume V_0. This probability is clearly (since the particles of the gas are supposed to be independently distributed):

(11) $$\frac{W}{W_0} = \left(\frac{V}{V_0}\right)^N$$

so that (10) and (11) indeed reproduce (9). Upon comparing (9) and (7), Einstein concludes that

$$\mathcal{E} = Nh\nu \tag{12}$$

for which he gave the following interpretation: "*Monochromatic radiation* of low density—i.e. as long as Wien's radiation formula is valid—*behaves in a thermodynamical sense as if it consisted of mutally independent energy quanta of magnitude* $R\beta\nu/N$" [$\equiv h\nu$ in the notation we use here; the underlining is ours and emphasizes that Einstein arrived here at a conclusion which is more general than that he had initially stated.] He was further encouraged in his interpretation by noticing that, with the particle density φ_ν defined from the energy density u_ν by

$$u_\nu = h\nu\, \varphi_\nu \tag{13}$$

one obtains (in the same Wien approximation):

$$< h\nu > \equiv \int_0^\infty d\nu\, u_\nu \Big/ \int_0^\infty d\nu \varphi_\nu = 3kT \tag{14}$$

Notice $RHS = 2.3(\frac{1}{2}kT)$, where 3 is the number of dimensions of the space in which radiation takes place, and 2 is the number of independent polarizations of light.

So far, the analogy is evidently just that: a mere analogy. However, Einstein discusses in the same paper several applications of his interpretation, among which the so-called photo-electric effect, which earned him the Nobel prize (1922). It is here, indeed, that "Einstein's extraordinary boldness lies" (Pais, 1982). The physical phenomena can be described as follows. When an electromagnetic wave of short enough wave-length impiges on a metal surface, electrons are ejected. The effect was discovered by Hertz in 1877. The puzzling feature, discovered by Lenard between 1899 and 1902 was that the energy of the ejected electron is independent of the intensity of the incident light but depends on its frequency. Maxwell electromagnetism does not explain this since in that theory the energy carried by a light wave is measured by the light intensity. Hence, as the intensity of the incident light is increased, more energy becomes available to be imparted to the electrons and so it would seem that their energy should then increase, which it does not: only the number of electrons which are ejected increases. In contrast to Maxwell theory, Einstein's view of light led him to a description of the phenomena (see Figure 1) whereby

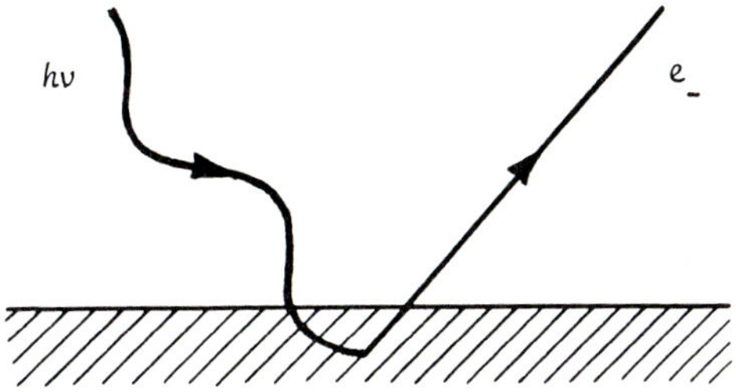

Fig. 1

the energy E, of the ejected electron is given by

$$E = h\nu - E_0 \tag{15}$$

here ν is the frequency of the incident light, and E_0 is the energy required to remove the electron from the metal.

The prediction (15) was confirmed experimentally by Millikan who, ironically enough, had started out thinking the prediction was "unreasonable". From the point of view of classical physics, Millikan was right indeed: Einstein's prediction and the phenomenon it turns out to describe accurately were so unreasonable that a drastically new extension of classical physics had to be invented. It was not that just a new elementary particle—later to be called the *photon*—had been found but, more significantly, that *a new concept of particles was called for*, particles which—as we shall presently see—obeyed their own kind of statistics.

Einstein (1909) returned to this problem ostensibly to patch up an obvious weakness of his (1905, PE) paper, namely that this paper did not use the full strength of Planck's formula, but only its Wien asymptotic behaviour for $h\nu \gg kT$. In this second paper, Einstein proved what is now known as the *Einstein fluctuation formula* (see (26) below) which established that the photon is not just a particle, nor is it just a wave, but, in fact, is both a wave and a particle; this was the germ of what we today refer to as either the wave-particle duality or, somewhat improperly, the complementarity principle.

Einstein's argument runs as follows; here again it is a demonstration, not a proof. He starts with the Planck version of Boltzmann's interpretation of entropy, but applied now to the radiation, namely (compare with (10) above):

$$S = k\, \ell n W \tag{16a}$$

or equivalently

$$W = \exp(S/k) \tag{16b}$$

Up to then this formula was used, in the black-body radiation context, for equilibrium situation only. Einstein extends it to the study of fluctuations. Consider, indeed, two vessels of volume V (resp.v), entropy Σ (resp.σ) and energy H (resp.η), with (V, Σ, H) large enough to serve as reservoir and (v, σ, η) being the system of interest, weakly coupled to (V, Σ, H). Schematically, the situation is represented in Figure 2.

The total system is close to equilibrium, and its total entropy S is developed in Taylor expansion with respect to the small energy fluctuations of the energy η:

$$S = S_0 + \left(\frac{dS}{d\eta}\right)_0 (\eta - \eta_0) + \frac{1}{2}\left(\frac{d^2 S}{d\eta^2}\right)_0 (\eta - \eta_0)^2 + \cdots \tag{17}$$

Since, according to Boltzmann's general ideas, S is maximum at equilibrium

$$\left(\frac{dS}{d\eta}\right)_0 = 0 \text{ and } \left(\frac{d^2 S}{d\eta^2}\right)_0 < 0 \tag{18}$$

Moreover, the fact that the big system (V, Σ, H) acts as a reservoir for the small system (v, σ, η) is expressed as the approximation

$$\left(\frac{d^2 S}{d\eta^2}\right)_0 \cong \left(\frac{d^2 \sigma}{d\eta^2}\right)_0 \tag{19}$$

Upon inserting (17), (18) and (19) in (16b) one receives

$$W \cong W_0 \ \exp\{\frac{1}{2k}\left(\frac{d^2\sigma}{d\eta^2}\right)_0 (\eta - \eta_0)^2\} \tag{20}$$

which one can rewrite as the Gaussian:

$$W \cong W_0 \exp\{-\frac{1}{2} \cdot \frac{(u - < u >)^2}{< (\Delta u)^2 >}\} \tag{21a}$$

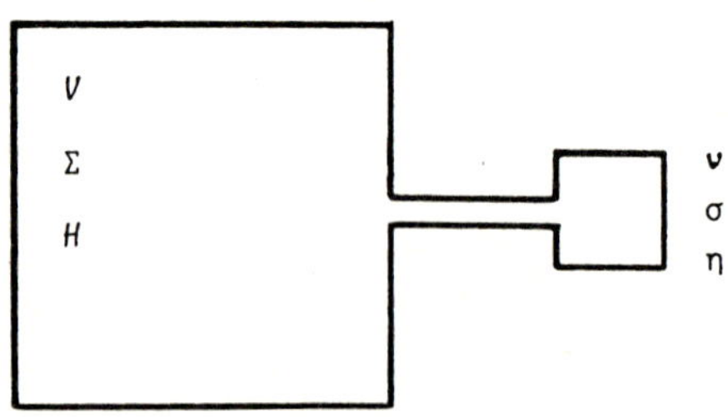

Fig. 2

with the identifications

(21b) $\eta \equiv u\,;\ \eta_0 \equiv < u >;\ < (\Delta u)^2 > \equiv < (u - < u >)^2 >$

(where $< \quad >$ denotes the equilibrium value), and with the approximations

(21c) $$< (\Delta u)^2 > \cong -\frac{k}{d^2 < \sigma > / d < u >^2}\,;\ \left(\frac{d^2\sigma}{d\eta^2}\right)_0 \cong \frac{d^2 < \sigma >}{d < u >^2}$$

Upon using now the thermodynamical relation $Td < \sigma > = d < u >$ we have

(22a) $$\frac{d < \sigma >}{d < u >} = \frac{1}{T}$$

and from there:

(22b) $$\frac{d^2 < \sigma >}{d < u >^2} = \frac{d}{d < u >}\left(\frac{1}{T}\right) = -\frac{1}{T^2}\frac{1}{d < u > / dT}$$

which, when fed into (21c) gives

(23) $$< (\Delta u)^2 > = kT^2 \ \frac{d < u >}{dT}$$

This is the central formula (which, incidentally, is not in Einstein's paper). At this point, Einstein finally specializes his discussion to the black-body radiation, using the full Planck radiation law (1) in the form

(24) $$< u > = Z\, h\nu\, [\exp(h\nu/kT) - 1]^{-1} \quad \text{with } Z \equiv \frac{8\pi}{c^3}\nu^2$$

Upon feeding this in (22b) and in (23) gives

(25) $$\frac{d^2 < \sigma >}{d < u >^2} = -k \cdot \frac{1}{< u > [h\nu + \frac{1}{Z} < u >]}$$

(26a) $$< (\Delta u)^2 > \ = < (\Delta u)^2 >_p + < (\Delta u)^2 >_w$$

with

(26b) $$< (\Delta u)^2 >_p \ \equiv < u > h\nu$$

(26c) $$< (\Delta u)^2 >_w \ \equiv < u >^2 Z^{-1}$$

The reader will notice that (25) corresponds to Planck's ansatz (7.1.23); here again, however, the difference in interpretation must be emphasized: Planck's formula (7.1.23) concerns the resonators, while Einstein's formula (25) is a property of the radiation itself; hence, the general thread is still with us. The new result is to be looked for in (26). From (24) we see that in the Wien approximation $h\nu \gg kT$ the dominant term in $< (\Delta u)^2 >$ is $< (\Delta u)^2 >_p$ which reduces then to the contribution of the photons as *particles* (see in particular (12) above); hence the notation $< \dots >_p$. On the contrary, we see again from (24) that in the classical approximation $h\nu \ll kT$ the dominant term in $< (\Delta u)^2 >$ is $< (\Delta u)^2 >_w$ which reduces then to the Rayleigh-Jeans contribution where the radiation is seen (Section 1) as an electromagnetic *wave* in equilibrium in a resonant cavity; hence the notation $< \dots >_w$. We see, therefore, that the fluctuation formula (26) contains contributions from *both* the *particle* and the *wave* aspects of the radiation. This *particle-wave duality*, first perceived in the case of the black-body radiation, was to play a central role in the later developments of what became quantum mechanics.

The modern view of formula (26) is that the photons are not independent particles in the classical sense—this would only give the first term in this formula—but obey their own kind of statistics, the Bose-Einstein statistics to be described in detail later—which then accounts for the correction given by the second term in the Einstein fluctuation formula (26).

The energy-fluctuations discussed above are accompanied by momentum fluctuations (with $p \equiv h\nu/c$), and Einstein's paper contains an application to the radiation pressure exercised on a mirror suspended in a black-body cavity: under the bombardment of the photons, the mirror is shown to perform a Brownian motion. Recall that it was Einstein (1905, BM) who had recognized Brownian motion to be one of the best available experimental proofs of the existence of microscopic particles.

The photon emerged from the theory of the black-body radiation as an object, the particle nature of which was made manifest experimentally by the prediction this concept allowed to make on the photoelectric effect, see formula (15). Still we saw through formula (26) that this was not a particle in the classical sense.

The question was then of whether these existed in nature some other waves which could similarly be interpreted as particles. One of the most common waves to come to mind are the sound waves in solid. These were studied from the present point of view by Einstein (1907, 1911) and Debye (1912). The argument runs as follows.

Suppose that the vibrational energy of a solid is given by

$$U = \int_0^\infty d\nu\, g(\nu)\, U_\nu \tag{27}$$

where, in analogy with the black-body radiation formula (1) (see also 7.1.20), we *assume*

$$U_\nu = h\nu\, [\exp(h\nu/kT) - 1]^{-1} \tag{28}$$

and

$$g(\nu) = \begin{cases} 12\pi\nu^2 V/s^3 & 0 \leq \nu \leq \nu_0 \\ 0 & \nu > \nu_0 \end{cases} \tag{29}$$

with V denoting the volume of the solid, and ν_0 being a "cut-off" frequency, to be interpreted in physical terms as the fact that in a real crystal the vibrations have a minimum wavelength comparable to the interatomic distance. Incidentally, one may note here that the idea that crystalline solids are regular lattices, with atoms sitting at their vertices, had been the subject of numerous speculations for about a century; X-rays had been discovered by Roentgen (1895); Ewald (1911) was studying the propagation of electromagnetic waves in space-lattices, in Sommerfeld's institute in Munich; and at the suggestion of von Laue, Friedrich and Knipping (1912) actually obtained in the laboratory the diffraction patterns of X-rays by lattice crystals. This discovery is thus essentially contemporary with the Debye paper we are now reviewing. In Debye's paper the cut-off frequency ν_0 is introduced to eliminate the ν^2 divergence in $g(\nu)$, and is thus determined by the condition

$$\int_0^{\nu_0} d\nu \, g(\nu) = 3N \tag{30}$$

where N is the number of 3-dimensional oscillators forming the crystalline solid. The term $12\pi V/s^3$ in (29) plays for this problem the role played by $8\pi/c^3$ in the black-body radiation (see 7.1.20); the differences are: (i) the introduction of V due to the fact that we are now considering the total energy and not a spatial energy density; (ii) the substitution of s, the speed of sound in the solid, for the light velocity c; (iii) $3 \cdot 4\pi$ instead of $2 \cdot 4\pi$, coming from the fact that sound waves have, in addition to the two transverse polarizations also present in light, a third degree of freedom, namely longitudinal waves.

From (29) and (30) we read:

$$\nu_0 = s\left(\frac{3}{4\pi} \cdot \frac{N}{V}\right)^{1/3} \tag{31}$$

Upon inserting (28), (29) and (31) in (27), we receive:

$$U = 9Nk \, \frac{T^4}{\Theta^3} \int_0^{x_0} dx \cdot \frac{x^3}{e^x - 1} \tag{32a}$$

with

$$\Theta = \frac{h\nu_0}{k} \quad \text{and} \quad x_0 = \frac{h\nu_0}{kT} \tag{32b}$$

Note that Θ has the dimension of a temperature: it is called the *Debye temperature*, and it is characteristic of the crystalling solid considered, a few hundred $°K$ for metals like aluminum, copper, silver and gold, and significantly higher for diamond. These are determined experimentally by measuring the specific heat

$$C_V \equiv \frac{1}{N} \cdot \frac{dU}{dT} \tag{33}$$

which can be computed from (32); the agreement with experiment is extremely good over the whole range of temperatures for which data are available (and for monoatomic crystalline structures). It is instructive to compute (33) from (32) in two extreme cases which we now examine briefly.

The high temperature limit $h\nu_0 \ll kT$ gives for the integral in (32a) an asymptotic expansion, the first term of which is $(x_0)^3/3$; hence we find, upon using (32b), an asymptotic expansion for U, the first term of which is $3NkT$. Consequently, as $T \to \infty$, the specific heat C_V tends to a constant, namely $3k$, independent of Θ and thus independent of the solid considered. We thus have recovered the classical "law" of Dulong-Petit (1819). By the turn of the Century, it was known that this empirical rule holds only at high temperatures, C_V becoming considerably smaller as T decreases; diamond is a notorious case in point. No satisfactory explanation, however, was available for the observed dependence of C_V on temperature; this is precisely what the quantum formula (32) provides, not only qualitatively, but quantitatively as well.

To see this, let us observe what happens in the other extreme case, the low temperature limit $kT \ll h\nu_0$. An asymptotic expansion for the integral in (32c) can be written again for this case, the first term of which is simply obtained by letting the upper limit of the integral go to infinity; in this limit the integral reduces to the constant $\pi^4/15$, so that

$$U \cong \frac{3\pi^4}{5} Nk \cdot \frac{T^4}{\Theta^3} \tag{34a}$$

and thus

$$C_V \cong \frac{12\pi^4}{5} k \left(\frac{T}{\Theta}\right)^3 \tag{34b}$$

In particular $C_V \to 0$ as $T \to 0$; this prediction was first made by Einstein (1907) on a basis of a simpler assumption, namely that g was a δ-distribution instead of (29); Einstein (1911), however, recognized that his assumption was responsible for his getting a wrong functional dependence of the specific heat on the temperature in the very low temperature range. It was Debye (1912) who derived (32), and thus (34), predicting the T^3-dependence which seemed to be in much better agreement with experiment. Nevertheless, it is the simpler prediction that C_V vanishes as the temperature approaches absolute zero which drew the attention of the physicists on the definite possibility that the quantum theory may have a domain of application much wider than the rather specialized field of black-body radiation. Among the famous conversions this prediction prompted, Nernst's reaction, although still guarded, is often quoted (e.g. Jammer, 1966, p.59 or Pais, 1982, p.399).

Finally, we should mention that the modern way to look at the Einstein-Debye theory of specific heats is to regard sound waves as particles, the *phonons*, which like the photons are not really classical particles as they, too, satisfy the statistics given by (26). These, at long last, could provide a realistic model to substitute for the somewhat mythical resonators Planck had introduced in the beginning of his theory of black-body radiation.

The cycle thus seems to have run its course. Still, on the strength of the wave-particle duality now recognized for photons and phonons, one might ask whether it should be possible to travel the same route in the opposite direction and thus be led to the conjecture that what are usually considered as particles could perhaps be reinterpreted as waves. It is precisely this question that L. de Broglie (1924) addressed in a series of papers culminating in his thesis.

Maurice de Broglie, himself a physicist of note, had been one of the scientific secretaries, and subsequently an editor, for the first Solvay Congress which had met at Nernst's instigation, in 1911, to discuss the new quanta. It is through his brother Maurice that Louis de Broglie became aware of the peculiar nature of X-rays, of the black-body radiation theory, and of the particle-wave duality Since his brother "not being a theoretician ... did not have particularly clear ideas on this subject", Louis de Broglie first set out to produce a "synthetic theory of radiation". He then gave a universal touch to his investigations by postulating that to every particle of mass m, moving with velocity v, should be associated a wave of wavelength λ given by the formula

$$\lambda = \frac{h}{mv} \quad (h \text{ is the Planck constant }) \tag{35a}$$

Note that this formula can be rewritten, through $\lambda \cdot \nu = v$ and $p = mv$, as

$$p = \frac{h\nu}{v} \tag{35b}$$

in close analogy with the formula $p = h\nu/c$ Einstein had used for the momentum of photons in his 1909 paper (see above).

When in the course of his thesis defense L. de Broglie was asked (by J. Perrin) whether his matter-waves could have any experimental reality, he suggested that one should investigate the diffraction of electrons by crystals. Such experiments were, in fact, underway—without de Broglie being aware of them. It appears to be true that empirical evidence could have been marshalled in support of de Broglie's thesis from experiments which had been conducted several years before; this remarkable story is told very well in Jammer (1966) pp.247–254. As it is, the 1937 Nobel prize recognized Davisson and Thomson "for the experimental discovery of the interference phenomena in crystals irradiated by electrons"; see e.g. Davisson & Germer (1927) and Thomson

(1927). This early experimental confirmation contributed most clearly to the general acceptance of de Broglie's relation (35) and the underlying extension of the wave-particle duality, from photons and phonons, to electrons and beyond. One should also add that, at the same time, strong theoretical evidences, stemming from the interpretation of the Schroedinger equation, were pointing in the same direction, as we shall see in Chapter Eight.

Section 3. THE STRUCTURE OF THE ATOM.

This topic's history is rich in contradictions, the first of which may very well be that, soon after—and, in fact, even before—the very existence of atoms and molecules had been established, physicists started speculating on how the atom was built, i.e. speculated on the parts of that which, etymologically, should have had no parts.

Fortunately, two experimental facts set the tone of the discussion, preventing it to go too far astray. The first (see Section 2 above) was the discovery of the electron and the fact that its mass is about 1/2000 the estimated mass of the lightest of all atoms, hydrogen. The second was the discovery of spectroscopy reported in the beginning of this chapter (see the introductory notes on Kirchhoff's work, in Section 1 above): by the turn of the Century, the characteristic "spectral lines" of many elements had been correctly identified.

The wealth of information thus obtained by spectroscopists must have looked as a nightmarish jungle to any scientist endowed with a sense that nature somehow ought to be understandable, or at the very least systematizable. The first glimpse of an organized order was proposed by an outsider. J. J. Balmer, often characterized as a Swiss schoolteacher which he was, happened also to be teaching geometry at the University of Basel as a privat docent, and to be a great lover of numerology. The story goes that a friend of his challenged him in one of his idle moments to find a simple mathematical algorithm fitting the spectral lines of Hydrogen. From Balmer's papers (1885), we learn that he was well aware of the latest available data and that he fitted to better than 1 part in 40.000 Ångström's observations on four lines in the Hydrogen spectrum. He further predicted the position of a fifth and went on gathering information on at least a dozen others which he obtained from Vogel, Huggins and Hagenbach.

The formula that Balmer found for the wavelengths of a collection of lines, known today as the Balmer series, is:

(1a) $$\lambda_m = h \cdot \frac{m^2}{m^2 - 4}, \quad m = 3, 4, \ldots$$

in which Balmer empirically attributed to h the value $3645.6 \cdot 10^{-7}$ millimeters. For our purpose in this section, it is convenient to rewrite Balmer's formula in term of frequencies, namely as:

(1b) $$\nu_m = K\Big(\frac{1}{2^2} - \frac{1}{m^2}\Big)$$

The formula was too good to be passed over by the professionals; in a series of papers, Rydberg (1890) confirmed Balmer's result, claimed priority, and generalized it to other "series" which, for our purpose, can be written in the form

$$\nu_{n,m} = K\left(\frac{1}{n^2} - \frac{1}{m^2}\right), \quad n \text{ and } m \in \mathbb{Z}^+, n < m \tag{2}$$

The challenge was then to explain these formulas from first principles, and, in particular, to give a theoretical explanation for the empirical value of K.

In 1913, Bohr published a series of three papers of seminal importance; the argumentation is boldly original and emphasizes that some drastic departure from traditional wisdom had to be dared if an explanation were ultimately to emerge. For more than a decade, considerable efforts were spent to boil down the apparent contradictions of these papers to some manageable principles leading to a consistent generalization of the classical theory; the result of these efforts will be presented in subsequent chapters. We first proceed here by outlining the core features of Bohr's first paper, restricting our attention mainly to the Hydrogen atom, with some incidental comments on He^+; for this purpose, we find it convenient to isolate five main assumptions in Bohr's argument.

ASSUMPTION 1. *The atom is constituted of a positively charged nucleus surrounded by orbiting electrons. The total negative charge of the electrons is equal in magnitude to the positive charge of the nucleus. The radius of the nucleus is supposed to be negligibly small in comparison with the linear dimension of the atom. The mass M of the nucleus is supposed to be much larger than the mass m of the electron (if, however, a quantitative estimate of this is to be taken into account, replace in the sequel m by the reduced mass $m' \equiv m[1 + (m/M)]^{-1}$*

Bohr, with some justice, refers to this as the Rutherford model, although some historical precedents can be traced to Perrin and to Nagaoka; we shall not need those references in the sequel.

This model is first discussed completely classically. Specifically, let a be the radius of the circular orbit of an electron around the nucleus; v be the velocity of the electron on its orbit, and ω be its frequency, m its mass, e its charge; and ϵ be the charge of the nucleus. We have then, by definition:

$$\omega \equiv v/2\pi a \tag{3}$$

Upon balancing the centrifugal force and the Coulomb force, we have:

$$m(2\pi\omega)^2 a = \frac{e\epsilon}{a^2} \tag{4}$$

Upon writing the total energy E of the electron as the sum of its kinetic energy and its potential energy, we have:

$$E = \frac{1}{2}mv^2 - \frac{e\epsilon}{a} \tag{5}$$

Upon using (3), (4) and (5) we find, by straightforward algebraic manipulations

$$W \equiv -E = \frac{e\epsilon}{2a} \tag{6}$$

$$\omega = \frac{\sqrt{2}\, W^{3/2}}{\pi e\epsilon\sqrt{m}} \tag{7}$$

Bohr notices, as he starts from (6) and (7) that, by Kepler's law, these formulas hold in fact for an electron on an elliptic orbit of principal axis $2a$ with the nucleus sitting at one of the foci; see (46) below. For later use, notice that W can be interpreted as the energy which would be required to remove the electron from its orbit and place it infinitely far away from the nucleus.

Bohr then emphasizes—as was well-known at the time to be a serious objection against Rutherford's model—that in this classical framework the electron on its orbit is accelerated; since it is electrically charged it should, according to Maxwell equations, emit an electromagnetic radiation. Consequently, its energy E should decrease in the course of time, and thus $W = -E$ should increase. From (6) and (7), this would imply that a decreases and ω increases as time elapses, i.e. the electron should spiral more and more rapidly towards the nucleus: the orbit is not stable! Bohr thus remarks: "It is obvious that the behavior of such a system will be very different from that of an atomic system occurring in nature." To correct for this, Bohr makes the following ansatz.

ASSUMPTION 2. *Regardless of classical theory, there are stable orbits of energy $W(n)$ with n running over $\mathbb{Z}^+$, characterized by the stability condition*

$$W(n) = h\,\omega(n)\,f(n) \tag{8}$$

where h is Planck constant, $\omega(n)$ is the frequency of the electron on the orbit $W(n)$, and $f : n \in \mathbb{Z}^+ \to \mathbb{R}^+$ is a function to be determined later on.

Upon feeding the stability conditon (8) in the classical equation (7) we obtain:

$$W(n) = \frac{\pi^2 me^2\epsilon^2}{2h^2} \cdot \frac{1}{f(n)^2} \tag{9a}$$

or equivalently

$$\omega(n) = \frac{\pi^2 me^2\epsilon^2}{2h^3} \cdot \frac{1}{f(n)^3} \tag{9b}$$

Note that it is one thing to select some particular solutions among all the solutions which are a priori possible, but that it is quite another thing to ignore part of a theory in order to decree that the solutions which one chooses are stable when in fact the original, full theory prevents this to be the case. Bohr was aware of this, but he seems to have believed from the beginning of his investigations that this departure from classical theory was necessary; he did not seem to worry whether any explanation could ultimately be possible within the accepted principles of the classical theory; quite to the contrary, he appears to have felt that such an explanation should not even be sought after.

While Bohr started his enquiry with the aim to describe the structure of the atom irrespective of what spectroscopy may do to help, the latter turned out to provide the key he was looking for. He once commented: "As soon as I saw Balmer's formula, the whole thing was immediately clear to me." The next three assumptions appear to be the essential steps he took to arrive at his conclusion.

ASSUMPTION 3. *Atoms produce spectral lines one at a time, i.e. individual atoms are not responsible for the whole spectrum simultaneously. Spectral lines are emitted, or absorbed, when an electron transits from an orbit with energy $W(n)$ to an orbit with energy $W(n')$, and the frequency $\nu_{n,n'}$ of the radiation emitted, or absorbed, in this transition is given by:*

$$h\nu_{n,n'} = W(n') - W(n) \tag{10}$$

where h is the Planck constant.

Note that this assumption embodies two important facts. The first is the observation, by Kirchhoff & Bunsen (see Section 1), that absorption and emission lines coincide. The second is Einstein's interpretation of $h\nu$ as the energy of a photon of frequency ν. Upon introducing now (10) into (9a), we obtain:

$$\nu_{n,n'} = \frac{\pi^2 m e^2 \epsilon^2}{2h^3}\left[\frac{1}{f(n')^2} - \frac{1}{f(n)^2}\right] \tag{11}$$

Now comes the link to Balmer's formula (1b).

ASSUMPTION 4. *The Balmer spectral series is given by (11) with $n' = 2$ and $n \geq 3$.*

Upon comparing (11) [with $n' = 2$ and $n \geq 3$] to (1b), we obtain:

$$K = \frac{\pi^2 m e^2 \epsilon^2}{2h^3 k^2} \tag{12a}$$

$$f : n \in \mathbb{Z}^+ \to kn \in \mathbb{R}^+ \tag{12b}$$

In these formulas, k is a positive constant, as yet undetermined; this will be taken care of presently. (*Stricto sensu,* Assumption 4 gives (12b) for $n \in$

$\mathbb{Z}^+ \setminus \{1\}$, but this is a detail we do not want to comment upon here beyond saying that the extension to $\mathbb{Z}^+$ is a negligible leap of faith in regards to the boldness of each of the assumptions we are making explicit here).

The final assumption will allow to determine the constant k appearing in (12).

ASSUMPTION 5. *For large values of n and n' the predictions of the theory coincide with the predictions of the classical theory.*

This assumption is the first appearance in Bohr's writings of his famous *correspondence principle*, which he later on was to refine and sharpen until he made it a touchstone for the emerging quantum theories. Here it simply appears as a transposition of a fact already noticed by Einstein in his work on black-body radiation: when a large number of quanta are simultaneously involved one recovers the classical theory.

In the specific case at hand here, assumption 5 is used as follows.

On the one hand, the theory so far outlined gives for large values of n:

$$\nu_{n,n-1} = K\left[\frac{1}{(n-1)^2} - \frac{1}{n^2}\right] \cong K \cdot \frac{2}{n^3} \tag{13}$$

and, upon using (9b):

$$\omega(n) = K \cdot \frac{1}{kn^3} \tag{14}$$

On the other hand, the classical theory of the orbiting electron leads to:

$$\nu_{n,n-1} = \omega(n) \tag{15}$$

Then Assumption 5 tells us to insert (13) and (14) into (15) from which one obtains $k = \frac{1}{2}$, so that (9a) and (11) become

$$W(n) = \frac{2\pi^2 m e^2 \epsilon^2}{h^2} \cdot \frac{1}{n^2} \qquad n \in \mathbb{Z}^+ \tag{16}$$

$$\nu_{n,n'} = K\left[\frac{1}{(n')^2} - \frac{1}{n^2}\right] \tag{17a}$$

with

$$K = \frac{2\pi^2 m e^2 \epsilon^2}{h^3} \tag{17b}$$

Note that (16) and (17a) hold for all positive integers n and n'; the classical limit of large n was only used to determine the constant k. For $n' < n$, $\nu_{n,n'}$ is positive and is the frequency of a line in the emission spectrum; when on the contrary $n' > n$, $\nu_{n,n'}$ is negative and $-\nu_{n,n'}$ is the frequency of a line in the absorption spectrum. Notice, further that K is now fully expressed in terms of fundamental constants: the charge and (reduced) mass of the electron, the charge of the nucleus, and the Planck constant. Upon specializing (17b) to the Hydrogen atom, for which $\epsilon = e$, one can, for instance, compute the value of Balmer's constant (see (1a)) which now comes out at $3646 \cdot 10^{-7}$ millimeters indeed, thus providing an excellent experimental check for the theory. Bohr further noticed that upon introducing (16) into (6) one obtains

$$a(n) = n^2 \cdot \frac{h^2}{4\pi^2 me\epsilon} \tag{18}$$

giving the radius (or more generally, the half-principal axis) of the n-th orbit. In particular, for the deepest orbit, that which corresponds to $n = 1$, and for the Hydrogen atom, where $\epsilon = e$, Bohr computed the quantity $a(1)$, now known as the *Bohr radius*, to be $a \cong 0.5 \cdot 10^{-10}m$ (within 5% of its modern value) thus giving an estimate of the size of the atoms.

Formula (6), which we just used, links the Bohr radius a to the ionization energy W (i.e. the energy necessary to expel the electron from its deepest lying orbit all the way to infinity); this was evaluated by Bohr, from the fundamental constants m, e and h, and confirmed directly from scattering data in experiments carried out by Franck and Hertz between 1914 and 1919. In the spectroscopic arena the predictions based on (17) were also met with spectacular success: having already used the Balmer series $n' = 2$, Bohr identified the series suspected by Ritz (1908) and confirmed by Paschen (1908) corresponding to $n' = 3$; he predicted from $n' = 1$ a series in the ultraviolet which was observed by Lyman (1914); and he predicted from $n' = 4$ and 5 two series in the infrared which were observed by Brackett (1922) and by Pfund (1924). He went further to consider the ionized Helium atom $He^+(\epsilon = 2e)$ and he identified, as pertaining to it, two series found earlier by Fowler and Pickering; his prediction was confirmed by Evans (1913), including the role of the factor $[1+(m/M)])^{-1}$ in the expression of the reduced mass of the electron (see Assumption 1). For atoms with several electrons, the theory needs some further refinements due to the fact, ignored at the time of Bohr's contribution, that electrons although they are not classical particles (see Section 2) are non-classical in a way different from that in which photons and phonons are non-classical: the latter satisfy the Bose-Einstein statistics, whereas the electrons satisfy the Fermi-Dirac statistics with the effect, in particular, that no two electrons can be in the same state; more on that later, as we would rather, in this section, concentrate on the difficulties pertaining to Bohr's theory itself. Clearly some efforts had first to be spent to extract some general principle from Bohr's seemingly ad hoc assumptions; and to elicit the connection which might exist between the quantum ansatz made in the theory of black-body radiation and of the specific heat of solids, and the quantum ansatz now made by Bohr in his theory of the structure of the atom. A major step in this direction was accomplished by Sommerfeld (1915, 1916; see also 1931). He came up with an ingenious condition which, in its simplest form, can be expressed as follows.

SOMMERFELD RULE. *The quantum-theoretically stable orbits of a Hamiltonian system $H(p_1, \ldots, p_n; q_1, \ldots, q_n)$ are characterized by the conditions:*

$$\oint dq_k\, p_k = h\, n_k \quad k = 1, \ldots, n \tag{19}$$

where h is the Planck constant and n_k $(k = 1, \ldots, n)$ are positive integers; the integral is taken along the orbit considered and extends over one period.

Let us first examine briefly what this condition says for a one-dimensional harmonic oscillator

$$H = \frac{1}{2m}p^2 + \frac{1}{2}kq^2 \tag{20}$$

The general solution of the Hamilton equation

$$\dot{q} = \frac{\partial H}{\partial p} \quad \text{and} \quad \dot{p} = -\frac{\partial H}{\partial q} \tag{21}$$

is

$$\begin{aligned} q(t) &= a\ \cos(2\pi\nu t + \varphi) \\ p(t) &= -2\pi\nu\ ma\ \sin(2\pi\nu t + \varphi) \\ &\text{with } \nu = \frac{1}{2\pi}\sqrt{\frac{k}{m}} \end{aligned} \tag{22}$$

Notice that the energy along each such orbit is a constant (since $\dot{H} = 0$) given by:

$$E \equiv H(p(t), q(t)) = 2\pi^2\nu^2 ma^2 \tag{23}$$

Sommerfeld's integral (the *LHS* of 19) is

$$\oint dq \cdot p = \int_0^T dt\, (2\pi\nu a)^2\ m\ \sin^2(2\pi\nu t + \varphi) \tag{24a}$$

i.e.

$$\oint dq \cdot p = 2\pi^2 \nu m a^2 \tag{24b}$$

Upon inserting (23) in (24b) we see that the Sommerfeld condition (19) reads

$$E = nh\nu \quad n \in \mathbb{Z}^+ \tag{25}$$

which is exactly the *Planck-Einstein relation* between the possible energies E of a harmonic oscillator and its fundamental frequency ν (see e.g. 7.1.21 and

Planck's "most essential point" quoted above that formula; see also Section 2 for Einstein's interpretation of this point). Hence Sommerfeld's rule seems indeed to go straight at the core of the arguments presented in the first two sections of this chapter.

Let us now study the Bohr atom from the same point of view; it is a Hamiltonian system with

$$H = \frac{1}{2}m\nu^2 - \frac{e\epsilon}{r} \tag{26}$$

Since we know from classical celestial mechanics that the orbits are planar we can consider this as a 2-dimensional problem, and the Hamiltonian (26) expressed in polar coordinates is

$$H = \frac{1}{2}m(\dot{r}^2 + r^2\dot{\varphi}^2) - \frac{e\epsilon}{r} \tag{27}$$

This suggests introducing

$$p_r \equiv m\dot{r} \quad \text{and} \quad p_\varphi \equiv mr^2\dot{\varphi} \tag{28}$$

so that H can be rewritten in the form

$$H = \frac{1}{2m}(p_r^2 + \frac{1}{r^2}p_\varphi^2) - \frac{e\epsilon}{r} \tag{29}$$

The first two Hamilton equations

$$\dot{\varphi} = \frac{\partial H}{\partial p_\varphi} \quad \text{and} \quad \dot{r} = \frac{\partial H}{\partial p_r} \tag{30}$$

together with (28) are tautologies, which only inform us that we picked the right canonical variables.

The second pair of Hamilton equations

$$\dot{p}_\varphi = -\frac{\partial H}{\partial \varphi} = 0 \tag{31}$$

$$\dot{p}_r = -\frac{\partial H}{\partial r} = \frac{1}{m}p_\varphi^2\frac{1}{r^3} - e\epsilon\frac{1}{r^2} \tag{32}$$

contains the information we need. We first note that (31) gives us that p_φ is a constant of the motion; this is nothing but Kepler's second law of planetary motion, a consequence of the invariance of the Hamiltonian H under the rotation group. Upon using the fact that the Hamiltonian itself is a constant of the motion, we rewrite (29) in the form

$$E = \frac{p_\varphi^2}{2m}\left[\left(\frac{p_r}{p_\varphi}\right)^2 + \frac{1}{r^2}\right] - \frac{e\epsilon}{r} \tag{33}$$

Notice further from (28) that

$$\frac{p_r}{p_\varphi} = \frac{1}{r^2} \cdot \frac{dr}{d\varphi} \tag{34}$$

With the change of variable

$$s = \frac{1}{r} \tag{35}$$

and (34), we can now rewrite (33) in the form

$$E = \frac{p_\varphi^2}{2m}\left[\left(\frac{ds}{d\varphi}\right)^2 + s^2\right] - e\epsilon s \tag{36}$$

Upon differentiating (36) with respect to φ, recalling that E and p_φ are constants, we obtain

$$\frac{d^2 s}{d\varphi^2} + s - \frac{me\epsilon}{p_\varphi^2} = 0 \tag{37}$$

This equation is equivalent to (32) and its solutions are

$$s = s_0 \cos(\varphi - \varphi_0) + \frac{1}{L} \tag{38}$$

with

$$L = (me\epsilon)^{-1}(p_\varphi)^2 \tag{39}$$

We have thus obtained (see 35) elliptic orbits of period 2π. Up to here everything was classical mechanics. We now introduce Sommerfeld conditions.

For the conjugate pair (p_φ, φ), the condition

$$\oint d\varphi\, p_\varphi = n_\varphi h \quad , \quad n_\varphi \in \mathbb{Z}^+ \tag{40}$$

simply gives, since p_φ is a constant of the motion:

$$p_\varphi = \frac{h}{2\pi} n_\varphi \quad , \quad n_\varphi \in \mathbb{Z}^+ \tag{41}$$

This is known as the *angular momentum quantization*; this result was already in Bohr (1913), but it is obtained here in a much simpler way.

For the conjugate pair (p_r, r) the Sommerfeld condition

$$\oint dr\, p_r = n_r h \quad , \quad n_r \in \mathbb{Z}^+ \tag{42}$$

gives, upon substituting $-W$ for E in (33):

$$\oint dr\left[-2mW+2me\epsilon\frac{1}{r}-p_\varphi^2\frac{1}{r^2}\right]^{\frac{1}{2}}=n_r h \tag{43}$$

which can be integrated (e.g. by contour integration) to give, upon taking (41) into account:

$$W(n)=\frac{2\pi^2 me^2\epsilon^2}{h^2}\cdot\frac{1}{n^2} \quad \text{with } n=(n_r+n_\varphi)\in\mathbb{Z}^+ \tag{44}$$

which gives back the formula (16) found by Bohr's method. We can actually see that this is no mere accident, if we go through the following elementary computation.

From (38), in which we can assume without loss of generality $s_0\geq 0$, we see that at the perihelion $(\varphi-\varphi_0\equiv\pi, r\equiv r_+)$ and at the aphelion $(\varphi-\varphi_0\equiv 0, r\equiv r_-)$ p_r vanishes so that the energy reduces to:

$$E=\frac{1}{2m}p_\varphi^2\cdot\frac{1}{r_\pm^2}-e\epsilon\frac{1}{r_\pm} \tag{45}$$

from which we conclude that the semi-major axis of the ellipse is given by

$$a\equiv\frac{1}{2}(r_++r_-)=\frac{1}{2}\frac{e\epsilon}{W} \tag{46}$$

Kepler's third law states that the periods T (resp. T_1) of two orbits with semi-major axis a (resp. a_1) are related by:

$$\left(\frac{T}{T_1}\right)^2=\left(\frac{a}{a_1}\right)^3 \tag{47}$$

i.e. upon taking (46) into account

$$\frac{1}{T}=\frac{1}{T_1}\cdot\left(\frac{W}{W_1}\right)^{3/2} \tag{48}$$

For every circular orbit (recall $n_r=0$) we have from (39) and (41), upon using $L=a=r(t)\ \forall\, t\in\mathbb{R}$:

$$a=\frac{h^2}{4\pi^2 me\epsilon}\cdot n_\varphi^2 \tag{49}$$

which again coincide with the formula we found by Bohr's method (see 18); in particular for $\epsilon=e$ and $n_\varphi=1$ we find back the formula for the Bohr radius.

Moreover, we have for these circular orbits

(50) $$\frac{1}{T} = \frac{1}{2\pi m a^2} p_\varphi$$

and thus, upon using (39) and (41):

(51) $$\frac{1}{T(n_\varphi)} = \frac{4\pi^2 m e^2 \epsilon^2}{h^3} \cdot \frac{1}{n_\varphi^3}$$

In particular, we obtain from (51) and (44) with $n_r = 0$ and $n_\varphi = 1$:

(52) $$W_1 = \frac{1}{2} h \cdot \frac{1}{T_1}$$

a formula we can now generalize to all stable elliptic orbits. Indeed from (44), (48) and (52), we finally obtain for arbitrary $n = n_r + n_\varphi$, $n_r = 0, 1, 2, \ldots$, $n_\varphi = 1, 2, \ldots$; and $\omega(n) \equiv T(n)^{-1}$:

(53) $$W(n) = h\,\omega(n) \cdot \frac{1}{2} n$$

which is exactly the Bohr stability condition (8) supplemented by the precise form of the function $f : n \in \mathbb{Z}^+ \to \frac{1}{2} n$ which we derived from the other conditions that Bohr imposed; see in particular (12b)and (15-16).

We have, therefore, recovered all the essential points of Bohr's treatment from a straightforward application of one simple stability condition, namely the Sommerfeld rule. In that sense, *Sommerfeld's theory contains Bohr's which it places in a more general context;* recall that Sommerfeld's rule also led to the Einstein-Plank formula (25) for the harmonic oscillator. Note, nevertheless, that Bohr's Assumptions 1 and 3 remain necessary to a physical interpretation of the theory.

The Bohr-Sommerfeld method can be extended one step further to include relativistic corrections. The need for such corrections is indicated by the following preliminary computation. According to the results we just obtained, the velocity v of an electron on a circular orbit is given by (3) in which we can now substitute the value of a (see 49) and $\omega \equiv T^{-1}$ (see 50) to find, with $Z \equiv \epsilon/e$:

(54) $$\frac{v}{c} = \alpha Z (n_\varphi)^{-1}$$

where c is the velocity of light, and α is the pure number

(55) $$\alpha \equiv \frac{2\pi e^2}{hc} \cong \frac{1}{137}$$

An anecdotic note might be in order here. For a while α^{-1} was suspected to be exactly an integer, thus rising a bizarre theoretical challenge: why would nature chose to link, through the integer 137 such disparate fundamental constants as the charge of the electron, the Planck constant, and the velocity of light? Subsequent measurements of e, h and c, however, gave for α^{-1} the modern value 137.0388... an unremarkable number which one feels less of an urge to explain! What, however, *is* relevant for us here is that, for the deepest orbit ($n = 1$) of hydrogen atom ($Z = 1$), or the ionized helium atom $He^+(Z = 2)$, the velocity of the electron is not quite negligible against the velocity of light. The question is whether this could be experimentally detected, and theoretically accounted for.

With m_0 denoting the reduced mass of the electron at rest, i.e. $m_0 = m[1+(m/M)]^{-1}$, the effective mass of the electron moving with velocity v is given in the special theory of relativity (see 5.2.38) by:

$$\mu = m_0(1-\beta^2)^{-1/2} \tag{56}$$

with $\beta \equiv v/c$. To take this into account, we must replace the non-relativistic formula (26) by:

$$H = (\mu - m_0)c^2 - \frac{e\epsilon}{r} \tag{57}$$

which reduces indeed to (26) in the non-relativistic approximation (see Chapter Five, in particular 5.2.39). We can rewrite (57) as:

$$\left[1 + \frac{1}{m_0 c^2}(H + \frac{e\epsilon}{r})\right]^2 = \frac{1}{1-\beta^2} \tag{58}$$

Notice further that

$$\beta^2 = \frac{v^2}{c^2} = \frac{1}{c^2}(\dot{r}^2 + r^2\dot{\varphi}^2) \tag{59}$$

Upon replacing now (28) by

$$p_r = \mu\dot{r} \quad \text{and} \quad p_\varphi = \mu r^2\dot{\varphi} \tag{60}$$

we obtain (see 56 and 59):

$$\frac{\beta^2}{1-\beta^2} = \frac{1}{m_0^2 c^2}\left(p_r^2 + \frac{1}{r^2}p_\varphi^2\right) \tag{61}$$

Upon inserting this in (58) we receive:

$$\left[1 + \frac{1}{m_0 c^2}(H + \frac{e\epsilon}{r})\right]^2 = 1 + \frac{p_\varphi^2}{m_0^2 c^2}\left[(\frac{p_r}{p_\varphi})^2 + \frac{1}{r^2}\right] \tag{62}$$

We now use again the change of variable $s = r^{-1}$ (see 35) to rewrite (62) as

$$\left[1 + \frac{1}{m_0 c^2}(H + se\epsilon)\right]^2 = 1 + \frac{p_\varphi^2}{m_0^2 c^2}\left[\left(\frac{ds}{d\varphi}\right)^2 + s^2\right] \tag{63}$$

which, incidentally, reduces indeed to (36) in the limit $c \to \infty$. Notice, further that H and p_φ are still constants of the motion, so that the differentiation of (63) with respect to φ gives:

$$\left[1 + \frac{1}{m_0 c^2}(E + se\epsilon)\right] \cdot \frac{e\epsilon}{m_0 c^2} = \frac{p_\varphi^2}{m_0^2 c^2}\left[\frac{d^2 s}{d\varphi^2} + s\right] \tag{64}$$

i.e.

$$\frac{d^2 s}{d\varphi^2} + \gamma^2 (s - \Lambda) = 0 \tag{65a}$$

with

$$\gamma^2 = 1 - \frac{e^2 \epsilon^2}{p_\varphi^2 c^2} \tag{65b}$$

and

$$\Lambda = \frac{m_0 e\epsilon}{p_\varphi^2 \gamma^2}\left(1 + \frac{E}{m_0 c^2}\right) \tag{65c}$$

which indeed reduces to (37) as $c \to \infty$; instead of (38), the equation (65) admits for solutions

$$s = s_0 \cos[\gamma(\varphi - \varphi_0)] + \Lambda \tag{66}$$

which are now periodic of period $2\pi/\gamma$. Hence, since $\gamma < 1$, the perihelion precesses by the amount

$$\Delta\varphi = 2\pi\left(\frac{1-\gamma}{\gamma}\right) \text{ radians per revolution} \tag{67}$$

Here again, an incidental remark is called for. If the electromagnetic potential $e\epsilon/r$ were replaced by a gravitational potential, the above computation when carried out for the orbit of Mercury around the Sun would give a precession of 7 seconds of arc per century, a far cry from the result (6.3.94) we obtained from general relativity. Back to the atom, where gravitation plays a most negligible role as compared to the effect of the electromagnetic forces at work here, special relativity is all what we need. The fact, however, that the perihelion does precess (see 67) seems to preclude a straightforward use of Sommerfeld's rule. It is, however, easy to overcome this: in order to have integration variables which have the period of our system, we introduce $\psi = \gamma\varphi$, i.e. we choose a polar coordinate system (r, ψ) which precesses with the perihelion. With this change of variable, the Sommerfeld rule reads

$$\oint d\psi\, p_\psi = n_\psi h \quad n_\psi \in \mathbb{Z}^+ \tag{68}$$

$$\oint dr\, p_r = n_r h \quad n_r \in \mathbb{Z}^+ \tag{69}$$

(68) gives immediately

$$p_\psi = \frac{h}{2\pi} n_\psi \quad n_\psi \in \mathbb{Z}^+ \tag{70}$$

while (69), together with (62), gives

$$\oint dr \left[A + 2B\frac{1}{r} + C\frac{1}{r^2}\right]^{\frac{1}{2}} = n_r h \tag{71a}$$

with

$$A = 2m_0 E + \frac{E^2}{c^2} \tag{71b}$$

$$B = m_0 e\epsilon + \frac{e\epsilon E}{c^2} \tag{71c}$$

$$C = -p_\psi^2 + \frac{e^2\epsilon^2}{c^2} \tag{71d}$$

Here again, we notice that (71) reduces to (43) as $c \to \infty$. (71) can then be integrated as (43) was, and gives (after a few straightforward manipulations):

$$E = m_0 c^2\{(1+\Xi)^{-\frac{1}{2}} - 1\} \tag{72a}$$

with

$$\Xi = \alpha^2 Z^2 \{n_r + (n_\psi^2 - \alpha^2 Z^2)^{\frac{1}{2}}\}^{-2} \tag{72b}$$

This formula is the exact solution to our problem. For the hydrogen atom ($Z = 1$) or the ionized helium atom $He^+(Z = 2)$, we see from (55) that $\alpha Z \ll n_\psi(\geq 1)$, and we can, therefore, expand E in power series in (αZ) to obtain for $W = -E$:

$$W = \frac{2\pi^2 m_0 e^4 Z^2}{h^2} \cdot \frac{1}{n^2} \cdot \chi(n, n_\psi) \tag{73a}$$

with

$$n = n_r + n_\psi \; ; \; n_r = 0, 1, 2, \ldots ; \; n_\psi = 1, 2, \ldots \tag{73b}$$

and

$$\chi(n, n_\psi) = 1 + \frac{\alpha^2 Z^2}{n^2}\left(\frac{n}{n_\psi} - \frac{3}{4}\right) + \ldots \tag{73c}$$

If we keep only the first term in χ, (73) reduces to (44) and thus (16): we have recovered the Bohr-Sommerfeld results as a non-relativistic approximation.

The second term in χ gives the first relativistic correction; this was first proposed by Bohr (1915) in the particular case of circular orbits, where $n_r = 0$ so that $n = n_\psi$, $\frac{n_r}{n_\chi} - \frac{3}{4} = \frac{1}{4}$, and thus W appears to depend on n only. It was Sommerfeld (1916) who first found the exact expression (72) and its expansion (73) in power series of α^2. It is quite remarkable that (73) depends not only on n, but also on n_ψ separately, thus lifting the degeneracy still present in (44). This fact was immediately seized as the solution of an already old problem in spectroscopy. Indeed, it was known since Michelson (1891) that the lines in the Balmer series of the hydrogen atom were not strictly single, well-defined, lines but rather had a fine structure of their own, which (44), and thus (16), could not explain; the second term in χ was seen as a possible explanation of this phenomena, and for this reason α is known in the physical literature as the *fine structure constant*. Since, however, this correction is proportional to $\alpha^2 \cong 1/18780$ it is a rather minute one; the theory was, nevertheless, satisfactorily confirmed by Paschen (1916) in the precise measurements he made on He^+ for which $Z = 2$; notice that Z appears at the power four in W for this correction, which makes the experimental detection somewhat more manageable. As is typical of much of the "old quantum theory" described in the present chapter, (73) was later given quite a different interpretation.

Our account of Sommerfeld's theory will be essentially complete when we finally notice that the structure of the atom is to be expressed as a three-dimensional problem rather than a two-dimensional one as we presented it so far. The introduction of this third degree of freedom leads to the introduction of a third quantum number entering through the condition

$$\oint d\theta\, p_\theta = n_\theta h \tag{74a}$$

i.e.

$$p_\theta = \frac{h}{2\pi} n_\theta \tag{74b}$$

This implies only that we should replace n_ψ by $(n_\psi + n_r)$ in (73) so that the prediction of the theory remains qualitatively unaffected. However, an external magnetic field can be used to lift this degeneracy, and Sommerfeld saw a further confirmation of his theory in the Stern-Gerlach (1922) experiment.

We should also remark that what we called the Sommerfeld rule had been formulated independently by Wilson (1915) and Ishiwara (1915); these authors, however, did not push the consequences of their formulation with the technical strength that allowed Sommerfeld to include the relativistic idea leading to (73); it is the latter formula which is most probably responsible for the primacy given, in the physics literature, to Sommerfeld's formulation.

Largely because of the early successes the quantum theory had in making experimental predictions, it was the object of much activity in the next ten years (1915–1925), the history of which is vividly recounted in Jammer (1966). This period is marked by three developments the significance of which is, however, better understood within the framework of the "new quantum mechanics"; the latter evolved immediately after 1925 and we will present its main tenets in Chapter Eight. We, therefore, conclude the present chapter with only a few brief allusions to these developments.

The first development we want to mention is the rather long-winded attempt to consolidate the Bohr-Sommerfeld theory in better harmony with the general framework of classical mechanics. The problem here was to trace in classical mechanics, the precise origin of the quantization rules. In particular, it was noted that some ambiguity remained in the Sommerfeld theory, concerning the proper choice of the canonical variables $\{(p_k, q_k)|k = 1, 2, \ldots, n\}$, as different choices were shown to lead to different predictions. The answer was sought in an aspect of classical mechanics which had been developed by astronomers and mathematicians in their study of multiply periodic systems. The situation of that and allied problems at the turn of the Century can be found in Poincaré (1905). The importance of this kind of ideas in the realm of quantum theory was recognized by Ehrenfest (1913), Epstein (1916) and Schwarzschild (1916c), and it led to a vast literature on the use of "action-angle variables" and of "adiabatic invariants"; one of its first applications to quantum theory was a theoretical account of the Stark effect. As far as the conceptual foundations of quantum mechanics are concerned, the main lesson to be drawn from these investigations can be summarily described by saying that the Hamilton-Jacobi formulation emerged as that part of classical mechanics which lended itself most naturally to a quantum generalization; more on that in Chapter Eight.

The second development that occured in the decade 1915–1925 and which we will only mention here in passing, is the fact that the emerging quantum theory became able to give predictive rules on the relative intensities (and not only the positions) of the spectral lines. A collection of "selection rules" resulted, which had their own origin in a better understanding of the mechanisms by which matter interacts with electromagnetic radiation. Here again, this can be more economically understood in terms of the "matrix elements" which appear in the "new quantum mechanics" to be described in Chapter Eight.

Finally, the third development which we want to mention in this chapter is the conjecture that the electron orbiting around the atom is carrying a fourth quantum number—besides Sommerfeld's n_r, n_ψ and n_τ—known today under the name of spin, and assuming only two values: $s = \pm\frac{1}{2}$. This conjecture is associated with the names of Pauli (1925) and of Uhlenbeck & Goudsmit (1925), although it is established that in these turbulent times, some significant precedents can be found. Pauli was motivated by the question, initially asked by Bohr, as to why all electrons, for an atom in its ground state, were not all lying in the orbit of lowest energy W (i.e. $n = 1$). To account for this, he postulated the existence of the quantum number s, insisting on its essential two-valuedness, and he complemented this with the further postulate that, in an atom, any two electrons must differ at least in one of the four quantum numbers characterizing their individual states. By a simple enumeration, filling up the atom levels starting with those of the lowest energy, Pauli accounted for the periodic table of elements well-known to every chemistry student. This proposal, for which Pauli received the Nobel prize in 1945, is known today as the Pauli exclusion principle, and it is seen as a consequence of the fact that the electrons satisfy the Fermi-Dirac statistics. Stimulated in turn by Pauli's proposal, Uhlenbeck and Goudsmit interpreted s as a state of rotation of the electron on itself; this "spinning" electron was soon realized to be impossible to account for within the classical theory: its size and mass were much too large and the velocity at the surface of the electron was computed to be larger than the velocity of light. It was realized later that a proper account of the spin can only be given in term of the "projective" representations of the group O_3^+; to understand this, we will need the Hilbert space formulation of quantum mechanics to be presented in Chapter Eight. It turned out that the spin of the electron allowed a simpler reinterpretation of the Sommerfeld formula (73); for this reason, among others, the spin was often said, in the popular physics literature, to be of relativistic origin. While it is true that O_3^+ appears as a subgroup of the relativistic, inhomogeneous Lorentz group, and that it plays a distinguished role in the theory of the representations of the latter (see Wigner, 1939), it is known today (see e.g. Galindo & Sanchez del Rio, 1961; and Levy-Leblond, 1965) that a perfectly consistent, non-relativistic theory of the spin of the electron is possible as well.

Such were the highlights of the "old quantum mechanics". The situation, as it presented itself just before the birth of the "new quantum mechanics", in contrast with what it became afterwards, is quite properly illustrated by two quotes Jammer (1966) excerpts. The first is due to Lorentz (1923): "All this has great beauty and extreme importance, but, unfortunately we do not understand it ... quantum mechanics ... has still to be made." The other, contrasting quote is from Mackey (1963): "We should like to emphasize that in quantum mechanics the quantum rules [i.e. in particular, Sommerfeld's rule] are a deduction ... The laws of quantum mechanics ... give differential equations ... Everything else is deductible from these laws." The great divide occurred in 1925–1926 with the publication of the papers of Heisenberg, Born, Jordan and Schroedinger; a few years later, von Neumann gave the theory a firm conceptual and mathematical foundation, which allows then a consistent description of the quantum mechanics of systems with finitely many degrees of freedom. These developments are presented in the next chapter.

CHAPTER 8.
THE HILBERT SPACE FORMULATION OF QUANTUM MECHANICS

SYNOPSIS

A physical system is characterized by: its observable properties (or simply "observables"); its modes of preparation (the result of a preparation being called a "state" of the system); and a prediction rule giving, for every state ρ, the expected value of every observable A.

For instance, in the classical description of a system of N distinguishable particles in $\mathbb{R}^3$, the observables are identified with the smooth, real-valued functions on $\mathbb{R}^{6N} = T^*\mathbb{R}^{3N}$; the states are identified with probability distributions on $\mathbb{R}^{6N}$; and the prediction rule is

$$
(0) \qquad \begin{aligned} &< \rho; A > \equiv \int dq\, dp\ \rho(q, p)\, A(q, p) \\ &\text{with } q \equiv (q_1, \ldots, q_N)\ ,\ p \equiv (p_1, \ldots, p_N) \\ &\text{and } dq\, dp \equiv d^3q_1 \ldots d^3q_N\, d^3p_1 \ldots d^3p_N \end{aligned}
$$

In the Hilbert space formulation of quantum mechanics, a system of N distinguishable particles in $\mathbb{R}^3$ is described as follows: the observables are identified with the self-adjoint operators A acting within the Hilbert space $\mathcal{H} = \mathcal{L}^2_{\mathbb{C}}(\mathbb{R}^{3N}, d^{3N}q)$; the states are identified with the density-matrices ρ on $\mathcal{H}$, i.e. with the positive, trace-class operators ρ satisfying $tr\rho = 1$; and the prediction rule is:

$$
(1) \qquad < \rho; A > = tr\rho A
$$

Among the states of particular importance are the pure states, i.e. those states which cannot be decomposed into a convex combination of other states. In the classical description these are the Dirac measures, whereas in the quantum description the pure states are the one-dimensional projectors $P = P^* = P^2$. In a time-honored abuse of language any vector $\Phi \in \mathcal{H}$ satisfying $\| \Phi \| = 1$

is also called a pure state, the corresponding density-matrix being the projector

$$P_\Phi : \Psi \in \mathcal{H} \mapsto (\Psi, \Phi)\Phi \in \mathcal{H}; \tag{2a}$$

for these states (1) reduces to

$$< P_\Phi; A >= (A\Phi, \Phi). \tag{2b}$$

We shall show that behind the choice involved in writing $\mathcal{H}$ as $\mathcal{L}^2_{\mathbb{C}}(\mathbb{R}^{3N}, d^{3N}\mathbf{q})$ lies the identification of the observable "position of the n-th particle along the j-th coordinate" with the operator Q^j_n defined by (leaving aside domain questions):

$$(Q^j_n\Phi)(q^1_1, \ldots, q^j_n, \ldots, q^3_N) \equiv q^j_n\Phi(q^1_1, \ldots, q^j_n, \ldots, q^3_N). \tag{3}$$

For any $\Phi \in \mathcal{H}$ with $\| \Phi \|= 1$ and any function A of the position-observable $\mathbf{Q}_1, \ldots, \mathbf{Q}_N$ we have then:

$$< P_\varphi; A(\mathbf{Q}_1,, \ldots, \mathbf{Q}_N) >= \int d^3\mathbf{q}_1 \ldots d^3\mathbf{q}_N\, \rho(\mathbf{q}_1, \ldots, \mathbf{q}_N)\, A(\mathbf{q}_1, \ldots, \mathbf{q}_N) \tag{4a}$$

where:

$$\rho(\mathbf{q}_1, \ldots, \mathbf{q}_N) \equiv |\Phi(\mathbf{q}_1, \ldots, \mathbf{q}_N)|^2 \tag{4b}$$

can thus be interpreted (compare with equation (0) above) as a probability distribution in configuration space.

We shall also see that, in conformity with the identification (3) above, the generator of the translation of the n-th particle in the j-th direction, interpreted as the observable "momentum of the n-th particle along the j-th coordinate" is the operator defined by (leaving again aside domain questions):

$$(P^j_n\Phi)(q^1_1, \ldots, q^j_n, \ldots, q^3_N) \equiv (-i\partial_{q^j_n}\Phi)(q^1_1, \ldots, q^j_n, \ldots, q^3_N) \tag{5}$$

Three remarks should be made at this point of our preliminary overview, since they will be the object of much discussion in this chapter. Firstly, we chose in the above definition (5) natural units which are particularly well-adapted to the description of microscopic phenomena, namely units where the Planck constant $\hbar \equiv h/2\pi = 1$. Secondly, the Fourier transform $\mathcal{F}$ transforms the momentum observables into multiplication operators so that

$$|\tilde{\Phi}(\mathbf{p}_1, \ldots, \mathbf{p}_N)|^2 \equiv |(2\pi)^{-3N/2} \int d^3\mathbf{q}_1 \ldots d^3\mathbf{q}_N \exp\{-i\mathbf{p}\cdot\mathbf{q}\}\Phi(\mathbf{q}_1, \ldots, \mathbf{q}_n)|^2 \tag{6}$$

can be interpreted, in analogy with (4) above, as a probability distribution in momentum space. The absence, in quantum mechanics, of a (point-wise positive) joint probability distribution for both position and momentum (in contrast with the classical formula (0) above) will be one of the essential differences between the classical and quantum descriptions, a fact which will have to be accounted for when we will discuss the sense in which classical mechanics can be recovered from the quantum theory. Thirdly, the position- and momentum-operators defined by (3) and (5), satisfy the canonical commutation relations (with the notation $[A, B] \equiv AB - BA$):

(7a)
$$[Q^j_n, Q^k_m] = 0 = [P^j_n, P^k_m]$$
$$[P^j_n, Q^k_m] = -i\delta_{n,m}\,\delta_{j,k} I$$

which are strongly reminiscent of the classical Poisson brackets

(7b)
$$\{q^j_n, q^k_m\} = 0 = \{p^j_n, p^k_m\}$$
$$\{p^j_n, q^k_m\} = -\delta_{n,m}\,\delta_{j,k} 1$$

The Heisenberg uncertainty relations then follow immediately from the mathematical identity:

(8)
$$< \rho; (A- < \rho; A >)^2 >< \rho; (B- < \rho; B >)^2 > \geq < \rho; i[A, B] >^2$$

We shall also argue in this chapter that if the system is conservative the time-evolution is described by a one-parameter group $\{U_t | t \in \mathbb{R}\}$ of unitary operators acting on $\mathcal{H}$; the generator H of this group is interpreted as the "energy" observable of the system. In case the particles are free, except for their mutual interactions:

(9)
$$H = \sum_{n=1}^{N} \frac{1}{2M_n} |\mathbf{P}_n|^2 + V(\mathbf{Q}_1, \ldots, \mathbf{Q}_N)$$

Hence, upon writing

(10a)
$$\Phi_t(\mathbf{q}_1, \ldots, \mathbf{q}_N) \equiv (U_t\Phi)(\mathbf{q}_1, \ldots, \mathbf{q}_N)$$

we will obtain the Schroedinger equation

(10b)
$$i\partial_t\, \Phi_t(\mathbf{q}_1, \ldots, \mathbf{q}_N) = (-\sum_{n=1}^{N} \frac{1}{2M_n}\Delta_n + V(\mathbf{Q}_1, \ldots, \mathbf{Q}_N))(\Phi_t(\mathbf{q}_1, \ldots, \mathbf{q}_N)$$

Among the most convincing successes of the Hilbert space formalism of quantum mechanics are applications to the structure of atoms and molecules, and to the scattering of a particle by a potential, as we shall see towards the end of this chapter which will, however, describe first the main conceptual and mathematical aspects of the formalism itself.

Section 1. THE HEISENBERG MATRIX FORMULATION.

The three seminal papers on the matrix formulation of quantum mechanics appeared within a span of less than four months. They were written from Göttingen by Heisenberg (1925 P), Born & Jordan (1925) and Born, Heisenberg & Jordan (1925); together with the papers of Heisenberg (1925 M) and of Born & Wiener (1925–1926), they record the beginnings of the process in the course of which Hilbert space concepts came to be recognized as the proper language of quantum theory. While it is true that these authors were guided by an analogy with the way Fourier analysis is used to trace the time evolution in classical theory, we should emphasize that this is by way of analogy only: indeed one of the main successes of the emerging Hilbert space formalism was to make quantum theory much more independent from the classical computational support previously required (see Chapter Seven) by the Bohr-Sommerfeld quantization rules. This advance was accomplished through an almost accidental detour via the theory of matrix algebras. Until then, the latter had essentially been of interest to pure mathematicians only; there were, however, a few notable exceptions amongst mathematical physicists. The first of these exceptionally broad mathematical physicists was Born himself, who, some twenty years earlier, had learned about the theory from his teacher J. Rosanes, who in turn had been a student of Frobenius. Another mathematical physicist who knew about matrix algebras was Courant; he had been writing Hilbert's lecture notes for what became a classic, the *Methoden der Mathematischen Physik* which had just appeared in 1924 and was soon recognized to contain some of the most pertinent tools necessary for the elaboration of the new quantum theory. One of Courant's assistants was Jordan who accidentally learned of quantum mechanics from Born in a train trip from Göttingen to Hanover (see M. Jammer, 1966)! He is the Jordan of the "Jordan algebras" which he invented a few years later, again in connection with an attempt towards a cleaner axiomatization of quantum theory. More will be said on the latter in Chapter Nine, a development Hilbert did not foresee: he was fond of pointing out that the detour via matrix algebras could have been considerably shortened had Born and his colleagues recognized, as Hilbert did, that matrices had previously appeared in connection with differential equations (see Section 2). It is, therefore, worth looking closer at this historically fortunate detour, the object of this section.

In his 1925 paper, Heisenberg was already very explicit on his conceptual motivations. *Firstly,* he stressed the importance—already hinted to by Bohr—of restricting to *observable* properties all the formal assumptions of the theory. The language then had to become much sharper concerning such things as the position and velocity of an electron; in particular, the use of putative concepts, such as the orbit of an electron in the atom, was to be banished in so far as the corresponding properties could not be observed. It is in connection with this requirement that Heisenberg arrived at his famous *multiplication law,* which we shall describe below. *Secondly,* Heisenberg wanted to bring the

Bohr-Sommerfeld quantization rule into the formalism itself; in particular, Heisenberg wanted to find a way to avoid the two-step process consisting (as we saw in Chapter Seven) in having *first* to solve a problem classically and having *then* to impose an ad hoc quantization rule, the application of which required subtle variations from one problem to the next. For this part of the argument we shall follow here the presentation of Born & Jordan, although the core of it is sometimes called the Heisenberg *canonical commutation relation. Thirdly,* Heisenberg aimed at building up a formalism in which complex systems could be handled according to some fixed rules established once and for all; for instance, the "complex systems" Heisenberg mentioned as being on his mind were atoms with several electrons, and atoms placed in arbitrary electric and magnetic fields. The latter problem was solved by Pauli (1926) for the hydrogen atom using the formalism and methods just proposed by Heisenberg, Born & Jordan. The historical impact of this result is underscored in Born (1955): "The first non-trivial and physically important application of quantum mechanics was made ... by Pauli ... From this moment there was no longer any doubt about the correctness of the theory." In his first paper, Heisenberg discussed simpler problems: the "anharmonic oscillator" $\ddot{x} + \omega_0^2 x^2 + \lambda x^3 = 0$, and the "rotator" which mimics, quantum mechanically, the classical picture of an electron revolving at constant distance a and with constant angular velocity ω around the nucleus. These examples are discussed today in every introductory physics text on quantum mechanics, and there is thus little point repeating these computations here, although the reader should try solving these two problems as exercises while he proceeds through this section. We mention now only the results. The energy levels of the oscillator, up to second order in λ are given by

$$(1) \qquad W_n = \frac{h}{2\pi}\omega_0(n + \frac{1}{2}) \quad \text{with} \quad n \in \mathbb{Z}^+$$

The energy levels of the rotator are given by:

$$(2) \qquad W_n = \frac{h^2}{8\pi^2 m a^2}(n^2 + n + \frac{1}{2}) \quad \text{with} \quad n \in \mathbb{Z}^+$$

Beyond these three general requirements, Heisenberg was motivated by a more precise problem: for the prediction of the relative intensities of the spectral lines, it is necessary to have a proper quantum-theoretical description of the multipole expansion of the radiation emitted by a moving electron. He recognized that a solution of this problem depends on the answer to the following more general question, which he termed "kinematic". Given a quantum-theoretical description $\{x(t)\}$ of a (classically) observable quantity $x(t)$, what should be the quantum-theoretical description of $x(t)^2$? To explore this question, Heisenberg restricted his attention to the radiation emitted by the electron as it passes—according to Bohr's theory of the hydrogen atom (see Chapter Seven)—from a stationary state, labeled n, to another stationary state, labeled

$(n-\alpha)$. Two observable properties characterize this radiation: its frequency and its amplitude, both depending on two variables, n and $(n-\alpha)$.

According to Bohr's theory (see 7.3.10) the frequency is given by

$$\nu(n,n-\alpha)=\frac{1}{h}[W(n)-W(n-\alpha)] \tag{3}$$

and evidently satisfies

$$\nu(n,n-\alpha)+\nu(n-\alpha,n-\beta)=\nu(n,n-\beta) \tag{4}$$

In analogy with the expression

$$\mathbb{R}e\{x_\alpha(n)e^{2\pi i\nu(n)\alpha t}\} \tag{5}$$

obtained from a classical Fourier analysis of the radiation, Heisenberg proposed the following "quantum-theoretical" representation for the amplitude of the radiation in order to emphasize the separate dependence on n and $(n-\alpha)$:

$$\mathbb{R}e\{x(n,n-\alpha)e^{2\pi i\nu(n,n-\alpha)t}\} \tag{6}$$

where x describes both polarization and phase and is thus a vector in $\mathbb{C}^3$.

Addressing now the question of how to represent the square of the amplitude, Heisenberg proposed that the "simplest and most natural" way to represent this quantity while preserving (4) is to write for it:

$$\mathbb{R}e\{\xi(n,n-\beta)e^{2\pi i\nu(n,n-\beta)t}\} \tag{7a}$$

with

$$\begin{aligned}&\xi(n,n-\beta)e^{2\pi i\nu(n,n-\beta)t}=\\&\Sigma_\alpha[x(n,n-\alpha)e^{2\pi i\nu(n,n-\alpha)t}]\times[x(n-\alpha,n-\beta)e^{2\pi i\nu(n-\alpha,n-\beta)t}]=\\&\Sigma_\alpha\, x(n,n-\alpha)x(n-\alpha,n-\beta)e^{2\pi i\nu(n,n-\beta)t}\end{aligned} \tag{7b}$$

and thus

$$\xi(n,n-\beta)=\sum_\alpha x(n,n-\alpha)x(n-\alpha,n-\beta) \tag{7c}$$

The content of Heisenberg's *kinematic postulate* is that this multiplication rule is universal, i.e. can be iterated to give all powers of any observable quantity; moreover, Heisenberg's proposal extended to seeing in (7c) the germ of a rule to multiply any two observable quantities. He, however, pointed out the difficulty that, whereas in the classical theory $x(t)y(t)=y(t)x(t)$, his multiplication rule would, in general, give different results for the quantum descriptions

$\{x(t)y(t)\}$ and $\{y(t)x(t)\}$. This difficulty, however, did not affect his subsequent analysis since he only needed, in that paper, a quantum description of classical observables of the form $x(t)\dot{x}(t)$; for such observables he had a remarkable intuition, namely, that they appear as time derivative of $\frac{1}{2}x(t)^2 = \frac{1}{2}x(t)x(t)$ and can thus be seen as symmetric products of the form $\frac{1}{2}(x(t)\dot{x}(t) + \dot{x}(t)x(t))$. We shall exploit this remark later on in this section and in Chapter Nine.

To Born belongs the credit of recognizing that Heisenberg's "inspired" multiplication law is in fact the multiplication law of matrices. With that insight, he set out to stream-line Heisenberg's whole argument and to present it in a mathematically more acceptable form; this he did in his paper with Jordan (loc. cit.) and we shall follow here their presentation. In that paper Born & Jordan distinguish three separate steps in Heisenberg's programme; these are: a "kinematical postulate", a "dynamical postulate" and a "quantization postulate".

1. Kinematical postulate. *A physical system (with one degree of freedom) is to be described by representing the position $q(t)$ and the momentum $p(t)$ at time t by Hermitian matrices over $\mathbb{C}$*

$$q(t) = \{q(n,m)e^{2\pi i\nu(n,m)t}\} \tag{8a}$$
$$p(t) = \{p(n,m)e^{2\pi i\nu(n,m)t}\} \tag{8b}$$

where the real-valued function ν satisfies

$$\nu(n,m) + \nu(m,k) + \nu(k,n) = 0 \quad \forall \quad n,m,k \tag{9}$$
$$\nu(n,m) \neq 0 \quad \textit{whenever} \quad n \neq m \tag{10}$$

Remarks. (i) We will later have to make this postulate more precise, both physically and mathematically. First of all, the range of the indices $n, m, k, \ldots$ will have to be specified; as in Heisenberg's paper they are here to be thought of as labels for the "stationary states" of the system, a notion which will have to be made more precise. Secondly, Born & Jordan were aware of the fact that their "matrices" are somewhat peculiar: not only have they infinitely many entries (i.e. $n, m, k \ldots$ range over an infinite index set) but they are, moreover, unbounded (in a sense which we will specify later on). At this exploratory stage of the theory, Born & Jordan decided to ignore the mathematical problems that occur when one deals with such unbounded objects, and to proceed, nonetheless, at a heuristic level. We now follow them in this endeavor, with, however, the understanding that things will have to be tightened up; we shall show how this can be done in Sections Two and Three below.

(ii) In the spirit of Bohr's theory, the real numbers $\nu(n,m)$ are meant to be the quantum-theoretical frequencies belonging to the transition between stationary states labeled n and m. Condition (9) implies (upon writing $n =$

$m = k$)

(11) $$\nu(n,n) = 0 \quad \forall \quad n$$

and thus (upon writing $m = k$)

(12) $$\nu(n,m) = -\nu(m,n) \quad \forall \quad n,m$$

With n_0 fixed but arbitrary, define

(13) $$\varphi(n) \equiv \nu(n, n_0)$$

Condition (9), together with its consequence (12), then implies

(14) $$\nu(n,m) = \varphi(n) - \varphi(m)$$

Anticipating on things to come, Born & Jordan wrote this relation in the form

(15) $$\nu(n,m) = \frac{1}{h}[W_n - W_m]$$

where h is the Planck constant. (15) has the mathematical form of Bohr's assumption (7.3.10), with $\nu_{nm} = \nu(m,n)$. From an epistemological point of view, however, the content of these two formulas is different. Bohr starts with the W_n as the energy of the stationary state (or "orbit") labeled n, while here (15) appears only as a definition of $W_n = h\varphi(n)$ (up to an irrelevant additive constant, linked to our arbitrary choice of n_0); the interpretation of W_n as the energy of a stationary state will appear here as a consequence of the theory.

(iii) There is a slight redundancy in the above kinematic postulate. Indeed (12) implies that $q(t)$ and $p(t)$ are Hermitian for all times t if and only if they are Hermitian at one time, say t_0. The condition that these matrices be Hermitian is now equivalent (through 12) to the condition

(16) $$\left.\begin{array}{l} q(n,m)^* = q(m,n) \\ p(n,m)^* = p(m,n) \end{array}\right\} \quad \forall \quad n,m$$

where * denotes the complex conjugation. It should also be pointed out that Born & Jordan assumed that both q *and* p are Hermitian matrix valued functions of t, whereas Heisenberg initial paper concentrated on q only.

(iv) Define now the matrix

(17) $$\Phi \equiv \{\delta(n,m)\varphi(n)\}$$

where δ is the Kronecker symbol. For any matrix-valued function g on $\mathbb{R}$, of the form

(18) $$g(t) \equiv \{g(n,m)e^{2\pi i\nu(n,m)t}\}$$

we obtain, upon using (14),

(19) $$\dot{g}(t) = 2\pi i\,[\Phi, g(t)]$$

where the commutator $[A,B]$ of any two matrices A and B is defined by:

(20) $$[A,B] \equiv AB - BA$$

This concludes the Born-Jordan formalization of Heisenberg's original "kinematical postulate".

The dynamics then enters in a second postulate, with the introduction of a "Hamiltonian function" H of p and q, from which equations of motion can be defined. To do this, Born & Jordan needed a few computational preliminaries, which we now briefly review.

First of all, Born & Jordan needed a notion of a matricial function of several matricial variables. At the heuristic level at which they worked, one simply gets by with formal polynomials. Two problems are temporarily ignored. The first one is again that one works with unbounded objects; for that we defer again to Sections Two and Three below. The second problem is that the product AB of two Hermitian matrices A and B is in general not Hermitian, unless A and B commute, i.e. unless $[A, B] = 0$; we shall come back to this problem later on in the present section.

Secondly, Born & Jordan needed the notion of a partial derivative of a matricial function of several matricial variables, so that they could write a quantum analog for the classical Hamilton equations of motion. Let A be a matrix and "define" $tr\, A = \Sigma_n A(n,n)$; here again this presents some mathematical difficulties: in particular, the sum has to be absolutely convergent for the sequel to make proper sense. Let us, nevertheless, proceed formally, and suppose A to be a function (e.g. a polynomial) of the n matrices $x_1, x_2, \ldots, x_n$. The formal partial derivative of A with respect to x_k is defined as the matrix

$$\frac{\partial A}{\partial x_k} \equiv \{\frac{\partial A}{\partial x_k}(m,n)\} \equiv \{\frac{\partial\, trA}{\partial x_k(n,m)}\} \tag{21}$$

At first sight this may seem to be a strange definition; it is, therefore, useful to explore some of its formal properties. With λ and μ two complex numbers; A and B two matrices, we have

$$\frac{\partial}{\partial x_k}(\lambda A + \mu B) = \lambda \frac{\partial A}{\partial x_k} + \mu \frac{\partial B}{\partial x_k} \tag{22}$$

and

$$\frac{\partial}{\partial x_k}(AB) = \frac{\partial}{\partial x_k}(BA) \tag{23}$$

The following two properties are verified by inspection from the definition (21) and the product rule for matrices. With N a positive integer and x a matrix

$$\frac{\partial}{\partial x}(x^N) = N x^{N-1} \tag{24}$$

With $x_1, x_2, \ldots, x_N$ denoting N (independent!) matrices:

(25) $$\frac{\partial}{\partial x_k}(x_1 x_2 \ldots x_N) = x_{k+1} \ldots x_N x_1 \ldots x_{k-1}$$

In fact, the reader who has checked explicitly the above two formulas will recognize that his argument can be generalized to check the following property of which these two formulas are particular cases. Let M and N be two positive integers; $I \equiv \{i \in \mathbb{Z} | 1 \leq i \leq M\}$ and $J = \{j \in \mathbb{Z} | 1 \leq j \leq N\}$; $\ell : i \in I \mapsto \ell_i \in J$; $\{x_j | j \in J\}$ be a collection of (independent!) matrices; and

(26a) $$A \equiv \prod_{i=1}^{M} x_{\ell_i} \equiv x_{\ell_1} x_{\ell_2} \ldots x_{\ell_M}$$

Then

(26b) $$\frac{\partial A}{\partial x_k} = \sum_{i=1}^{M} \delta_{\ell_i, k} \Big(\prod_{j=i+1}^{M} x_{\ell_j} \Big) \Big(\prod_{j'=1}^{i-1} x_{\ell_{j'}} \Big)$$

Note that, together with (22), (26) could be taken as an alternate definition of the partial derivative. The following consequence of (26) will be useful in the sequel. For any polynomial A in two matrices x and y we have (formally, at least!)

(27) $$[x, \frac{\partial A}{\partial x}] + [y, \frac{\partial A}{\partial y}] = 0$$

To see this, remark first that the linearity (22) makes it sufficient to prove (27) for monomials, i.e. for A of the form

(28) $$A \equiv \prod_{j=1}^{N} (x^{m_j} y^{n_j})$$

From (26) one obtains

(29a) $$\frac{\partial A}{\partial x} = \sum_{j=1}^{N} \sum_{k=0}^{m_j - 1} x^{m_j - 1 - k} y^{n_j} A_j x^k$$

(29b) $$\frac{\partial A}{\partial y} = \sum_{j=1}^{N} \sum_{k=0}^{n_j - 1} y^{n_j - 1 - k} A_j x^{m_j} y^k$$

with

(29c) $$A_j = \prod_{\ell=j+1}^{N} (x^{m_\ell} y^{n_\ell}) \prod_{\lambda=1}^{j-1} (x^{m_\lambda} y^{n_\lambda})$$

From (29a) one then obtains

(30a) $$[x, \frac{\partial A}{\partial x}] = \sum_{j=1}^{N} (x^{m_j} y^{n_j} A_j - y^{n_j} A_j x^{m_j})$$

and similarly from (29b):

(30b) $$[y, \frac{\partial A}{\partial y}] = \sum_{j=1}^{N} (y^{n_j} A_j x^{m_j} - A_j x^{m_j} y^{n_j})$$

From (30a), (30b) and (29c), (27) follows for every monomial of the form (28) and thus, as we already remarked, for all polynomial in x and y.

Recall now (see Chapter One) that in classical mechanics, the Lagrangean L is defined by

$$L \equiv p\dot{q} - H(p,q) \tag{31}$$

and that the classical Hamiltonian equations of motion

$$\dot{q} = \frac{\partial H}{\partial p} \quad ; \quad \dot{p} = -\frac{\partial H}{\partial q} \tag{32}$$

are obtained from the variational principle consisting in imposing that for all $t_2 > t_1$:

$$S(t_1, t_2) \equiv \int_{t_1}^{t_2} dt\, L \tag{33}$$

be an extremum, a condition written symbolically

$$\delta \int dt\, L = 0 \tag{34}$$

In analogy with this classical variational principle Born & Jordan proposed the following (formal) quantum variational principle; the quantum kinematical postulate formulated above is presupposed in the statement of this second postulate.

2. Dynamical postulate. *A function H of the matrices p and q is said to be a Hamiltonian for the system if the equations of motions can be obtained by imposing that trL be an extremum, where*

$$L \equiv p\dot{q} - H(p,q) \tag{35}$$

Remarks.

(i) This postulate could be seen as a definition of the Hamiltonian H and of the Lagrangean L; what one has in mind then is that q and p are observables, so that one is able to determine empirically the differential equation their evolution satisfies, and thus decide whether the system is Hamiltonian, i.e. whether a function H exists, which satisfies the postulate. What Born & Jordan, however, seem to have had in mind—as one can appreciate by reading the examples they gave in their paper—is that a physical system is characterized by its Hamiltonian. For instance, a harmonic (resp. anharmonic) oscillator is characterized by the prescription that $H(p,q) = \frac{1}{2m}p^2 + \frac{1}{2}\omega_0^2 q^2$ (resp. $\frac{1}{2m}p^2 + \frac{1}{2}\omega_0^2 q^2 + \frac{1}{3}\lambda q^3$), with m, ω_0 and λ real numbers.

(ii) This postulate is formal in the sense that, given H, it may very well happen that $tr\,L$ is not properly defined. Following Born & Jordan, we temporarily ignore this difficulty and proceed to explore the formal consequences of the postulates, postponing to Section Three the presentation of a mathematically consistent theory.

(iii) The first formal consequence of the dynamical postulate is that, since it is a variational principle, we can compute where the extrema of $tr\,L$ occur by separately equating to zero the partial derivative with respect to each matrix element. From (35) and the form of p and q given by the kinematical postulate, we have:

$$tr\,L = \sum_{m,n} p(m,n)\, 2\pi i\, \nu(n,m)\, q(n,m) - tr\,H \tag{36}$$

The variational principle gives thus

$$2\pi i\, \nu(n,m)\, p(m,n) - \frac{\partial\, tr\,H}{\partial q(n,m)} = 0 \tag{37a}$$

$$2\pi i\, \nu(n,m)\, q(n,m) - \frac{\partial\, tr\,H}{\partial p(m,n)} = 0 \tag{37b}$$

Recalling now (21) and (12) we can rewrite (37) in the form

$$(-\dot{p})(m,n) - \frac{\partial H}{\partial q}(m,n) = 0 \tag{38a}$$

$$(\dot{q})(n,m) - \frac{\partial H}{\partial p}(n,m) = 0 \tag{38b}$$

i.e., since these equations hold for all n and m

$$\dot{q} = \frac{\partial H}{\partial p} \quad ; \quad \dot{p} = -\frac{\partial H}{\partial q} \tag{39}$$

Hence, the quantum equations of motion are formally identical with the classical Hamilton equations (32). This formal analogy was, in fact, the motivating reason behind the a priori mysterious definition (21) of the partial derivatives of a matricial function with respect to its matricial variables. In practical applications—such as those which were discussed in the early days of this new quantum mechanics—the partial derivatives appearing in the RHS of equations (39) are more comfortably computed via the alternate definition of the partial derivative stemming from (26); in most of the cases then discussed, the particular cases (24) and (25) of (26) were already adequate. We shall see below an even more effective way to compute the equations of motion. Nevertheless, the dynamical postulate stated above throws some light on the original motivation for the formalism to which we will arrive.

The third, and last, building block in the Born & Jordan analysis of Heisenberg's paper was the realization that the Bohr-Sommerfeld quantum condition, as used by Heisenberg, amounts to the following prescription.

3. QUANTIZATION POSTULATE. *With q and p represented as prescribed by the kinematical postulate, one has*

$$[p, q](n, n) = \frac{h}{2\pi i} \quad \forall \quad n \tag{40}$$

where $[p, q] \equiv pq - qp$ and h is the Planck constant.

Remarks.

(i) With the matrix multiplication rule written explicitly, (40) reads

$$\sum_k \{p(n, k)\, q(k, n) - q(n, k)\, p(k, n)\} = \frac{h}{2\pi i} \tag{41}$$

which is the form in which Born & Jordan wrote their postulate.

(ii) In the particular, but physically important, case where $H(p, q) = \frac{1}{2m}p^2 + V(q)$, (39) gives $p = m\dot{q}$. From the form of q assumed in the kinematical postulate, (41) reduces then to

$$\sum_k \nu(k, n)\, |q(n, k)|^2 = \frac{h}{8\pi^2 m} \tag{42}$$

which is the original form in which Heisenberg wrote his version of the Bohr-Sommerfeld condition.

(iii) Born & Jordan recognized immediately that their condition (40) rules out the possibility that q and p be finite matrices. Suppose indeed that they were finite matrices, i.e. that the cardinality of the index set $\{n\}$ were finite, say N; since $tr\, AB = tr\, BA$, we would then have $0 = tr[p, q] = \Sigma_n [p, q](n, n) = Nh/2\pi i$. In a footnote, they even made a remark to the effect that p and q could not, moreover, be represented by bounded operators; we shall prove this in Section Three.

(iv) The statement of the above "quantization" postulate is free from any explicit reference to classical mechanics, a distinctive improvement over the earlier versions of the Sommerfeld quantization rule.

(v) We shall see in Section Three that (40) can be interpreted as a consequence of the definition of the momentum p as the generator of space translations. In the present context, however, this quantization postulate must only be regarded as a seminal axiom, the value of which is to be judged by the fruit it will bear. It is nonetheless interesting, from an epistemological point of view, to see how the Sommerfeld quantization rule provided an inductive guide for the formulation of this new postulate. It is, however, to be emphasized that the former is less general than the latter, so that no true mathematical derivation of the new quantization postulate can really be presented starting only from the Sommerfeld rule. Here is how Born & Jordan proceeded.

The Sommerfeld rule (7.3.19) reads

$$\oint dq\, p = hn \tag{43}$$

where h is the Planck constant, and n is an integer. Classically, one could write p and q as Fourier series

$$p = \sum_k p_k \, e^{2\pi i\nu kt} \qquad q = \sum_k q_k \, e^{2\pi i\nu kt} \tag{44a}$$

and thus

$$J \equiv \oint dq\, p = \int_0^{\frac{1}{\nu}} dt\, \dot{q}\, p = 2\pi i \sum_k k\, q_k\, p_{-k} \tag{45}$$

i.e.

$$1 = 2\pi i \sum_k k \frac{\partial}{\partial J}(q_k\, p_{-k}) \tag{46}$$

Taking stock of (43) and recalling that, via the kinematical postulate, to the classical observables p and q correspond the matrices $\{p(n,k)\}$ and $\{q(n,k)\}$, Born & Jordan boldly postulated *by fiat* that (41) was to be the quantum formula corresponding to the classical formula (46). They did not give the details, some of which were suggested in Heisenberg's paper merely by way of analogy to earlier works of Thomas (1925) and Kuhn (1925). For this correspondence, inducing a quantum formula from a classical one, one can only propose some very loose a priori justifications, rooted ultimately in Bohr's Assumption 5 (see Section 7.3); such arguments, however, are not compelling, and they cannot be, would it only be for the simple reason that (41) is more general than its antecedents: a truly new principle was postulated here.

This concludes the remarks we wanted to make specifically on the third postulate of Born & Jordan. We next present some (formal) consequences of the three postulates taken together as the germ of a theory.

Proposition. *Assume the three postulates stated above. Then (i) we have*

$$[p, q] = -i\hbar I \tag{47}$$

where $\hbar = h/2\pi$, h is the Planck constant, and I is the unit matrix $\{\delta(n,m)\}$.

(ii) With g any polynomial in p and q, and H of the form

$$H(p,q) = H_1(p) + H_2(q) \tag{48}$$

where H_1 and H_2 are polynomials, we have

(49) (a) $\dot{g} = \frac{i}{\hbar}[H, g]$

(b) H is a constant of the motion

(c) H is a diagonal matrix

(50) (d) $h\nu(n,m) = H(n,n) - H(m,m)$

Proof: As before, the argument is only formal: we do not worry about convergence when we write down the infinite sums involved in computing products of infinite matrices; in modern language, this amounts to ignoring all the domain questions which appear when one deals with unbounded operators (see Sections Two and Three).

To prove (i), let $A \equiv [p,q] \equiv pq - qp$. We have then $\dot{A} = \dot{p}q + p\dot{q} - q\dot{p} - \dot{q}p = [p,\dot{q}] - [q,\dot{p}]$. From the equations of motion (39), this can be rewritten

$$\dot{A} = [p, \frac{\partial H}{\partial p}] + [q, \frac{\partial H}{\partial q}] \tag{51}$$

We now use (27) to conclude that $\dot{A} = 0$. Recall now from (19) that $\dot{A} = 2\pi i\,[\Phi, A]$. We, therefore, have $[\Phi, A] = 0$. We now use—for the first time in the theory—the condition (10) of the kinematical postulate, which says that Φ is non-degenerate. Since Φ is, by its definition (17) diagonal, we conclude from the above argument that $A \equiv [p,q]$ is diagonal. The desired conclusion (47) then follows from (40): we have a diagonal matrix, the diagonal elements of which are all equal to one another.

From (47) one obtains by recursion

$$[p^n, q] = -in\hbar p^{n-1} \tag{52a}$$
$$[q^n, p] = +in\hbar q^{n-1} \tag{52b}$$

Upon using (24) we can rewrite (52) in the form

$$[p^n, q] = -i\hbar \frac{\partial}{\partial p}(p^n) \tag{53a}$$
$$[q^n, p] = +i\hbar \frac{\partial}{\partial q}(q^n) \tag{53b}$$

With H as in (48) we conclude, by linearity, from (53):

$$[H, q] = [H_1(p), q] = -i\hbar \frac{\partial H_1}{\partial p} = -i\hbar \frac{\partial H}{\partial p} \tag{54a}$$

and

$$[H, p] = [H_2(q), p] = +i\hbar \frac{\partial H_2}{\partial q} = +i\hbar \frac{\partial H}{\partial q} \tag{54b}$$

From (54) and the equations of motion (39) we obtain

$$\dot{q} = \frac{i}{\hbar}[H, q] \tag{55a}$$
$$\dot{p} = \frac{i}{\hbar}[H, p] \tag{55b}$$

Upon using the fact that the operations $g \to \dot{g}$ and $g \mapsto [H, g]$ are linear, and that they are *derivations*, i.e. that they satisfy

$$D(g_1 g_2) = D(g_1)\, g_2 + g_1\, D(g_2) \tag{56}$$

we obtain (49) from (55).

In the particular case where $g = H$, (49) implies that $\dot{H} = 0$, i.e. H is a constant of the motion.

In the course of proving (i), we already showed that every constant of the motion is diagonal, so H is indeed a diagonal matrix.

As a consequence of this, we can rewrite (49) in the form

$$2\pi i\, \nu(n, m)\, g(n, m) = \frac{i}{\hbar}\{H(n, n) - H(m, m)\}\, g(n, m) \tag{57}$$

and thus, since $g(n, m)$ are arbitrary (50) follows from (57). □

Remarks.

(i) The conclusions (ii b-d) of the proposition confirm Heisenberg's original interpretation of the index set $\{n\}$ as labelling the stationary energy states, a remark which we will make more precise in Section Two. We should already notice, however, that with $W_n \equiv H(n, n)$ one recovers, as a consequence of the present postulates, the old quantum condition (7.3.10) proposed by Bohr.

(ii) Born & Jordan immediately recognized the central importance of (47) which they called the "sharpened" or the "exact" quantum condition. Today we refer to this condition as the *canonical commutation relation* (for one degree of freedom), or also the "'Heisenberg commutation relation"; the latter attribution, however, is placed in proper historical perspective by Born (1956) who reminisced: "I shall never forget the thrill I experienced when I succeeded in condensing Heisenberg's ideas on quantum conditions in the mysterious equation $pq - qp = h/2\pi i$." While this condition was derived here upon using explicitly the restriction (10), Born & Jordan properly pointed out in their paper that "all further conclusions rest on this condition", namely (47). The subsequent developments of the theory fully vindicated this foresight: (47) holds true even when (10) does not.

(iii) The restriction (48) imposed in the statement of the proposition is often physically meaningful. Mathematically (49) cannot be true if one takes for H any arbitrary polynomial in p and q. Consider for instance, the monomial

$$H(p, q) = p^2 q \tag{58}$$

We have then

$$\dot{q} = \frac{\partial H}{\partial p} = pq + qp \tag{59}$$

but

$$\frac{i}{\hbar}[H, q] = 2pq \tag{60}$$

One may immediately object that the matrix (58) is not Hermitian; this is easily cured by replacing (58) by symmetrized products

$$(61)\qquad \begin{aligned} H(p,q) &= \frac{1}{2}(p^2q+qp^2) \\ &= \frac{1}{3}(p^2q+pqp+qp^2) \end{aligned}$$

Since the partial derivative is a cyclic operation (see 23) the Hamiltonian (61) gives, through (39), the same equations of motion as those one would obtain from the function (58); however, the Hamiltonian (61) now satisfies (49).

(iv) The above remark suggests immediately the following question: What is the matrix representing, in the quantum theory, a given function of the classical variables p and q ? We should warn the reader that this question has been the object of perennial debates, as it truly lies at the core of the interpretation of the theory. It is, therefore, interesting to examine here how Born & Jordan went about this problem; in fact, we shall later in this section use some of their computations to give a proof of a "no-go" result. The solution which they proposed can be described as follows.

One first notices, by recursion from (53), that

$$(62a)\qquad [p^m,q^n] = -i\hbar m \sum_{k=0}^{n-1} q^{n-1-k}\, p^{m-1}\, q^k$$

$$(62b)\qquad [p^m,q^n] = -i\hbar n \sum_{\ell=0}^{m-1} p^{m-1-\ell}\, q^{n-1}\, p^\ell$$

so that

$$(62c)\qquad \begin{aligned} &\frac{1}{n}\sum_{k=0}^{n-1} q^{n-1-k}\, p^{m-1}\, q^k \\ &= \frac{1}{m}\sum_{\ell=0}^{m-1} p^{m-1-\ell}\, q^{n-1} p^\ell \\ &= \frac{i}{\hbar mn}[p^m,q^n] \end{aligned}$$

Born & Jordan proposed then to associate to the classical monomial p^mq^m the following Hermitian matrix as its quantum representative:

$$(63)\qquad H(p,q) \equiv \frac{i}{\hbar}\frac{1}{(m+1)(n+1)}[p^{m+1},q^{n+1}]$$

On the one hand, since the partial derivative is a cyclic operation, we can use (62) to conclude that

$$(64a)\qquad \dot q \equiv \frac{\partial H}{\partial p} = \frac{\partial}{\partial p}(p^mq^n)$$

$$(64b)\qquad \dot p \equiv -\frac{\partial H}{\partial q} = -\frac{\partial}{\partial q}(p^mq^n)$$

The RHS of (64) can be computed from (29) and gives

$$\dot{q} = \sum_{\ell=0}^{m-1} p^{m-\ell-1}\, q^n\, p^\ell \tag{65a}$$

$$\dot{p} = -\sum_{k=0}^{n-1} q^{n-k-1}\, p^m q^k \tag{65b}$$

Upon using again (62) one can rewrite (65) in the form

$$\dot{q} = \frac{i}{\hbar}\frac{1}{(n+1)}[p^m, q^{n+1}] \tag{66a}$$

$$\dot{p} = -\frac{i}{\hbar}\frac{1}{(m+1)}[p^{m+1}, q^n] \tag{66b}$$

On the other hand, we can use (63) to compute, via the Jacobi identity and (53):

$$\frac{i}{\hbar}[H, q] = \frac{i}{\hbar}\frac{1}{n+1}[p^m, q^{n+1}] \tag{67a}$$

$$\frac{i}{\hbar}[H, p] = \frac{i}{\hbar}\frac{1}{m+1}[q^n, p^{m+1}] \tag{67b}$$

Hence, upon comparing (66) and (67):

$$\dot{q} = \frac{i}{\hbar}[H, q] \quad \text{and} \quad \dot{p} = \frac{i}{\hbar}[H, p] \tag{68}$$

Finally, by the same reasoning as used in the proof of the proposition, we conclude that the Hamiltonian (63) satisfies (49) again. This conclusion extends by linearity to a formal correspondence rule giving the quantum Hamiltonian that corresponds to an arbitrary polynomial function of the classical variables p and q.

With this generalization, we close our discussion of the proposition.

Before proceeding any further, let us construct explicitly an example for the matrices p, q and H we are considering. Consider the linear space over $\mathbb{C}$, the elements of which are the sequences $\Psi : n \in \mathbb{Z}^+ \mapsto \Psi(n) \in \mathbb{C}$ where $\Psi(n) = 0$ except for a finite number of values of n. A basis $\{\Psi_m | m \in \mathbb{Z}^+\}$ for this space is given by the sequences of the form $\Psi_m(n) = \delta(m, n)$. On this space, define the linear operator a by

$$(a\Psi)(n) \equiv [(n+1)\hbar]^{1/2}\Psi(n+1) \tag{69}$$

Notice in particular that

$$a\Psi_0 = 0 \tag{70}$$

and

(71) $$a\Psi_n = (n\hbar)^{1/2}\Psi_{n-1} \quad \forall \quad n \in \mathbb{Z}^+ \setminus \{0\}$$

Define similarly the linear operator a^* by

(72) $$(a^*\Psi)(n) \equiv \begin{cases} 0 & \text{for } n = 0 \\ (n\hbar)^{1/2}\Psi(n-1) & \text{for } n \in \mathbb{Z}^+ \setminus \{0\} \end{cases}$$

Notice also

(73) $$a^*\Psi_n = [(n+1)\hbar]^{1/2}\Psi_{n+1} \quad \forall \quad n \in \mathbb{Z}^+$$

From (69) and (72), or equivalently from (70), (71) and (73) one obtains then

(74) $$[a, a^*] = \hbar I$$

where I is the identity operator.

From the operators a and a^* we construct now the linear operators

(75a) $$p \equiv \frac{i}{\sqrt{2}}(a^* - a)$$

(75b) $$q \equiv \frac{1}{\sqrt{2}}(a^* + a)$$

From (74) and (75) we then obtain

(76) $$[p, q] = -i\hbar I$$

Finally for every operator A on our space (of finite sequences) we define the matrix $A \equiv \{A(n, m)\}$ by

(77) $$A\Psi_n = \sum_m A(n, m)\Psi_m$$

We obtain in this manner two matrices $p \equiv \{p(n, m)\}$ and $q \equiv \{q(n, m)\}$ which satisfy (47) as a direct consequence of (76). The reader should write down these matrices explicitly, and verify that *neither* $|p(n, m)|$ *nor* $|q(n, m)|$ is bounded uniformly in n and m.

The physical significance of this representation is that the Hamiltonian of the harmonic oscillator

(78) $$H \equiv \frac{1}{2}(p^2 + q^2) = a^*a + \frac{1}{2}\hbar$$

satisfies

(79) $$H\Psi_n = \epsilon_n \Psi_n$$

with

$$\epsilon_n = (n + \frac{1}{2})\hbar \tag{80}$$

for all $n \in \mathbb{Z}^+$. (79) states that the Hamiltonian is diagonal in the basis $\{\Psi_n | n \in \mathbb{Z}^+\}$ we have chosen, and (80) states that its eigenvalues are non-degenerate.

Mathematically, the significance of this representation can be seen as follows. Let X be a matrix such that

$$[X, p] = 0 = [X, q] \tag{81}$$

From this, and from (78) we conclude that

$$[X, H] = 0 \tag{82}$$

Since H is diagonal and non-degenerate (see 79 and 80) we already know (see the proof of the proposition) that X is diagonal. Moreover, (81) implies (and, in fact, is equivalent to)

$$[X, a] = 0 = [X, a^*] \tag{83}$$

From the fact that X is diagonal, we conclude then, upon making use of (73) (or of 70–71) that all diagonal entries of X are equal, i.e. since X is diagonal, X is a multiple of the identity matrix, that is to say

$$X = \{\lambda\delta(n, m)\} \equiv \lambda I \tag{84}$$

for some $\lambda \in \mathbb{C}$. This property expresses the fact that the representation of p and q, which we just constructed, is *irreducible*. We shall see in Section Three that: (i) this property is equivalent to the fact that every matrix A can be expressed as a "function" of p and q, in a sense to be made precise; and (ii) up to unitary equivalence, there exists exactly one irreducible representation of p and q, also in a sense to be made precise there. Our representation is, therefore, canonical; in particular, its relevance extends very much beyond the case of the harmonic oscillator (78).

From this example one gets the feeling that the three postulates of the theory are—at the core—self-consistent even though some mathematical tightening up still remains to be done before one could claim to have really isolated that essential core.

The main achievement realized in the formulation of these three postulates was that the emerging quantum theory was becoming less dependent on the classical substrate on which it had originally be grafted; in particular, much of the earlier contradictions of the theory, stemming as they were from too close an adhesion to classical notions, were now losing their virulence.

This progress, however, still left open an important question of principle, namely whether and how one may still speak of a correspondence between the classical and the quantum theories. In a series of papers, Dirac (1925, 1926) remarked that there is a strong analogy between the quantum expression (49), namely

$$\dot{g} = \frac{i}{\hbar}[H, g] \tag{85a}$$

where

$$[H, g] \equiv Hg - gH \tag{85b}$$

and the classical expression

$$\dot{g} = -\{H, g\} \tag{86a}$$

where

$$\{H, g\} \equiv \frac{\partial H}{\partial q} \cdot \frac{\partial g}{\partial p} - \frac{\partial H}{\partial p} \cdot \frac{\partial g}{\partial q} \tag{86b}$$

This suggested that the quantum commutator $(i/\hbar)[.,.]$ should correspond to the classical Poisson bracket $-\{.,.\}$. Notice in particular that both operations are linear in each of their variables that they satisfy the Jacobi identity

$$[[a, b], c] + [[b, c], a] + [[c, a], b] = 0 \tag{87}$$

$$\{\{a, b\}, c\} + \{\{b, c\}, a\} + \{\{c, a\}, b\} = 0\ ; \tag{88}$$

and that both $[a, \cdot]$ and $\{a, \cdot\}$ are derivations:

$$[a, bc] = [a, b]c + b[a, c] \tag{89}$$

$$\{a, bc\} = \{a, b\}c + b\{a, c\}\,. \tag{90}$$

This analogy suggests the following question, known as the *Dirac problem.* Does there exist two matrices P and Q and a correspondence φ which, to every polynomial g in the classical variables p and q, associates a matrix $\varphi(g)$ in such a manner that: (i) $\varphi(p) = P$ and $\varphi(q) = Q$; (ii) to the unit function $1 : (p, q) \in \mathbb{R}^2 \mapsto 1$, φ associates the unit matrix $I = \{\delta(n, m)\}$; (iii) φ is linear; (iv) $(i/\hbar)[\varphi(f), \varphi(g)] = \varphi(-\{f, g\})$ for every polynomials f and g in p and q; (v) the matrices P and Q form an irreducible system, i.e. the only matrices X satisfying $[X, P] = 0 = [X, Q]$ are of the form $X = \lambda I$ where $\lambda \in \mathbb{C}$ and I is the unit matrix.

SCHOLIUM. *The Dirac problem admits no solution.*

Proof: We proceed by contradiction. Suppose thus that such a correspondence φ exists. From $\{p,q\} = -1$ we immediately conclude that P and Q satisfy

$$[P,Q] = -i\hbar I \tag{91}$$

which is the relation (47) of Born & Jordan. From this relation alone follows (as we have seen above) that

$$[P^m,Q^n] = -i\hbar m \sum_{k=0}^{n-1} Q^{n-1-k}P^{m-1}Q^k \tag{92a}$$

and

$$[P^m,Q^n] = -i\hbar n \sum_{\ell=0}^{m-1} P^{m-1-\ell}Q^{n-1}P^\ell \tag{92b}$$

From the definition (86b) of the Poisson bracket, we have

$$-\{p^2,p\} = 0 \quad \text{and} \quad -\{p^2,q\} = 2p \tag{93}$$

From (92) we have in particular

$$\frac{i}{\hbar}[P^2,P] = 0 \quad \text{and} \quad \frac{i}{\hbar}[P^2,Q] = 2P \tag{94}$$

The putative correspondence φ, together with the irreducibility assumption give then, upon comparing (93) and (94):

$$\varphi(p^2) = P^2 + \lambda I \quad \text{with } \lambda \in \mathbb{C} \tag{95}$$

Similarly, we obtain

$$\varphi(q^2) = Q^2 + \mu I \quad \text{with } \mu \in \mathbb{C} \tag{96}$$

From the definition (86b) of the Poisson bracket, we have

$$-\{p^2,q^2\} = 4pq \tag{97}$$

From (92) we have in particular

$$\frac{i}{\hbar}[P^2,Q^2] = 2(PQ+QP) \tag{98}$$

From (95), (96), (97) and (98) we obtain

$$\varphi(pq) = \frac{1}{2}(PQ+QP) \tag{99}$$

We next prove that φ, if it exists, must satisfy

(100) $$\varphi(p^n) = P^n \quad \forall \quad n \in \mathbb{Z}^+$$

Clearly (100) is satisfied, by definition, for $n = 1$; we can, therefore, proceed by recursion: suppose (100) holds for $n-1$ with $n > 1$. From the definition (86b) of the Poisson bracket, we have

(101) $$-\{p^n, p\} = 0 \quad \text{and} \quad -\{p^n, q\} = np^{n-1}$$

From (92) we have

(102) $$\frac{i}{\hbar}[P^n, P] = 0 \quad \text{and} \quad \frac{i}{\hbar}[P^n, Q] = nP^{n-1}$$

The putative correspondence φ, together with the irreducibility assumption, gives then upon comparing (101) and (102):

(103) $$\varphi(p^n) = P^n + \lambda_n I \quad \text{with} \quad \lambda_n \in \mathbb{C}$$

We still have to show $\lambda_n = 0$. To see this, we notice from (86b) that

(104) $$-\{p^n, pq\} = np^n$$

From (92) we compute

(105) $$\frac{i}{\hbar}[P^n, \frac{1}{2}(PQ + QP)] = nP^n$$

The desired formula (100) follows then from (103), (104) and (105). We prove similarly that φ, if it exists, must satisfy

(106) $$\varphi(q^n) = Q^n$$

Finally, the definition of the Poisson bracket (see 86b) gives

(107) $$-\{p^{m+1}, q^{n+1}\} = (m+1)(n+1)p^m q^n$$

from which we conclude, upon using (100) and (106) that φ, if it exists, must satisfy

(108) $$\varphi(p^m q^n) = \frac{i}{\hbar}\frac{1}{(m+1)(n+1)}[P^{m+1}, Q^{n+1}]$$

Notice, incidentally, that (108) coincides with the Born & Jordan proposal (63).

Up to this point, everything thus seems to be in order. However, we now compute from (86b):

$$-\{p^3, q^3\} + 3\{p^2q, pq^2\} = 0 \tag{109}$$

The putative correspondence φ would then give, according to (100), (106), (108) and (92), when applied to the LHS of (109):

$$\begin{aligned} &\frac{i}{\hbar}[\varphi(p^3), \varphi(q^3) - 3\frac{i}{\hbar}[\varphi(p^2q), \varphi(pq^2)] = \\ &\frac{i}{\hbar}[P^3, Q^3] - 3\frac{i}{\hbar}[\frac{1}{2}(P^2Q + QP^2), \frac{1}{2}(PQ^2 + Q^2P)] \end{aligned} \tag{110}$$

One straightforward way to compute the RHS of (110) is to use (91) and (92) repeatedly, and to rewrite each term appearing in the commutator expansion as a sum of ordered monomials of the form Q^nP^m. We then find that the coefficients of these monomials cancel, except for the lowest monomial which gives, for the RHS of (110), the matrix $(-3\hbar^2 I)$. This differs from the result obtained by applying the correspondence φ to the RHS of (109): $\varphi(0) = 0$ by linearity. This is the desired contradiction. □

Remarks.

(i) A result of this type was first obtained by Groenewold (1946); see also van Hove (1951); thms 3.4.9 and 3.4.13 in Abraham & Marsden (1978); and references quoted therein, in particular Chernoff (1981). From these papers one sees that this scholium, stated here in its simplest form, does generalize in two directions briefly described in remarks (ii) and (iii) below.

(ii) We stated our scholium for one degree of freedom, i.e. for $(p, q) \in \mathbb{R}^2 = T^*\mathbb{R}$. A similar result holds for finitely many degrees of freedom, i.e. for $(p, q) \in \mathbb{R}^{2n} = T^*\mathbb{R}^n$, with n any finite positive integer.

(iii) We imposed, in our statement of the Dirac problem, that the matrices P and Q form an irreducible system. The scholium is still true if this assumption is weakened to the requirement that P and Q form a system with finite multiplicity, i.e. to the case where one has a direct sum of finitely many irreducible representations of the canonical commutation relations.

(iv) The finite multiplicity condition just mentioned is a necessary condition for the truth of the scholium, as was first noticed by van Hove (1951). Consider indeed the map φ which, to every polynomial g in the classical variables p and q, associates the operator

$$\varphi(g) \equiv i\hbar(\partial_q g)\partial_p - i\hbar(\partial_p g)\partial_q - p(\partial_p g) + g \tag{111}$$

acting on the Schwartz space $S(\mathbb{R}^2)$ defined as follows. The elements of $S(\mathbb{R}^2)$ are C^∞-functions $\Psi : \mathbb{R}^2 \to \mathbb{C}$ satisfying

$$\sup_{(p,q)\in\mathbb{R}^2} \| p^m q^n (\partial_p^k \partial_q^\ell \Psi)(p, q) \| < \infty \tag{112}$$

for all $m, n, k, \ell \in \mathbb{Z}^+$; we equip $S(\mathbb{R}^2)$ with the structure of a prehilbert space by considering it as a (dense) subspace of $\mathcal{L}^2(\mathbb{R}^2, dpdq)$. To make contact with the Dirac problem we need to establish that $S(\mathbb{R}^2)$ contains an orthonormal basis. For this, and later purposes, it is useful to construct such a basis as follows. Notice first that

$$\mathcal{L}^2(\mathbb{R}^2, dpdq) \cong \mathcal{L}^2(\mathbb{R}, dp) \otimes \mathcal{L}^2(\mathbb{R}, dq) \tag{113a}$$

$$\mathcal{L}^2(\mathbb{R}, dx) \cong \ell^2 \tag{113b}$$

Identify now the sequence $\Psi_0 \equiv (1, 0, 0 \ldots) \in \ell^2$ with the function

$$\Psi_0 : x \in \mathbb{R} \mapsto \pi^{-1/2} \exp(-x^2/4) \tag{114}$$

Through (113b), we obtain from (73) an orthonormal basis in $\mathcal{L}^2(\mathbb{R}, dx)$; notice incidentally that this identification produces, via the algebraic relation (75), a representation of the canonical commutation relation by the operators

$$(q\Psi)(x) = x\Psi(x) \tag{115a}$$

$$(p\Psi)(x) = -i\hbar(\frac{\partial\Psi}{\partial x})(x) \tag{115b}$$

With the orthonormal basis $\{\Psi_n | n \in \mathbb{Z}^+\}$ in $\mathcal{L}^2(\mathbb{R}, dx)$ just obtained, define for every $(m, n) \in \mathbb{Z}^+ \times \mathbb{Z}^+$

$$\Psi_{m,n} : (p, q) \in \mathbb{R}^2 \mapsto \Psi_m(p)\Psi_n(q) \in \mathbb{C} \tag{116}$$

It is then easy to check that these functions $\Psi_{m,n}$ belong to $S(\mathbb{R}^2)$ and that $\{\Psi_{m,n} | (m, n) \in \mathbb{Z}^+ \times \mathbb{Z}^+\}$ is an orthonormal basis.

Upon taking the matrix elements of the operators (111) with respect to any orthonormal basis in $S(\mathbb{R}^2)$—we just proved that such an orthonormal basis exists—we obtain a correspondence which satisfies the requirements (i)–(iv) of the Dirac problem. The last requirement—namely irreducibility—of the Dirac problem is, however, not satisfied. Indeed, (111) gives in particular

$$\varphi(p) = -i\hbar\partial_q \quad \text{and} \quad \varphi(q) = i\hbar\partial_p + q \tag{117}$$

To see that the corresponding matrices do not form an irreducible system, notice that the space $S(\mathbb{R}^2)$ harbours another representation of the canonical commutation relation, obtained by writing

$$\psi(p) = -i\hbar\partial_p \quad \text{and} \quad \psi(q) = i\hbar\partial_q + p \tag{118}$$

Clearly the operators $\psi(p)$ and $\psi(q)$ are not scalar multiples of the identity operator, and yet they commute with the operators $\varphi(p)$ and $\varphi(q)$.

(v) The map (111) is called the *prequantization map* and it plays an important role in the *geometric quantization programme* of Souriau (1966) and Kostant (1970); see also Simms & Woodhouse (1976), Guillemin & Sternberg (1977), Abraham & Marsden (1978), and Sniatycki (1980). The existence of the "other" representation (118) was pointed out by Segal (1960) and Streater (1966). The mathematical structure underlying the simultaneous presence of the two representations (117) and (118) has been investigated by the author in connection with the theory of modular algebras; see Emch (1981).

(vi) Physically, the negative content of the scholium can be interpreted in a variety of ways. At one extreme one could view Dirac's remark as a mere mnemonic analogy. At the other extreme, one could insist that Dirac's analogy between the classical Poisson bracket $-\{.,.\}$ and the quantum commutator $(i/\hbar)[.,.]$ must be laid at the foundation of quantum theory, even at the cost of removing—at least temporarily—one of the requirements (usually the fifth one) of the Dirac problem; this is in essence the starting point of the geometric quantization programme which can then be viewed as a geometric prescription to extract irreducible representations from the reducible prequantization representation (111). Somewhere between these two extreme views, one finds the attitude of most physicists. Recognizing that a true quantization map φ, satisfying all five requirements of the Dirac problem, does not exist, they work within the framework given by the irreducible representation (115), and they adopt one ordering rule or another to form products of the operators p and q. One such rule, for instance, consists in writing the operators p and q in term of the operators a and a^* one can uniquely define from p and q using (75). One then replaces any polynomial in the classical variables p and q by the same polynomial in the quantum variables p and q; next one expands this polynomial in term of the variables a and a^*. Finally one replaces every monomial M in a and a^* appearing in this expansion by a monomial of the form $(a^*)^m a^n$ where m (resp. n) is the number of times the factor a^* (resp. a) appeared in the monomial M; see also Section 10.2. Another "ordering" procedure consists in associating, to a function g of the classical variables p and q, the operator $\varphi(g)$ obtained by the following generalization of Fourier analysis

(119) $$\varphi(g) \equiv \iint da db\, \tilde{g}(a,b) E(\hbar; a, b)$$

with $\tilde{g}$ and $E(\hbar; \cdot, \cdot)$ defined by the following identities

(120) $$g(p,q) \equiv \iint da db\, \tilde{g}(a,b) e^{-i(ap+bq)}$$

where (p,q) runs over $\mathbb{R}^2$, while $E(\hbar; a, b)$ is defined for every $(a,b) \in \mathbb{R}^2$ by

(121) $$E(\hbar; a, b) = \exp\{-i(\hbar^{-1/2} ap + \hbar^{1/2} bq)\}$$

where p and q are now the operators defined from the representation (115). The reader will notice that some care should be exercised when writing (119)–(121).

First of all, for the integrals (119) and (120) to make sense, $\tilde{g}$ should be tame enough: $\tilde{g} \in \mathcal{L}^1(\mathbb{R}^2, da\, db)$ would do, but one can do better than that; see Segal (1963a) and Grossman, Loupias & Stein (1968). Secondly, the operator-exponential in (121) must be properly defined; this can be achieved in $\mathcal{L}^2(\mathbb{R}, dx)$ by noticing that the operators $(\hbar^{-\frac{1}{2}}ap + \hbar^{\frac{1}{2}}bq)$ are essentially self-adjoint on $\mathcal{S}(\mathbb{R})$, so that (121) can be properly defined from the spectral theorem for the corresponding self-adjoint operators (see Section 2). Equivalently, $E(\hbar; a, b)$ can be defined explicitly by its action on $\mathcal{L}^2(\mathbb{R}, dx)$:

$$(122)\qquad \begin{gathered}[E(\hbar; a, b)\Psi](x) \equiv \\ \exp\{-i\hbar^{\frac{1}{2}}b(x - \hbar^{\frac{1}{2}}a/2\}\Psi(x - \hbar^{\frac{1}{2}}a)\end{gathered}$$

Notice for completeness that $\mathcal{S}(\mathbb{R})$ is stable under the action of $E(\hbar; a, b)$, and that when the physical units are chosen in such a manner that $\hbar = 1$, $E(\hbar; a, b)$ coincides with the *Weyl operator*:

$$(123)\qquad W(a, b) \equiv e^{-i(ap+bq)}$$

With these technicalities out of the way, we can now come to the main point of this remark. The scholium tells us that if a correspondence φ between classical and quantum observables is chosen in such a manner that the requirements (i), (ii), (iii) and (v) of the Dirac problem are satisfied, then the correspondence cannot be a Lie isomorphism from the classical theory, with its Poisson bracket $-\{.,.\}$, to the quantum theory, with its commutator $(i/\hbar)[.,.]$. The proof of the scholium, nevertheless, already suggests that φ could preserve these Lie structures to first order in $\hbar$, so to speak. This can be illustrated with the two correspondences just described. More precisely—see Scholium 8.3.3 and its Corollary 8.3.4 below—one can start from a quantum theory where the RHS of (119), together with (121), becomes a definition of a quantum observable; one can then control mathematically the limit $\hbar \to 0$ in such a way that (120) follows, together with the classical phase space itself, equipped with its symplectic structure. This symplectic structure, so essential to the Hamiltonian formulation of classical mechanics appears in this manner as a consequence of the quantum theory.

This remark suggests viewing classical mechanics as a derived theory, obtained as a result of an approximation of a more fundamental theory, namely quantum mechanics. This is to be conceptually compared with the way Newtonian physics can be viewed as an approximation of relativistic physics. In both cases, the question then is whether the particular circumstances of a given physical situation do warrant considering only the classical approximation.

For this suggestion to be acceptable, however, we still have to give quantum mechanics the free-standing structure which would make it a true theory. This process had been started in the work of Heisenberg, Born and Jordan reviewed in this section. Another approach towards the same end was proposed by

Schroedinger, as we shall see in Section Two. Building on these two preliminary steps, von Neumann presented a synthesis that qualifies as a well-defined theory for systems with finitely many degrees of freedom; this will be reviewed in Section Three. The extension to more general theories, covering cases where infinitely many degrees of freedom are involved, will be discussed in the last two chapters of this book.

Section 2. THE SCHROEDINGER WAVE MECHANICS

The aims of this section are: (a) to review the conceptual background behind the formulation of the Schroedinger equation; (b) to discuss the connection between this equation and the Heisenberg matrix formalism reviewed in the preceding section; and (c) to show how the Born statistical interpretation of quantum theory emerged from these premises.

The *Schroedinger equation* comes in two avatars: an *eigenvalue equation*

$$(1) \qquad [-\frac{h^2}{8\pi^2 m}\Delta + V(x)]\Psi(x) = E\Psi(x)$$

and a *wave equation*

$$(2) \qquad [-\frac{h^2}{8\pi^2 m}\Delta + V(x)]\Psi(x,t) = i\frac{h}{2\pi}\frac{\partial\Psi}{\partial t}(x,t)$$

In these two equations, h is the Planck constant, and Δ is the Laplace operator in $\mathbb{R}^n$ (the case $n = 3$ is particularly important in physical applications, but the theory and its applications are more general than that):

$$(3) \qquad \Delta \equiv \sum_{k=1}^{n}(\frac{\partial}{\partial x_k})^2 \equiv \sum_{k=1}^{n}\partial_k^2$$

Equation (1) (resp. eqn. 2) is also referred to as the *time-independent* (resp. *time-dependent*) *Schroedinger equation.* This nomenclature will be made more precise later on in this section.

In the third paper of the series where he first wrote these equations, Schroedinger (1926) pointed out that his equations could be obtained from—and that their physical content was in fact equivalent to—the Heisenberg, Born & Jordan formalism. This, however was *not* how Schroedinger arrived at these equations. Perhaps nowhere in this book is it more proper to recall with Klein (1967) a quote from Benjamin Farrington: "History is the most fundamental science, for there is no human knowledge which cannot lose its scientific character when men forget the conditions under which it originated, the questions which it answered, and the functions it was created to serve."

The questions which Schroedinger's theory was designed to answer were of a quite different nature than those addressed by Heisenberg, and so were his methods and tastes. Schroedinger indeed wrote in connection with the

Heisenberg, Born & Jordan approach: "I was discouraged, if not repelled by what appeared to me a rather difficult method of transcendental algebra,defying visualization." On the positive side, Schroedinger's direct motivation had its roots in the ideas of Einstein and of de Broglie. A summary of Einstein's view of the old quantum mechanics can be found in a letter he wrote to Born in January 1920: "I do not believe that the solution to the quanta has to be found by giving up the continuum ... I believe now, as before, that one has to look for redundancy in determination by using differential equations so that the solutions themselves no longer have the character of a continuum. But how?" (see Born, 1971).

Schroedinger found the answer to the "how" of Einstein's question in the Einstein-de Broglie particle-wave duality. While Heisenberg, Born & Jordan used concepts—such as position and momentum—which primarily pertain to particle mechanics, Schroedinger concentrated on the wave aspect: if there was to be a wave, there must be a wave equation. In Schroedinger's words: "Our classical mechanics is perhaps the complete analogy of geometrical optics and as such is wrong and not in agreement with reality; it fails whenever the radii of curvature and the dimensions of the path are no longer large compared with a certain wavelength ... Then an undulatory mechanics has to be established, and the most obvious approach to it is the elaboration of the Hamiltonian analogy into a wave theory." For the Hamiltonian analogy Schroedinger is referring to, see Hamilton (1833).

With the historical context and motivations now in place, we can proceed to give a derivation of Schroedinger's equation.

We start with the usual wave equation

$$[\Delta - \frac{1}{v^2}\frac{\partial^2}{\partial t^2}]\Psi(x,t) = 0 \tag{4}$$

where Δ is the Laplace operator (3) and v is the velocity of the wave. The core of the argument consists in inserting in (4) the following four ingredients.

The first ingredient is to focus—temporarily—on monochromatic solutions, i.e. on waves of the form

$$\Psi(x,t) = \Psi(x)e^{-2\pi i\nu t} \tag{5}$$

The second ingredient is the de Broglie expression (7.2.35) of the particle-wave duality which attributes, to a particle of momentum $p \equiv mv$, a wave of wavelength λ (i.e. of frequency $\nu = v/\lambda$) given by

$$\lambda = \frac{h}{mv} \quad \text{i.e.} \quad p = \frac{h\nu}{v} \tag{6}$$

The third ingredient is the Einstein relation (see 7.2.12) between the energy E and the frequency ν of a quantum wave, namely

$$E = h\nu \tag{7}$$

The fourth ingredient is the classical expression of the energy E of a particle as the sum of its kinetic energy $p^2/2m$ and its potential energy $V(x)$, i.e.

$$E = \frac{1}{2m} p^2 + V(x) \tag{8}$$

From (5) and (7) we obtain

$$\frac{\partial^2 \Psi}{\partial t^2}(x,t) = (-2\pi i E/h)^2 \Psi(x,t) \tag{9}$$

From (6), (7) and (8) we have

$$v = E[2m(E - V)]^{-1/2} \tag{10}$$

Upon inserting (9) and (10) into (4) and taking (5) into account, we receive the eigenvalue equation (1), i.e. the time-independent Schroedinger equation.

Notice further that if $\{E_n, \Psi_n | n = 1, 2, \ldots\}$ is any (finite) collection of solutions of (4), then Ψ defined by

$$\Psi(x,t) \equiv \Sigma_n \Psi_n(x) e^{-2\pi i E_n t/h} \tag{11}$$

satisfies the wave-equation (2), i.e. the time-dependent Schroedinger equation.

The genius of Schroedinger was to consider the two equations (1) and (2) as the fundamental equations of a new theory, where (1) was to describe the "stationary states" of the system, while (2) was to describe the time-evolution of "arbitrary states".

Three remarks are in order at this point. The first remark will precise the mathematical form of these equations. The second remark will be concerned with the predictive value of the theory. The third remark will address the conceptual questions of finding: (a) a physical interpretation for the wave-function $\Psi(x,t)$, and (b) the relations which may exist between this theory and that proposed by Heisenberg, Born & Jordan.

Thus, our first remark is that equations (1) and (2) are well-defined mathematical problems only when one agrees on what would constitute a solution, i.e. when one agrees on an "admissible" class of functions, to which the solutions should belong. One usually accepts today that the Schroedinger equation is solved when the following steps have been carried out.

The *first step* in the solution of the Schroedinger equation is to define the differential operator

$$-\frac{h^2}{8\pi^2 m}\Delta + V(x) \tag{12}$$

on a dense linear subspace of the Hilbert space $\mathcal{H} = \mathcal{L}^2(\mathbb{R}^n, d^n x)$ in such a manner that it admits a unique self-adjoint extension H. The reader must note here that we distinguish between Hermitian operators and self-adjoint operators. Specifically, let $D(A)$ be a linear subspace of $\mathcal{H}$, and A be a linear operator $A : D(A) \to \mathcal{H}$. Let $\{B, D(B)\}$ be similarly defined. A is said to be an *extension* of B, or equivalently B is said to be a *restriction* of A, if $D(A) \supseteq D(B)$ and $A\Psi = B\Psi$ for all $\Psi \in D(B)$. Two operators $\{A, D(A)\}$ and $\{C, D(C)\}$ are said to be *Hermitian conjugate* to one another if $(A\Psi, \Phi) = (\Psi, C\Phi)$ for all $(\Psi, \Phi) \in D(A) \times D(C)$. A is said to be *densely defined* if $D(A)$ is dense in $\mathcal{H}$, i.e. if given any $\Phi \in \mathcal{H}$ and any $\epsilon > 0$ there exists $\Psi \in D(A)$ such that $\| \Phi - \Psi \| < \epsilon$; for instance, the Schwartz space $\mathcal{S}(\mathbb{R}^n)$ (see 8.1.112) is dense in $\mathcal{L}^2(\mathbb{R}^n, d^n x)$, and the operators

(13a) $$(p_k \Psi)(x) = -i\hbar(\partial_k \Psi)(x)$$

and

(13b) $$(q_k \Psi)(x) = x_k \Psi(x)$$

with $D(p_k) = D(q_k) = \mathcal{S}(\mathbb{R}^n)$, $k = 1, 2, \ldots, n$, are densely defined. Now if $\{A, D(A)\}$ is a densely defined linear operator, one can show that there exists a unique linear operator $\{A^*, D(A^*)\}$, called the *adjoint* of $\{A, D(A)\}$, such that every $\{C, D(C)\}$ Hermitian conjugate to $\{A, D(A)\}$ is a restriction of $\{A^*, D(A^*)\}$. Finally $\{A, D(A)\}$ is said to be *Hermitian* if $\{A^*, D(A^*)\}$ is an extension of $\{A, D(A)\}$; and in particular, $\{A, D(A)\}$ is said to be *self-adjoint* if it is Hermitian *and* $D(A^*) = D(A)$.

The second step in the solution of the Schroedinger equation is to identify the spectral family associated to H. Here again this statement calls for a definition and an appeal to a general theorem. A *spectral family* on $\mathcal{H}$ is a function

(14) $$P : \lambda \in \mathbb{R} \mapsto P_\lambda \in \mathcal{B}(\mathcal{H})$$

(where $\mathcal{B}(\mathcal{H})$ is the algebra of all bounded linear operators from $\mathcal{H}$ into itself) satisfying conditions (15)–(19) below

(15) $$\text{for all } \lambda \in \mathbb{R} : P_\lambda^* = P_\lambda$$

(16) $$\text{for all } \lambda \text{ and } \mu \text{ in } \mathbb{R}, \text{ with } \lambda \leq \mu : P_\lambda P_\mu = P_\mu P_\lambda = P_\lambda$$

(17) $$\text{for all } \lambda \in \mathbb{R} : \quad s - \lim_{\epsilon \to +0} P_{\lambda+\epsilon} = P_\lambda$$

and finally:

(18) $$s - \lim_{\lambda \to -\infty} P_\lambda = 0$$

(19) $$s - \lim_{\lambda \to +\infty} P_\lambda = I$$

Conditions (15) and (16) imply in particular that every P_λ is a projector, i.e. $P_\lambda = P_\lambda^2 = P_\lambda^*$. Conditions (17–19) call for two notational precisions. First, $\epsilon \to +0$ means that ϵ tends to zero through positive values. Second, $s - \lim$ means the limit in the *strong-operator topology*. A basis of neighbourhood in this topology is obtained by considering all sets of the form

(20) $$N(A, S, \epsilon) \equiv \{B \in \mathcal{B}(\mathcal{H}) | \parallel (B - A)\Psi_k \parallel < \epsilon \quad \forall \Psi_k \in S\}$$

where $A \in \mathcal{B}(\mathcal{H})$, S is a finite set of vectors in $\mathcal{H}$, and $\epsilon > 0$. This strong-operator topology is the weakest topology on $\mathcal{B}(\mathcal{H})$ for which the maps $B \in \mathcal{B}(\mathcal{H}) \mapsto B\Psi \in \mathcal{H}$ are continuous when $\mathcal{H}$ is equipped with its strong topology (i.e. with the topology of the convergence of vectors in the norm).

As an example of a spectral family, consider $P : \lambda \in \mathbb{R} \mapsto P_\lambda \in \mathcal{B}(\mathcal{H})$ with P_λ defined for every $\Psi \in \mathcal{H} \equiv \mathcal{L}^2(\mathbb{R}, dx)$ by

(21a) $$(P_\lambda \Psi)(x) \equiv \chi_{(-\infty,\lambda]}(x)\Psi(x)$$

with

(21b) $$\chi_{(-\infty,\lambda]}(x) \equiv \begin{cases} 1 & \text{for } -\infty < x \leq \lambda \\ 0 & \text{otherwise} \end{cases}$$

For another example, let $\{\Psi_n | n \in \mathbb{Z}^+\}$ be any orthonormal basis in a Hilbert space $\mathcal{H}$, and define

(22) $$P_\lambda : \Psi \in \mathcal{H} \mapsto \sum_{n \leq \lambda} (\Psi, \Psi_n)\Psi_n$$

Let us now come back to the general case of an arbitrary spectral family $\{P_\lambda | \lambda \in \mathbb{R}\}$ on some Hilbert space $\mathcal{H}$. Denote by $B(\mathbb{R})$ the σ-algebra of Borel subsets of $\mathbb{R}$. Notice then that for every $\Psi \in \mathcal{H}$ with $\parallel \Psi \parallel = 1$,

(23) $$\mu_\Psi : \Delta \in B(\mathbb{R}) \mapsto \int_\Delta d(P_\lambda \Psi, \Psi)$$

is a probability measure on $\mathbb{R}$. Consider now

(24) $$D(H) \equiv \{\Psi \in \mathcal{H} | \int_{\mathbb{R}} \lambda^2 d(P_\lambda \Psi, \Psi) < \infty\}$$

One can then prove that there exists a unique self-adjoint operator H, with domain $D(H)$ given by (24), such that

(25) $$(H\Psi, \Phi) = \int_{\mathbb{R}} \lambda \, d(P_\lambda \Psi, \Phi) \qquad \forall \Psi \in D(H) \quad \text{and} \quad \forall \Phi \in \mathcal{H}$$

Conversely, one can prove that to every self-adjoint operator H corresponds a unique spectral family $\{P_\lambda | \lambda \in \mathbb{R}\}$ such that (24) and (25) are satisfied. These two results are known as the *spectral theorem* for self-adjoint operators. A proof of this theorem can be found in every textbook on functional analysis, e.g. Stone (1932), Riesz & Nagy (1955), Akhiezer & Glazman (1961), Kato (1966), Yosida (1971), Reed & Simon (1972), Amrein, Jauch & Sinha (1977), Kadison & Ringrose (1983).

The reader who desires concrete examples will check that the spectral family (21) defines, through (24), the domain

(26a) $$D(q) \equiv \{\Psi \in \mathcal{L}^2(\mathbb{R}, dx) | \int_{\mathbb{R}} dx\, x^2 |\Psi(x)|^2 < \infty\}$$

and that (25) defines then the self-adjoint operator q as the map

(26b) $$q : \Psi \in D(q) \mapsto q\Psi \in \mathcal{L}^2(\mathbb{R}, dx)$$

with

(26c) $$(q\Psi)(x) = x\Psi(x)$$

Furthermore, this operator is the unique self-adjoint extension of its restriction $(q|\mathcal{S}(\mathbb{R}))$ to $\mathcal{S}(\mathbb{R})$; compare with (13b). Note further that the Fourier transform

(27) $$(\mathcal{F} f)(k) \equiv (2\pi)^{-1/2} \int dx\, e^{-ikx} f(x)$$

defined originally on $(\mathcal{L}^2 \cap \mathcal{L}^1)(\mathbb{R}, dx)$, extends uniquely to a unitary operator $\mathcal{F}$ from $\mathcal{L}^2(\mathbb{R}, dx)$ onto itself; recall that $\mathcal{F}$ is said to be unitary whenever $\mathcal{F}\mathcal{F}^* = I = \mathcal{F}^*\mathcal{F}$. We can now define from (21) a new spectral family

(28) $$\tilde{P} : \lambda \in \mathbb{R} \mapsto \tilde{P}_\lambda \equiv \mathcal{F}^* P_\lambda \mathcal{F} \in \mathcal{B}(\mathcal{H})$$

For this spectral family, (24) gives

(29a) $$D(p) \equiv \{\Psi \in \mathcal{L}^2(\mathbb{R}, dx) | \int_{\mathbb{R}} dk\, k^2 |(\mathcal{F}\Psi)(k)|^2 < \infty\}$$

(25) can then be used to define the self-adjoint operator p as the map

(29b) $$p : \Psi \in D(p) \mapsto p\Psi \in \mathcal{L}^2(\mathbb{R}, dx)$$

with

(29c) $$(\mathcal{F} p\Psi)(k) = \hbar k (\mathcal{F}\Psi)(k)$$

with

(29d) $$(p\Psi)(x) = -i\hbar(\partial_x \Psi)(x)$$

Here again, this operator is the unique self-adjoint extension of its restriction $(p|S(\mathbb{R}))$ to $S(\mathbb{R})$; compare with (13a). The two self-adjoint operators p and q just defined constitute together what is called the *Schroedinger representation of the canonical commutation relation* for one-degree of freedom; the generalization to n degrees of freedom is straightforward from (13). Notice in particular, that $S(\mathbb{R}) \subset D(p) \cap D(q)$, and is stable under the action of p and q. Hence on $S(\mathbb{R})$ the expression

$$[p,q] \equiv pq - qp = -i\hbar I \tag{30}$$

is mathematically well-defined. In addition Foias, Geher & Sz-Nagy (1960) noticed that the operator

$$\frac{1}{2}(p^2 + q^2) : S(\mathbb{R}) \to \mathcal{L}^2(\mathbb{R}, dx) \tag{31}$$

admits exactly one self-adjoint extension H. The spectral family of this H is of the form (22), where the orthonormal basis $\{\Psi_n | n \in \mathbb{Z}^+\}$ is given by recursion as follows. On $S(\mathbb{R})$ define the operators

$$a \equiv 2^{-1/2}(q + ip) \tag{32a}$$

$$a^* \equiv 2^{-1/2}(q - ip) \tag{32b}$$

From our previous remark, $S(\mathbb{R})$ is stable under a and a^*. Let $\Psi_0 \in S(\mathbb{R})$ be the function

$$\Psi_0(x) = \pi^{-1/2} \exp(-x^2/4) \tag{33}$$

Then $\Psi_{n+1} \in S(\mathbb{R})$ is defined for every $n \in \mathbb{Z}^+$ by

$$\Psi_{n+1} = [(n+1)\hbar]^{-1/2} a^* \Psi_n \tag{34}$$

From (30)–(34) one obtains for every $n \in \mathbb{Z}^+$

$$H\Psi_n = \epsilon_n \Psi_n \tag{35}$$

with

$$\epsilon_n = (n + \frac{1}{2})\hbar \tag{36}$$

We have thus recovered the result (8.1.79–80). From (22) and (24) we obtain now the further information that H is a self-adjoint operator with domain

$$D(H) = \{\Psi \in \mathcal{L}^2(\mathbb{R}, dx) \mid \sum_{n=0}^{\infty} {\epsilon_n}^2 \mid (\Psi, \Psi_n) \mid^2 < \infty\} \tag{37}$$

Moreover, upon using (26c) and (29d), we observe that the restriction of H to $\mathcal{S}(\mathbb{R}) \subset D(H)$ takes the form of the Schroedinger operator (12) with $V(x) = x^2/2$. We have, therefore, solved the time-independent Schroedinger equation (1) for this particular potential V corresponding physically to the harmonic oscillator.

In general, we can now state unambiguously what is meant by solving the *eigenvalue-equation* (1): a solution (E, Ψ) is deemed admissible if and only if $\Psi \in D(H), E \in \mathbb{R}$, and $H\Psi = E\Psi$.

The harmonic oscillator is particular in that the solutions of the Schroedinger eigenvalue equation (1) span, by linearity, a dense subspace of $\mathcal{L}^2(\mathbb{R}, dx)$. This is in general not the case. For instance, for H defined on

$$D(H) \equiv \{\Psi \in \mathcal{L}^2(\mathbb{R}, dx) \mid \int_{\mathbb{R}} dk\, k^4 \mid (\mathcal{F}\Psi)(k)|^2 < \infty\} \tag{38a}$$

by

$$(\mathcal{F}H\Psi)(k) = \hbar^2 k^2 (\mathcal{F}\Psi)(k) \tag{38b}$$

the eigenvalue equation (1) admits no solution. Note that H, given by (38), corresponds physically to the case $V \equiv 0$, i.e. to a free particle.

In the general case, solving the *time-independent Schroedinger equation* means more than solving it as an eigenvalue equation: it means determining explicitly the spectral family $\{P_\lambda | \lambda \in \mathbb{R}\}$ canonically associated to the self-adjoint operator H. As for the eigenvalue problem, this problem is well-posed if and only if the differential operator (12) has been defined on a domain where it admits a unique self-adjoint extension H. Finding such a domain, and then determining explicitly the spectral family $\{P_\lambda | \lambda \in \mathbb{R}\}$ of H has been, and continues to be a rich field of mathematical research; see e.g. Kato (1966), Reed & Simon (1972), or Amrein, Jauch & Sinha (1977). We shall see several applications, now, and later in this chapter.

The *third step* in the solution of the Schroedinger equation is to consider the time-dependent equation (2). The main mathematical tool here is the Stone theorem which establishes a one-to-one correspondence between self-adjoint operators $\{H, D(H)\}$ acting in a Hilbert space $\mathcal{H}$, and strongly continuous, one-parameter groups $\{U_t | t \in \mathbb{R}\}$ of unitary operators acting on $\mathcal{H}$. Several proofs of this theorem are available, see e.g. any of the classical texts in functional analysis mentioned earlier in this section. Let us briefly sketch how this theorem comes about in the present context. From the preceding two steps we obtained the self-adjoint operator $\{H, D(H)\}$ and its spectral family $\{P_\lambda | \lambda \in \mathbb{R}\}$. From the latter we constructed, for every $\Psi \in \mathcal{H}$ with $\| \Psi \| = 1$ the probability measure μ_Ψ, see (23). Clearly the function

$$f_\Psi : t \in \mathbb{R} \mapsto \int_{\mathbb{R}} \exp\{-\frac{i}{\hbar}\lambda t\} \mu_\Psi(d\lambda) \in \mathbb{C} \tag{39}$$

is continuous and satisfies

$$| f_\Psi(t) | \leq 1 = f_\Psi(0) = \| \Psi \|^2 \quad \forall\, t \in \mathbb{R} \tag{40}$$

It further satisfies

(41) $$\sum_{j,k=1}^{N} a_k a_j^* f_\Psi(t_k - t_j) \geq 0$$

for every finite positive integer N, every sequence $\{t_k | k = 1, 2, \ldots, N\} \subset \mathbb{R}$, and every sequence $\{a_k | k = 1, 2, \ldots, N\} \subset \mathbb{C}$; i.e. f_Ψ is a *function of positive-type*. Note that the classical Bochner theorem claims that the converse is also true, namely every such function can be obtained as the Fourier transform of a probability measure. Consider now for every $t \in \mathbb{R}$ the map

(42) $$\begin{aligned} &\varphi_t : (\Psi, \Phi) \in \mathcal{H} \times \mathcal{H} \mapsto \\ &\varphi_t(\Psi, \Phi) \equiv \int_{\mathbb{R}} \exp\{-\frac{i}{\hbar}\lambda t\} d(P_\lambda \Psi, \Phi) \in \mathbb{C} \end{aligned}$$

Clearly:

(43) $$\varphi_t(\Psi, \Psi) = \| \Psi \|^2 f_{\Psi/\|\Psi\|}(t) \quad \forall \Psi \in \mathcal{H}$$

(44a) $$| \varphi_t(\Psi, \Phi) | \leq 2 \| \Psi \| \cdot \| \Phi \| \quad \forall \Psi, \Phi \in \mathcal{H}$$
(44b) $$\varphi_t(\,\cdot\,, \Phi) \text{ is linear on } \mathcal{H} \quad \forall \Phi \in \mathcal{H}$$
(44c) $$\varphi_t(\Psi, \cdot\,) \text{ is antilinear on } \mathcal{H} \quad \forall \Psi \in \mathcal{H}$$

From (42) and (44) we conclude, via the Riesz representation theorem, that for each $t \in \mathbb{R}$, there exists a bounded operator U_t from $\mathcal{H}$ into itself such that

(45) $$(U_t \Psi, \Phi) = \int_{\mathbb{R}} \exp\{-\frac{i}{\hbar}\lambda t\} d(P_\lambda \Psi, \Phi) \quad \forall \Psi, \Phi \in \mathcal{H}$$

From this formula, we read immediately that

(46a) $$U_0 = I$$
(46b) $$U_t^* = U_{-t} \qquad \forall t \in \mathbb{R}$$
(46c) $$U_t U_s = U_{t+s} \qquad \forall s, t \in \mathbb{R}$$
and thus
(46d) $$U_t U_t^* = I = U_t^* U_t \quad \forall t \in \mathbb{R}$$

Finally, from the continuity of f_Ψ we obtain:

(46e) $$s - \lim_{\epsilon \to 0} U_{t+\epsilon} = U_t \qquad \forall t \in \mathbb{R}$$

The formulas (46) express precisely what is meant by saying that $\{U_t | t \in \mathbb{R}\}$ is a *strongly continuous, one-parameter group of unitary operators* acting in

$\mathcal{H}$. We have thus obtained the first part of Stone's theorem: the construction of $\{U_t | t \in \mathbb{R}\}$ from the self-adjoint operator H. Conversely given any strongly continuous, one-parameter group of unitary operators acting in $\mathcal{H}$, consider for every $\Psi \in \mathcal{H}$ the function

$$F_\Psi : t \in \mathbb{R} \mapsto (U_t\Psi, \Psi) \in \mathbb{C} \tag{47}$$

These functions are continuous, of positive-type, and satisfy $|F_\Psi(t)| \leq \| \Psi \|^2 = F_\Psi(0)$. The classical Bochner theorem (see above, and in particular eqn. 39) thus allows us to reverse the above construction and to obtain a spectral family $\{P_\lambda | \lambda \in \mathbb{R}\}$ for which (45) is satisfied. From this spectral family, the self-adjoint operator H is constructed by using (24) and (25), thus giving the second part of Stone's theorem. Finally, from the point of view of partial differential equation, it is interesting to notice, and easy to prove, that

$$D(H) = \{\Psi \in \mathcal{H} \mid s - \lim_{t\to 0} \frac{i\hbar}{t}(U_t - I)\Psi \text{ exists } \} \tag{48a}$$

$$H\Psi = s - \lim_{t\to 0} \frac{i\hbar}{t}(U_t - I)\Psi \quad \forall\, \Psi \in D(H) \tag{48b}$$

and that $D(H)$ is stable under $\{U_t | t \in \mathbb{R}\}$, i.e.

$$\Psi \in D(H) \Rightarrow U_t\Psi \in D(H) \quad \forall\, t \in \mathbb{R} \tag{49}$$

The equality

$$HU_t = i\hbar\partial_t U_t \tag{50}$$

therefore holds on $D(H)$; and thus, for every $\Psi \in D(H)$ $\Psi_t \equiv U_t\Psi$ *solves the time-dependent Schroedinger equation* (2) in a sense now made mathematically precise.

The validity of the equality (50) on $D(H)$ is the reason why the self-adjoint operator H is sometimes referred to as the *generator* of $\{U_t | t \in \mathbb{R}\}$. This name, however, is alternatively used for the antiself-adjoint operator $(-iH/\hbar)$ and one writes formally

$$U_t = \exp\{-\frac{i}{\hbar}Ht\} \tag{51}$$

Whenever the latter formula is used in this book it is meant as a shorthand notation for (45): the bounded operator U_t is defined on the whole of $\mathcal{H}$ whereas the expansion of the exponential as a power series only makes sense when the domain of convergence is carefully controlled.

The Schroedinger equation being now defined as a well-posed mathematical problem, we can address the question of whether it is a physical theory.

We already mentioned that the solution (35–36) of the eigenvalue equation (1) with the potential $V(x) = x^2/2$ coincides with the Heisenberg, Born & Jordan treatment of the harmonic oscillator with Hamiltonian $(p^2+q^2)/2$. For an arbitrary harmonic oscillator, i.e. for $V(x) = \omega^2 x^2/2$, one obtains similarly an orthonormal basis $\{\Psi_n | n \in \mathbb{Z}^+\}$ of eigenvectors $\Psi_n \in \mathcal{H}$ corresponding to the non-degenerate eigenvalue

(52) $$\epsilon_n = (n + \frac{1}{2})h\nu$$

with $\nu = \omega/2\pi$. To make contact with the physics of the "old quantum theory" of Chapter Seven, we first notice that the infinite sums involved in the following expressions converge for every strictly positive number $\beta > 0$

(53a) $$Z_\beta \equiv \sum_{n=0}^{\infty} e^{-\beta\epsilon_n}$$

(53b) $$< H >_\beta \equiv Z_\beta^{-1} \sum_{n=0}^{\infty} \epsilon_n e^{-\beta\epsilon_n}$$

(53c) $$< H^2 >_\beta \equiv Z_\beta^{-1} \sum_{n=0}^{\infty} \epsilon_n{}^2 e^{-\beta\epsilon_n}$$

Furthermore:

(54) $$< H^2 >_\beta - < H >_\beta^2 = \partial_\beta{}^2 \ell n Z_\beta$$

from which we compute

(55) $$< H^2 >_\beta - < H >_\beta^2 = (h\nu)^2 e^{\beta h\nu}(e^{\beta h\nu} - 1)^{-2}$$

For $\beta \equiv (kT)^{-1}$, this expression coincides, up to the usual factor $(8\pi\nu^2/c^3)$, with the Einstein fluctuation formula (7.2.26) which itself was seen (in Section 7.2) to contain much of the physics of the Planck radiation law (7.1.22). Recall in particular that the Einstein fluctuation formula (7.2.26) is the most compelling statement of the Einstein particle-wave duality and of the meaning of the classical limit. These were two of the essential conceptual ingredients in Schroedinger's approach, and it is, therefore, interesting that they can be recovered in (55) from the Schroedinger equation: the baby has not been thrown away with the water of the bath! Moreover, the manner in which (55) has been obtained points the way to the full-fledged quantum theory to be developed in Section Three below, where we will see that (54), through (53), can be interpreted as the dispersion of the energy of the harmonic oscillator in canonical thermal equilibrium at temperature T.

The less compelling part of the argument presented so far in this section is perhaps the derivation (4)–(11) of the equations (1) and (2). Indeed, Dirac (1961) recalls in his obituary for Schroedinger, quoted in Jammer (1966), that after having derived his equation Schroedinger had abandoned it for a few months, since he could not at first satisfactorily obtain from it the energy levels of the (relativistic) hydrogen atom. He only believed in his equation, strongly enough to publish it, when he succeeded in showing that the eigenvalues E_n solving equation (1) for the potential

$$V : x \in \mathbb{R}^3 \mapsto -e^2 r^{-1} \in \mathbb{R} \tag{56}$$

with $r = (\Sigma_{k=1}^3 x_k^2)^{1/2}$ and e the charge of the electron, coincide with the treatment of the (non-relativistic) hydrogen atom by Bohr. Since the latter was vindicated by experimental verification—except for the effect of the spin of the electron, which had not yet been recognized—a physical theory was born, which could compete with the others. First amongst these was the new theory of Heisenberg, Born & Jordan, presented in Section One.

The reaction to Schroedinger's first paper was mixed, ranging from great enthusiasm (Einstein, Planck) to downright hostility (Heisenberg).

Among the enthusiastic reactions, a curious anecdote is worth mentioning here, since it points out an important property of the Schroedinger equation. Having been told about Schroedinger's work by Planck, Einstein understood Schroedinger's equation to be of the form

$$F(\Psi, E) = 0 \tag{57a}$$

with

$$F(\Psi, E) \equiv [\Delta + \frac{E^2}{h^2(E - V)}]\Psi \tag{57b}$$

On April 16, 1926, he wrote to Schroedinger objecting that such an equation did not satisfy a fundamental prerequisite, namely that for two independent systems, [1] and [2], the solutions (Ψ_i, E_i) of $F(\Psi, E) = 0$, $i = 1, 2$, should satisfy

$$F(\Psi_1 \cdot \Psi_2, E_1 + E_2) = 0 \tag{58}$$

which (57b) does not satisfy. Einstein went on to suggest an alternate form for F, namely

$$F(\Psi, E) = [\Delta + \frac{1}{h^2}(E - V)]\Psi \tag{59}$$

which, we know, is Schroedinger's original equation (1), except for a numerical factor which one can take care of, by adjusting the physical units. Einstein nevertheless added in his letter to Schroedinger: "The idea of your article shows real genius." Schroedinger was elated by Einstein's rediscovery of his eigenvalue equation (1), as he wrote to Einstein on April 23, 1926. In the meantime, Einstein had checked Schroedinger's paper and had sent him a note, dated April 22, 1926, to the effect that his letter "was superfluous". Not so to us, since this initial letter by Einstein contains (58) which is indeed an essential property of the Schroedinger equation. In a further letter to Schroedinger, dated April 26, 1926, Einstein wrote: "I am convinced that you have made a decisive advance with your formulation of the quantum condition, just as I am equally convinced that the Heisenberg-Born route is off the track." This last comment illustrates the fact that, at the time of their almost simultaneous publication, the contributions of Heisenberg, Born & Jordan and of Schroedinger appeared to present conflicting views of the essence of quantum theory.

Heisenberg was "deeply disturbed". In a letter to Pauli—see Fierz & Weisskopf (1960)—he wrote: "The more I reflect on the physical part of the Schroedinger theory, the more detestable [or disgusting] I find it. Schroedinger really simply throws overboard everything in quantum theory: namely the photoelectric effect, the Franck collisions, the Stern-Gerlach effect, etc. Then it is not hard to make a theory."

In this climate of heated controversy, Schroedinger (1926) in his paper "On the relation of the Heisenberg-Born-Jordan quantum mechanics to mine" established that the two theories were, in act, formally equivalent. He indeed recognized—see also C. Eckart (1926)—that the operators q and p, with

$$(q\Psi)(x) = x\Psi(x) \quad \text{and } (p\Psi)(x) = -i\hbar(\partial_x\Psi)(x) \tag{60}$$

(for a precise definition of q and p as self-adjoint operators see (26) and (29) above) satisfy the canonical commutation relation (30) which he compared to the Heisenberg-Born-Jordan relation (8.1.47) as follows. Let $\{\Psi_n | n \in \mathbb{Z}^+\}$ be an orthonormal basis in $\mathcal{L}^2(\mathbb{R}, dx)$ (we should add: with $\Psi_n \in D(p) \cap D(q) \quad \forall n \in \mathbb{Z}^+$), e.g. the orthonormal basis (34) consisting of the solutions of the eigenvector equation (1) for the harmonic oscillator (this basis satisfies $\Psi_n \in S(\mathbb{R}) \subset D(p) \cap D(q) \quad \forall n \in \mathbb{Z}^+$ indeed). Then the matrices $\{q_{n,m} | n, m \in \mathbb{Z}^+\}$ and $\{p_{n,m} | n, m \in \mathbb{Z}^+\}$ satisfy (8.1.47) which appears thus as a special expression of (30).

Upon identifying the operators q and p with the quantum theoretical representative of the position and momentum, Schroedinger obtained that his differential operator (12) could be rewritten in the form

$$\frac{1}{2m}p^2 + V(q) \tag{61}$$

which is precisely the form of the Heisenberg, Born & Jordan Hamiltonian (8.1.48).

The generalization from $n = 1$ to $n(\in \mathbb{Z}^+)$ degrees of freedom is straightforward.

Thus the Heisenberg, Born & Jordan matrix mechanics now appeared to be formally contained in the Schroedinger wave mechanics. The fact that the two theories are in fact mathemaically *equivalent* is the content of a theorem proven by von Neumann (1931) to the effect that, up to unitary equivalence, there exists only one irreducible representation of the canonical commutation relation (30), subject to some auxiliary technical condition necessary to its proper physical interpretation; see Section Three below.

Since two physical theories which are mathematically equivalent are expected to give the same predictions, the controversy subsided. This, however, is a somewhat fallacious argument, and it seems proper to mention here Einstein's warning: "Momentary success carries more power of conviction for most people than reflections on principle"; letter from Einstein to M. Besso, dated July 24, 1949, and quoted in Pais (1982). On principle, both theories were genial, but lucky guesses, and very different ones at that! Which of these inductive guesses one prefers is a matter of taste, and on that point the controversy could have continued unabated. We shall come back on this in Chapter Nine. A contributing factor to the acceptance of a formal reconciliation was Born's own attitude on the role of mathematics in physical thinking.

First of all, Born was ready for the introduction of linear differential operators in quantum theory; this is documented by a paper he had written with Wiener independently of Schroedinger's idea. Indeed Born-Wiener (1925–6), considering a space of functions of the time variable t, had introduced the differential operator

(62a)
$$D = \frac{d}{dt}$$

and had identified it with

(62b)
$$D = \frac{2\pi i}{h} W$$

where W represented the energy. Now (62a & b) give

(62c)
$$W = -i\frac{h}{2\pi}\frac{d}{dt}$$

Upon comparing this expression with the time-dependent Schroedinger equation, written in the form

(63)
$$H = i\frac{h}{2\pi}\partial_t$$

we notice two differences. The first one is the choice of the convention $(-i)$ instead of its complex conjugate $(+i)$; being nothing but a convention, this difference is irrelevant. The second difference is essential; in the RHS of (63) appears a partial differential operator, indicative of Schroedinger's intuition that a wave equation was involved. Born & Wiener missed that. Nevertheless, it is remarkable that they computed the commutation relation between an arbitrary polynomial function (with coefficients in $\mathbb{C}$) of their differential operator D and an arbitrary multiplication operator on their space. From their formula, or directly if one so wishes, one computes in particular, for the multiplication operator T formally defined by $(Tf)(t) = tf(t)$:

$$[D, T] = 1 \tag{64}$$

Born later reminisced—as quoted by Jammer (1966)—that (64) "was absolutely the same as for q and p. But we did not see that. And I never will forgive myself, for if we had done this, we would have had the whole wave mechanics from quantum mechanics at once, a few months before Schroedinger." Their failure to identify (62c) with

$$p = -i\frac{h}{2\pi}\partial_x \tag{65}$$

was, however, fortunate: H and p have in general very different spectral properties, as is in particular the case for the harmonic oscillator which Born & Wiener also solved correctly. Incidentally, we should mention here the following corollary of von Neumann's uniqueness theorem (loc. cit.). Let H be a self-adjoint operator acting on $\mathcal{H} = \mathcal{L}^2(\mathbb{R}^n, d^n x)$, and $\{U_t | t \in \mathbb{R}\}$ be the unitary evolution group generated by H. Suppose, furthermore, that $H - aI \geq 0$ for some $a \in \mathbb{R}$, as is so often the case; e.g. harmonic oscillator, hydrogen atom, etc. Then there exists no "time-operator" T, i.e. there exists no self-adjoint operator $\{T, D(T)\}$ acting on $\mathcal{H}$ and such that $U_{-t}TU_t = T + tI$ on $D(T)$, for all $t \in \mathbb{R}$ (so that one would have, at least formally $[H, T] = i\hbar I$), i.e. no operator T for which anything like (64) would hold.

Having thus recognized his oversight, Born was ready for a closer examination of Schroedinger's theory; in fact he was soon to propose for this theory a probabilistic interpretation which is still very much a part of quantum theory as we know it today.

The problem which Born addressed was to elucidate the meaning of the wave-function Ψ appearing as a solution of the Schroedinger wave-equation.

Schroedinger himself ascribed, at first erroneously to Ψ, and then correctly to

$$\rho(x, t) \equiv e|\Psi(x, t)|^2 \tag{66}$$

the meaning of an electric charge distribution; in (66) e is the charge of the electron, and $|\Psi(x, t)|$ is the absolute value of the complex number $\Psi(x, t)$; recall that Schroedinger's purpose was to solve the problem of the hydrogen atom, where the Hamiltonian H is given by the LHS of (2) with V defined in (56), so that Ψ was to describe the wave to be associated to the electron of this atom.

In particular, upon computing the time-derivative of ρ given by (66) for any solution Ψ of the time-dependent equation (2), Schroedinger found back the Maxwell continuity equation

$$\dot{\rho} + \text{div } J = 0 \tag{67}$$

with

$$J = -i\frac{eh}{4\pi m}\{\Psi^* \cdot \nabla\Psi - (\nabla\Psi)^*\Psi\} \tag{68}$$

Upon substituting into (68) the form

$$p = -i\hbar\nabla \tag{69}$$

of (65), he obtained

$$J = e\frac{1}{2}\{\Psi^*(\frac{1}{m}p\Psi) + (\frac{1}{m}p\Psi)^*\Psi\} \tag{70}$$

in close analogy with the classical formula

$$J = \eta e v = \eta e(\frac{1}{m}p) \tag{71}$$

In a series of papers Born (1926) consolidated this tentative result, while reflecting on Einstein's particle-wave duality and the experimental data on electron scattering recently obtained by J. Franck, Born's Göttingen colleague. This series of papers is the basis of what is called today the *Born statistical interpretation of quantum mechanics*; their lasting influence was recognized by the award of the 1954 Nobel prize to Born.

One particular aspect of Born's contribution was that he proposed to interpret $|\Psi(x)|^2$ as the *probability density* for a *particle* to be at the position x when it is described by the *wave-function* Ψ. Notice that the requirement that the probability density $|\Psi(x)|^2$, defined through

$$\mu_\Psi(dx) = |\Psi(x)|^2 d^n x, \tag{72}$$

be a finite measure μ_Ψ on $\mathbb{R}$, is equivalent to the requirement that the wave-function Ψ be a square-integrable function of x, i.e. that Ψ belong to the *Hilbert space* $\mathcal{H} = \mathcal{L}^2(\mathbb{R}^n, d^n x)$; this gives a physical interpretation to our initial requirement on the class of admissible solutions to the Schroedinger equation.

Born, however, went much further, and what he did is best explained using the spectral theorem. We already mentioned (see 23–25 above) that to every self-adjoint operator $\{A, D(A)\}$ acting in a Hilbert space $\mathcal{H}$, corresponds a unique spectral family $\{P_\lambda | \lambda \in \mathbb{R}\}$. We also pointed out that, given this spectral family, one can define for every $\Psi \in \mathcal{H}$ with $\| \Psi \| = 1$ a probability measure on $\mathbb{R}$, namely (see 23)

$$\mu_\Psi : \Delta \in \mathcal{B}(\mathbb{R}) \mapsto \int_\Delta d(P_\lambda \Psi, \Psi)$$

We finally remarked that with this probability measure, one has, provided that $\Psi \in D(A)$:

$$(A\Psi, \Psi) = \int_{\mathbb{R}} \lambda \, \mu_\Psi(d\lambda) \tag{73}$$

(compare with 25). The heart of the Born statistical interpretation of quantum mechanics is to *interpret* the expression (73) as the *expectation value* $< A >_\Psi$ *of the observable A when the system is described by the wave function* Ψ.

In particular for the operator $\{q, D(q)\}$ acting in $\mathcal{L}^2(\mathbb{R}, dx)$ and defined by (26), we obtain from the spectral family (21), that the corresponding probability measures μ_Ψ are absolutely continuous with respect to the Lebesgue measure on $\mathbb{R}$ and are given by (72) (with $n = 1$; the generalization to $n \in \mathbb{Z}^+$ is straightforward); we have, therefore, recovered the particular case with which we started our account of the Born interpretation.

Let us now examine the connection of this interpretation with Schroedinger's. The interpretation of ρ (see 66) in Born's perspective is clear; moreover, this also accounts for J (see 68). Let $\mathcal{H} \equiv \mathcal{L}^2(\mathbb{R}^n, d^n x)$, $\tilde{\Psi}$ denote the Fourier transform of Ψ, and $p_j (1 \le j \le n)$ be the j-th component of the momentum, i.e. the operator $p_j = -i\hbar\partial_j$. The expression

$$(p_j \Psi, \Psi) = \int d^n k (\hbar k_j) |\tilde{\Psi}(k)|^2 \tag{74}$$

is well defined for every $\Psi \in D$ where

$$D \equiv \bigcap_{j=1}^{n} D(p_j) \tag{75a}$$

with

$$D(p_j) \equiv \{\Psi \in \mathcal{H} | \int d^n k \; k_j^2 \; |\tilde{\Psi}(k)|^2 < \infty\} \tag{75b}$$

Upon taking the inverse Fourier transform of (74), we obtain

$$\begin{aligned} (p\Psi, \Psi) &= \int d^n x \; \Psi^*(x) \, (-i\hbar\nabla\Psi)(x) \\ &= \int d^n x \, (-i\hbar\nabla\Psi)^*(x) \, \Psi(x) \end{aligned} \tag{76}$$

(where we used the vector notation $\nabla = \{\partial_j | 1 \leq j \leq n\}$). For any $(e/m) \in \mathbb{R}$ and every $\Psi \in D$ we have, therefore, that the expected value

$$<\frac{e}{m}p>_\Psi \equiv (\frac{e}{m}p\Psi, \Psi) \tag{77}$$

of the current $J \equiv ep/m$, when the system is described by the wave-function Ψ, is given by:

$$<\frac{e}{m}p>_\Psi = \int d^n x\, J(x) \tag{78}$$

with $J(x)$ precisely given by (68). Hence Schroedinger's interpretation appears indeed as a particular case of Born's.

Let us now turn to another particular case, also considered by Born. Let $\Psi \in \mathcal{H}$ with $\| \Psi \| = 1$, and A be any self-adjoint operator acting in $\mathcal{H}$, and such that

$$A\Psi = a\Psi \tag{79}$$

for some $a \in \mathbb{R}$. Note that such an operator always exists, e.g.

$$P_\Psi : \Phi \in \mathcal{H} \mapsto (\Phi, \Psi)\Psi \in \mathcal{H} \tag{80}$$

satisfies (79). Notice then that for any A satisfying (79)

$$<(A - <A>_\Psi I)^2>_\Psi = 0 \tag{81}$$

i.e. A shows no dispersion for Ψ. Conversely, if Ψ and A satisfy (81), then they also satisfy (79), and the corresponding probability measure $\mu_\Psi(d\lambda)$ is a Dirac measure concentrated at $a = <A>_\Psi$.

Up to this point of our presentation of Born's interpretation we only encountered straight, classical probability theory. The first indication that something might be amiss is given by the following remark.

Let A and B be two self-adjoint operators acting in a Hilbert space $\mathcal{H}$, and suppose for sake of simplicity that $D(A) = \mathcal{H} = D(B)$. Define then for every $\Psi \in \mathcal{H}$

$$A_0 \equiv A - <A>_\Psi I \tag{82}$$

$$\Delta_\Psi A \equiv (<A_0{}^2>_\Psi)^{\frac{1}{2}} \tag{83}$$

Clearly $\Delta_\Psi A$ is to be interpreted as the dispersion of A with respect to Ψ. Define similarly B_0 and $\Delta_\Psi B$ and notice that:

$$[A_0, B_0] = [A, B] \tag{84}$$

and that:

(85a) $$f : \lambda \in \mathbb{R} \mapsto \| (A_0 - i\lambda B_0)\Psi \|^2 \in \mathbb{R}^+$$

is a positive definite quadratic function of λ which can be rewritten in the form

(85b) $$f(\lambda) = (\Delta_\Psi A)^2 + \lambda < i[A, B] >_\Psi + \lambda^2(\Delta_\Psi B)^2$$

Hence

(86) $$\Delta_\Psi A \cdot \Delta_\Psi B \geq \frac{1}{2} \| < i[A, B] >_\Psi \|$$

Some care must be used if one wants to extend this inequality to the case when A or B, or both, are unbounded. Consider for instance the case where $\mathcal{H} = \mathcal{L}^2([0,1], d\theta)$; Θ is the bounded operator $(\Theta\Psi)(\theta) = \theta\Psi(\theta)$; and L is the operator $-i\hbar d/d\theta$ defined on the dense subspace $D(L)$ the elements of which are the absolutely continuous functions on $[0,1]$ such that $\Psi(0) = \Psi(1)$ and $\Psi' \in \mathcal{L}^2([0,1], d\theta)$. It is easy to find a dense subspace D in $\mathcal{H}$, on which $[L, \Theta]$ is well-defined and equal to $-i\hbar I$; hence $[L, \Theta]$ admits a unique extension, by continuity, to a bounded operator on $\mathcal{H}$. Note now that for Ψ defined by $\Psi(\theta) = \exp\{2\pi i\theta\}$, both sides of (86) are well-defined and violate the inequality: $LHS = 0$, $RHS = 1/2$. A quick survey of the proof will show why. This warning having been sounded, the reader will check that the proof, as we gave it, does apply to the case where A (resp. B) is replaced by p (resp. q) defined by (29) (resp. 26), *and* Φ is restricted to belong to the dense subspace $S(\mathbb{R}) \subset D(p) \cap D(q)$, which is stable under both p and q. The canonical commutation relation (30) then implies, through (86), that

(87) $$\Delta_\Psi p \cdot \Delta_\Psi q \geq \frac{1}{2}\hbar$$

This is the so-called *Heisenberg uncertainty relation* which states that there is *no* wave-function $\Psi(\in S(\mathbb{R}))$ with respect to which the product of the dispersions of p and q is smaller than $\hbar/2$. Although (87) is in fact a result which would have naturally belonged to classical Fourier analysis, the first time a relation of this type appeared, in the context of quantum mechanics, was in a paper by Heisenberg (1927) where this author wrote $\delta p \cdot \delta q = h/2\pi \equiv \hbar$ (the equality and the missing factor $\frac{1}{2}$ are not typographical oversights here). Kennard (1927) pointed out that the lower limit in (87) is reached for Gaussian wave-functions Ψ; the lower eigenfunction of the harmonic oscillator provides an immediate physical example of this situation; in this case $< p >_\Psi = 0 = < q >_\Psi$; similarly, for every $(\alpha, \beta) \in \mathbb{R}^2$ a wave-function $\Psi \in S(\mathbb{R})$ can be found such that $< p >_\Psi = \alpha$, $< q >_\Psi = \beta$ and the equality sign in (87) holds; these are the "Schroedinger coherent states"; they will appear in a more general context towards the end of Section Three. At Pauli's suggestion, Weyl

(1928) derived (87) from the Cauchy-Schwartz inequality. The proof of (86–87) we gave here—which is indeed reminiscent of one of the classical proofs of the Cauchy-Schwartz inequality—is essentially that of Robertson (1929u). As Kennard had already suggested, (87) soon became recognized as lying at "the core of the new theory" and we find this relation at the very beginning of Pauli's epoch-marking *Principles of Wave Mechanics*; Pauli (1933).

The Born interpretation and outlook—evidently including (86–87)—were part and parcel of the lectures given in Göttingen by Hilbert during the Winter 1926–1927; see Hilbert, von Neumann & Nordheim (1927). This first attempt to present quantum mechanics in an axiomatic manner most probably played an important stimulating role in the approach of von Neumann (1932) which we shall review in Section Three.

In view of the close connection between Schroedinger's interpretation of his wave-function, and the more general interpretation given by Born—see in particular (77–78)—it might appear surprising that Schroedinger was not satisfied. Nor was Einstein. In a letter to Schroedinger, dated May 31, 1928, Einstein indeed wrote: "The Heisenberg-Born tranquilizing philosophy—or religion?—is so delicately contrived that, for the time-being, it provides a gentle pillow for the true believer from which he cannot very easily be aroused. So let him lie there. But this religion has ... damned little effect on me."

Einstein's lasting dissatisfaction came to focus on the claim, made by the orthodox "believers", that the new quantum mechanics is a *complete theory*, in the same manner as the classical mechanics of the Nineteenth Century could be considered to be a complete theory. This disagreement has been sharpened into a far-reaching debate on "determinism" and "reality". As a simple matter of facts, one should notice that in classical mechanics there exists a large class of states which are dispersion-free for *all* observables *simultaneously*, namely the Dirac measures on phase space, i.e. on the cotangent bundle T^*Q of the configuration space Q, e.g. $\mathbb{R}^n$. Even if one does not like the use of Dirac measures in classical mechanics, one can at least always approximate them by completely additive, smooth measures in such a way that the dispersion of all observables can be made, simultaneously, as small as one wants. This is precisely what seems to be precluded, in quantum theory, by the *uncertainty relations* (86)–(87): this was a hard feature to accept from a "complete theory".

This last statement needs to be made mathematically more precise, and this will be done when we shall have a mathematical structure which is precise enough. Such a structure is provided by the work of von Neumann (1932) to be discussed in the next section.

Section 3. VON NEUMANN'S HILBERT SPACE FORMALISM

A formal consolidation of the quantum theories presented in the previous two sections took place in the late 1920's and early 1930's. This development is witnessed by the impressive number of treatises—some of them having now become true classics—which appeared on the subject during this period, e.g. Schroedinger (1926), Biggs (1927), Birtwistle (1928), de Broglie (1928), Landé (1928), Weyl (1928), Frenkel (1929), Sommerfeld (1929), Born & Jordan (1930), Dirac (1930), Wigner (1931), von Neumann (1932), and Pauli (1933). Among these, the mathematical synthesis offered by von Neumann, in his *Mathematical Foundations of Quantum Mechanics*, seems to be the closest to our purposes here which are: (a) to present the theory in a form germane to the tastes of a mathematician, and (b) to open the road to an axiomatic treatment of the subject.

The way by which von Neumann captured the mathematical essence of the quantum theories known at the time can be summarized as follows—although this concise formulation is not to be found verbatim in von Neumann's book.

VON NEUMANN'S POSTULATES. *(1) A physical system is characterized by a triple* $\{S, \mathcal{A}, < . ; . >\}$ *where: S is the set of its possible states (or "modes of preparation"); $\mathcal{A}$ is the set of its observables (or measurable attributes); and* $< . ; . >$ *is a prediction rule which associates to every pair* $(\phi, A) \in S \times \mathcal{A}$ *a real number* $< \phi; A >$ *to be interpreted as the expectation value of the observable A when the system is in the state ϕ. (2) A quantum system is a physical system characterized by a separable Hilbert space $\mathcal{H}$ which allows to make the following identifications: S is identified with the set of all positive, trace-class operators ρ on $\mathcal{H}$ such that $tr(\rho) = 1$; $\mathcal{A}$ is identified with the set $\mathcal{B}_{sa}(\mathcal{H})$ of all bounded self-adjoint operators acting on $\mathcal{H}$; and $< \rho; A >$ is given by $tr(\rho A)$.*

We already indicated, in the synopsis of this chapter, how the physical systems encountered in classical mechanics provide a particular realization of the first postulate. In this section we shall examine the mathematical structures involved in the second postulate, then describe a few consequences of this postulate, and finally point out some of its limitations.

Earlier in this chapter we have used particular models of separable Hilbert spaces, such as ℓ^2 and $\mathcal{L}^2(\mathbb{R}^n, d^n x)$. von Neumann realized that, behind Schroedinger's correct claim that his formalism is equivalent to the Heisenberg-Born & Jordan formalism, lies the fact that the theory does not depend on the choice of a particular model for the Hilbert space, but only on the underlying abstract Hilbert space structure. For sake of completeness, we first briefly state the abstract definitions relevant to this structure.

An *inner product* on a vector space $\mathcal{H}$ over $\mathbb{C}$ is a map $(. \, , \, .) : \mathcal{H} \times \mathcal{H} \to \mathbb{C}$ satisfying:

(1a) (i) $(\lambda\Phi + \lambda'\Phi', \Psi) = \lambda(\Phi, \Psi) + \lambda'(\Phi', \Psi)$ for all $\Phi, \Phi', \Psi \in \mathcal{H}$, and all $\lambda, \lambda' \in \mathbb{R}$

(1b) (ii) $(\Psi, \Phi) = (\Phi, \Psi)^*$ for all $\Phi, \Psi \in \mathcal{H}$ (where * denotes the complex conjugation in $\mathbb{C}$)

(1c) (iii) $(\Phi, \Phi) \geq 0$ for all $\Phi \in \mathcal{H}$

(1d) (iv) $(\Phi, \Phi) = 0$ only when $\Phi = 0$

A *prehilbert space* $\{\mathcal{H}, (.\,,\,.)\}$ over $\mathbb{C}$ is a vector space $\mathcal{H}$ over $\mathbb{C}$, equipped with an inner product $(.\,,\,.)$. Note that

(2) $$\|\,.\,\| : \Phi \in \mathcal{H} \mapsto +(\Phi, \Phi)^{1/2} \in \mathbb{R}^+$$

defines a norm on $\mathcal{H}$. If $\{\mathcal{H}, (.\,,\,.)\}$ is complete with respect to this norm, it is said to be a *Hilbert space* over $\mathbb{C}$.

A Hilbert space is said to be *separable* if it admits a countable orthonormal basis $\{\Psi_i | i \in I\}$, i.e. a countable collection of vectors Ψ_i such that

(3a) $$(\Psi_i, \Psi_j) = \delta_{ij} \quad \text{for all } i, j \in I$$

(3b) $$(\Phi, \Psi_i) = 0 \quad \text{for all } i \in I \Rightarrow \Phi = 0$$

We should note in passing that quantum theories can be built by replacing, in the above definitions, the field $\mathbb{C}$ of complex numbers by the field $\mathbb{R}$ of real numbers, or the field $\mathbb{Q}$ of quaternions. The case of $\mathbb{R}$ has been investigated by Stueckelberg (1960) and his collaborators. The first suggestion that $\mathbb{Q}$ may be used in quantum theory is due to P. Jordan (1934); the algebra of operators on a quaternionic Hilbert space as been studied as early as Teichmueller (1936); the possibilities open by a quaternionic quantum mechanics were investigated by Finkelstein et al. (1959–1963), by Dyson (1962) and, in connection with the classification of the irreducible projective representations of the Lorentz group, by Emch (1963). We shall confine ourselves here to the usual case where $\mathbb{C}$ is the field over which the Hilbert space $\mathcal{H}$ is defined.

While the mathematical structures involved in the three elements S, $\mathcal{A}$ and $<\,.\,;\,.\,>$ are intimately linked, we review them here in sequence.

a. The states

We first turn our attention to the set S appearing in von Neumann's postulate for quantum systems.

A bounded linear operator A, acting on a Hilbert space $\mathcal{H}$, is said to be *positive* whenever there exists a bounded linear operator B, acting on $\mathcal{H}$, such that $A = B^*B$ (where B^* denotes the Hermitian adjoint of B). Note that every positive operator A is self-adjoint and admits a unique positive square-root, denoted $A^{1/2}$. Note then that: a bounded linear operator A is positive if and only if $(A\Phi, \Phi) \geq 0 \quad \forall \Phi \in \mathcal{H}$; A, B positive implies $(A + B)$ positive; and A positive, $\lambda \in \mathbb{R}^+$ implies (λA) positive.

We denote by $\mathcal{B}(\mathcal{H})$ the set of all bounded linear operators acting on a Hilbert space $\mathcal{H}$; $\mathcal{B}^+(\mathcal{H})$ denotes then the set of all positive elements in $\mathcal{B}(\mathcal{H})$; we further denote by $\mathcal{U}(\mathcal{H}) \equiv \{U \in \mathcal{B}(\mathcal{H}) | U^*U = I = UU^*\}$ the set of all *unitary* operators acting on $\mathcal{H}$. Note then $A \in \mathcal{B}^+(\mathcal{H})$ and $\mathcal{U} \in (\mathcal{H})$ implies $U^*AU \in \mathcal{B}^+(\mathcal{H})$.

Let now $\{\Psi_i\}$ be an orthonormal basis in $\mathcal{H}$. The map

(4) $$tr : A \in \mathcal{B}^+(\mathcal{H}) \mapsto \Sigma_i (A\Psi_i, \Psi_i) \in [0, \infty]$$

satisfies the following properties

(5a) $$tr\ (A+B) = tr\ A + tr\ B \quad \forall A, B \in \mathcal{B}^+(\mathcal{H})$$

(5b) $$tr\ (\lambda A) = \lambda tr A \quad \forall A \in \mathcal{B}^+(\mathcal{H}), \forall \lambda \in \mathbb{R}^+$$

(5c) $$tr\ (BB^*) = tr\ (B^*B) \quad \forall B \in \mathcal{B}(\mathcal{H})$$

(5d) $$tr\ (U^*AU) = tr\ A \quad \forall A \in \mathcal{B}^+(\mathcal{H}), \forall U \in \mathcal{U}(\mathcal{H})$$

Note that (5d) is equivalent to the statement that the map tr, given by (4), is independent of the orthonormal basis $\{\Psi_i\}$ chosen to define it. To see that this alternate statement is true, let $\{\Phi_k\}$ be another orthonormal basis in $\mathcal{H}$; by Parseval equality we have indeed:

$$\begin{aligned} tr\ A &\equiv \Sigma_i(A\Psi_i, \Psi_i) \equiv \Sigma_i \parallel A^{1/2}\Psi_i \parallel^2 \\ &= \Sigma_{i,k}|(A^{1/2}\Psi_i, \Phi_k)|^2 = \Sigma_{k,i}|(A^{1/2}\Phi_k, \Psi_i)|^2 \\ &= \Sigma_k \parallel A^{1/2}\Phi_k \parallel^2 = \Sigma_k(A\Phi_k, \Phi_k). \end{aligned}$$

This property (5d) shows why we define the *trace* tr on $\mathcal{B}^+(\mathcal{H})$ and not on $\mathcal{B}(\mathcal{H})$ itself. Suppose indeed that B is a self-adjoint operator with (non-degenerate) discrete spectrum consisting of the terms of a conditionally convergent series; then on rearranging the order of the corresponding (normalized) eigenvectors $\{\Psi_i\}$ of B (which can be achieved by a unitary transformation acting on $\mathcal{H}$), we can make the sum in the RHS of (4) converge to any arbitrary real number. Hence, the requirement that the trace be independent of the basis chosen to define it precludes any extension of its domain of definition to the whole of $\mathcal{B}(\mathcal{H})$; this is one reason behind our choice of the domain of tr in (4); there are other reasons as well, e.g.

(6) $$tr\ A = 0 \quad \text{implies} \quad A = 0.$$

A bounded linear operator T is said to be of *trace-class* whenever

(7) $$\parallel T \parallel_{\mathcal{T}} \equiv tr\{(T^*T)^{1/2}\} < \infty$$

We denote by

(8) $$\mathcal{T}(\mathcal{H}) \equiv \{T \in \mathcal{B}(\mathcal{H}) \mid \parallel T \parallel_{\mathcal{T}} < \infty\}$$

the set of all trace-class (bounded linear) operators on $\mathcal{H}$. Three properties of $\mathcal{T}(\mathcal{H})$ are important in connection with von Neumann's postulate.

Firstly, with $T, S \in \mathcal{T}(\mathcal{H})$ and $\lambda, \mu \in \mathbb{C}$, the linear operator

(9) $$(\lambda T + \mu S) : \Psi \in \mathcal{H} \mapsto \lambda(T\Psi) + \mu(S\Psi) \in \mathcal{H}$$

belongs again to $\mathcal{T}(\mathcal{H})$, and this composition law equips $\mathcal{T}(\mathcal{H})$ with the structure of a vector space over $\mathbb{C}$. Furthermore, the norm (7) then equips $\mathcal{T}(\mathcal{H})$ with the structure of a Banach space; in particular, every Cauchy sequence $\{T_n\} \subset \mathcal{T}(\mathcal{H})$ with respect to this norm converges to an element $T \in \mathcal{T}(\mathcal{H})$.

Secondly, with $T \in \mathcal{T}(\mathcal{H})$ and $A \in \mathcal{B}(\mathcal{H})$, the linear operator

$$AT : \Psi \in \mathcal{H} \mapsto A(T\Psi) \in \mathcal{H} \tag{10}$$

belongs again to $\mathcal{T}(\mathcal{H})$, and we have:

$$\| AT \|_{\mathcal{T}} \leq \| A \| \cdot \| T \|_{\mathcal{T}} \tag{11}$$

where

$$\| A \| \equiv \sup_{\Psi \in \mathcal{H}, \|\Psi\| \leq 1} \| A\Psi \| \tag{12}$$

Similarly $TA \in \mathcal{T}(\mathcal{H})$. Hence $\mathcal{T}(\mathcal{H})$ is a two-sided ideal in $\mathcal{B}(\mathcal{H})$; it happens also to be a *-ideal, i.e. $T \in \mathcal{T}(\mathcal{H})$ implies $T^* \in \mathcal{T}(\mathcal{H})$ (where T^* denotes the Hermitian adjoint of T.

Thirdly, with

$$\mathcal{F}(\mathcal{H}) \equiv \{F \in \mathcal{B}(\mathcal{H}) | \dim(F\mathcal{H}) < \infty\} \tag{13}$$

denoting the set of all *finite-rank operators* acting on $\mathcal{H}$, it is easy to check that $\mathcal{T}(\mathcal{H})$ is the completion of $\mathcal{F}(\mathcal{H})$ with respect to the norm (7). Let $\mathcal{C}(\mathcal{H})$ be the Banach space obtained as the completion of $\mathcal{F}(\mathcal{H})$ with respect to the norm (12); the elements of $\mathcal{C}(\mathcal{H})$ are called *compact* (or *completely continuous*) *operators* on $\mathcal{H}$. One can prove—see Schatten (1950)—that

$$\mathcal{T}(\mathcal{H}) = \mathcal{C}(\mathcal{H})^* \quad \text{and} \quad \mathcal{T}(\mathcal{H})^* = \mathcal{B}(\mathcal{H}) \tag{14}$$

where $\mathcal{X}^*$ denotes the Banach space *dual* to the Banach space $\mathcal{X}$, i.e. $\mathcal{X}^*$ is the set of all bounded linear maps $\varphi : \mathcal{X} \to \mathbb{C}$, equipped with the natural vector space structure over $\mathbb{C}$, namely

$$< \lambda\varphi + \mu\psi; X > = \lambda < \varphi; X > + \mu < \psi; X > \quad \forall X \in \mathcal{X}, \tag{15}$$

and the norm

$$\| \varphi \| = \sup_{X \in \mathcal{X}, \|X\| \leq 1} | < \varphi; X > | \,. \tag{16}$$

It is to be noted—see Dixmier (1950)—that

$$\mathcal{B}(\mathcal{H})^* = \mathcal{T}(\mathcal{H}) \oplus \mathcal{C}(\mathcal{H})^{\perp} \quad \text{where} \tag{17a}$$

$$\mathcal{C}(\mathcal{H})^{\perp} \equiv \{\varphi \in \mathcal{B}(\mathcal{H})^* | < \varphi; X > = 0 \ \forall X \in \mathcal{C}(\mathcal{H})\} \tag{17b}$$

Finally—see Schatten (1957)—$\mathcal{C}(\mathcal{H})$ is not the dual of any Banach space, unless $\mathcal{H}$ is finite-dimensional (in which case: $\mathcal{F}(\mathcal{H}) = \mathcal{T}(\mathcal{H}) = \mathcal{C}(\mathcal{H}) = \mathcal{B}(\mathcal{H}) = \mathcal{B}(\mathcal{H})^*$).

These mathematical preliminaries were not all available at the time of the original formulation of the von Neumann postulate; the hindsight they provide allows us, however, to bring in perspective, and thus better understand, the true nature and possible limitations of the original formulation.

We first notice that *the set S of states* considered by von Neumann is now mathematically well-defined as

$$(18) \qquad S \equiv \{\rho \in \mathcal{T}(\mathcal{H}) | \rho \text{ positive, and } tr\, \rho = 1\}$$

In the physical literature, the elements ρ of S are referred to as *density matrices*, or following von Neumann (1927), as "statistical operators".

The algebraic structure of S is that of a *convex set*, i.e. for every $\rho_1, \rho_2 \in S$ and every $\lambda \in [0,1]$

$$(19) \qquad \lambda\rho_1 + (1-\lambda)\rho_2 \in S$$

A state $\rho \in S$ is said to be *pure* if it is an extreme point of S, i.e. if ρ cannot be written in the form (19) with $\rho_1, \rho_2 \in S$, $\rho_1 \neq \rho_2$ and $\lambda \in (0,1)$. We denote by $\mathcal{E}$ the set of all pure states.

This algebraic structure of S is in line with the interpretation of each element ρ of S as a summary of a mode of preparation for the systems under consideration: a mode of preparation is intuitively characterized by the ensemble of all systems which are prepared according to a certain prescription. Then $\rho \equiv \lambda\rho_1 + (1-\lambda)\rho_2$ is the mode of preparation corresponding to an ensemble where a fraction λ of the systems is prepared according to the prescription ρ_1, *and* a fraction $(1-\lambda)$ of the systems is prepared according to the prescription ρ_2. So goes, at least, the conventional wisdom.

Let us now continue with the analysis of the mathematical structure of the set S defined by (18). Topologically, S is a closed subset of $\mathcal{T}(\mathcal{H})$ in the $\|\cdot\|_{\mathcal{T}}$ -norm topology. In fact, S is the closure, in this topology, of the convex set $S \cap \mathcal{F}(\mathcal{H})$; every $\rho \in S$ can be written in the form

(20a) $\qquad \rho = \Sigma_i\ \lambda_i\, P_{\Psi_i} \qquad$ where

(20b) $\qquad \lambda_i \in [0,1] \quad$ with $\Sigma_i \lambda_i = 1$

(20c) $\qquad P_{\Psi_i} : \Phi \in \mathcal{H} \mapsto (\Phi, \Psi_i)\Psi_i \in \mathcal{H}$

(20d) $\qquad \{\Psi_i\}$ is an orthonormal basis in $\mathcal{H}$.

Note that for every $\Psi \in \mathcal{H}$ with $\|\ \Psi\ \| = 1$, P_Ψ belongs to the set $\mathcal{E}$ of all pure states; and that every $\rho \in \mathcal{E}$ is of this form. Consequently $S \cap \mathcal{F}(\mathcal{H})$ is the set of all finite convext combinations of pure states, i.e. $S \cap \mathcal{F}(\mathcal{H})$ is the convex hull of $\mathcal{E}$ in $\mathcal{T}(\mathcal{H})$. Since S is the closure of this set, S is the *closed convex hull of its extreme points* in the norm-topology of $\mathcal{T}(\mathcal{H})$! We shall see below, however, that this might not be the most natural topology with respect to which one should construct the closed convex hull of $\mathcal{E}$, thus pointing to one of the possible limitations of the von Neumann postulate.

This topological structure of $\mathcal{S}$ is indicative of the ways in which von Neumann consolidated, and generalized, the quantum mechanical formalisms of Schroedinger and Heisenberg together with the synthesis provided by the Born interpretation. Firstly, to every $\Psi \in \mathcal{H}$ with $\| \Psi \| = 1$, i.e. to every Schroedinger wave-function, or to every Born probability amplitude, corresponds the pure state $P_\Psi \in \mathcal{E} \subset \mathcal{S}$ (see 20c); the pure states $\rho \in \mathcal{E}$ being necessarily of the form P_Ψ, they are often referred to as *vector-states*. Secondly, every state $\rho \in \mathcal{S}$ can be written as a convex combination of such pure states (see 20a); when this sum is infinite, the topology in which the sum converges is made explicit: it is the topology of the norm in the Banach space $\mathcal{T}(\mathcal{H})$.

b. The observables

The next object appearing in von Neumann's postulate (2) is the set $\mathcal{B}_{sa}(\mathcal{H})$ of all bounded, self-adjoint operators acting on $\mathcal{H}$.

One might a priori object to the restriction which consists in considering bounded observables only; we already know indeed some important physical observables which are *not* bounded, such as for instance the position Q and the momentum P. We also know, however, that every self-adjoint operator $A : D(A) \to \mathcal{H}$, bounded or not, determines a spectral family $P : \lambda \in \mathbb{R} \mapsto P_\lambda \in \mathcal{B}(\mathcal{H})$ see (8.2.14–19 and 8.2.25) such that

$$(21) \qquad A = \int \lambda \, d\, P_\lambda$$

Upon considering any bounded, continuous, monotonically increasing function $f : \lambda \in \mathbb{R} \mapsto f(\lambda) \in \mathbb{R}$, we can replace an unbounded, self-adjoint operator A by the bounded self-adjoint operator

$$(22) \qquad f(A) = \int f(\lambda) d\, P_\lambda$$

Empirically, this corresponds to the choice of a new (non-linear) scale in which one would now measure the property considered. No major objection of principle can therefore be raised against our restricting first our attention to the set of bounded observables, although we will ultimately want to consider unbounded observables as well since the above construction is rather unwieldy in applications.

$\mathcal{B}_{sa}(\mathcal{H})$ being the set of all the observables we want to consider, it is its mathematical structure that we should analyze with a view towards our ultimate aim, which is to extract from this study an abstract axiomatization of quantum theory. In the perspective of von Neumann's book, however—namely, the formulation of quantum mechanics as a concrete theory of operators acting on a Hilbert space—$\mathcal{B}(\mathcal{H})$ is the primary object. We thus first analyze some of its algebraic and topological properties; more will be said on the latter in Subsection d.

First of all, the composition laws of operators acting on a Hilbert space $\mathcal{H}$, namely

$$(\lambda A + \mu B) : \Psi \in \mathcal{H} \mapsto \lambda(A\Psi) + \mu(B\Psi) \in \mathcal{H} \tag{23}$$

and

$$(AB) : \Psi \in \mathcal{H} \mapsto A(B\Psi) \in \mathcal{H} \tag{24}$$

equip $\mathcal{B}(\mathcal{H})$ with the structure of an *algebra*. Moreover, the Hermitian conjugation

$$A \in \mathcal{B}(\mathcal{H}) \mapsto A^* \in \mathcal{B}(\mathcal{H}) \tag{25}$$

defined by

$$(A^*\Psi, \Phi) = (\Psi, A\Phi) \quad \forall\, \Psi, \Phi \in \mathcal{H} \tag{26}$$

satisfies

$$(A^*)^* = A \tag{27a}$$
$$(\lambda A + \mu B)^* = \lambda^* A^* + \mu^* B^* \tag{27b}$$
$$(AB)^* = B^* A^* \tag{27c}$$

i.e. * is an *involution*, and it equips $\mathcal{B}(\mathcal{H})$ with the structure of an *involutive algebra*.

From the topological point of view, the norm

$$\| \,.\, \| : A \in \mathcal{B}(\mathcal{H}) \mapsto \| A \| \equiv \sup_{\Psi \in \mathcal{H}, \|\Psi\| \leq 1} \| A\Psi \| \tag{28}$$

equips $\mathcal{B}(\mathcal{H})$ with the structure of a *Banach space*. Notice further that we have

$$\| AB \| \leq \| A \| \,.\, \| B \| \tag{29}$$

and

$$\| I \| = 1 \tag{30}$$

where I is the identity operator $I : \Psi \in \mathcal{H} \mapsto \Psi \in \mathcal{H}$, acting as a *unit* in $\mathcal{B}(\mathcal{H})$ (i.e. $IA = A = AI$). These two properties of the norm, together with the Banach space structure of $\mathcal{B}(\mathcal{H})$, characterize the fact that $\mathcal{B}(\mathcal{H})$ is a *Banach algebra* with unit. In fact, $\mathcal{B}(\mathcal{H})$ is a C^**-algebra,* i.e. it is an involutive, Banach algebra where the involution and the norm satisfy the additional condition

$$\| A^*A \| = \| A \|^2 \tag{31}$$

Note in particular that (31), together with (29), implies

$$\| A^* \| = \| A \| \tag{32}$$

Hence, as a consequence of (29) and (32), the multiplication and the involution are continuous operations in the topology induced by the norm.

A C^*-algebra $\mathcal{A}$ which, as a Banach space, is the dual of a Banach space $\mathcal{X}$ is called a *W^*-algebra*; the Banach space $\mathcal{X}$, of which $\mathcal{A}$ is then the dual, is called the *predual* of $\mathcal{A}$ and it is denoted $\mathcal{A}_*$.

We already mentioned—see (14)—that $\mathcal{B}(\mathcal{H})$ with its norm (28) is the dual of the Banach space $\mathcal{T}(\mathcal{H})$ of trace-class operators, equipped with its trace norm $\| T \|_{\mathcal{T}} = tr\{(T^*T)^{1/2}\}$. Hence $\mathcal{B}(\mathcal{H})$ is a W^*-algebra.

Another example, and a counterexample, will illustrate these notions. The space $\mathcal{L}^\infty(\mathbb{R}^n)$ of essentially bounded functions from $\mathbb{R}^n$ to $\mathbb{C}$, equipped with its usual vector space structure, the point-wise multiplication, the sup-norm and the involution $f \to f^*$, defined by $f^* : x \in \mathbb{R}^n \mapsto f(x)^* \in \mathbb{C}$, is clearly a C^*-algebra. The space $\mathcal{L}^1(\mathbb{R}^n)$ of absolutely integrable functions from $\mathbb{R}^n$ to $\mathbb{C}$, equipped with its natural vector space structure and its $\mathcal{L}^1$-norm is a Banach space, the dual of which is $\mathcal{L}^\infty(\mathbb{R}^n)$. Hence $\mathcal{L}^\infty(\mathbb{R}^n)$ is a W^*-algebra. Note that the product in $\mathcal{L}^\infty(\mathbb{R}^n)$ trivially satisfies $fg = gf$, i.e. $\mathcal{L}^\infty(\mathbb{R}^n)$ is *abelian* (one also says "commutative"); $\mathcal{B}(\mathcal{H})$ is not, and this is an essential difference between classical and quantum mechanics.

Let us now mention an example of a C^*-algebra which is not a W^*-algebra. The set $\mathcal{C}(\mathcal{H})$ of compact operators—which we obtained previously as the closure, in the norm (12)≡(28), of the set $\mathcal{F}(\mathcal{H})$ of finite-rank operators (see 13)—inherits from $\mathcal{B}(\mathcal{H})$ the structure of C^*-algebra, but it is not a W^*-algebra when $\mathcal{H}$ is infinite-dimensional since $\mathcal{C}(\mathcal{H})$ cannot then be obtained as the dual of any Banach space (Schatten, 1957).

A W^*-algebra with unit, realized as a subalgebra of the W^*-algebra $\mathcal{B}(\mathcal{H})$ for some Hilbert space $\mathcal{H}$, is called a *von Neumann algebra*. Another characterization of these objects will be given in Subsection d. Let us, nevertheless, note here that $\mathcal{B}(\mathcal{H})$ is obviously a von Neumann algebra; so is $\mathcal{L}^\infty(\mathbb{R}^n, d^n x)$ since it can be realized as the algebra of multiplication operators on $\mathcal{L}^2(\mathbb{R}^n, d^n x)$.

The mathematical theory of C^*- and W^*-algebras is well-developed: see Dixmier (1957 and 1964), Sakai (1971), Emch (1972), Pederson (1979), Bratteli & Robinson (1979 and 1980), Kadison & Ringrose (1983). We shall discuss some applications of the general theory in Chapters Nine and Ten. We only wanted here to indicate that these structures are already present in the physical situations covered by von Neumann postulates. We should add that this very fact played an important stimulating role in the development of the mathematical theory itself, starting with Murray & von Neumann (1936).

We now want to turn our attention to the set $\mathcal{B}_{sa}(\mathcal{H})$ of the observables which appear in von Neumann postulate 2, characterizing the quantum physical systems to be considered in this section.

The set

$$\mathcal{B}_{sa}(\mathcal{H}) \equiv \{A \in \mathcal{B}(\mathcal{H}) | A^* = A\} \tag{33}$$

inherits canonically from $\mathcal{B}(\mathcal{H})$—which is a Banach space over $\mathbb{C}$—the structure of a Banach space over $\mathbb{R}$. In fact, $\mathcal{B}(\mathcal{H})$ is the complexification of $\mathcal{B}_{sa}(\mathcal{H})$, since every $A \in \mathcal{B}(\mathcal{H})$ can be written uniquely in the form

$$A = A_1 + i A_2 \tag{34}$$

with $A_1, A_2 \in \mathcal{B}_{sa}(\mathcal{H})$ and $i = \sqrt{-1}$; indeed these conditions determine uniquely

$$A_1 = \frac{1}{2}(A + A^*) \quad \text{and} \quad A_2 = \frac{1}{2i}(A - A^*) \tag{35}$$

The algebraic structure, however, requires some adaptation since $A, B \in \mathcal{B}_{sa}(\mathcal{H})$ does *not* imply $(AB) \in \mathcal{B}_{sa}(\mathcal{H})$ unless $AB = BA$. Notice, nevertheless, that $A \in \mathcal{B}_{sa}(\mathcal{H})$ does imply $A^n \in \mathcal{B}_{sa}(\mathcal{H})$ for all $n \in \mathbb{Z}^+$, and that $A^n \cdot A^m = A^{n+m}$ does make sense within $\mathcal{B}_{sa}(\mathcal{H})$ since A^n and A^m obviously commute with one another. Furthermore, for all $A, B \in \mathcal{B}_{sa}(\mathcal{H})$

$$A \circ B \equiv \frac{1}{2}\{(A+B)^2 - A^2 - B^2\} \tag{36}$$

again belongs to $\mathcal{B}_{sa}(\mathcal{H})$ and satisfies

$$A \circ B \equiv \frac{1}{4}\{(A+B)^2 - (A-B)^2\} \tag{37}$$

This suggests the introduction of the following mathematical structure. An *abelian, real, Jordan algebra* $\mathcal{A}$ is a vector space over $\mathbb{R}$, equipped with a product $\circ : (a, b) \in \mathcal{A} \times \mathcal{A} \mapsto a \circ b \in \mathcal{A}$ satisfying the following four properties for all $a, b, c \in \mathcal{A}$ and all $\lambda \in \mathbb{R}$:

$$(\lambda a) \circ b = \lambda (a \circ b) \tag{38}$$

$$a \circ (b + c) = a \circ b + a \circ c \tag{39}$$

$$a \circ b = b \circ a \tag{40}$$

$$a^2 \circ (b \circ a) = (a^2 \circ b) \circ a \tag{41}$$

where the n-th power a^n (for $n \in \mathbb{Z}^+$) of an element $a \in \mathcal{A}$ is defined recursively by

$$a^n \equiv a^{n-1} \circ a \tag{42}$$

The map

$$\{.\,,\,.\,,\,.\}:(a,b,c)\in \mathcal{A}\times\mathcal{A}\times\mathcal{A} \mapsto \{a,b,c\}\equiv(a\circ b)\circ c - a\circ(b\circ c)\in\mathcal{A} \tag{43}$$

is called the *associator* of the Jordan algebra $\mathcal{A}$; the latter is said to be *associative* if $\{a,b,c\} = 0 \quad \forall\ a,b,c \in \mathcal{A}$. The name "abelian real Jordan algebra" has apparently been coined by Albert (1946) and the study of these algebras, together with some of their generalizations, has become a vigourous field of mathematics, see e.g. Braun & Koecher (1966). A brief review of the role and place of Jordan algebras in the general mathematical study of algebras has been written by Paige (1963); it can serve as a good introduction to this field. The concept of a Jordan algebra has its roots in attempts to axiomatize quantum mechanics, see in particular Jordan (1934) and Jordan, von Neumann & Wigner (1934).

Notice, in connection with the above definition, that from (39), (40) and (42) we obtain trivially:

$$(a+b)^2 = a^2 + 2a\circ b + b^2 \tag{44}$$

and thus

$$a\circ b = \frac{1}{2}\{(a+b)^2 - a^2 - b^2\} \tag{45}$$

which takes the same form as (36). Upon using (38) with $\lambda = -1$, in (45) we further obtain

$$(a-b)^2 = a^2 - 2\ a\circ b + b^2 \tag{46}$$

Upon subtracting (46) from (44), we get

$$a\circ b = \frac{1}{4}\{(a+b)^2 - (a-b)^2\} \tag{47}$$

which takes the same form as (37). Upon adding (46) to (44), we further obtain:

$$(a+b)^2 + (a-b)^2 = 2a^2 + 2b^2 \tag{48}$$

The role of (48) in the distributivity of the product $\circ$ has been recognized by von Neumann (1936). Let us indeed substitute $(a+c)$ and $(a-c)$ for a in (48), and then subtract the resulting equalities from one another; upon using (47), we obtain

$$(a+b)\circ c + (a-b)\circ c = 2\ a\circ c \tag{49}$$

Upon replacing $(a+b)$ and $(a-b)$ by a and b respectively, we then recover the distributive law (39). The relevance of this remark to the axiomatization of quantum mechanics comes from the fact that (47) and (48) involve only the composition laws consisting in taking squares and making sums, two composition procedures which are easier to interpret empirically; in particular (47) can be taken as an empirical procedure to construct the symmetric product $\circ$ appearing in the definition of a Jordan algebra.

The reader will now verify that (36), or equivalently (37), equips $\mathcal{B}_{sa}(\mathcal{H})$ with the structure of an abelian, real Jordan algebra. In particular, the symmetric product $\circ$ in $\mathcal{B}_{sa}(\mathcal{H})$ and the ordinary product in $\mathcal{B}(\mathcal{H})$ are linked by the relation

$$A \circ B = \frac{1}{2}(AB + BA) \tag{50}$$

The defining property (41), while straightforwardly satisfied in $\mathcal{B}_{sa}(\mathcal{H})$, might still appear somewhat mysterious as an empirical relation between observables. This property of the symmetric product in $\mathcal{B}_{sa}(\mathcal{H})$ is in fact related to another structure also present in this special Jordan algebra, namely the commutator

$$\{A, B\} \equiv [A, B]/i\hbar \equiv (AB - BA)/i\hbar \tag{51}$$

This suggests the following definition. A *Jordan-Lie algebra* $\mathcal{A}$, with *structure constant* $K \in \mathbb{R}$, is an abelian, real Jordan algebra on which a product $\{.\,,\,.\}$: $(a, b) \in \mathcal{A} \times \mathcal{A} \mapsto \{a, b\} \in \mathcal{A}$ is defined and satisfies the following five properties for all $a, b, c \in \mathcal{A}$ and all $\lambda \in \mathbb{R}$.

$$\{\lambda a, b\} = \lambda\{a, b\} \tag{52}$$
$$\{a, b + c\} = \{a, b\} + \{a, c\} \tag{53}$$
$$\{a, b\} = -\{b, a\} \tag{54}$$
$$\{a, b^2\} = 2\{a, b\} \circ b \tag{55}$$
$$\{a, b, c\} = K\{\{a, c\}, b\} \tag{56}$$

where $\{.\,,\,.\,,\,.\}$ is the "associator" of the Jordan structure.

Note that in every abelian, real Jordan algebra, the associator satisfies the condition

$$\{a, b, c\} + \{b, c, a\} + \{c, a, b\} = 0 \tag{57}$$

for all $a, b, c \in \mathcal{A}$. Consequently, in a Jordan-Lie algebra $\mathcal{A}$ with structure constant $K \neq 0$, the product $\{.\,,\,.\}$ satisfies, because of (56), the Jacobi identity

$$\{\{a, b\}, c\} + \{\{b, c\}, a\} + \{\{c, a\}, b\} = 0 \tag{58}$$

so that $\mathcal{A}$ is indeed a Lie algebra.

Note further that for every $a \in \mathcal{A}$, the map

$$\{a,\ .\} : b \in \mathcal{A} \mapsto \{a, b\} \in \mathcal{A} \tag{59}$$

is linear; moreover, upon replacing b by $(b+c)$ in (55), we obtain

$$\{a, b \circ c\} = \{a, b\} \circ c + b \circ \{a, c\} \tag{60}$$

Hence, $\{a, .\} : \mathcal{A} \to \mathcal{A}$ is a *derivation* with respect to the Jordan structure of $\mathcal{A}$.

We shall see later that this result is important for the interpretation of observables as generators of one-parameter continuous groups of symmetries and provides a quantum generalization of the classical Noether theorem.

The relation (60) is also important in the following connection. Let $\mathcal{X}$ be an associative algebra (over a field of characteristic different from 2); one can then define on $\mathcal{X}$ two bilinear products

$$[x, y]_+ = \frac{1}{2}(xy + yx) \tag{61}$$

$$[x, y]_- = (xy - yx) \tag{62}$$

Let $\mathcal{X}^+$ (resp. $\mathcal{X}^-$) denote the mathematical object consisting of the vector space $\mathcal{X}$ equipped with the product $[.,.]_+$ (resp. $[.,.]_-$). The Poincaré -Birkhoff-Witt theorem states that every Lie algebra is isomorphic to a subalgebra of some $\mathcal{X}^-$. It is known that the analog of this theorem is not valid for Jordan algebras; see for instance Jordan, von Neumann & Wigner (1934). For Jordan-Lie algebras, however, the map

$$(a, b) \in \mathcal{A} \times \mathcal{A} \mapsto ab = a \circ b + \mu\{a, b\} \tag{63}$$

is clearly bilinear. This fact, together with the relation (60) shows that the map (63) is associative, i.e.

$$(ab)c - a(bc) = 0 \tag{64}$$

if and only if

$$K\{\{a, c\}, b\} + \mu^2(\{\{a, b\}, c\} + \{\{b, c\}, a\}) = 0 \tag{65}$$

Whenever the Jacobi identity (58) is satisfied (which, as we saw, is in particular the case when $K \neq 0$), the condition (65) reduces to

$$(K + \mu^2)\{\{a, c\}, b\} = 0 \quad \forall\ a, b, c \in \mathcal{A} \tag{66}$$

Hence, when $K < 0$ we can always define on $\mathcal{A}$ a product, namely (63) with $\mu = \sqrt{-K}$, making $\mathcal{A}$ an associative algebra with

$$a \circ b = \frac{1}{2}(ab + ba) \tag{67a}$$

$$\{a, b\} = \frac{1}{2\mu}(ab - ba)\,. \tag{67b}$$

The case $K = 0$ occurs in classical mechanics where the Jordan product on $C_R^\infty(\mathbb{R}^{2N})$ is given by the point-wise multiplication of functions, while the Lie product is given by the Poisson-bracket

$$\{f, g\} \equiv \sum_{k=1}^{N} (\partial_{q_k} f \cdot \partial_{p_k} g - \partial_{q_k} g \cdot \partial_{p_k} f) \tag{68}$$

The case $K > 0$ occurs in quantum mechanics. Indeed, we already saw that $\mathcal{B}_{sa}(\mathcal{H})$ is an abelian, real Jordan algebra; (51) now makes it a Jordan-Lie algebra with structure constant $K = \hbar^2/4$. Condition (66) shows that if we want to imbed $\mathcal{B}_{sa}(\mathcal{H})$ in an associative algebra $\mathcal{X}$ compatible with its Jordan and Lie structures, we must complexify it. This is precisely what is achieved by embedding $\mathcal{B}_{sa}(\mathcal{H})$ into $\mathcal{B}(\mathcal{H})$.

With the Jordan-Lie structure now identified on $\mathcal{B}_{sa}(\mathcal{H})$ we better understand the meaning of condition (41) in the definition of the Jordan product, namely here

$$A^2 \circ (B \circ A) = (A^2 \circ B) \circ A \quad \forall\ B \in \mathcal{B}_{sa}(\mathcal{H}) \tag{69}$$

From (56) we see that with A, C fixed in $\mathcal{B}_{sa}(\mathcal{H})$

$$\{A, B, C\} = 0 \quad \forall\ B \in \mathcal{B}_{sa}(\mathcal{H}) \tag{70}$$

is equivalent to

$$[[A, C], B] = \quad \forall\ B \in \mathcal{B}_{sa}(\mathcal{H}) \tag{71}$$

Upon using this relation with $B \equiv P_\Phi : \Psi \in \mathcal{H} \mapsto (\Psi, \Phi)\Phi \in \mathcal{H}$ and letting Φ run over the unit ball in $\mathcal{H}$, we find:

$$[A, C] = i\lambda I \quad \text{with } \lambda \in \mathbb{R} \tag{72}$$

Recall now that $\mathcal{B}(\mathcal{H})$ is a Banach algebra with unit. As a consequence of this fact, we will now show that (72) implies

$$[A, C] = 0 \tag{73}$$

Recall indeed that an element X of a Banach algebra $\mathcal{A}$ with unit I is said to be *regular* if there exists an element $X^{-1} \in \mathcal{A}$ such that $XX^{-1} = I = X^{-1}X$; an element $Y \in \mathcal{A}$ which is not regular is said to be *singular*. The *spectrum* $\sigma(Z)$ of an arbitrary element $Z \in \mathcal{A}$ is the set $\sigma(Z) \equiv \{z \in \mathbb{C} | zI - Z \text{ is singular }\}$. Note that $\sigma(Z)$ is non-void and that $|\sigma(Z)| \equiv \sup_{z \in \sigma(Z)} |z| \leq \| Z \|$. With A as in (72), let $a \in \mathbb{R}$ with $a > \| A \|$, and form $A_a \equiv A + aI$; clearly A_a is a regular element in $\mathcal{B}(\mathcal{H})$. For any $\mu \in \mathbb{C}$ we have then:

$$A_a C - \mu I = A_a (C A_a - \mu I)(A_a)^{-1} \tag{74}$$

so that

$$\sigma(A_a C) = \sigma(C A_a) \tag{75}$$

We further notice, as a consequence of (72) that

$$A_a C - \mu I = C A_a - (\mu - i\lambda) I \tag{76}$$

so that

$$\mu \in \sigma(A_a C) \quad \text{iff} \quad (\mu - i\lambda) \in \sigma(C A_a) \tag{77}$$

Together with (75), this implies

$$\mu \in \sigma(A_a C) \quad \text{iff} \quad (\mu - in\lambda) \in \sigma(A_a C) \quad \forall\, n \in \mathbb{Z}^+ \tag{78}$$

Since $\sigma(A_a C)$ is non-void, and bounded (namely by $\| A_a \| \cdot \| C \|$) we must have $\lambda = 0$. Hence (72) implies (73) indeed.

Consequently (70) implies (73), and in particular we now see that (69) expresses precisely that $[A, A^2] = 0$. This gives a more immediate meaning to the defining property (41) of the Jordan product in the Jordan-Lie algebra $\mathcal{B}_{sa}(\mathcal{H})$ of observables appearing in von Neumann's postulate.

Finally, $\mathcal{B}_{sa}(\mathcal{H})$ inherits canonically the topological structures of $\mathcal{B}(\mathcal{H})$: when equipped with the structure it inherits from the C^*-algebraic (resp. W^*-algebraic) structure of $\mathcal{B}(\mathcal{H})$, the Jordan algebra $\mathcal{B}_{sa}(\mathcal{H})$ becomes what is called a JB-algebra (resp. a JW-algebra).

c. The expectation values

The last part of von Neumann's postulates for quantum mechanics is to associate to every pair $(\rho, A) \in \mathcal{S} \times \mathcal{B}_{sa}(\mathcal{H})$ the expectation value $< \rho; A >= tr(\rho A)$ of the observable A when the system is in the state ρ.

We immediately notice that the map

$$\rho : A \in \mathcal{B}_{sa}(\mathcal{H}) \mapsto < \rho; A > \equiv tr(\rho A) \;\in \mathbb{R} \tag{79}$$

satisfies, for all $A, B \in \mathcal{B}_{sa}(\mathcal{H})$ and all $\lambda, \mu \in \mathbb{R}$:

$$< \rho; \lambda A + \mu B > \; = \lambda < \rho; A > + \mu < \rho; B > \tag{80}$$

$$\| \rho \| = 1 \tag{81}$$

where

$$\| \rho \| \equiv \sup_{\substack{A \in \mathcal{B}_{sa}(\mathcal{H}) \\ \|A\| \leq 1}} | < \rho; A > | \tag{82}$$

and

(83) $$< \rho; A^2 > \geq 0$$

From an axiomatic point of view, these properties of the states can be considered as (part of) the empirical reasons for equipping the set of the observables with the structure of a JB-algebra. $\mathcal{B}_{sa}(\mathcal{H})$, however, is not just a JB-algebra: it is a JW-algebra. Let us now see how this additional structure comes about from the point of view of the states.

We first notice that the canonical decomposition $A = A_1 + iA_2$ of an arbitrary element $A \in \mathcal{B}(\mathcal{H})$ with $A_1, A_2 \in \mathcal{B}_{sa}(\mathcal{H})$ allows to extend ρ from the Jordan-Lie algebra $\mathcal{B}_{sa}(\mathcal{H})$, on which it is defined, to its complexification $\mathcal{B}(\mathcal{H})$. This unique extension of ρ to $\mathcal{B}(\mathcal{H})$ is obviously given by:

(84) $$\rho : A \in \mathcal{B}(\mathcal{H}) \mapsto < \rho; A > \equiv tr \ \rho A \in \mathbb{C}$$

and it satisfies, for all $A, B \in \mathcal{B}(\mathcal{H})$ and $\lambda, \mu \in \mathbb{C}$

(85) $$< \rho; A^* > \ = < \rho; A >^*$$

(86) $$< \rho; \lambda A + \mu B > \ = \lambda < \rho; A > + \mu < \rho; B >$$

(87) $$\| \rho \| \ = 1$$

where

(88) $$\| \rho \| \ \equiv \sup_{\substack{A \in \mathcal{B}(\mathcal{H}) \\ \|A\| \leq 1}} | < \rho; A > |$$

and

(89) $$< \rho; A^* A > \ \geq 0$$

These properties suggest to consider the set S^*, the elements of which are the maps

(90a) $$\phi : A \in \mathcal{B}(\mathcal{H}) \mapsto < \phi; A > \in \mathbb{C}$$

satisfying, for all $A, B \in \mathcal{B}(\mathcal{H})$ and $\lambda, \mu \in \mathbb{C}$:

(90b) $$< \phi; A^* A > \ \geq 0$$
(90c) $$< \phi; \lambda A + \mu B > = \lambda < \phi; A > + \mu < \phi; B >$$
(90d) $$< \phi; I > = 1$$

From the fact that $\mathcal{B}(\mathcal{H})$ is a C^*-algebra with unit (where $\| I \| = 1$), one verifies that the maps defined by (90 a–d) satisfy:

(90e) $$< \phi; A^* > = < \phi; A >^* \quad \forall \ A \in \mathcal{B}(\mathcal{H})$$
(90f) $$< \phi; I > = \sup_{\substack{A \in \mathcal{B}(\mathcal{H}) \\ \|A\| \leq 1}} | < \phi; A > | \equiv \| \phi \| = 1$$

and thus

(90g) $$| < \phi; A > | \leq \| A \| \quad \forall \ A \in \mathcal{B}(\mathcal{H})$$

Upon comparing now (90a–g) with (84)–(89), we see that $S \subseteq S^* \subseteq \mathcal{B}(\mathcal{H})^*$. It is, moreover, clear that every $\phi \in S^*$ is uniquely determined by its restriction to $\mathcal{B}_{sa}(\mathcal{H})$. Moreover, S^* shares with S the property of being a convex set. For all these reasons, we refer to S^* as the set of all C^**-states* on the C^*-algebra $\mathcal{B}(\mathcal{H})$. Note in particular, that the definition of S^* does not involve the W^*-algebra structure of $\mathcal{B}(\mathcal{H})$, nor even the fact that $\mathcal{B}(\mathcal{H})$ is an algebra of operators acting on some Hilbert space.

From (17) we conclude that, while $S \subset S^*$, we have $S \neq S^*$ exactly when the dimension of $\mathcal{H}$ is infinite. While this is a clear-cut mathematical result, the question of whether S^* rather than S should be considered as the set of *all* states on a physical system is debatable. As we shall now see, the debate in fact rests on a matter of taste, namely on a choice between two idealizations of the procedures one can actually perform and check in the laboratory.

The arguments in favor of S^* usually involve considering S as a subset of $S^* \subset \mathcal{B}(\mathcal{H})^*$, with $\mathcal{B}(\mathcal{H})^*$ equipped with its *weak* **-topology* (abbreviated w^*-*topology*); this is the weakest topology such that the maps

$$\text{(91a)} \qquad \phi \in \mathcal{B}(\mathcal{H})^* \mapsto <\phi; A> \in \mathbb{C}$$

are continuous for all $A \in \mathcal{B}(\mathcal{H})$; a basis of neighbourhoods for this topology is obtained by considering all sets of the form

$$\text{(91b)} \qquad N(\phi, S, \epsilon) \equiv \{\psi \in \mathcal{B}(\mathcal{H})^* | \ | <\psi; A> - <\phi; A> | < \epsilon \ \forall A \in S\}$$

where $\phi \in \mathcal{B}(\mathcal{H})^*$, S is a finite subset of $\mathcal{B}(\mathcal{H})$, and $\epsilon > 0$. One then argues that this is the correct empirical topology to be put on the set of states of a physical system. In this topology S^* is closed, S is not, and in fact S^* is the w^*-closure of S; more specifically (see e.g. Thm II.1.6 in Emch, 1972) S^* is the w^*-closed convex hull of the set $\mathcal{E}$ of the extreme points of S, i.e. of the set $\mathcal{E}$ of vector states on $\mathcal{B}(\mathcal{H})$. Note that this result is stronger than the result one would have obtained from the Krein-Milman theorem (see e.g. Thm. V.8.4 in Dunford & Schwartz, 1957), namely that S^* is the w^*-closed convex hull of the set $\mathcal{E}^*$ of *its* extreme points. Indeed, provided that $\dim \mathcal{H} = \infty$, $\mathcal{E}^*$ is strictly larger than $\mathcal{E}$, a result which can be established as follows (for details, as well as for the general theory behind this, see Kadison & Singer, 1959). Clearly $\mathcal{E} \subseteq \mathcal{E}^*$. Let now $\mathcal{A} = \mathcal{L}^\infty(\mathbb{R}^n, d^n x)$ be the von Neumann algebra of multiplication operators on $\mathcal{L}^2(\mathbb{R}^n, d^n x)$, and let $S^*_\mathcal{A}$ be the set of all C^*-states on $\mathcal{A}$; see (90), with now $\mathcal{B}(\mathcal{H})$ replaced by $\mathcal{A}$. Again by the Krein-Milman theorem, $S^*_\mathcal{A}$ is the w^*-closed convex hull of the set $\mathcal{E}^*_\mathcal{A}$ of *its* extreme points; so $\mathcal{E}^*_\mathcal{A} \neq \emptyset$. With $\psi \in \mathcal{E}^*_\mathcal{A}$, let $\mathcal{F}_\phi \equiv \{\phi \in S^* \text{ such that } \phi|\mathcal{A} = \psi\} \neq \emptyset$ by the Hahn-Banach theorem. $\mathcal{F}_\phi$ is convex and w^*-compact. From the Krein-Milman theorem, once more, one obtains $\mathcal{F}_\phi \cap \mathcal{E}^* \neq \emptyset$, while $\mathcal{F}_\phi \cap \mathcal{E} = \emptyset$ trivially. This proves our assertion. The above argument shows also how the plane-waves of the physicists can be formalized as C^*-states belonging to $\mathcal{E}^*$ and not to $\mathcal{E}$, thus providing an illustration of the kind of idealization involved when one considers states in S^* which are not in S.

The above argument in favor of S^* notwithstanding, most physics texts traditionally restrict their attention to S. When a case is made for this constraint, it usually has to do with the fact that the states in S, but not those in $S^*\setminus S$, generalize to the quantum realm an idealized property often assumed of the measures occuring in classical probability theory, namely complete additivity. We now extend this condition to the present context.

Let

(92) $$\mathcal{P}(\mathcal{H}) \equiv \{P \in \mathcal{B}(\mathcal{H}) | P = P^* = P^2\}$$

be the collection of all *projectors* acting in $\mathcal{H}$. Note that for every $P \in \mathcal{P}(\mathcal{H})$

(93) $$P\mathcal{H} \equiv \{P\Psi | \Psi \in \mathcal{H}\}$$

is a closed subspace of $\mathcal{H}$. For P and Q in $\mathcal{P}(\mathcal{H})$ we write

(94a) $$P \subseteq Q$$

whenever

(94b) $$P\mathcal{H} \subseteq Q\mathcal{H}$$

i.e. whenever

(94c) $$P\Psi = \Psi \quad \text{implies} \quad Q\Psi = \Psi$$

This relation is clearly a *partial ordering* of $\mathcal{P}(\mathcal{H})$. With respect to this relation $\mathcal{P}(\mathcal{H})$ is a *complete lattice*, i.e. given any collection $\{P_k | k \in K\}$ of elements of $\mathcal{P}(\mathcal{H})$, there always exists a lowest upper bound (resp. a largest lower bound) for $\{P_k | k \in K\}$, i.e. an element of $\mathcal{P}(\mathcal{H})$ denoted $\bigcup P_k$ (resp. $\bigcap P_k$) such that, with $Q \in \mathcal{P}(\mathcal{H})$:

(95a) $$P_k \subseteq Q \quad \forall\ k \in K \quad \text{implies} \quad \bigcup_{k\in K} P_k \subseteq Q$$

and

(95b) $$Q \subseteq P_k \quad \forall\ k \in K \quad \text{implies} \quad Q \subseteq \bigcap_{k\in K} P_k$$

The map

(96) $$P \in \mathcal{P}(\mathcal{H}) \mapsto P^\perp \equiv I - P \in \mathcal{P}(\mathcal{H})$$

further equips $\mathcal{P}(\mathcal{H})$ with the structure of an *orthocomplemented* lattice, i.e. this map satisfies the following four conditions for all $P \in \mathcal{P}(\mathcal{H})$:

(97a) $(P^\perp)^\perp = P$

(97b) $P \cup P^\perp = I$

(97c) $P \cap P^\perp = 0$

(97d) $Q \in \mathcal{P}(\mathcal{H})$ and $P \subseteq Q$ implies $Q^\perp \subseteq P^\perp$

This orthocomplementation is canonical in the following sense. For any $P, Q \in \mathcal{P}(\mathcal{H})$ with $P \subseteq Q$, let

(98) $$\mathcal{P}(P,Q) \equiv \{R \in \mathcal{P}(\mathcal{H}) | P \subseteq R \subseteq Q\}$$

Clearly $\mathcal{P}(P,Q)$ canonically inherits from $\mathcal{P}(\mathcal{H})$ the property of being a complete lattice; moreover, the map

(99) $$R \in \mathcal{P}(P,Q) \mapsto (P \cup R^{\perp}) \cap Q \ \in \ \mathcal{P}(P,Q)$$

now equips $\mathcal{P}(P,Q)$ with the structure of an orthocomplemented lattice.

Two elements P and Q in $\mathcal{P}(\mathcal{H})$ are said to be *mutually orthogonal* whenever $P \subseteq Q^{\perp}$ (and hence $Q \subseteq P^{\perp}$, see 97a and 97d).

Two elements P and Q in $\mathcal{P}(\mathcal{H})$ are said to be *mutually compatible* whenever any (and thus all) of the following equivalent conditions is satisfied:

(100a) $$(P \cap Q^{\perp}) \cup Q = (Q \cap P^{\perp}) \cup P$$

(100b) $$< \phi; P \cup Q > + < \phi; P \cap Q > = < \phi; P > + < \phi; Q > \quad \forall \phi \in S$$

(100c) $$< \phi; P \cup Q > + < \phi; P \cap Q > = < \phi; P > + < \phi; Q > \quad \forall \phi \in S^*$$

(100d) $$P \cup Q + P \cap Q = P + Q$$

(100e) $$P \circ Q \equiv \frac{1}{2}(PQ + QP) = P \cap Q$$

(100f) $$[P,Q] \equiv PQ - QP = 0$$

The reader will notice that (100a) only appeals to the structure of $\mathcal{P}(\mathcal{H})$ as an orthocomplemented lattice, whereas (100b) and (100c) use the restriction of the notion of state from $\mathcal{B}_{sa}(\mathcal{H})$ to $\mathcal{P}(\mathcal{H})$; (100d) involves the vector space structure of $\mathcal{B}_{sa}(\mathcal{H})$, (100e) its Jordan structure, and (100f) its Lie structure.

From (100c) we notice that if $\{P_k | k = 1, 2, \ldots, N\}$ is a *finite* collection of mutually orthogonal elements in $\mathcal{P}(\mathcal{H})$, then

(101) $$< \phi; \bigcup_{k=1}^{N} P_k > = \sum_{k=1}^{N} < \phi; P_k > \quad \forall \phi \in S^*$$

This relation generalizes to the present context the notion of *finitely additive measure* encountered in classical probability theory.

It is, furthermore, easy to check that for *any* collection $\{P_k | k \in K\}$ of mutually orthogonal elements in $\mathcal{P}(\mathcal{H})$ we have:

(102) $$< \phi; \bigcup_{k \in K} P_k > = \sum_{k \in K} < \phi; P_k > \quad \forall \phi \in S$$

This relation generalizes the notion of *completely additive measure*. (102) does in general *not* hold for $\phi \in S^*$. Indeed, let $\{\Psi_k | k \in \mathbb{Z}^+\}$ be an orthonormal basis in an infinite dimensional Hilbert space $\mathcal{H}$; $P_k : \Phi \in \mathcal{H} \mapsto (\Phi, \Psi_k)\Psi_k$;

and $\phi \in S^* \cap C(\mathcal{H})^\perp$ (see 17); then $<\phi; P_k>= 0$ while $\bigcup_{k\in\mathbb{Z}^+} P_k = I$ and thus $<\phi; \bigcup_{k\in\mathbb{Z}^+} P_k>= 1$. This counterexample can be used to show that (102) in fact characterizes S in S^*. For this reason, the elements of S are often referred to as *normal states*.

Note that $\phi \in S$ is uniquely determined by its restriction

$$\phi : P \in \mathcal{P}(\mathcal{H}) \mapsto <\phi; P> \in [0,1] \tag{103}$$

which, itself, is uniquely determined by its restriction to the set

$$\mathcal{P}_0(\mathcal{H}) \equiv \{P \in \mathcal{P}(\mathcal{H}) | P : \Phi \in \mathcal{H} \mapsto (\Phi, \Psi)\Psi \in \mathcal{H}\} \tag{104}$$

of all *atoms* (i.e. one-dimensional projectors) of the lattice $\mathcal{P}(\mathcal{H})$.

Note also that $\phi \in S$ is dispersion-free on a given $P \in \mathcal{P}(\mathcal{H})$, i.e.

$$<\phi; (P - <\phi; P> I)^2>= 0 \tag{105a}$$

if and only if:

$$<\phi; P> \in \{0, 1\}\,. \tag{105b}$$

With $\lambda \in \{0,1\}$ and $P \in \mathcal{P}(\mathcal{H})$, let

$$S_P^\lambda \equiv \{\phi \in S \,|\, <\phi; P>= \lambda\} \tag{106}$$

We have then

$$P \subseteq Q \quad \text{iff} \quad S_P^1 \subseteq S_Q^1 \tag{107}$$

$$S_P^0 = S_{P^\perp}^1 \tag{108}$$

and thus

$$P \perp Q \quad \text{iff} \quad S_P^1 \subseteq S_Q^0 \quad \text{iff} \quad S_Q^1 \subseteq S_P^0 \tag{109}$$

Furthermore, P and Q in $\mathcal{P}(\mathcal{H})$ are compatible (see 100) if and only if the sublattice $\mathcal{P}[P,Q]$ of $\mathcal{P}(\mathcal{H})$, generated by P and Q (upon taking all possible intersections, unions and complementations) is a Boolean lattice; this occurs if and only if $\phi \in S$, when restricted to $\mathcal{P}[P,Q]$, is a probability measure.

The above remarks lead to the following interpretation of $\mathcal{P}(\mathcal{H})$. The elements of $\mathcal{P}(\mathcal{H})$ are called *"elementary observables"* or *"propositions"*. The relation $P \subseteq Q$ is interpreted as P *implies* Q, or more precisely "P true with probability 1 implies Q true with probability 1". The orthocomplementation $P \to P^\perp$ is interpreted by saying that $P^\perp$ is the *negation* of P.

Finally, $\phi \in S$ is pure (i.e. $\phi \in \mathcal{E}$) if and only if there exists an atom $P \in \mathcal{P}_0(\mathcal{H})$ such that $\phi \in S_P^1$; in fact this atom is then unique, and the density matrix ρ associated to ϕ coincides with P. This relation establishes a bijective correspondence between $\mathcal{E}$ and $\mathcal{P}_0(\mathcal{H})$. For $\phi \in \mathcal{E}$ and $Q \in \mathcal{P}_0(\mathcal{H})$ we have

$$(110) \qquad < \phi; Q > \equiv tr\ PQ = |(\Psi, \Phi)|^2$$

where $\Psi, \Phi \in \mathcal{H}$ with $\| \Psi \| = 1 = \| \Phi \|$ and $P\Psi = \Psi$, $Q\Phi = \Phi$. In line with the above remarks, $< \phi, Q >$ is the probability that the system is in the pure state Q when it has been prepared in the pure state P. For this reason (110) is called the *transition-probability* between P and Q, and sometimes, by abuse of language, the transition probability between Ψ and Φ.

d. von Neumann algebras

Some other topological characterizations of $S \subset \mathcal{B}(\mathcal{H})^*$ can be given in terms of various topologies on $\mathcal{B}(\mathcal{H})$; since these topologies are also relevant to the original definition of a von Neumann algebra, we briefly review these mathematical notions here; for proofs, see Dixmier (1957) or Kadison & Ringrose (1983).

The *uniform topology* is the topology induced on $\mathcal{B}(\mathcal{H})$ by the operator norm (28); it is also referred to as the "operator-norm topology" or the "metric topology" of $\mathcal{B}(\mathcal{H})$. It is the topology we considered when we defined $\mathcal{B}(\mathcal{H})$ as a C^*-algebra.

The *strong (operator) topology* is defined as the weakest topology on $\mathcal{B}(\mathcal{H})$ for which the maps $A \in \mathcal{B}(\mathcal{H}) \mapsto A\Psi \in \mathcal{H}$, associated to every $\Psi \in \mathcal{H}$, are continuous when $\mathcal{H}$ is equipped with its norm-topology. A basis of neighbourhoods for this topology is obtained by considering all sets of the form

$$(111) \qquad N(A, S, \epsilon) \equiv \{B \in \mathcal{B}(\mathcal{H}) | \ \| (B - A)\Psi \| < \epsilon \quad \forall \ \Psi \in S\}$$

where $A \in \mathcal{B}(\mathcal{H})$, S is a finite subset of $\mathcal{H}$, and $\epsilon > 0$.

The *weak (operator) topology* is defined as the weakest topology on $\mathcal{B}(\mathcal{H})$ for which the maps $A \in \mathcal{B}(\mathcal{H}) \mapsto (A\Phi, \Psi) \in \mathbb{C}$, associated to every pair $\{\Phi, \Psi\} \in \mathcal{H} \times \mathcal{H}$, are continuous. A basis of neighbourhoods for this topology is obtained by considering all sets of the form

$$(112) \qquad N(A, S, \epsilon) \equiv \{B \in \mathcal{B}(\mathcal{H}) | \ |((B - A)\Phi, \Psi)| < \epsilon \ \forall \{\Phi, \Psi\} \in S\}$$

where $A \in \mathcal{B}(\mathcal{H})$, S is a finite subset of couples $\{\Phi, \Psi\} \in \mathcal{H} \times \mathcal{H}$ and $\epsilon > 0$.

To define the next two topologies on $\mathcal{B}(\mathcal{H})$ we introduce the set $\overline{\mathcal{H}}$ of all sequences $\overline{\Psi} \equiv \{\Psi_k\} \subset \mathcal{H}$ satisfying $\| \overline{\Psi} \|^2 \equiv \Sigma_k \ \| \Psi_k \|^2 < \infty$. This defines a norm, and thus a topology on $\overline{\mathcal{H}}$. For each $A \in \mathcal{B}(\mathcal{H})$, we denote by $\overline{A}$ the map $\overline{\Psi} \in \overline{\mathcal{H}} \mapsto \overline{A\Psi} \equiv \{A\Psi_k\} \in \overline{\mathcal{H}}$.

The *ultrastrong topology* is defined as the weakest topology on $\mathcal{B}(\mathcal{H})$ for which the maps $A \in \mathcal{B}(\mathcal{H}) \mapsto \overline{A\Psi} \in \overline{\mathcal{H}}$, associated to every $\overline{\Psi} \in \overline{\mathcal{H}}$, are continuous. A basis of neighbourhoods for this topology is obtained by considering all sets of the form

$$(113)\qquad N(A,S,\epsilon)\equiv\{B\in B(\mathcal{H}))|\ \ \|\overline{B-A}\,\overline{\Psi}\,\|^2<\epsilon\ \forall\ \overline{\Psi}\in S\}$$

where $A\in B(\mathcal{H})$, S is a finite subset of $\overline{\mathcal{H}}$, and $\epsilon>0$.

The *ultraweak topology* is defined as the weakest topology on $B(\mathcal{H})$ for which the maps $A\in B(\mathcal{H})\mapsto(\overline{A}\overline{\Phi},\overline{\Psi})\equiv\Sigma_k(A\Phi_k,\Psi_k)\in\mathbb{C}$, associated to every pair $\{\overline{\Phi},\overline{\Psi}\}\in\overline{\mathcal{H}}\times\overline{\mathcal{H}}$, are continuous. A basis of neighbourhoods for this topology is obtained by considering all sets of the form

$$(114)\qquad N(A,S,\epsilon)\equiv\{B\in B(\mathcal{H})|\ \ |(\overline{B-A}\,\overline{\Phi},\overline{\Psi})|<\epsilon\ \forall\ \{\overline{\Phi},\overline{\Psi}\}\in S\}$$

where $A\in B(\mathcal{H})$, S is a finite subset of couples, $\{\overline{\Phi},\overline{\Psi}\}\in\overline{\mathcal{H}}\times\overline{\mathcal{H}}$, and $\epsilon>0$.

The relations between these topologies on $B(\mathcal{H})$ are that the weak topology is weaker than the ultraweak and the strong topologies; the latter two topologies are weaker than the ultrastrong topology, which in turn, is weaker than the uniform topology. The ultraweak and the strong topologies are not similarly related. We therefore have, for any subset $\mathcal{A}\subseteq B(\mathcal{H})$

$$(115)\qquad {}^{u}\overline{\mathcal{A}}\subseteq{}^{us}\overline{\mathcal{A}}\subseteq{}^{uw}\overline{\mathcal{A}}\,(\text{ resp. }{}^{s}\overline{\mathcal{A}})\subseteq{}^{w}\overline{\mathcal{A}}$$

where ${}^{\tau}\overline{\mathcal{A}}$ denotes the closure of $\mathcal{A}$ in the topology τ.

Finally, on subsets of $B(\mathcal{H})$ which are bounded with respect to the norm (28), the strong and the ultrastrong topologies coincide; and the weak and ultraweak topologies coincide.

With B fixed in $B(\mathcal{H})$, the maps $A\in B(\mathcal{H})\mapsto AB\in B(\mathcal{H})$, and $A\in B(\mathcal{H})\mapsto BA\in B(\mathcal{H})$ are continuous in all five topologies; however, the maps $(A,B)\in B(\mathcal{H})\times B(\mathcal{H})\mapsto AB\in B(\mathcal{H})$ are not continuous in any of the above topologies, except in the uniform topology. Finally, the map $A\in B(\mathcal{H})\mapsto A^*\in B(\mathcal{H})$ is continuous for the uniform, the ultraweak and the weak-operator topologies, but not for the strong-operator and the ultrastrong topologies.

Let now $\mathcal{A}$ be any subset of $B(\mathcal{H})$, and define the *commutant* $\mathcal{A}'$ of $\mathcal{A}$ as the set

$$(116)\qquad \mathcal{A}'\equiv\{B\in B(\mathcal{H})|AB-BA=0\quad\forall\ A\in\mathcal{A}\}$$

Clearly: $\mathcal{A}'$ is a subalgebra of $B(\mathcal{H})$; $\mathcal{A}'$ contains the identity operator $I\in B(\mathcal{H})$; and $\mathcal{A}$ is contained in its *bicommutant* $\mathcal{A}''=(\mathcal{A}')'$. Note that if $\mathcal{A}$ and $\mathcal{B}$ are two subsets of $B(\mathcal{H})$, $\mathcal{A}\subseteq\mathcal{B}$ implies $\mathcal{B}'\subseteq\mathcal{A}'$. Hence $\mathcal{A}\subseteq\mathcal{A}''$ implies $\mathcal{A}'''\equiv(\mathcal{A}'')'\subseteq\mathcal{A}'\subseteq(\mathcal{A}')''=\mathcal{A}'''$, i.e. $\mathcal{A}'=\mathcal{A}'''$; thus $\mathcal{A}''=(\mathcal{A}'')''$. Hence the operation of taking the bicommutant resembles that of taking the topological closure; this is more than a formal analogy, as we shall now see. Since the product in $B(\mathcal{H})$ is separately continuous in the weak-operator topology, $\mathcal{A}'$ is closed in this topology; so is then $\mathcal{A}''$. We therefore have, as a consequence of $\mathcal{A}\subseteq\mathcal{A}''$:

$$(117)\qquad {}^{w}\overline{\mathcal{A}}\ \subseteq\ \mathcal{A}''$$

Notice further that if $\mathcal{A}$ is a *-subalgebra of $\mathcal{B}(\mathcal{H})$—i.e. $\mathcal{A}$ is a subalgebra of $\mathcal{B}(\mathcal{H})$ such that $A \in \mathcal{A}$ implies $A^* \in \mathcal{A}$—so is $\mathcal{A}'$ and thus $\mathcal{A}''$.

Let now $\mathcal{A}$ be a *-subalgebra of $\mathcal{B}(\mathcal{H})$ satisfying the condition that the subspace

$$\{A\Psi | A \in \mathcal{A}, \Psi \in \mathcal{H}\} \text{ is norm-dense in } \mathcal{H} \tag{118}$$

One can then prove that

$$\mathcal{A}'' = {}^{us}\overline{\mathcal{A}} \tag{119}$$

Together with (107) and (109), this implies that

$$\mathcal{A}'' = {}^{us}\overline{\mathcal{A}} = {}^{uw}\overline{\mathcal{A}} = {}^{s}\overline{\mathcal{A}} = {}^{w}\overline{\mathcal{A}} \tag{120}$$

and

$$\mathcal{A}'' \text{ is closed in the uniform topology of } \mathcal{B}(\mathcal{H}) \tag{121}$$

for any *-subalgebra of $\mathcal{B}(\mathcal{H})$ satisfying (118); notice that this condition is always trivially satisfied for a *-subalgebra of $\mathcal{B}(\mathcal{H})$ containing the identity operator $I \in \mathcal{B}(\mathcal{H})$.

These results justify the following definition. A *von Neumann algebra* is a *-subalgebra $\mathcal{A}$ of $\mathcal{B}(\mathcal{H})$ satisfying $\mathcal{A} = \mathcal{A}''$. Since the bicommutant of any subset of $\mathcal{B}(\mathcal{H})$ contains the identity operator $I \in \mathcal{B}(\mathcal{H})$, every von Neumann algebra satisfies (118); from (121) we conclude, therefore, that it is a C^*-subalgebra of $\mathcal{B}(\mathcal{H})$; and from (120), we see that it is closed in the weak-operator topology (and a fortiori in all the other topologies defined in this subsection; see 115). Conversely, let $\mathcal{A}$ be a $(C)^*$-subalgebra of $\mathcal{B}(\mathcal{H})$ that contains the identity $I \in \mathcal{B}(\mathcal{H})$ and is closed in any one of the ultrastrong, ultraweak, strong-operator or weak-operator topologies on $\mathcal{B}(\mathcal{H})$. Then the above argument shows that $\mathcal{A}$ is closed in all these topologies, and that it is a von Neumann algebra.

The von Neumann algebras are the "rings of operators" singled out, studied and classified by Murray & von Neumann (1936). These algebras provide, among other things, the most natural setting in which one could possibly present (see, notably, Kadison & Ringrose, 1983) the spectral theory of bounded operators acting on a Hilbert space, as well as several interesting aspects of the theory of unbounded operators. These algebras also appear naturally in the theory of representations of C^*-algebras and provide an effective tool to study systems involving infinitely many degrees of freedom such as those encountered in Statistical Mechanics and Quantum Field Theory (see Chapters Nine and Ten). They already throw some light on some aspects of von Neumann postulates, as we shall see below.

Clearly every A in a *-subalgebra $\mathcal{A} \subseteq \mathcal{B}(\mathcal{H})$ can be written uniquely as $A = A_1 + iA_2$ with A_1 and A_2 self-adjoint elements of A. The spectral theorem for self-adjoints operators on $\mathcal{H}$ (see in particular 8.2.25 which we only need here for $D(H) = \mathcal{H}$) shows then that if $\mathcal{P} \equiv \{P \in \mathcal{A} | P = P^* = P^2\}$ denotes the set of all projectors in a von Neumann algebra $\mathcal{A}$, then $\mathcal{A} = \mathcal{P}''$. Similarly, let $\mathcal{U} \equiv \{U \in \mathcal{A} | UU^* = I = U^*U\}$ denote the set of all unitarics in a von Neumann algebra $\mathcal{A}$. Let $A \in \mathcal{A}$ with $A = A^*$ and $\| A \| < 1$. Clearly $(I - A^2)$ is a positive operator belonging to $\mathcal{A}$. From the spectral theorem again, we see that $(I - A^2)^{1/2} \equiv B$ belongs to $\mathcal{A}$; so does then $U \equiv A + iB$. In fact, $U \in \mathcal{U}$. We conclude from this that $\mathcal{A} = \mathcal{U}''$. These two results are expressed by saying that a von Neumann algebra is generated by the set $\mathcal{P}$ (resp. $\mathcal{U}$) of its projectors (resp. its unitaries). They allow to construct immediately non-isomorphic examples of von Neumann algebras.

$\mathcal{B}(\mathcal{H})$ is trivially a von Neumann algebra, and so is $\mathbb{C}I \equiv \{\lambda I | \lambda \in \mathbb{C}\}$; in fact $(\mathbb{C}I)' = \mathcal{B}(\mathcal{H})$, and thus $\mathbb{C}I = \mathcal{B}(\mathcal{H})'$. A von Neumann algebra $\mathcal{A}$ is said to be a *factor* if $\mathcal{A} \cap \mathcal{A}' = \mathbb{C}I$; clearly both $\mathcal{B}(\mathcal{H})$ and $\mathbb{C}I$ are factors. Since a von Neumann algebra is generated by its projectors, we see that a von Neumann algebra is a factor if and only if 0 and I are the only projectors in $\mathcal{A}$ that commute with all projectors in $\mathcal{A}$. At the opposite of factors are the maximal abelian von Neumann subalgebras of $\mathcal{B}(\mathcal{H})$, which we are now going to define. Notice first that to say that a subalgebra $\mathcal{A}$ of $\mathcal{B}(\mathcal{H})$ is *abelian* is exactly to say that $\mathcal{A} \subseteq \mathcal{A}'$. Notice further that a von Neumann algebra is abelian if and only if the set $\mathcal{P}$ of its projectors is a Boolean sublattice of $\mathcal{P}(\mathcal{H})$. We now say that a *-subalgebra $\mathcal{A}$ of $\mathcal{B}(\mathcal{H})$ is a *maximal abelian* von Neumann subalgebra of $\mathcal{B}(\mathcal{H})$ whenever $\mathcal{A} = \mathcal{A}'$. Note indeed that $\mathcal{A}$ is then an abelian von Neumann algebra, and that every abelian subalgebra $\mathcal{B}$ of $\mathcal{B}(\mathcal{H})$ that contains $\mathcal{A}$ actually coincides with $\mathcal{A}$. In von Neumann's formalism for quantum mechanics, a subset $\mathcal{B}$ of $\mathcal{B}_{sa}(\mathcal{H})$ is said to be a *complete set of commuting* (or *compatible*) *observables*, abbreviated c.s.c.o., whenever $\mathcal{B}''$ is a maximal abelian von Neumann subalgebra of $\mathcal{B}(\mathcal{H})$.

Typical examples of abelian and maximal abelian von Neumann algebras can be obtained as follows.

Let A be an observable in the sense of von Neumann, i.e. $A \in \mathcal{B}_{sa}(\mathcal{H})$ the set of all bounded self-adjoint operators acting in a Hilbert space $\mathcal{H}$; let further $\{P_\lambda | \lambda \in \mathbb{R}\}$ be the spectral family of A. Clearly $\mathcal{A} \equiv \{A\}'' = \{P_\lambda | \lambda \in \mathbb{R}\}''$ and $\mathcal{A} \subseteq \mathcal{A}'$, i.e. $\mathcal{A}$ is an abelian von Neumann algebra. In particular, let $\{\Psi_k\}$ be an orthonormal basis in $\mathcal{H}$ and A be defined, by linearity and continuity, from

$$(122) \qquad A\Psi_k = a_k \Psi_k \quad \text{with} \quad a_k \in \mathbb{R} \quad \text{and} \quad |a_k| < a < \infty$$

Then the von Neumann algebra $\mathcal{A}_1$ generated by this observable is maximal abelian if and only if $k \neq \ell$ implies $a_k \neq a_\ell$. More generally, the spectrum of a bounded self-adjoint operator A is said to be *simple* whenever the von Neumann algebra $\mathcal{A} \equiv \{A\}''$ is maximal abelian in $\mathcal{B}(\mathcal{H})$. For instance, let $f : \mathbb{R} \to \mathbb{R}$ be a continuous, strictly increasing, bounded function. The operator

A defined on $\mathcal{H} \equiv \mathcal{L}^2(\mathbb{R}, dx)$ by

$$(123) \qquad (A\Psi)(x) = f(x)\Psi(x) \quad \forall \ \Psi \in \mathcal{H}$$

Then $\mathcal{A}_2 \equiv \{A\}'' \cong \mathcal{L}^\infty(\mathbb{R}, dx)$ is maximal abelian in $\mathcal{B}(\mathcal{H})$. Similarly, with $\mathcal{H}$ again $\mathcal{L}^2(\mathbb{R}, dx)$, let $U : a \in \mathbb{R} \mapsto U(a) \in \mathcal{B}(\mathcal{H})$ be defined by

$$(124) \qquad (U(a)\Psi)(x) = \Psi(x - a) \quad \forall \ \Psi \in \mathcal{H}$$

Note $U(-a) = U(a)^*$ and $U(a)U(b) = U(a+b) \quad \forall \, a, b \in \mathbb{R}$. Hence $U(a)U(a)^* = I = U(a)^*U(a)$. Again the von Neumann algebra $\mathcal{A}_3 \equiv \{U(a) | a \in \mathbb{R}\}''$ is maximal abelian in $\mathcal{B}(\mathcal{H})$ and isomorphic to $\mathcal{L}^\infty(\mathbb{R}, dx)$. Notice further that $\mathcal{A}_2 \cap \mathcal{A}_3 = \mathbb{C}I$ and thus $(\mathcal{A}_2 \cup \mathcal{A}_3)'' = \mathcal{B}(\mathcal{H})$.

These examples show that $\mathcal{B}(\mathcal{H})$ possesses one more interesting property not shared by all von Neumann algebras: the set $\mathcal{P}(\mathcal{H})$ of all its projectors is an *atomic* lattice, i.e. given any $P \in \mathcal{P}(\mathcal{H})$ with $P \neq 0$, there exists $Q \in \mathcal{P}(\mathcal{H})$ such that $Q \neq 0$, $Q \subseteq P$, and $R \in \mathcal{P}(\mathcal{H})$ together with $R \subseteq Q$ implies either $R = 0$ or $R = Q$. Indeed, for every $P \in \mathcal{P}(\mathcal{H})$ and $P \neq 0$, there exists $\Psi \in \mathcal{H}$ such that $\| \Psi \| = 1$ and $P\Psi = \Psi$; then $Q : \Phi \in \mathcal{H} \mapsto (\Phi, \Psi)\Psi \in \mathcal{H}$ satisfies the above conditions.

Amongst the three examples of (maximal) abelian von Neumann algebras just constructed, the reader will verify that $\mathcal{A}_1$ is atomic, while $\mathcal{A}_2$ and $\mathcal{A}_3$ are not atomic.

These three examples can further be used to illustrate the fact that all pure states on a (maximal) abelian von Neumann subalgebra $\mathcal{A}$ of $\mathcal{B}(\mathcal{H})$ can be extended to vector states of $\mathcal{B}(\mathcal{H})$ if and only if $\mathcal{A}$ is atomic. This remark brings us back to the paper of Kadison & Singer (1959) mentioned in the preceding subsection, and thus to the topological characterizations of $\mathcal{S}$ in $\mathcal{S}^*$, which we promised in the beginning of the present subsection.

Let $\mathcal{A}$ be a von Neumann algebra, acting on a Hilbert space $\mathcal{H}$. Let $\mathcal{A}^*$ be the Banach space dual of $\mathcal{A}$, i.e. $\mathcal{A}^*$ is the set of all linear maps $\phi : A \in \mathcal{A} \mapsto \, < \phi; A > \in \mathbb{C}$ such that $\| \phi \| \equiv \sup_{A \in \mathcal{A}; \|A\| < 1} \mid < \phi; A > \mid < \infty$. $\mathcal{A}^*$ thus comes equipped with the norm $\| \cdot \|$; we refer to the topology induced on $\mathcal{A}^*$ by this norm as the *uniform topology* of $\mathcal{A}^*$. $\mathcal{A}^*$ then enjoys the following three properties.

Firstly, for $\phi \in \mathcal{A}^*$ the following three conditions are equivalent:

(i) (resp. ii) $\phi : \mathcal{A} \to \mathbb{C}$ is continuous when $\mathcal{A}$ is equipped with the weak- (resp. strong-) operator topology it inherits from $\mathcal{B}(\mathcal{H})$; (iii) there exist a finite index set K, $\{\lambda_k | k \in K\} \subset \mathbb{R}^+$, $\{\Phi_k | k \in K\} \subset \mathcal{H}$, $\{\Psi_k | k \in K\} \subset \mathcal{H}$ such that

$$(125a) \qquad \| \phi \| = \sum_{k=1}^{N} \lambda_k$$

$$(125b) \qquad (\Phi_k, \Phi_\ell) = \delta_{k\ell} = (\Psi_k, \Psi_\ell) \quad \forall \ k, \ell \in K$$

$$(125c) \qquad < \phi; A > = \sum_{k=1}^{N} \lambda_k (A\Phi_k, \Psi_k) \quad \forall \ A \in \mathcal{A}$$

Secondly, for $\phi \in \mathcal{A}^*$ the following three conditions are equivalent: (i) (resp. ii) $\phi : \mathcal{A} \to \mathbb{C}$ is continuous when $\mathcal{A}$ is equipped with the ultraweak (resp. ultrastrong) topology it inherits from $\mathcal{B}(\mathcal{H})$; (iii) there exist $\{\Phi_k | k \in \mathbb{Z}^+\}$ and $\{\Psi_k | k \in \mathbb{Z}^+\}$ such that

$$\text{(126a)} \qquad \sum_{k=1}^{\infty} \| \Phi_k \|^2 < \infty \ , \ \sum_{k=1}^{\infty} \| \Psi_k \|^2 < \infty$$

$$\text{(126b)} \qquad < \phi; A > = \sum_{k=1}^{\infty} (A\Phi_k, \Psi_k)$$

Thirdly, let $\mathcal{A}_\sim$ (resp. $\mathcal{A}_*$) be the collection of all elements of $\mathcal{A}^*$ that satisfy (125) (resp. 126). Then $\mathcal{A}_*$ is the closure of $\mathcal{A}_\sim$ with respect to the uniform topology of $\mathcal{A}^*$. Furthermore, the bilinear form $(\phi, A) \in \mathcal{A}^* \times \mathcal{A} \mapsto < \phi; A > \in \mathbb{C}$, restricted to $\mathcal{A}_* \times \mathcal{A}$ is such that $\mathcal{A}$, when equipped with its norm (28) is the Banach space dual of $\mathcal{A}_*$, the latter being equipped with the norm it inherits from $\mathcal{A}^*$.

Notice now that since $\mathcal{A}$ is a C^*-algebra, every linear functional $\phi : \mathcal{A} \to \mathbb{C}$ that is positive—i.e. $< \phi; A^*A > \geq 0 \ \forall A \in \mathcal{A}$—necessarily belongs to $\mathcal{A}^*$. For such functional, one can assume, without loss of generality, that $\Psi_k = \Phi_k$ in (125) and (126). A positive linear functional $\phi : \mathcal{A} \to \mathbb{C}$ is said to be a *C^*-state* on the C^*-algebra $\mathcal{A}$ with unit I, whenever $< \phi; I > = 1$ (see 90). The first of the above three properties of the dual $\mathcal{A}^*$ of a von Neumann algebra $\mathcal{A}$ thus tells us that a C^*-state on $\mathcal{A}$ is weakly (resp. strongly) continuous whenever it can be extended to a state on $\mathcal{B}(\mathcal{H})$ belonging to the (algebraic) convex hull of $\mathcal{E}$, the collection of all vector states on $\mathcal{B}(\mathcal{H})$. Similarly, the second of these properties tells us that a C^*-state on $\mathcal{A}$ is ultraweakly (resp. ultrastrongly) continuous whenever it can be extended to a state on $\mathcal{B}(\mathcal{H})$ belonging to $\mathcal{S}$, the collection of all normal states on $\mathcal{B}(\mathcal{H})$.

Finally, the third of the above three properties of the dual $\mathcal{A}^*$ of a von Neumann algebra $\mathcal{A}$ tells us in particular that every von Neumann algebra is a W^*-algebra. Conversely, upon using the GNS construction to be discussed in Chapter Nine, one can prove that every W^*-algebra with unit is isomorphic, algebraically and topologically, to a von Neumann algebra, i.e. to a von Neumann subalgebra of $\mathcal{B}(\mathcal{H})$ for some $\mathcal{H}$.

e. Symmetries

We shall close this section by discussing three consequences of the von Neumann postulates for quantum theory. These are: the definition of symmetries, the von Neumann uniqueness theorem, and the classical limit. This subsection is devoted to the concepts of symmetry and of groups of symmetries.

Intuitively, a symmetry of a physical system is a transformation of this system that preserves its structure. One of the most economical ways to capture this notion, in the mathematical setting proposed by von Neumann for quantum theory, is first to restrict one's attention to $\mathcal{P}(\mathcal{H})$, the collection of "elementary observables" on the quantum system considered. Recall from Subsection c (see also d) that $\mathcal{P}(\mathcal{H})$ is an atomic, orthocomplemented lattice where the partial ordering $P \subseteq Q$ is interpreted as P implies Q, and the orthocomplementation $P \to P^{\perp}$ is interpreted as the negation. One then defines a symmetry of the quantum system $\{\mathcal{S}, \mathcal{B}(\mathcal{H}), < . \, ; \, . >\}$ as a bijective map

$$\alpha : P \in \mathcal{P}(\mathcal{H}) \mapsto \alpha[P] \in \mathcal{P}(\mathcal{H}) \tag{127a}$$

satisfying:

$$\alpha[P] \subseteq \alpha[Q] \quad \text{if} \quad P \subseteq Q \tag{127b}$$

$$\alpha[P^{\perp}] = \alpha[P]^{\perp} \tag{127c}$$

It immediately follows from this definition that

$$\alpha[0] = 0 \quad \text{and} \quad \alpha[I] = I \tag{127d}$$

$$\alpha[\bigcup P_i] = \bigcup \alpha[P_i] \quad \text{and} \quad \alpha[\bigcap P_i] = \bigcap \alpha[P_i] \tag{127e}$$

$$\alpha[P] \text{ compatible with } \alpha[Q] \quad \text{iff} \quad P \text{ compatible with } Q \tag{127f}$$

$$\alpha[P] \in \mathcal{P}_0(\mathcal{H}) \quad \text{iff} \quad P \in \mathcal{P}_0(\mathcal{H}) \tag{127g}$$

As a consequence of these properties, we see that α is uniquely determined by its restriction to $\mathcal{P}_0(\mathcal{H})$, the collection of all atoms of $\mathcal{P}(\mathcal{H})$. Note that the atoms of $\mathcal{P}(\mathcal{H})$ are one-dimensional projectors and are thus in one-to-one correspondence with the one-dimensional subspaces of $\mathcal{H}$. Suppose $\dim \mathcal{H} \geq 3$. One can then use the first fundamental theorem of projective geometry (see e.g. Baer, 1952 or Artin, 1957) to prove (see Emch & Piron, 1963) that there exists a bijective map

$$U : \Psi \in \mathcal{H} \mapsto U\Psi \in \mathcal{H} \tag{128a}$$

and a continuous automorphism

$$\chi : \lambda \in \mathbb{C} \mapsto \chi[\lambda] \in \mathbb{C} \tag{128b}$$

such that for all $\Psi, \Phi \in \mathcal{H}$, all $\lambda \in \mathbb{C}$, and all $P \in \mathcal{P}(\mathcal{H})$:

$$U(\Psi + \Phi) = U\Psi + U\Phi \tag{129a}$$

$$U(\lambda\Psi) = \chi[\lambda]U\Psi \tag{129b}$$

$$(U\Psi, U\Phi) = \chi[(\Psi, \Phi)] \tag{129c}$$

$$\alpha[P] = UPU^* \tag{129d}$$

Since there are only two continuous automorphism of $\mathbb{C}$—the identity and the complex conjugation—U is either unitary or antiunitary; hence

$$|(U\Psi, U\Phi)|^2 = |(\Psi, \Phi)|^2 \tag{130}$$

i.e. every symmetry preserves the transition probability (see 110) between pure states. This characterization of symmetries is due to Wigner (1931); and the associated theorem, asserting that a map $\alpha : \mathcal{P}_0(\mathcal{H}) \rightarrow \mathcal{P}_0(\mathcal{H})$ that satisfies (130) is necessarily of the form (128–129), is known as *Wigner's theorem*; for an elementary, but complete, proof of the latter result, given in the spirit of Wigner's original sketch, see Bargmann (1964) who also mentions somewhat earlier references to other complete, but less elementary, proofs of this Wigner theorem.

Note next that (129d) allows to extend naturally a symmetry to a map

$$\alpha : A \in \mathcal{B}_{sa}(\mathcal{H}) \mapsto \alpha[A] \equiv UAU^* \in \mathcal{B}_{sa}(\mathcal{H}) \tag{131a}$$

such that, for all $A, B \in \mathcal{B}_{sa}(\mathcal{H})$ and $\lambda, \mu \in \mathbb{R}$, we have

$$\alpha[\lambda A + \mu B] = \lambda\alpha[A] + \mu\alpha[B] \tag{131b}$$
$$\alpha[A^2] = \alpha[A]^2 \tag{131c}$$
$$\alpha[A \circ B] = \alpha[A] \circ \alpha[B] \tag{131d}$$

Hence every symmetry α can be extended, without loss of generality to a bijective map of $\mathcal{B}_{sa}(\mathcal{H})$ onto itself, that respects the Jordan algebra structure of $\mathcal{B}_{sa}(\mathcal{H})$. Another way to say this is to note that the Jordan structure of $\mathcal{B}_{sa}(\mathcal{H})$ does not allow to distinguish between those symmetries which are unitarily implemented and those which are antiunitarily implemented. The Lie algebra structure of $\mathcal{B}_{sa}(\mathcal{H})$, however, allows to make such a distinction.

To see this, note that we can extend α further to $\mathcal{B}(\mathcal{H})$ in two different ways which, a priori, appear to be equally natural. We can indeed either define

$$\alpha : A \in \mathcal{B}(\mathcal{H}) \mapsto \alpha[A] \equiv UAU^* \quad \in \quad \mathcal{B}(\mathcal{H}) \tag{132a}$$

or we can define

$$\alpha : A \in \mathcal{B}(\mathcal{H}) \mapsto \alpha[A] \equiv UA^*U^* \quad \in \quad \mathcal{B}(\mathcal{H}) \tag{132b}$$

Both of these definitions enjoy the following three properties, where A and B are arbitrary in $\mathcal{B}(\mathcal{H})$:

$$\alpha[A + B] = \alpha[A] + \alpha[B] \tag{133}$$
$$\alpha[A^*] = \alpha[A]^* \tag{134}$$
$$\alpha[AB + BA] = \alpha[A]\alpha[B] + \alpha[B]\alpha[A] \tag{135}$$

These definitions, however, differ in the following two pairs of properties, where A and B are arbitrary in $\mathcal{B}(\mathcal{H})$ and λ is arbitrary in $\mathbb{C}$; α defined according to

(132a) satisfies

(136a) $$\alpha[AB] = \alpha[A]\alpha[B]$$
(137a) $$\alpha[\lambda A] = \chi(\lambda)\alpha[A]$$
while α, defined according to (132b) satisfies
(136b) $$\alpha[AB] = \alpha[B]\alpha[A]$$
(137b) $$\alpha[\lambda A] = \chi(\lambda)^*\alpha[A]$$

where, in (137 a and b), χ is the identity (resp. the complex conjugation) on $\mathbb{C}$ when U is unitary (resp. when U is antiunitary).

Let now $\{. \ , \ .\}$ be the quantum Lie bracket defined in (51) and naturally extended to $\mathcal{B}(\mathcal{H})$ by:

(138) $$\{A, B\} \equiv (AB - BA)/i\hbar \quad \forall\, A, B \in \mathcal{B}(\mathcal{H}).$$

From (136) and (137) we conclude that both definitions of α (see 132 a and b) lead to

(139) $$\alpha[\{A, B\}] = +\{\alpha[A], \alpha[B]\} \text{ if } U \text{ is unitary}$$
(140) $$\alpha[\{A, B\}] = -\{\alpha[A], \alpha[B]\} \text{ if } U \text{ is antiunitary}$$

Hence, independently of which of the two definitions (132) one chooses, the Lie structure of $\mathcal{B}_{sa}(\mathcal{H})$ is preserved (resp. reversed) when U is unitary (resp. when U is antiunitary).

Since $\mathcal{B}_{sa}(\mathcal{H})$, rather than $\mathcal{B}(\mathcal{H})$, is the primary object of physical relevance, it is a matter of convention to choose between α linear or antilinear on $\mathcal{B}(\mathcal{H})$ (see 133 and 137). If we prefer, for reasons of mathematical convenience, to have α linear, then (136) and (137) imply that α preserves (resp. reverses) the order of the ordinary product in $\mathcal{B}(\mathcal{H})$ when U is unitary (resp. when U is antiunitary).

More generally, one can prove—see Kadison (1965)—that if α is a linear, adjoint-preserving map from a C^*-algebra $\mathcal{A}$ onto a C^*-subalgebra $\mathcal{B}$ of $\mathcal{B}(\mathcal{H})$ for some Hilbert space $\mathcal{H}$, then the following conditions are equivalent: (i) $\alpha[AB + BA] = \alpha[A]\alpha[B] + \alpha[B]\alpha[A]$; and (ii) there exists a projector $P \in \mathcal{B}'' \cap \mathcal{B}'$ such that $\alpha[AB]P = \alpha[A]\alpha[B]P$ and $\alpha[AB](I - P) = \alpha[B]\alpha[A](I - P)$ for all $A, B \in \mathcal{A}$. In particular, when α is a bijective map of $\mathcal{A}$ onto itself, the above conditions can serve as definition of a symmetry; note that condition (i) can then be replaced by the condition $\alpha[A^2] = \alpha[A]^2 \ \forall\, A \in \mathcal{A}$ with $A = A^*$. When $\mathcal{A}$ is a von Neumann algebra this definition is equivalent to our definition (127) of a symmetry, with $\mathcal{P}(\mathcal{H})$ replaced by $\mathcal{P} = \{Q \in \mathcal{A} | Q = Q^2 = Q^*\}$. If, furthermore, $\mathcal{A} = \mathcal{B}(\mathcal{H})$ note that the P of Kadison's theorem can only be 0 or I.

Upon returning to von Neumann's scheme for quantum theory, we note that if α is a symmetry

(141a) $$< \nu[\phi]; P > \equiv < \phi; \alpha[P] > \qquad P \in \mathcal{P}(\mathcal{H})$$

defines a map

$$\nu : \phi \subset S \mapsto \nu[\phi] \in S \tag{141b}$$

which is affine. This passage from α to ν is referred to, in the physics literature, as the passage from the *Heisenberg picture* to the *Schroedinger picture*. The study of the general equivalence between these two pictures was one of the main purposes of the paper by Kadison (1965) mentioned above.

Two more remarks on the implementation (129d) of a symmetry are in order here.

Firstly, if $\alpha = \beta \circ \beta \equiv \beta^2$ where β is a symmetry, then the symmetry α must be unitarily implemented. This is the reason why unitarily implemented symmetries play a paramount role in the study of connected Lie groups of symmetries.

Secondly, if U is a unitary (or antiunitary) operator implementing a symmetry α, then ωU, with $\omega \in \mathbb{C}$ and $|\omega| = 1$ implements the same symmetry. This remark is innocuous when one deals with only one symmetry, but it becomes serious when one has to deal with an infinite group of symmetries.

Let indeed $Sym(\mathcal{H})$ denote the group of all symmetries of a quantum system $\{S, \mathcal{B}_{sa}(\mathcal{H}), < \, . \, ; \, . \, >\}$, and suppose that G is a connected Lie group of symmetries of this system, i.e. that a map

$$\alpha : g \in G \mapsto \alpha(g) \in \text{Sym}\ (\mathcal{H}) \tag{142a}$$

is given, satisfying for all $g, g' \in G$, $\phi \in S$ and $A \in \mathcal{B}_{sa}(\mathcal{H})$:

$$g \in G \mapsto < \phi; \alpha(g)[A] > \in \mathbb{R} \text{ continuous} \tag{142b}$$

$$\alpha(g)\alpha(g') = \alpha(gg') \tag{142c}$$

If one chooses arbitrarily a representative $U(g)$ in the class $\mathcal{U}(g)$ of unitary operators implementing $\alpha(g)$, one has

$$U(g_1)U(g_2) = \omega(g_1, g_2)U(g_1g_2) \tag{143}$$

with $\omega(g_1, g_2) \in S^1 \equiv \{z \in \mathbb{C} |\ |z| = 1\}$, for all $g_1, g_2 \in G$. If another choice of representative, say $\tilde{U}$, is made, we have

$$\tilde{U}(g) = \nu(g)U(g) \quad \text{with} \quad \nu(g) \in S^1 \tag{144}$$

$$\tilde{U}(g_1)\tilde{U}(g_2) = \tilde{\omega}(g_1, g_2)\tilde{U}(g_1g_2) \tag{145}$$

$$\tilde{\omega}(g_1, g_2) = \nu(g_1)\nu(g_2)\omega(g_1, g_2)\nu(g_1g_2)^{-1} \tag{146}$$

The question is then whether a judicious choice of

$$\nu : g \in G \mapsto \nu(g) \in S^1 \tag{147}$$

would be possible in such a manner that

$$\tilde{U} : g \in G \mapsto \tilde{U}(g) \in \mathcal{U}(\mathcal{H}) \tag{148}$$

becomes a continuous unitary representation of G, i.e. satisfies for all $g, g' \in G$, and $\Phi, \Psi \in \mathcal{H}$:

$$g \in G \mapsto (\tilde{U}(g)\Phi, \Psi) \in \mathbb{C} \text{ continuous} \tag{149b}$$
$$\tilde{U}(g)\tilde{U}(g') = \tilde{U}(gg') \tag{149c}$$

In other words, is it possible to reduce the study of continuous actions of a connected Lie group, as symmetries on a quantum system, to the study of continuous unitary representations of such groups? This problem has been studied systematically by Bargmann (1954). The analysis is done in two steps: one first studies the problem locally in G, and then one extends to the whole of G the solutions obtained in the first step.

Let us sketch the kind of mathematics involved in these two steps. Let first $N_0(e)$ be a neighbourhood of the identity e in G, and for each $g \in N_0(e)$, select an element $U(g)$ in the class $\mathcal{U}(g)$ of unitary operators implementing $\alpha(g)$. $\{U(g)|g \in N_0(e)\}$ is said to be a *local lifting* if for every $\epsilon > 0$, every $\Psi \in \mathcal{H}$, and every $g \in N_0(e)$, there exists a neighbourhood $N(g)$ of g, with $N(g) \subset N_0(e)$ such that $\| U(g')\Psi - U(g)\Psi \| < \epsilon$ for every $g' \in N(g)$. From the continuity (142b) of α, one shows that a local lifting always exists. Two local liftings $\{U(g)|g \in N_0(e)\}$ and $\{\tilde{U}(g)|g \in \tilde{N}_0(e)\}$ are said to be *equivalent* if there exists a continuous function $\nu : N_0(e) \cap \tilde{N}_0(e) \to S^1$ such that $\tilde{U} = \nu U$ on $N_0(e) \cap \tilde{N}_0(e)$. The local problem thus consists in determining whether there exists one or several equivalence classes of local liftings. We can clearly assume, without loss of generality, that we choose in every equivalence class a local lifting that satisfies $U(e) = I$. For any such local lifting, (143) now restricted to the pairs $\{g_1, g_2\} \in G \times G$ such that $g_1, g_2, g_1 g_2 \in N_0(e)$, defines a function ω satisfying

(150a) ω continuous in each of its factors

$$\omega(g, e) = 1 = \omega(e, g) \quad \forall\ g \in N_0(e) \tag{150b}$$
$$\omega(g_1, g_2)\omega(g_1 g_2, g_3) = \omega(g_1, g_2 g_3)\omega(g_2, g_3) \tag{150c}$$

with (150c) valid for all pairs $\{g_1, g_2\} \in G \times G$ such that $g_1, g_2, g_3, g_1 g_2, g_2 g_3, g_1 g_2 g_3 \in N_0(e)$. The relations (150) define what is called a *local factor*. If $\{U(g)|g \in N_0(e)\}$ and $\{\tilde{U}(g)|g \in \tilde{N}_0(e)\}$ are two local liftings, they are equivalent if and only if their associated local factors ω and $\tilde{\omega}$ are linked by the relation (146) where now $g_1, g_2, g_1 g_2 \in N_0(e) \cap \tilde{N}_0(e)$ and ν is a continuous function from $N_0(e) \cap \tilde{N}_0(e)$ to S^1.

Consider now for every local lifting $\{U(g)|g \in N_0(e)\}$ the set $H \equiv \{(\nu, g)|\nu \in S^1, g \in N_0(e)\}$ equipped with the direct product topology and the local group law

$$(\nu_1, g_1)(\nu_2, g_2) \equiv (\nu_1\nu_2\omega(g_1, g_2), g_1g_2) \tag{151}$$

where ω is the local factor associated to the local lifting considered. As a consequence of the solution of the fifth Hilbert problem (see e.g. Montgomery & Zippin, 1955), H is a local Lie group; it is, furthermore, a local Lie group extension

$$e \to N_0(e) \to H \to S^1 \to 1 \tag{152}$$

of $N_0(e) \subset G$ by S^1. It is then easy to check that two local liftings $\{U(g)|g \in N_0(e)\}$ and $\{\tilde{U}(g)|g \in \tilde{N}_0(e)\}$ are equivalent if and only if their local Lie groups H and $\tilde{H}$ are equivalent local Lie group extensions of $N_0(e) \cap \tilde{N}_0(e) \subset G$ by S^1, i.e. if there is a local Lie group isomorphism π between H and $\tilde{H}$ such that the following diagram is commutative.

$$\begin{array}{ccccccc} e \to & N_0(e) \to & H \to & S^1 \to & 1 \\ & \downarrow id & \downarrow \pi & \downarrow id & \\ e \to & \tilde{N}_0(e) \to & \tilde{H} \to & S^1 \to & 1 \end{array} \tag{153}$$

Upon using next the usual fonctorial relation between the category of local Lie groups with their local morphisms, and the category of Lie algebras with their morphisms, the problem of finding all equivalence classes of local liftings is reduced to the problem of finding all equivalence classes of Lie algebra extensions of the Lie algebra of G by the Lie algebra of S^1. This extension problem is a particular case of the general theory of extensions and their equivalence classes discussed in Cartan & Eilenberg (1956) for all kinds of mathematical structures, including Lie groups and Lie algebras; for the latter, see also Bourbaki (1960).

One fringe benefit of this reformulation of the problem is that if one replaces the complex Hilbert space of von Neumann's quantum mechanics by a Hilbert space constructed over the (non-commutative!) field of quaternions, one needs only substitute in the final result above the Lie algebra of S^1 by the Lie algebra of $SU(2, \mathbb{C})$; see Emch (1963). In this case, the Lie algebra extension problem admits only one solution and, therefore, it is always possible to arrange things in such a way that the "phase" $\omega(g_1, g_2) \equiv 1$ locally.

The situation is not so simple in the case of interest to us here, namely the case where the Hilbert space is constructed over $\mathbb{C}$, i.e. where one has to classify the Lie algebra extensions by S^1. This is the problem that Bargmann (1954) solved explicitly for many Lie groups G of interest to physicists. One of the general results obtained by Bargmann can be read off directly from the above formulation of the problem. Indeed, as a corollary (see Bourbaki, 1960) to the Levi-Malcev theorem every extension of a semi-simple Lie algebra is inessential, i.e. is equivalent to the trivial, direct product, extension. Hence, if the Lie algebra of the symmetry group G is semi-simple, the "phase" ω can always be eliminated locally, to give $\omega(g_1, g_2) \equiv 1$ on some neighbourhood of the identity e in G.

Let, for instance, $M^{n,m}$ be the (n, m)-Minkowski space defined at the beginning of Section Two in Chapter Four. Recall that physicists are mostly interested in the cases $M^{n,o} = \mathbb{R}^n$ with $n = 3$ or 4, and $M^{n,1}$ with $n = 3$. Let further $P^{n,m}$ be the group of all isometries of $M^{n,m}$ (see 4.2.1b); $P_e^{n,m}$ be the component of the identity e in $P^{n,m}$; T^{n+m} be the group of translations $z \in M^{n,m} \mapsto \hat{z} \equiv z + \varsigma \in M^{n,m}$ (with ς running over $M^{n,m}$); clearly $T^{n+m} \subset P_e^{n,m}$. Let finally $L^{n,m}$ be the subgroup of $P^{n,m}$ consisting of all homogeneous transformations $z \in M^{n,m} \mapsto \hat{z} \equiv \Lambda z \in M^{n,m}$; and $L_e^{n,m} \equiv L^{n,m} \cap P_e^{n,m}$ be the component of the identity in $L^{n,m}$. For $n + m > 2$, $L_e^{n,m}$ are connected, simple Lie groups. Hence, for these groups, the "phase" ω can always be eliminated locally. This result was obtained by Bargmann (1954), and in the particular cases of $O_+^3 \equiv L_e^{3,0}$ and $L^{3,1}$ by Wigner (1931 and 1939).

Upon playing with the commutators of the Lie algebra of $P_e^{n,m}$, with $n + m > 2$, one can further show that, for these groups again, the "phase" ω can be eliminated locally; see Bargmann and Wigner (loc. cit.). Note, in particular, that $P^{n,m} \supset T^{n+m}$.

The reader should *not* surmize, from the above examples, that the "phase" ω can always be eliminated locally for groups of physical interest, and we shall give three counterexamples later in this subsection.

Suppose for a moment that G is a connected Lie group such that every local Lie group extension of G by S^1 is inessential, i.e. is locally equivalent to the trivial, direct product, extension. For any continuous group action $\alpha : g \in G \mapsto \alpha(g) \in \text{Sym}\ (\mathcal{H})$, the "phase" ω can then be eliminated locally, i.e. one can find a local lifting $\{U(g) | g \in N_0(e)\}$ of α such that, for every $g_1, g_2 \in N_0(e)$ with $g_1 g_2 \in N_0(e)$, we have:

$$U(g_1)U(g_2) = U(g_1 g_2) \tag{154}$$

It is then easy to verify, by patching up neighbourhoods, that (154) can be extended to a continuous unitary representation U of the universal covering group $\tilde{G}$ of G in such a manner that U implements $\alpha : G \mapsto \text{Sym}\ (\mathcal{H})$. Hence the study of all continuous group actions α of such a group G is reduced to the study of the continuous unitary representations of $\tilde{G}$. Evidently, if G is, in addition, simply connected i.e. $G = \tilde{G}$, then the study of its continuous group actions reduces to the study of its own continuous unitary representations. There are, however, several connected Lie groups, of interest to physicists, where the "phase" ω can be eliminated locally, but which are not simply connected. The most obvious example is O_+^3 —for which we have already seen that ω can be eliminated locally; this group is doubly connected with covering group $\tilde{G} = SU(2,\mathbb{C}) : O^3_+ \simeq SU(2,\mathbb{C})/Z_2$ where Z_2 is the group

(155) $$Z_2 \equiv \{\begin{pmatrix} 1 & 0 \\ 0 & 1 \end{pmatrix}, \begin{pmatrix} -1 & 0 \\ 0 & -1 \end{pmatrix}\}$$

The Lie algebra $su(2,\mathbb{C})$ of $SU(2,\mathbb{C})$, and thus O^3_+, is

(156a) $$su(2,\mathbb{C}) \equiv \{\sum_{k=1}^{3} a^k J_k | a^k \in \mathbb{R}\}$$

with the Lie bracket $\{. \;,\; .\} = [. \;,\; .]/i$ (where $[A,B] = AB - BA$ is the usual commutator in $\mathcal{B}(\mathbb{C}^2)$); and with $J_k = \sigma_k/2$ where

(156b) $$\sigma_1 \equiv \begin{pmatrix} 0 & 1 \\ 1 & 0 \end{pmatrix}; \sigma_2 \equiv \begin{pmatrix} 0 & -i \\ i & 0 \end{pmatrix}; \sigma_3 \equiv \begin{pmatrix} 1 & 0 \\ 0 & -1 \end{pmatrix}$$

are the so-called *Pauli matrices.* Note, in particular, that for all cyclic permutations (j,k,ℓ) of the indices $(1,2,3)$ we have

(156c) $$\{J_j, J_k\} = J_\ell$$

(Note in passing, the analogy between (156c) and the Poisson brackets $\{L_j, L_k\} = L_\ell$ where L_i are the three components of the classical *angular momentum* $L = \mathbf{x} \times \mathbf{p}$; compare with 8.3.227–8).

Since $\tilde{G} \equiv SU(2,\mathbb{C})$ is compact, it admits an invariant measure μ which is finite, the so-called Haar measure; and $\mathcal{L}^2(\tilde{G}, d\mu)$ is separable. The Peter-Weyl theorem (see e.g. Pontrjagin, 1958) then asserts that the regular representation of $\tilde{G}$ on $\mathcal{L}^2((\tilde{G}, d\mu)$ is a direct sum of irreducible continuous unitary representations (or i.c.u.-representations) of $\tilde{G}$ which are finite-dimensional, each i.c.u.-representation occuring in this sum with a finite multiplicity equal to its dimension; moreover, every i.c.u.-representation of $\tilde{G}$ occurs in this decomposition (i.e. is unitarily equivalent to an i.c.u.-representation appearing in the decomposition of the regular representation). In fact, for $SU(2,\mathbb{C})$, one can show (see e.g. Wigner, 1931) that there exists exactly one i.c.u.-representation U_n of $\tilde{G}$ for every $n \in \mathbb{Z}^+$. Let further u_n be the representation of $su(2,\mathbb{C})$ obtained from U_n. Then upon writing $n = 2s+1$, one finds that the i.c.u.-representation U_n is uniquely characterized by the value $s(s+1)$ of the Casimir operator

(157) $$J_n^2 = \sum_{k=1}^{3} u_n(J_k)^2$$

As for the phase reduction problem for the continuous actions of O^3_+, the representations $\{U_n | n \in \mathbb{Z}^+\}$ fall in two classes. The first (resp. second) class contains all i.c.u.-representations where n is odd (resp. even), i.e. where s is an integer (resp. a "half-integer"). The representations of the first class are continuous unitary representations of O^3_+. The representations of the second class, however, are only "up to a sign" continuous unitary representations of O^3_+, i.e. for these representations, one cannot eliminate the "phase" ω globally to better than

$$U(g_1)U(g_2) = \pm U(g_1 g_2) \quad \forall\ g_1, g_2 \in\ O^3_+ \tag{158}$$

In particular, one obtains $U(2\pi) = -I$ (compare with 155).

The same dichotomy occurs, for instance, for the proper Poincaré group $P^{3,1}_e$ of special relativity; see Wigner (1939).

Before leaving the subject of the continuous actions of O^3_+ we want to make one more observation. Let

$$U : g\ \in\ SU(2, \mathbb{C}) \mapsto U(g)\ \in\ \mathcal{U}(\mathcal{H}) \tag{159}$$

be any (i.e. not necessarily irreducible) continuous unitary representation of $SU(2,\ \mathbb{C})$, and let $U(2\pi)$ denote the unitary operator in this representation corresponding to the element $-I \in\ Z_2 \subset SU(2, \mathbb{C})$ (see 155). Since, moreover $U(2\pi)^2 = I$ the identity operator in $\mathcal{H}$, its spectrum belongs to $\{-1, +1\}$, and the two closed subspaces $\mathcal{H}_+$ and $\mathcal{H}_-$, defined by

$$\mathcal{H}_\pm \equiv \{\Psi_\pm \in \mathcal{H} | U(2\pi)\Psi_\pm = \pm\Psi_\pm\} \tag{160}$$

provide a decomposition of $\mathcal{H}$ into two orthogonal subspaces, each of which is stable under $\{U(g)|g \in SU(2, \mathbb{C}\}$ since $[U(2\pi), U(g)] = 0$ for all $g \in\ SU(2, \mathbb{C})$. It is then easy to verify that U, when restricted to $\mathcal{H}_+$ (resp. $\mathcal{H}_-$) is a continuous, unitary representation (resp. a continuous, unitary, up to a sign, representation) of O^3_+; and that

$$\alpha_\pm : g\ \in\ O^3_+ \mapsto \alpha(g) \in\ \text{Sym}\ (\mathcal{H}_\pm) \tag{161a}$$

defined by

$$\alpha_\pm(g)[A] = U(g)AU(g)^* \quad \forall\ A \in \mathcal{B}_{sa}(\mathcal{H}_\pm) \tag{161b}$$

(through the identification $O_3^+\ \simeq\ SU(2,\ \mathbb{C})/Z_2$), are continuous actions (see 142) of O^3_+; in particular

$$\alpha_\pm(2\pi)[A] = A \quad \forall\ A \in \mathcal{B}_{sa}(\mathcal{H}_\pm) \tag{161c}$$

We can evidently extend (161a & b) to a continuous action $\alpha\ :\ SU(2,\ \mathbb{C}) \mapsto \mathcal{B}_{sa}(\mathcal{H})$. Let then

$$\mathcal{A} \equiv \{A \in \mathcal{B}_{sa}(\mathcal{H}) | \alpha(2\pi)[A] = A\} \tag{162a}$$

It is easy to see that $A \in \mathcal{B}_{sa}$ belongs to $\mathcal{A}$ if and only if

(162b) $$(A\Psi_+, \Psi_-) = 0 \quad \forall \{\Psi_+, \Psi_-\} \in \mathcal{H}_+ \times \mathcal{H}_-$$

Hence, if $\mathcal{H}_+ \neq \emptyset \neq \mathcal{H}_-$:

(163a) $$\mathcal{A} = \mathcal{B}_{sa}(\mathcal{H}_+) \oplus \mathcal{B}_{sa}(\mathcal{H}_-) \neq \mathcal{B}_{sa}(\mathcal{H})$$

or, equivalently $\mathcal{A} = \mathcal{A}''$ and

(163b) $$\mathcal{A}' = \{P_+, P_-\}' \neq \{\lambda I | \lambda \in \mathbb{C}\}$$

where P_+ (resp. P_-) is the projector from $\mathcal{H}$ to $\mathcal{H}_+$ (resp. $\mathcal{H}_-$). Since a physical rotation by 2π in $\mathbb{R}^3$ is the identity, we should want to impose (162 a) as a condition to be satisfied by every observable. This shows that von Neumann's postulates have to be enlarged to allow for the algebra $\mathcal{A}$ of observables not to be necessarily identical with $\mathcal{B}_{sa}(\mathcal{H})$. This was first recognized (in an only slightly different context) by Wick, Wightman & Wigner (1952), and the phenomena just described is characterized in the physics literature by saying that there is a *superselection rule* (or "no observable quantity or coherence") between systems (or "particles") with integer spin s, and those with half-integer spin s. A brief axiomatization, in the framework of the quantum proposition-calculus, can be found in Emch & Piron (1963); for a didactic account, see Jauch (1968). Note that a quantum theory with superselection rules, i.e. where $\mathcal{A}' \neq \mathbb{C}$ and actually $\mathcal{A}'' \cap \mathcal{A}' \neq \mathbb{C}I$, appears as some intermediary between two extremes: on the one hand, the quantum mechanics defined by von Neumann's postulates, where $\mathcal{A} = \mathcal{B}_{sa}(\mathcal{H})$ and thus $\mathcal{A}' = \mathbb{C}I$ and in particular $\mathcal{A}'' \cap \mathcal{A}' = \mathbb{C}I$; and on the other hand, classical mechanics where the algebra of observables is abelian, i.e. $\mathcal{A} \subseteq \mathcal{A}'$, and thus $\mathcal{A}'' \cap \mathcal{A}' = \mathcal{A}''$.

We now come back to the problem of whether the "phase" ω can always be eliminated, at least locally; the answer is negative, even in some cases of physical interest, and we show this by presenting three counterexamples, the first of them with all the details necessary to constitute a proof.

Let us consider $\mathbb{R}^{2n}$, equipped with its natural structure of abelian, $2n$-dimensional Lie group; for notational convenience, we identify $\mathbb{R}^{2n}$ with $\mathbb{C}^n \equiv \{z = (a, \hat{a}) | (a, \hat{a}) \in \mathbb{R}^n \times \mathbb{R}^n\}$. With $\lambda \in (0, \infty)$ fixed, let $\{E_\lambda(z) | z \in \mathbb{C}^n\}$ be defined on $\mathcal{H} \equiv \mathcal{L}^2(\mathbb{R}^n, d^n x)$ by

(164a) $$[E_\lambda(z)\Psi](x) \equiv \exp\{-i\lambda\, \hat{a}\,(x - \lambda a/2)\}\Psi(x - \lambda a)$$

where Ψ runs over $\mathcal{H}$. A straightforward computation shows that for every $z, z' \in \mathbb{C}^n$:

(164b) $E_\lambda(z)$ is unitary

(164c) $$E_\lambda(z)E_\lambda(z') = \omega_\lambda(z, z')E_\lambda(z + z')$$

(164d) $$\omega_\lambda(z, z') = \exp\{i\lambda^2\sigma(z, z')/2\}$$

(164e) $$\sigma(z, z') = a \cdot \hat{a}' - a' \cdot \hat{a} = -\sigma(z', z)$$

where $\cdot$ denotes the usual pairing between $\mathbb{R}^n$ and its dual. Define now

$$\text{(165a)} \qquad \alpha_\lambda : z \in \mathbb{C}^n \mapsto \alpha(z) \in \text{Sym}\,(\mathcal{H})$$
$$\text{(165b)} \qquad \alpha_\lambda(z)[A] = E_\lambda(z) A\, E_\lambda(z)^* \quad \forall\; A \in \mathcal{B}_{sa}(\mathcal{H})$$

Clearly, α_λ satisfies (142 b & c), so that α_λ describes a continuous action of $\mathbb{C}^n$ on $\mathcal{B}_{sa}(\mathcal{H})$. Moreover, $\{E_\lambda(z)|z \in \mathbb{C}^n\}$ is a (local) lifting, and ω_λ is a (local) factor. It is, furthermore, easy to check that this local factor is not locally equivalent to the trivial factor $\omega \equiv 1$. Hence, there exist group actions α_λ of $\mathbb{C}^n$ where the "phase" ω cannot be eliminated locally.

Remark further that two (local) factors ω_λ and ω_μ obtained in the above manner are locally inequivalent, unless $\lambda = \mu$. This generalizes from $\mathbb{C}^n$ to any m-dimensional, connected and simply connected abelian Lie group G, by replacing in (164 d & e) σ by any bilinear, antisymmetric (not necessarily non-degenerate) form on the Lie algebra of G; see Bargmann (1954), and already Weyl (1931).

In the perspective of quantum mechanics, the interest of $\{E_\lambda(z)|z \in \mathbb{C}^n\}$, defined as in (164a), with $\lambda = \hbar^{1/2}$, is that it generalizes to arbitrary n a formula (see 8.1.122) we obtained for $n = 1$ in connection with the description of the canonical commutation relation for one degree of freedom. Indeed, define the self-adjoint operators $\{P_\lambda^{(k)}, Q_\lambda^{(k)}|k = 1, 2, \ldots, n\}$ by

$$\text{(166)} \qquad E_\lambda(a, \hat{a}) \equiv \exp\{-i(a \cdot P_\lambda + \hat{a} \cdot Q_\lambda)\}$$

Let $D(P_\lambda^{(k)})$, $D(Q_\lambda^{(k)})$ denote the domain of definition of the operators $P_\lambda^{(k)}$ and $Q_\lambda^{(k)}$. Then there exists a dense subspace D—namely $\mathcal{S}(\mathbb{R}^n)$—of $\mathcal{H}$ such that:

$$\text{(167a)} \qquad D \subseteq \bigcap_{k=1}^{n} \{D(P_\lambda^{(k)}) \cap D(Q_\lambda^{(k)})\}$$
$$\text{(167b)} \qquad P_\lambda^{(k)} D \subseteq D \;;\; Q_\lambda^{(k)} D \subseteq D$$
$$\text{(167c)} \qquad P_\lambda^{(k)}|D \text{ and } Q_\lambda^{(k)}|D \text{ are essentially self-adjoint}$$
$$\text{(167d)} \qquad [P_\lambda^{(k)}, P_\lambda^{(\ell)}] = 0 = [Q_\lambda^{(k)}, Q_\lambda^{(\ell)}]$$
$$\text{(167e)} \qquad [P_\lambda^{(k)}, Q_\lambda^{(\ell)}] = -i\lambda^2\;\delta^{k\ell} I$$

with (167 d & e) holding on D. With $\lambda^2 = \hbar$, the relations (167) are interpreted as the quantum analog of the classical Poisson brackets between momentum and position coordinates. We shall see towards the end of this section that this is more than a mere analogy. Already here, the physics behind these relations (167) can be caught through the following computations.

From (164) we define, for every $(a, \hat{a}) \in \mathbb{R}^n \times \mathbb{R}^n$:

$$\text{(168a)} \qquad U_\lambda(a) \equiv E_\lambda(a/\lambda^2, 0)$$
$$\text{(168b)} \qquad V_\lambda(\hat{a}) \equiv E_\lambda(0, \hat{a}/\lambda^2)$$

Notice that $\{U_\lambda(a)|a \in \mathbb{R}^n\}$ and $\{V_\lambda(\hat{a})|\hat{a} \in \mathbb{R}^n\}$ are continuous unitary representations of $\mathbb{R}^n$, and satisfy for all $(a, \hat{a}) \in \mathbb{R}^n \times \mathbb{R}^n$:

$$U_\lambda(a)V_\lambda(\hat{a})U_\lambda(a)^* = exp(i\, a \cdot \hat{a}/\lambda^2)V_\lambda(\hat{a}) \tag{169}$$

from which we read

$$U_\lambda(a)Q_\lambda^{(k)}U_\lambda(a)^* = Q_\lambda^{(k)} - a^k I \tag{170}$$

holding over $D(Q_\lambda^{(k)})$, for all $a \in \mathbb{R}^n$. To express this relation in the framework of von Neumann's postulates, we introduce two objects, namely

$$\mathcal{A} \equiv \{V_\lambda(\hat{a})|\hat{a} \in \mathbb{R}^n\}'' \tag{171}$$

$$\alpha_\lambda : a \in \mathbb{R}^n \mapsto \alpha_\lambda(a) \in \text{ Sym } (\mathcal{H}) \tag{172a}$$

$$\alpha_\lambda(a)[A] \equiv U_\lambda(a)AU_\lambda(a)^* \;\; \forall\, A \in \mathcal{B}_{sa}(\mathcal{H}) \tag{172b}$$

Notice first that $\mathcal{A}$ is the von Neumann algebra of multiplication operators in $\mathcal{B}(\mathcal{H})$, and thus

$$\mathcal{A} = \{F_\lambda(\Delta)|\Delta \in B(\mathbb{R}^n)\}'' \tag{173}$$

where $B(\mathbb{R}^n)$ is the σ-algebra of Borel subsets of $\mathbb{R}^n$, and F is the projection-valued measure defined by

$$F_\lambda : \Delta \in B(\mathbb{R}^n) \mapsto F_\lambda(\Delta) \in \mathcal{P}(\mathcal{H}) \tag{174a}$$

$$(F_\lambda(\Delta)\Psi)(x) = \chi_\Delta(\lambda x)\Psi(x) \tag{174b}$$

$$\chi_\Delta(\xi) = \begin{cases} 1 & \text{if } \xi \in \Delta \\ 0 & \text{if } \xi \notin \Delta \end{cases} \tag{174c}$$

Notice then that α_λ defines a continuous action of $\mathbb{R}^n$ on $\mathcal{B}_{sa}(\mathcal{H})$ (see 142) and that (see 169), for every $a \in \mathbb{R}^n$:

$$\alpha_\lambda(a) : A \in \mathcal{A}_{sa} \mapsto \alpha_\lambda(a)[A] \in \mathcal{A}_{sa} \tag{175}$$

Finally, for every $a \in \mathbb{R}^n$ and every $\Delta \in B(\mathbb{R}^n)$, denote by $a[\Delta]$ the Borel subset of $\mathbb{R}^n$ defined by:

$$a[\Delta] = \{x \in \mathbb{R}^n |\, x - a \in \Delta\} \tag{176}$$

Then, as an action on $\mathcal{A}_{sa}$, α_λ is determined by

$$\alpha_\lambda(a)[F_\lambda(\Delta)] = F_\lambda(a[\Delta]) \;\; \forall\, (a, \Delta) \in \mathbb{R}^n \times B(\mathbb{R}^n) \tag{177}$$

In summary, we have extracted from (164) a projection-valued measure (174) F_λ on $\mathbb{R}^n$, and a continuous action (172) α_λ of $\mathbb{R}^n$ on $\mathcal{B}_{sa}(\mathcal{H})$ such that (177) is satisfied. With $\lambda^2 = \hbar$, this relation allows to interpret consistently Q as a quantum "position observable" and P as a quantum "momentum observable". For this to make sense canonically, it is indispensible that the factor ω_λ, appearing in (164), cannot be changed, and a fortiori eliminated, even locally, by a continuous change of the form (144) in the lifting $\{E_\lambda(a,\hat{a})|(a,\hat{a}) \in \mathbb{R}^n \times \mathbb{R}^n\}$. This is probably the most fundamental way in which the *Planck constant* arises in the von Neumann formulation of quantum mechanics.

We only mention two other cases where the "phase" ω cannot be eliminated locally. For the group G of Galilean relativity, Bargmann (1954) has exhibited a continuum of inequivalent local extensions of G by S^1, parametrized by the mass. Finally, the case $P_e^{n,m}$ with $n+m=2$ was left open in our previous discussion; Bargmann (1954) has shown that in this particular case, the phase can indeed not be eliminated locally.

f. The CCR for n degrees of freedom

In the previous subsection, we obtained (see 164–177) special realizations of the "canonical commutation relations" for n degrees of freedom. The content of von Neumann's uniqueness theorem, to be stated in the present subsection, is that these realizations are canonical in a sense which we are going to make precise. In last analysis, this result of von Neumann (1931), based on an earlier formulation by Weyl (1928), is the mathematical reason behind the equivalence of the Heisenberg quantum mechanics (recall Section One of this chapter) and the Schroedinger wave mechanics (recall Section Two). Moreover, we shall see that this Weyl-von Neumann formulation allows to understand better the relation between quantum and classical theories.

First of all, we should notice that, given n and λ, the realization E_λ is not unique in an obvious way. Let indeed X be a bijective linear isometry from $\mathcal{H} = \mathcal{L}^2(\mathbb{R}^n, d^n x)$ onto some Hilbert space $\overline{\mathcal{H}}$. Then $\{\overline{E}_\lambda(a,\hat{a})\} \equiv X\, E_\lambda(a,\hat{a})X^{-1}|(a,\hat{a}) \in \mathbb{R}^n \times \mathbb{R}^n\}$ satisfies again (164 b–e) and its consequences. We can, in particular, define from this $\overline{P}_\lambda$ and $\overline{Q}_\lambda$ as in (166), and these operators will satisfy (167). Similarly, we can define $\overline{U}_\lambda$ (resp. $\overline{V}_\lambda$) from $\overline{E}_\lambda$ as we did in (168), and these continuous unitary group representations of $\mathbb{R}^n$ (resp. $\mathbb{R}^n$) will satisfy (169) and (170). Finally, we can define $\overline{\alpha}_\lambda$ from $\overline{U}_\lambda$ as in (172), and $\{\overline{F}(\Delta) \equiv XF(\Delta)X^{-1}|\Delta \in B(\mathbb{R}^n)\}$; these objects will evidently again satisfy (177), with the same physical interpretation. Hence, a theory based on $\overline{E}$ will give exactly the same physical predictions as those obtained from the original realization E.

In particular, for any fixed $\mu \in (0,\infty)$, the map X from $\mathcal{L}^2(\mathbb{R}^n, d^n x) \equiv \mathcal{H}$ onto itself, defined by

$$(178) \qquad (X\Psi)(x) = \mu^n\, \Psi(\mu^2 x) \quad \forall\ \Psi \in \mathcal{H},$$

and interpreted as a consistent change of scale (i.e. of physical units), will give with $\mu^{-4} = \lambda^2 = \hbar$:

(179a) $$(\overline{Q}^{(k)}\Psi)(x) = x^k \ \Psi(x)$$

(179b) $$(\overline{P}^{(k)}\Psi)(x) = -i\hbar(\partial_k\Psi)(x)$$

This is the generalization, to n degrees of freedom, of the *Schroedinger representation* (compare to 8.2.60; and see also 8.2.69 or 8.2.74–75). The integral form of (179):

(180a) $$(\overline{V}(\hat{a})\Psi(x) = \exp(-i\hat{a}\cdot x/\hbar)\Psi(x)$$

(180b) $$(\overline{U}(a)\Psi)(x) = \Psi(x-a)$$

is known as the *Weyl form* of the Schroedinger reprsentation, for $\lambda^2 = \hbar$.

The above remarks, complementing the discussion of the previous subsection, suggest the following formalized definition. A *representation of the CCR for n degrees of freedom* is a triple $\{\mathcal{H}, F, U\}$ consisting of a separable Hilbert space $\mathcal{H}$ over $\mathbb{C}$; a projection-valued measure $F : \Delta \in B(\mathbb{R}^n) \mapsto F(\Delta) \in \mathcal{P}(\mathcal{H})$; and a unitary, continuous, group representation $U : a \in \mathbb{R}^n \mapsto U(a) \in \mathcal{U}(\mathcal{H})$ such that for every $a \in \mathbb{R}^n$ and every $\Delta \in B(\mathbb{R}^n)$:

(181) $$U(a)F(\Delta)U(a)^* = F(a[\Delta])$$

where $a[\Delta] \equiv \{x \in \mathbb{R}^n | x - a \in \Delta\}$.

This is the basic definition, from which everything else will follow.

From F, we define for every $k = 1, 2, \ldots, n$, the spectral family $\{F_\rho^{(k)} | \rho \in \mathbb{R}\}$ by

(182) $$F_\rho^{(k)} \equiv F(\Delta_\rho^{(k)}) \quad \text{where} \quad \Delta_\rho^{(k)} \equiv \{x \in \mathbb{R}^n | x^{(k)} \leq \rho\}$$

From each of these spectral family, we define the self-adjoint operator

(183) $$Q^{(k)} \equiv \int \rho \, dF_\rho^{(k)}$$

These definitions, together with (181), imply

(184) $$U(a)Q^{(k)}U(a)^* = Q^{(k)} - a^{(k)}I$$

on the domain $D(Q^{(k)})$ of $Q^{(k)}$. We refer to $Q \equiv \{Q^{(k)} | k = 1, 2, \ldots, n\}$ as the *position*, and to $U(a)$ as the *translation* operators for the representation $\{\mathcal{H}, F, U\}$.

We now *choose* $\lambda \in (0, \infty)$ and define the self-adjoint operators $\{P^{(k)} | k = 1, 2, \ldots, n\}$ by

(185) $$U(a) \equiv \exp(-ia \cdot P/\lambda^2)$$

We refer to $P \equiv \{P^{(k)} | k = 1, 2, \ldots, n\}$ as the *momentum* operator for the representation $\{\mathcal{H}, F, U\}$ with respect to λ. The choice of a particular value for λ corresponds to a choice of the physical units for the "momentum" (once the physical units for the "position" are fixed).

With the *same* choice of λ, we finally introduce the auxiliary objects $\{V(\hat{a}) | \hat{a} \in \mathbb{R}^n\}$ and $\{E(z) | z \equiv (a, \hat{a}) \in \mathbb{C}^n \equiv \mathbb{R}^n \times \mathbb{R}^n\}$ by

$$V(\hat{a}) \equiv \exp(-i\hat{a} \cdot Q/\lambda^2) \tag{186}$$

$$E(z) \equiv \exp(-i\lambda^2 a \cdot \hat{a}/2) U(\lambda^2 a) V(\lambda^2 \hat{a}) \tag{187}$$

Note that $V : \hat{a} \in \mathbb{R}^n \mapsto V(\hat{a}) \in \mathcal{U}(\mathcal{H})$ is a unitary, continuous group representation of $\mathbb{R}^n$. Moreover, as a consequence of the defining condition (181) we have:

$$U(a)V(\hat{a}) = \exp(ia \cdot \hat{a}/\lambda^2) V(\hat{a}) U(a) \tag{188}$$

$$E(z)E(z') = \exp\{i\lambda^2(a \cdot \hat{a}' - a' \cdot \hat{a})/2\} E(z + z') \tag{189}$$

Upon using the continuity of U, V and E, one can follow again the route travelled in the previous subsection and show that each of these two relations separately implies (181); the reader will verify indeed that this proof is independent of the identification of $\mathcal{H}$ with $\mathcal{L}^2(\mathbb{R}^n, d^n x)$. Either of the two relations (188) or (189) could, therefore, have been used to define a representation of the CCR for n degrees of freedom. We choose (181) for two reasons: first, its direct physical interpretation, and second, the elucidation it allows for the role of λ.

A representation $\{\mathcal{H}, F, U\}$ of the CCR for n degrees of freedom is said to be *irreducible* if $\{0\}$ and $\mathcal{H}$ are the only two closed subspaces of $\mathcal{H}$ that are stable under the action of the collection of operators $\{F(\Delta), U(a) | \Delta \in \mathcal{B}(\mathbb{R}^n), a \in \mathbb{R}^n\}$.

Upon using the tools provided in Subsection d, the reader will verify the following three results, where $\{\mathcal{H}, F, U\}$ is a representation of the CCR for n degrees of freedom, and $\mathcal{N}(\mathcal{H}, F, U)$ is the von Neumann algebra $\{F(\Delta), U(a) |\ \Delta \in \mathcal{B}(\mathbb{R}^n), a \in \mathbb{R}^n\}''$.

Firstly, $\mathcal{N}(\mathcal{H}, F, U) = \{U(a), V(\hat{a}) | a \in \mathbb{R}^n, \hat{a} \in \mathbb{R}^n\}'' = \{E(z) | z \in \mathbb{C}^n\}''$.

Secondly, $(\mathcal{H}, F, U)$ is irreducible if and only if any (and thus all) of the following equivalent conditions is satisfied: (i) $\mathcal{N}(\mathcal{H}, F, U)' = \{\varsigma I | \varsigma \in \mathbb{C}\}$; (ii) $\mathcal{N}(\mathcal{H}, F, U) = \mathcal{B}(\mathcal{H})$; (iii) every observable $A \in \mathcal{B}_{sa}(\mathcal{H})$ can be approximated in the weak-operator topology, by finite sums of the form $\Sigma_k \varsigma_k E(z_k)$ with $\varsigma_k \in \mathbb{C}$ and $z_k \in \mathbb{C}^n$; (iv) same as (iii) with the weak-operator topology replaced by any of the following three topologies: strong-operator, ultraweak or ultrastrong. Conditions (iii) and (iv) are paraphrased by saying that every observable is a "function" of the position and momentum observables.

Thirdly, the representation $\{E_\lambda(z) | z \in \mathbb{C}^n\}$ constructed in the previous subsection—and thus the Schroedinger representation constructed in the beginning of the present subsection—is irreducible.

Finally, two representations $\{\mathcal{H}, F, U\}$ and $\{\overline{\mathcal{H}}, \overline{F}, \overline{U}\}$ of the CCR for n degrees of freedom are said to be *unitarily equivalent* if there exists a bijective, linear isometry X from $\mathcal{H}$ onto $\overline{\mathcal{H}}$ such that $\forall\, a \in \mathbb{R}^n$ and $\forall\, \Delta \in \mathcal{B}(\mathbb{R}^n)$

$$\overline{F}(\Delta) = XF(\Delta)X^{-1} \quad \text{and} \quad \overline{U}(a) = XU(a)X^{-1} \tag{190}$$

PROPOSITION 1 (VON NEUMANN UNIQUENESS THEOREM). *For each $n \in \mathbb{Z}^+$ there exists exactly one (up to unitary equivalence) irreducible representation of the CCR for n degrees of freedom.*

Proof: This proposition is a statement about $\{\mathcal{H}, F, U\}$, and thus does not involve λ. For the purpose of the proof of this proposition, we are, therefore, free to construct the auxiliary objects V and E (see 186 and 187) for any convenient choice of the value of $\lambda \in (0, \infty)$; we choose $\lambda = 1$. We next define the map

$$E : f \in \mathcal{L}^1(\mathbb{C}^n, d^n z) \mapsto E(f) \in \mathcal{B}(\mathcal{H}) \tag{191a}$$

(recall $\mathbb{C}^n = \mathbb{R}^n \times \mathbb{R}^n$ and $d^n z \equiv d^n a \cdot d^n \hat{a}$) by the integral

$$E(f) \equiv \int d^n z\, f(z)\, E(z) \tag{191b}$$

understood in the weak-operator topology. Note that $E(f)$ belong indeed to $\mathcal{B}(\mathcal{H})$:

$$\| E(f) \| \leq \| f \|_1 \tag{192}$$

From the unitarity of $E(z)$ and the relation (189) we obtain

$$E(f)^* = E(f^*) \text{ with } f^* : z \in \mathbb{C}^n \mapsto f(-z)^* \in \mathbb{C} \tag{193}$$

$$E(\varsigma)^* E(f) E(\varsigma) = \int d^n z\, \exp\{i(a\hat{\alpha} - \alpha\hat{a})\} f(z)\, E(z) \tag{194}$$

We shall further need the fact that

$$E(f) = 0 \quad \text{with} \quad f \in \mathcal{L}^1(\mathbb{C}^n, d^n z) \quad \text{imply} \quad f = 0 \tag{195}$$

Clearly if $E(f) = 0$, (194) implies that for any polynomial $P : z \equiv (a, \hat{a}) \in \mathbb{C}^n \mapsto P(z) \in \mathbb{C}$, of the form

$$P(z) = \sum_n \xi_n \exp\{i(a\hat{\alpha}_n + \hat{a}\alpha_n)\} \tag{196}$$

where the sum is finite, $\{\xi_n\} \subset \mathbb{C}$, $\{(\alpha_n, \hat{\alpha}_n)\} \subset \mathbb{C}^n$, we have for all Φ and $\Psi \in \mathcal{H}$:

$$\int d^n z\; P(z)\, f(z)\, (E(z)\Phi, \Psi) = 0 \tag{197}$$

and thus, upon approximating the characteristic function of any Borel subset $S \subset \mathbb{C}^n$, we obtain from (197):

$$\int_S d^n z \;\; f(z)\,(E(z)\Phi, \Psi) = 0\,. \tag{198}$$

Consequently, we have for almost all $z \in \mathbb{C}^n : (f(z)E(z)\Phi, \Psi) = 0$, and thus $f(z)E(z)\Phi = 0$. Since $E(z)$ is unitary, this proves (195).

Consider now in particular the function

$$f_0 : z \in \mathbb{C}^n \mapsto (2\pi)^{-n} \exp(-|z|^2/4) \tag{199}$$

and the operator $A \equiv E(f_0)$; a straightforward computation shows that

$$A\;\; E(z)\,A = \exp(-|z|^2/4)\,A \tag{200}$$

With $z = 0$, this implies in particular $A^2 = A$. Since f_0 is real and symmetric, (193) implies $A^* = A$. Finally since $f_0 \neq 0$, (195) implies $A \neq 0$. Hence A is a non-zero projector. Consequently, there exists at least one $\Phi \in \mathcal{H}$ such that $\| \Phi \| = 1$ and $A\Phi = \Phi$. For any vector satisfying these conditions, (200) implies, for all $z \in \mathbb{C}^n$:

$$(E(z)\Phi, \Phi) = \exp(-|z|^2/4) \tag{201}$$

Clearly the closed linear span

$$\mathcal{H}_0 \equiv \overline{\mathrm{Span}}\{E(z)\Phi | z \in \mathbb{C}^n\} \tag{202}$$

is a closed subspace of $\mathcal{H}$, stable under the action of the von Neumann algebra $\mathcal{N} = \{E(z) | z \in \mathbb{C}\}''$. Since $\{\mathcal{H}, F, U\}$ is supposed to be irreducible, this implies that $\mathcal{H}_0$ coincides with $\mathcal{H}$, i.e. Φ is "cyclic" in $\mathcal{H}$ under the action of $\mathcal{N}$.

Let us now repeat this construction for any other irreducible representation $\{\overline{\mathcal{H}}, \overline{F}, \overline{U}\}$ of the CCR for n degrees of freedom; define then the map

$$X : \sum_{k=1}^{K} a_k E(z_k)\Phi \in \mathcal{H} \mapsto \sum_{k=1}^{K} a_k \overline{E}(z_k)\overline{\Phi} \in \overline{\mathcal{H}} \tag{203}$$

where $\{a_k | k = 1, 2, \ldots, K\} \subset \mathbb{C}$, $\{z_k | k = 1, 2, \ldots, K\} \subset \mathbb{C}^n$, and $K \in \mathbb{Z}^+$. From the above argument (see in particular (201) and the cyclicity of Φ), one concludes easily that X extends to a bijective linear isometry from $\mathcal{H}$ onto $\overline{\mathcal{H}}$ such that $\overline{E}(z) = XE(z)X^{-1}$ for all $z \in \mathbb{C}^n$. Consequently X implements a unitary equivalence between $\{\mathcal{H}, F, U\}$ and $\{\overline{\mathcal{H}}, \overline{F}, \overline{U}\}$. This concludes the proof of von Neumann's uniqueness theorem. □

Since the Schroedinger representation is irreducible, von Neumann's uniqueness theorem asserts that every irreducible representation of the CCR for n degrees of freedom is unitarily equivalent to the Schroedinger representation (179–180) acting in $\mathcal{L}^2(\mathbb{R}^n, d^n x)$, and thus gives the same physical predictions. In particular, both Schroedinger and Heisenberg assumed—at least implicitly—in their proposals for quantum mechanics, that every observable is a function of the position and momentum observables; together with von Neumann's postulate, this means that both Schroedinger and Heisenberg considered only irreducible representation of the CCR. Thus von Neumann's uniqueness theorem sets on a firm mathematical foundation the equivalence between the quantum mechanical formalisms of Heisenberg and of Schroedinger.

Upon returning to the proof of the proposition and applying it to the particular case where $\overline{E} = E$, one observes that the irreducibility of $\{E(z)|z \in \mathbb{C}\}$ implies that $\Phi \in \mathcal{H}$ with $\| \Phi \| = 1$ is determined by (201), uniquely up to a phase $\omega \in \mathbb{C}$ with $|\omega| = 1$. Hence the projector A is one-dimensional. Upon comparing with 8.2.33, the reader will verify that for every irreducible representation of the CCR, the range of A is spanned by the vector Φ corresponding to the ground state of the harmonic oscillator

$$H = \frac{1}{2}\sum_{k=1}^{n}(P^{(k)})^2 + \frac{1}{2}\sum_{k=1}^{n}(Q^{(k)})^2 \tag{204}$$

The above argument on the uniqueness of Φ can be extended to a proof of the following result on non-necessarily irreducible representations.

Corollary 2. *Every representation $\{\mathcal{H}, F, U\}$ of the CCR for n degrees of freedom is a direct sum of copies of the Schroedinger representation for n.*

Proof: From the proof of the proposition, we see that the definition of A, and the fact that it is a non-zero projector, do not depend on the irreducibility of the representation; the only thing we lose by considering a general representation $\{\mathcal{H}, F, U\}$ is that A is not a one-dimensional projector in general. Let thus $\{\Phi_j | j = 1, 2, \ldots\}$ be an o.n. basis in A. Then

$$\mathcal{H}_j = \overline{\text{Span}}\,\{E(z)\Phi_j | z \in \mathbb{C}\} \tag{205}$$

is invariant under the action of $\mathcal{N} = \{F(\Delta), U(a) | \Delta \in \mathcal{B}(\mathbb{R}^n), a \in \mathbb{R}^n\}''$. The proof of the proposition shows that the restriction of $\{\mathcal{H}, F, U\}$ to $\mathcal{H}_j$ is unitarily equivalent to the Schroedinger representation. From (200), one verifies next that $\mathcal{H}_j$ and $\mathcal{H}_\ell$ are orthogonal for $j \neq \ell$. From this it is easy to conclude that $\mathcal{H} = \bigoplus_j \mathcal{H}_j$ provides the desired decomposition of $\{\mathcal{H}, F, U\}$. □

Let us now make a few brief comments on the assumptions entering in the above Proposition and Corollary (for details, see Emch, 1972).

In our definition of a representation of the CCR, we imposed that $\mathcal{H}$ be separable. While this condition is used in the proof of the Corollary, it is not necessary to the proof of the Proposition itself. The latter can therefore be used to show that if $\{\mathcal{H}, F, U\}$ is irreducible, then $\mathcal{H}$ must be separable.

Note that if we had started with an (irreducible) triple $\{\mathcal{H}, U, V\}$ satisfying (188), the condition that both U and V are continuous representations of $\mathbb{R}^n$ is necessary to the proofs; there exists indeed an irreducible representation of that system (even with $n = 1$) where U is continuous, but V is not, and where $\mathcal{H}$ is not separable. Our assumptions guarantee that both U and V are continuous.

One could also have been inclined to define a representation of the CCR by assuming only the commutation relations (167) between the operators $P^{(k)}$ and $Q^{(\ell)}$ $(k, \ell = 1, 2, \ldots, n)$. However, a counterexample to the Proposition is then easy to produce in $\mathcal{L}^2([0,1], dx)$,—see, for a hint, the argument between 8.2.86 and 8.2.87—thereby showing that these commutation relations must be supplemented by some auxiliary conditions such as, for instance, the fact that the spectrum of each operator $Q^{(k)}$ covers $\mathbb{R}$. This condition is guaranteed by our definition of $\{\mathcal{H}, F, U\}$ (see in particular 184); moreover the Proposition then implies that the spectral families of each $P^{(k)}$ and $Q^{(k)}$ are absolutely continuous with respect to Lebesgue measure.

g. The classical limit

Remembering the difficulties encountered in the early formulations of quantum theories (see Chapter Seven and the first two sections of the present chapter) concerning their possible, or necessary, connections with classical mechanics, we find it proper to present some precise comments on this problem at the close of this section on von Neumann's formulation of quantum mechanics.

We start with an irreducible representation $\{\mathcal{H}, F, U\}$ of the CCR for n degrees of freedom. We know from Proposition 1 that we can assume, without loss of generality, that $\mathcal{H}$ is $\mathcal{L}^2(\mathbb{R}^n, d^n x)$ and that $E_\lambda : (a, \hat{a}) \in \mathbb{R}^n \times \mathbb{R}^n \to \mathcal{B}(\mathcal{H})$ is given by (164). We now make use of the freedom we have in choosing $\lambda \in (0, \infty)$, i.e. in choosing the physical units in which we measure momenta (and positions).

For every $f \in \mathcal{L}^1(\mathbb{C}^n, d^n z)$, we now define the operator

$$\hat{f} \equiv E_\lambda(f) = \int d^n z \; f(z)\, E_\lambda(z) \tag{206}$$

(compare with 191–3; see also Segal, 1963a; and for generalizations, Lavine, 1965 and Grossmann, Loupias & Stein, 1968). We further equip

$$\mathcal{A}_\lambda \equiv \{\hat{f} \mid f \in \mathcal{L}^1(\mathbb{C}^n, d^n z)\} \tag{207}$$

with the structure of *-subalgebra of $\mathcal{B}(\mathcal{H})$ by defining

$$(\lambda f)^\wedge = \lambda \hat{f} \quad ; (f+g)^\wedge = \hat{f} + \hat{g} \tag{208a}$$

$$(\hat{f})^* = (f^*)^\wedge \quad \text{with} \quad f^* : z \in \mathbb{C}^n \mapsto f(-z)^* \in \mathbb{C} \tag{208b}$$

and the product $\hat{f} *_\lambda \hat{g}$:

(208c) $$\hat{f} *_\lambda \hat{g} : \varsigma \in \mathbb{C}^n \mapsto \int d^n z \; f(z)\, g(\varsigma - z)\, \omega_\lambda(z, \varsigma)$$

with ω_λ given by (164 d–e). A norm is further defined on $\mathcal{A}_\lambda$ by $\| \hat{f} \| = \| E_\lambda(f) \| \leq \| f \|_1$. A Jordan- and a Lie-structure are defined on $\mathcal{A}_\lambda$ by

(209a) $$\hat{f} \circ_\lambda \hat{g} \equiv \frac{1}{2}(\hat{f} *_\lambda \hat{g} + \hat{g} *_\lambda \hat{f})$$

(209b) $$\{\hat{f}, \hat{g}\}_\lambda \equiv (\hat{f} *_\lambda \hat{g} - \hat{g} *_\lambda \hat{f})/i\lambda^2$$

Note that these structures are compatible with the Jordan-Lie algebra structure of $\mathcal{B}_{sa}(\mathcal{H})$ when $\lambda^2 = \hbar$ (see Subsection b above); and that, for each $\lambda \in (0, \infty)$, $\mathcal{A}_\lambda'' = \mathcal{B}(\mathcal{H})$, so that $\mathcal{A}_\lambda$ is dense in $\mathcal{B}(\mathcal{H})$ (see Subsection d above).

Upon taking a hint from the properties of the functions

(210) $$\phi : z \in \mathbb{C}^n \mapsto tr\{\rho \; E_\lambda(z)\} \in \mathbb{C}$$

associated to each state $\rho \in S$, we define a *state* on $\mathcal{A}_\lambda$ as a continuous function

(211a) $$\phi_\lambda : z \in \mathbb{C}^n \mapsto \phi_\lambda(z) \in \mathbb{C}$$

which satisfies $\phi(0) = 1$ and

(211b) $$\sum_{j,k=1}^{K} \varsigma_j \; \varsigma_k^* \;\; \phi_\lambda(z_j - z_k)\, \omega_\lambda(z_j, z_k) \geq 0$$

for all $K \in \mathbb{Z}^+$, all $\{\varsigma_k | k = 1, 2, \ldots, K\} \subset \mathbb{C}$, and all $\{z_k | k = 1, 2, \ldots, K\} \subset \mathbb{C}^n$. We refer to (211b) by saying that ϕ_λ is of λ-positive type; this condition implies in particular that ϕ_λ is bounded. Note also that for any $\phi \in \mathcal{L}^\infty(\mathbb{C}^n)$

(212) $$\phi : \hat{f} \in \mathcal{A}_\lambda \mapsto < \phi; \hat{f} > \equiv \int d^n z \; \phi(z)\, f(z) \in \mathbb{C}$$

is positive if and only if it is locally almost everywhere equal to a continuous function of λ-positive type.

A family $\{\phi_\lambda | \lambda \in (0, \infty)\}$ is said to be a *classical state* if, for each $\lambda \in (0, \infty)$, ϕ_λ is a state on $\mathcal{A}_\lambda$, *and* ϕ_λ converges uniformly on compacta as $\lambda \to 0$. This is equivalent to the condition that ϕ_λ converge, as $\lambda \to 0$, in the $\sigma(\mathcal{L}^\infty, \mathcal{L}^1)$-topology, i.e. in the topology of convergence of expectation values. For each classical state $\{\phi_\lambda | \lambda \in (0, \infty)\}$ there exists then a continuous function

(213a) $$\phi_0 : z \in \mathbb{C}^n \mapsto \phi_0(z) \in \mathbb{C}$$

satisfying $\phi_0(0) = 1$,

(213b) $$\sum_{j,k=1}^{K} \varsigma_j \varsigma_k^* \ \phi_0(z_j - z_k) \geq 0$$

(with $\{\varsigma_k\}$, $\{z_k\}$ and K as in (211b)), and

(213c) $$\phi_\lambda \to \phi_0 \text{ as } \lambda \to 0 \quad \text{uniformly on compacta}$$

Since the notion of classical state will play a central role in the forthcoming Scholium 3, a few examples are called for.

Our first example is the ground state of the harmonic oscillator

(214) $$H_\lambda = \frac{1}{2} P_\lambda^2 + \frac{\omega^2}{2} Q_\lambda^2$$

In the representation (164) this state is represented by the vector $\Phi^0 \in \mathcal{L}^2(\mathbb{R}, dx)$ given by

(215a) $$\Phi^0(x) = (\omega/\pi)^{1/4} \exp(-\omega^2 x^2/2)$$

and the corresponding state on $\mathcal{A}_\lambda$ is the function (see 210)

(215b) $$\phi_\lambda^0 : z \in \mathbb{C}^n \mapsto \exp\{-\lambda^2(\omega a^2 + \omega^{-1}\hat{a}^2)/4\}$$

Clearly $\{\phi_\lambda^0 | \lambda \in (0, \infty)\}$ converges, uniformly on compacta, as $\lambda \to 0$, to the function

(215c) $$\phi_0^0 : z \ \in \ \mathbb{C}^n \mapsto \phi_0^0(z) = 1$$

Hence $\{\phi_\lambda^0 | \lambda \in (0, \infty)\}$ is a classical state. Note that the Fourier transform of ϕ_0^0 is the Dirac measure that is concentrated at the origin.

Our second example is a generalization of the previous one, and is obtained from the remark that, for every $\hat{z} = (\hat{p}, q) \in \mathbb{R}^n \times \mathbb{R}^n$, the transformation $E_\lambda \to E_\lambda^{\hat{z}}$ given by

(216a) $$E_\lambda^{\hat{z}} : z \in \mathbb{C}^n \mapsto e^{\hat{z}}(z) \ E_\lambda(z) \ \in \ \mathcal{U}(\mathcal{H})$$

where

(216b) $$e^{\hat{z}}(z) \equiv \exp\{-i(a \cdot \hat{p} + \hat{a} \cdot q)\} \quad \forall \ z \in \mathbb{C}^n$$

leads again to an irreducible representation of the CCR. From Proposition 1, we know that there exists a unitary operator $X_\lambda(\hat{z})$ such that

(216c) $$E_\lambda^{\hat{z}}(z) = X_\lambda(\hat{z})^* \ E_\lambda(z) \ X_\lambda(\hat{z}) \quad \forall \ z \in \ \mathbb{C}^n$$

Let now ρ be any density matrix and define

(217) $$\phi_\lambda^{\hat z} : z \in \mathbb{C}^n \mapsto tr\{X_\lambda(\hat z)\,\rho\, X_\lambda(z)^* E_\lambda(z)\}$$

It is easy to check that $\{\phi_\lambda^{\hat z} | \lambda \in (0,\infty)\}$ is a classical state, and that $\phi_0^{\hat z} = e^{\hat z}$. Note again that the Fourier transform of $\phi_0^{\hat z} = e^{\hat z}$ is the Dirac measure on $\mathbb{R}^n \times \mathbb{R}^n$, now concentrated at $\hat z = (\hat p, q)$.

In the particular case where ρ is a one-dimensional projector, $\phi_\lambda^{\hat z}$ is called a coherent state in the sense of Perelomov (1972). We have indeed for all $A \in \mathcal{B}_{sa}(\mathcal{H})$:

(218) $$\begin{gathered} tr\{X_\lambda(\hat z)\,\rho\, X_\lambda(\hat z)^* A\} = (A\Phi_\lambda^{\hat z}, \Phi_\lambda^{\hat z}) \equiv < A >_\lambda^{\hat z} \\ \text{with} \quad \Phi_\lambda^{\hat z} \equiv X_\lambda(\hat z)\Phi\ ,\ \Phi \in \mathcal{H}\ ,\ \| \Phi \| = 1 \end{gathered}$$

The further particularization to the case where $\Phi = \Phi^0$ (see 215a) and $\lambda^2 = \hbar$ gives the coherent states in the sense of Schroedinger (1926N). These states give

(219a) $$< Q >_\lambda^{\hat z} = q \qquad < P >_\lambda^{\hat z} = \hat p$$

(219b) $$< (Q - qI)^2 >_\lambda^{\hat z} \cdot < (P - pI)^2 >_\lambda^{\hat z} = \hbar^2/4$$

These are, therefore, states of minimal quantum dispersion (see 8.2.82–87) centered at $\hat z = (\hat p, q) \in \mathbb{R}^n \times \mathbb{R}^n$.

Our third example generalizes the first example in still another direction. With the harmonic oscillator Hamiltonian (214) consider the density matrix

(220) $$\rho_\lambda^\beta \equiv \exp(-\beta H_\lambda)/tr\{\exp(-\beta H_\lambda)\}$$

Physically the state

(221a) $$\phi_\lambda^\beta : z \in \mathbb{C}^n \mapsto tr\{\rho_\lambda^\beta E_\lambda(z)\}$$

is interpreted as the canonical equilibrium state for the harmonic oscillator at natural temperature $\beta = 1/k\,T$ (compare with 8.2.53–55). A straightforward computation gives for every $z = (a, \hat a) \in \mathbb{R}^n \times \mathbb{R}^n$

(221b) $$\phi_\lambda^\beta(z) = \exp\{-\lambda^2 \Theta_\lambda^\beta(\omega a^2 + \omega^{-1}\hat a^2)/4\}$$

(221c) $$\Theta_\lambda^\beta = \coth\,(\beta\omega\lambda^2/2)$$

Clearly $\{\phi_\lambda^\beta | \lambda \in (0,\infty)\}$ is a classical state for every $\beta \in (0,\infty)$. It converges, uniformly on compacta as $\lambda \to 0$, to the function

(222a) $$\phi_0^\beta : z \in \mathbb{C}^n \mapsto \exp\{-\beta^{-1}(a^2 + \omega^{-2}\hat a^2)/2\}$$

the Fourier transform of which is the measure μ_β on $\mathbb{R}^n \times \mathbb{R}^n$ given by

$$d\mu_\beta(\hat{p}, q) = Z^{-1} \exp\{-\beta H(\hat{p}, q)\}\, d^n\hat{p}\, d^n q \tag{222b}$$

where

$$Z = \int d^n\hat{p}\, d^n q \;\; \exp\{-\beta H(\hat{p}, q)\} \tag{222c}$$

$$H(\hat{p}, q) = \frac{1}{2}\hat{p}^2 + \frac{\omega^2}{2} q^2 \tag{222d}$$

Note that μ_β is the canonical equilibrium measure, at natural temperature β for the classical harmonic oscillator (see 222d).

All these examples suggest that classical states define distributions on $\mathbb{R}^n \times \mathbb{R}^n$. This is indeed the case in general since the continuous functions ϕ_0 of positive type, obtained as limit of classical states are, by the classical Bochner theorem, Fourier transforms of normalized, but not necessarily completely additive, measures on the space $\mathbb{R}^n \times \mathbb{R}^n$. From this remark, one then obtains the following two results.

Scholium 3. *To every quantum observable*

$$\hat{f} \equiv \int d^n z\; f(z) \exp\{-i(aP_\lambda + \hat{a}Q_\lambda)\} \tag{223}$$

with $f \in \mathcal{L}^1(\mathbb{C}^n, d^n z)$ corresponds a unique function $\tilde{f} : \mathbb{C}^n \to \mathbb{C}$ given by

$$\tilde{f}(\hat{z}) \equiv \int d^n z\; f(z) \exp\{-i(a\hat{p} + \hat{a}q)\} \tag{224}$$

with $\hat{z} \equiv (\hat{p}, q)$ running over $\mathbb{C}^n \equiv \mathbb{R}^n \times \mathbb{R}^n$) such that for every classical state $\{\phi_\lambda | \lambda \in (0, \infty)\}$:

$$\lim_{\lambda \to 0} < \phi_\lambda;\, \hat{f} > = \int d\mu_\phi(\hat{z})\; f(\hat{z}) \tag{225}$$

where μ_ϕ is the Bochner measure corresponding to the continuous function ϕ_0 of positive-type determined by $\{\phi_\lambda | \lambda \in (0, \infty)\}$.

Corollary 4. *For every pair (f, g) of continuous functions, with compact support, from $\mathbb{R}^n \times \mathbb{R}^n$ to $\mathbb{C}$, and for every classical state $\{\phi_\lambda | \lambda \in (0, \infty)\}$*

$$\lim_{\lambda \to 0} < \phi_\lambda; \hat{f} \circ_\lambda \hat{g} > \; = \int d\mu_\phi(\hat{z})\; (\tilde{f} \cdot \tilde{g})(\hat{z}) \tag{226}$$

$$\lim_{\lambda \to 0} < \phi_\lambda; \{\hat{f}, \hat{g}\}_\lambda > \; = \int d\mu_\phi(\hat{z})\; \{\tilde{f}, \tilde{g}\}(\hat{z}) \tag{227}$$

where $\hat{f} \circ_\lambda \hat{g}$ (resp. $\{\hat{f}, \hat{g}\}_\lambda$) is the Jordan-(resp. Lie) product defined on $\mathcal{A}_\lambda$ by (209); μ_ϕ is the measure associated to $\{\phi_\lambda | \lambda \in (0, \infty)\}$ by Bochner's theorem, as in Scholium 3; $\tilde{f}$ (and similarly $\tilde{g}$) is the function associated to $\hat{f}$ (resp. $\hat{g}$) by Scholium 3; $\tilde{f} \cdot \tilde{g}$ is the usual point-wise product of functions, i.e. $(\tilde{f} \cdot \tilde{g})(\hat{z}) = \tilde{f}(\hat{z})\tilde{g}(\hat{z})$; and finally $\{\tilde{f}, \tilde{g}\}$ is the Poisson bracket:

(228) $$\{\tilde{f},\tilde{g}\} \equiv \sum_{k=1}^{n}(\partial_{q_k}\tilde{f}\cdot\partial_{p_k}\tilde{g} - \partial_{q_k}\tilde{g}\cdot\partial_{p_k}\tilde{f}).$$

These two results can be paraphrased by saying that the phase-space formalism of classical mechanics follows from the von Neumann formalism of quantum mechanics. Specifically, the phase-space $T^*\mathbb{R}^n \simeq \mathbb{C}^n \equiv \mathbb{R}^n \times \mathbb{R}^n$ of classical mechanics appears as the space of classical pure space; classical states determine measures on this phase space; to every quantum observable $\hat{f}$ corresponds a well-defined, essentially bounded, function $\hat{f}$ on the phase-space; this correspondence is known as the Wigner (1932)—Moyal (1949) rule, and it is derived here from first principles; this correspondence is clearly linear; moreover, the Jordan-(point-wise multiplication) and Lie-(Poisson bracket) structures of classical mechanics follow respectively from the Jordan-(symmetrized product) and Lie-(quantum commutator) structures of the algebra of quantum observables. The limit $\lambda \to 0$ corresponds to a "macroscopic" choice of physical units to mesure momentum and position: in these units the numerical value of the Planck constant becomes arbitrarily small.

The detailed proofs of the above results are straightforward and they can be found in Emch (1982 & 1983), where a generalization from $\mathbb{R}^n$ to curved homogeneous Riemannian manifolds is also indicated.

In the light of these results, quantum theory can be seen as the fundamental theory from which another theory—namely classical mechanics—is derived as a limiting case, where a fundamental constant—namely the Planck constant h—is formally let go to zero. This is reminiscent of the situation encountered in special relativity where Galilean relativity is obtained as a limiting theory when another fundamental constant—namely the inverse c^{-1} of the velocity of light c—is formally let go to zero.

Upon remembering the theory of (local) factors presented in Subsection e above, the reader will notice that, underneath the classical limit $h \to 0$, lies a concept of contraction of Lie groups (or deformation of their group algebra) that is quite analogous to the phenomena discovered by Inonu & Wigner (1953) in their discussion of the passage from Einstein's to Galilean relativity; see also Saletan (1961) and for an up-dated review with interesting references, Arnal et al. (1981).

The link just established between quantum and classical mechanics explain the similarity of methods used, in both theories, to solve some specific problems. This is the case, for instance, of the group theoretical considerations developed, for any Galilean-invariant system of N particles, to show that the time evolution is generated by a Hamiltonian of the form

(229) $$H = \sum_{k=1}^{N}\frac{1}{2m_k}|\mathbf{p}_k| + V(\mathbf{q}_1,\ldots,\mathbf{q}_N)$$

with V depending only on the interparticle distances; see e.g. Mackey (1963), Jauch (1968) and Abraham & Marsden (1978). For a comparison between the quantum and classical evolutions corresponding to such Hamiltonians, see Hepp (1974).

Section 4. QUANTUM SCATTERING THEORY

There are few—if any—areas of mathematical physics where von Neumann's Hilbert space formalism has been more successful than in its applications to quantum scattering theory. From the early days of atomic and nuclear physics to their current activities around the giant accelerators of high energy physics or in the quantum optics laboratories, physicists have used scattering experiments as a method of choice to probe the fundamental interactions between the elementary constituents of matter and radiation. Mathematically, quantum scattering theory, focusing as it does on a specific class of partial differential equations, has provided functional analysis with a wealth of precise problems that have helped sharpen our knowledge of self-adjoint operators and their spectral properties.

A mathematician interested in detailed accounts of what has been accomplished in this field should consult the books by Amrein, Jauch & Sinha (1977) and by Reed & Simon (1979); see also Kato (1966), Simon (1971), and Amrein (1981). To keep the presentation of the basic ideas at a reasonably elementary level, we shall concentrate here on the particular case of the scattering of a non-relativistic quantum particle by a potential.

Note that this particular case also covers another problem of physical interest, namely the mutual scattering of two interacting quantum particles A and B described individually in the Hilbert spaces $\mathcal{H}_A = \mathcal{L}^2(\mathbb{R}^3, d^3x_A)$ and $\mathcal{H}_B = \mathcal{L}^2(\mathbb{R}^3, d^3x_B)$. The Hilbert space of the joint system is then $\mathcal{H} = \mathcal{H}_A \otimes \mathcal{H}_B$ where the joint evolution is described by a strongly continuous one-parameter group $U : \mathbb{R} \to \mathcal{U}(\mathcal{H})$ of unitary operators. By Stone's theorem, we have $U(t) = \exp\{-iHt\}$. The Hamiltonian H is supposed to be of the form $H_0 + V$ with H_0 and V defined from

$$
\begin{aligned}
(H_0\Phi)(x_A, x_B) &= [-(\frac{1}{2m_A}\Delta_A + \frac{1}{2m_B}\Delta_B)\Phi](x_A, x_B) \\
(V\Phi)(x_A, x_B) &= V(x_B - x_A)\Phi(x_A, x_B)
\end{aligned}
\tag{1}
$$

where Δ denotes the Laplace operator. Informed by classical mechanics, one introduces the new set of canonical observables

$$
\begin{aligned}
&Q \equiv \frac{1}{M}(m_A q_A + m_B q_B) \;\; ; \;\; P \equiv (p_A + p_B) \\
&q \equiv (q_B - q_A) \;\; ; \;\; p \equiv \frac{1}{M}(m_A p_B - m_B p_A) \\
&\text{with } M \equiv (m_A + m_B) \;\; ; \;\; m \equiv \frac{1}{M} m_A m_B
\end{aligned}
\tag{2}
$$

that suggests the introduction of the unitary operator $U : \mathcal{H}_A \otimes \mathcal{H}_B \to \mathcal{H}_{CM} \otimes \mathcal{H}_{\text{rel}}$ [with $\mathcal{H}_{CM} \equiv \mathcal{L}^2(\mathbb{R}^3, d^3X)$ and $\mathcal{H}_{\text{rel}} = \mathcal{L}^2(\mathbb{R}^3, d^3x)$] defined by

$$\begin{aligned} &(U\Phi)(x_A, x_B) = \Phi(X, x) \quad \text{with} \\ &X = \frac{1}{M}(m_A x_A + m_B x_B) \ \text{and} \ x = x_B - x_A \end{aligned} \tag{3}$$

This unitary transformation allows to rewrite $U(\mathbb{R})$ in the form

$$\begin{aligned} &U(t) = U_{CM}(t) \otimes U_{\text{rel}}(t) \quad \text{with} \\ &U_{CM}(t) = \exp\{-iH_{CM}t\} \ \text{and} \ U_{\text{rel}}(t) = \exp\{-iH_{\text{rel}}t\} \end{aligned} \tag{4}$$

where H_{CM} and H_{rel} are defined from

$$\begin{aligned} (H_{CM}\Psi)(X) &= -\frac{1}{2M}(\Delta_X \Psi)(X) \\ (H_{\text{rel}}\psi)(x) &= -\frac{1}{2m}(\Delta_x \psi)(x) + V(x)\psi(x) \end{aligned} \tag{5}$$

Hence, as in classical mechanics, the motion of the center of mass is free and can be factored out, leaving the problem of a single quantum particle of mass m moving in the potential V.

We will assume, throughout this section, that we are given two self-adjoint oeprators H_0 and H_1 acting in a separable Hilbert space $\mathcal{H}$. We say that $(\mathcal{H}, H_0, H_1)$ is *standard* if $\mathcal{H} = \mathcal{L}^2(\mathbb{R}^3, d^3x)$; H_0 is defined as a self-adjoint extension of the second-order differential operator $(-\Delta_x/2m)$; and H_1 is defined as a self-adjoint extension of $(-\Delta_x/2m) + V$, where V is a multiplication operator.

Recall that, to every self-adjoint operator H acting in $\mathcal{H}$ is canonically associated a spectral family $\{P_\lambda | \lambda \in \mathbb{R}\}$ and a one-parameter group of unitary operators acting on $\mathcal{H}$, continuous in the strong-operator topology, namely

$$U(t) \equiv \exp\{-iHt\} = \int \exp(-i\lambda t)\, dP_\lambda \tag{6}$$

Recall further that for every $\Psi \in \mathcal{H}$

$$\mu_\Psi : \Delta \in \mathcal{B}(\mathbb{R}) \mapsto \int_\Delta d(P_\lambda \Psi, \Psi) \tag{7}$$

defines a probability measure on $\mathbb{R}$. $\{P_\lambda | \lambda \in \mathbb{R}\}$ defines an orthogonal decomposition

$$\mathcal{H} = \mathcal{H}_p(H) \oplus \mathcal{H}_{ac}(H) \oplus \mathcal{H}_{sc}(H) \tag{8}$$

where $\mathcal{H}_p(H)$ is the subspace of $\mathcal{H}$ spanned by the eigenvectors of H; $\mathcal{H}_{ac}(H)$ is the subspace of H defined by the condition $\Psi \in \mathcal{H}_{ac}(H)$ if and only if μ_Ψ is absolutely continuous with respect to Lebesgue measure. Note that, for every $\Psi \in \mathcal{H}_{ac}(H)$, the Riemann-Lebesgue lemma implies that

$$f_\Psi(t) \equiv (U(t)\Psi, \Psi) \tag{9}$$

satisfies

$$\lim_{|t|\to\infty} f_\Psi(t) = 0 \tag{10}$$

This is interpreted as follows. Let $\Psi \in \mathcal{H}$ with $\| \Psi \| = 1$; $\Psi_t \equiv U(t)\Psi$;

$$\psi_t : B \in \mathcal{B}(\mathcal{H}) \mapsto < \psi_t ; B > = (B\Psi_t, \Psi_t) \in \mathbb{C} \tag{11}$$

be the state of the particle at time t;

$$P_\psi : \Phi \in \mathcal{H} \mapsto (\Phi, \Psi)\Psi \in \mathcal{H} \tag{12}$$

be the proposition asserting that the particle is in the state ψ; then

$$p_\psi(t) \equiv < \psi_t ; P_\psi > = |(U(t)\Psi, \Psi)|^2 \tag{13}$$

is the probability to find the particle in the state ψ at time t, when we know that it is in that same state at time $t = 0$. The relation (10) therefore means that for every $\Psi \in \mathcal{H}_{ac}(H)$, one has

$$\lim_{|t|\to\infty} p_\psi(t) = 0 \tag{14}$$

If, instead of $\Psi \in \mathcal{H}_{ac}(H)$, we now assume that $\Psi \in \mathcal{H}_{sc}(H) \equiv [\mathcal{H}_p(H) \oplus \mathcal{H}_{ac}(H)]^\perp$, μ_Ψ is singular continuous—the Cantor measure is an example illustrating how nasty this situation can be—and the best one can guarantee on the behaviour of $p_\psi(t)$ at infinity is that

$$\lim_{T\to\infty} \frac{1}{2T} \int_{-T}^{T} dt\, p_\psi(t) = 0\,. \tag{15}$$

Even for the standard situations of scattering theory, where $\mathcal{H} = \mathcal{H}_{ac}(H_0 = -\Delta/2m)$, and thus $\mathcal{H}_{sc}(H_0) = 0$, one of the hard questions of the theory is to find conditions on V such that $\mathcal{H}_{sc}(H_1) = 0$; for a systematic introduction to the different techniques deployed for the control of this condition, see Reed & Simon (1979). Counterexamples are known, but they have been pushed now into a rather contrived corner of the theory; it thus seems better, in an elementary exposition such as that of the present section, to assume $\mathcal{H}_{sc}(H_0) = 0 = \mathcal{H}_{sc}(H_1)$, unless explicit mention is made to the contrary. For instance, these conditions are satisfied for $H_0 = -\Delta/2m$ acting in $\mathcal{L}^2(\mathbb{R}^3, d^3x)$, and

$H_1 = H_0 + V$ if there exists $\epsilon > 0$ such that

(16a) $$W(x) \equiv (1+ \| x \|)^{1+\epsilon} V(x)$$

can be written in the form

(16b) $$\begin{aligned} &W(x) = W_1(x) + W_2(x) \ \text{ with} \\ &W_1 \in \mathcal{L}^2(\mathbb{R}^3, d^3x) \ \text{ and } \ W_2 \in \mathcal{L}^\infty(\mathbb{R}^3, d^3x) \end{aligned}$$

This result often allows one to control separately the behaviour of V at infinity and its local behaviour, for instance by defining $W_1 = \chi_\Delta W$ where χ_Δ is the characteristic function of a compact $\Delta \subset \mathbb{R}^3$.

In its simplest form, scattering theory is concerned with the time-behaviour of pure normal states on $\mathcal{B}(\mathcal{H})$. Let ψ be such a state; $U : \mathbb{R} \to \mathcal{U}(\mathcal{H})$ be a unitary group evolution on $\mathcal{H}$, continuous with respect to the strong operator topology; H be its generator; ψ_t be defined as in (11); P_r be the projector defined by

(17a) $$(P_r\Phi)(x) \equiv \chi_{S(r)}(x)\Phi(x)$$

where $\chi_{S(r)}$ is the characteristic function of the ball

(17b) $$S(r) \equiv \{x \in \mathbb{R}^3 \mid \| x \| \leq r\} ;$$

and

(18) $$p_\psi(t,r) \equiv < \psi_t ; P_r > = \| P_r U(t)\Psi \|^2$$

denotes the probability to find the particle in $S(r)$ at time t when we know that it is in the state ψ at $t = 0$. We say that ψ is a *scattering state* for $U(\mathbb{R})$ if

(19) $$\lim_{|t|\to\infty} p_\psi(t,r) = 0 \quad \forall \ r < \infty$$

Alternate definitions have been considered. For instance, if we had been willing to consider situations where $\mathcal{H}_{sc}(H) \neq 0$, it would have made sense to replace (19) by the weaker condition

(20) $$\lim_{T\to\infty} \frac{1}{2T} \int_{-T}^{T} dt \ p_\psi(t,r) = 0 \quad \forall r < \infty .$$

Moreover, situations can be produced analytically—see, e.g. Pearson (1975)—where it is necessary to distinguish in (19) the limits corresponding to the remote past ($t \to -\infty$) and the distant future ($t \to +\infty$). We shall not make this distinction here, and we will mention below a class of standard situations where this is justified. We shall thus denote by $M_\infty(H)$ the set of pure normal states on $\mathcal{B}(\mathcal{H})$ that satisfy (19), and we will refer to its elements as the scattering states for the evolution $U(\mathbb{R})$ generated by H.

Similarly, we denote by $M_0(H)$ the set of pure normal states on $\mathcal{B}(\mathcal{H})$ that satisfy

$$\lim_{r\to\infty} \sup_{t\in\mathbb{R}} < \psi_t ; (I - P_r) >= 0 \tag{21}$$

and we will refer to the elements of $M_0(H)$ as the *bound states* for the evolution $U(\mathbb{R})$ generated by H.

Let us further denote by $E_\infty(H)$ [resp. $E_0(H)$] the linear subspace of $\mathcal{H}$ generated by the vectors $\Psi \in \mathcal{H}$ such that the corresponding ψ belong to $M_\infty(H)$ [resp. $M_0(H)$]. Note that $E_\infty(H)$ and $E_0(H)$ are closed subspaces of $\mathcal{H}$, that are stable under $U(\mathbb{R})$, and mutually orthogonal. In general, one has only $\mathcal{H}_p(H) \subseteq E_0(H)$ and $E_\infty(H) \subseteq \mathcal{H}_p(H)^\perp$; the following particular cases should nevertheless be noted. In the standard case $(\mathcal{H}, H_0, H_1)$ described earlier, $\mathcal{H}_{ac}(H_0) = \mathcal{H} = E_\infty(H_0)$ and $\mathcal{H}_p(H_0) = 0 = E_0(H_0)$; moreover, if $\mathcal{H}_{sc}(H_1) = 0$ [see e.g. (16)], and if V can be written in the form

$$\begin{aligned} &V(x) = V_1(x) + V_2(x) \ \text{with} \\ &V_1 \in \mathcal{L}^2(\mathbb{R}^3, d^3x) \ \text{ and } \ V_2 \in \mathcal{L}^\infty(\mathbb{R}^3, d^3x) \end{aligned} \tag{22}$$

then $\mathcal{H}_{ac}(H_1) = E_\infty(H_1)$ and $\mathcal{H}_p(H_1) = E_0(H_1)$.

Additional information on scattering states is obtained upon introducing the notion of *sojourn time*, either in a compact $\Delta \subset \mathbb{R}^3$ or in a state ψ, namely

$$\tau_\psi(\Delta) \equiv \int_{-\infty}^{+\infty} dt < \psi_t ; P_\Delta > \quad \text{or} \quad \tau_\psi \equiv \int_{-\infty}^{+\infty} dt \, p_\psi(t) \tag{23}$$

where $p_\psi(t)$ is defined in (13). These notions have been studied—see e.g. Lavine (1978)—in connection with the concepts of resonances and their line-width. Lavine has further work in progress on finding realistic bounds for the situation where a quantum particle is "almost" trapped in a rigid cavity of reasonably general geometric shape; this work is to be distinguished from the asymptotic time rescaling procedures developed for the "tunneling" situations where the rigid cavity is replaced by a soft potential-wall of finite but very large height, and where detailed information (of a somewhat different kind) is available—see e.g. Emch & Sinha (1979), and references quoted therein on the related theories of spectral concentration that grew out of the work by Kato (1966). The notion of sojourn time should also not be confused with the scattering theory concept of time delay to be introduced later on in this section.

We now come to the study of scattering situations where we want to compare two unitary group evolutions $U_k : \mathbb{R} \to \mathcal{U}(\mathcal{H})$ $(k = 0, 1)$, continuous with respect to the strong operator topology. For convenience, we refer to $U_0(\mathbb{R})$ [resp. $U_1(\mathbb{R})$] as the free evolution [resp. the evolution with interaction], keeping in mind the standard situation where the generator H_0 [resp. H_1] of $U_0(\mathbb{R})$ [resp. $U_1(\mathbb{R})$] is defined from $-\Delta/2m$ [resp. $-\Delta/2m + V(x)$] acting in $\mathcal{L}^2(\mathbb{R}^3, d^3x)$; the reader will verify that the mathematical concepts of the theory are more general than this particular interpretation, but that the role of this specific case comes into play when one starts trying to implement the axioms.

In a typical scattering situation, the way an experimenter sees the effects of the potential $V = H_1 - H_0$ can be intuitively described as follows. Far away from the scattering region, i.e. far away from the region where V is appreciably different from zero, particles behave as if they were free. As such an asymptotically free particle enters the scattering region, it is deflected and ultimately comes out of this region in a state which is again asymptotically free. The aims of scattering theory are therefore: (a) to establish a correspondence between the scattering states that belong to the description pertaining to $U_0(\mathbb{R})$, and those that belong to the description pertaining to $U_1(\mathbb{R})$; (b) to use this to establish a correspondence between the states of the incoming and outgoing particles; (c) to relate these correspondences to V itself.

A systematic programme to give an abstract formalization of this situation was started by Jauch (1958); the emphasis there was on providing the theory with a set of general, mathematically well-formulated axioms, while mathematical existence proofs had already begun to appear for various concrete, but special, realizations; see e.g. Cook (1957). To bring order into what has to be proven, it is convenient to isolate three separate conditions.

The first condition is that for every vector Ψ_- [resp. Ψ_+] in $E_\infty(H_0)$ the following limits exist

(24a) $$s - \lim_{t \to -\infty} U_1(t)^* U_0(t) \Psi_-$$

(24b) $$s - \lim_{t \to +\infty} U_1(t)^* U_0(t) \Psi_+$$

We denote by $\Omega_-(H_1, H_0)$ [resp. $\Omega_+(H_1, H_0)$] the linear operator acting from $E_\infty(H_0)$ to $\mathcal{H}$ defined by (24a) [resp. (24b)]; when no confusion is likely, we simply write Ω_- for $\Omega_-(H_1, H_0)$, and similarly Ω_+ for $\Omega_+(H_1, H_0)$. Note that the convention to denote by Ω_+ [resp. Ω_-] the operator we call Ω_- [resp. Ω_+] is also used in the literature. $\Omega_\pm$ are referred to as the *wave operators* of the theory.

The second condition, known as *weak asymptotic completeness*, is that the ranges of Ω_+ and Ω_- coincide, i.e.

(25) $$\text{Ran } \Omega_+ = \text{Ran } \Omega_-$$

When this condition is satisfied, we refer to this subspace as Ran $\Omega_\pm$.

The third condition, known as *asymptotic completeness*, is

(26) $$\text{Ran } \Omega_\pm = E_\infty(H_1)$$

The physical meaning of these three conditions can be described as follows. To assume that the wave operator Ω_- exists is to assume that for every asymptotic vector state $\Psi_- \in E_\infty(H_0)$ there exists a vector $\Psi \in \mathcal{H}$, the evolution of which through $U_1(\mathbb{R})$ approximates the free evolution of Ψ_- as $t \to -\infty$ in the sense that

$$\lim_{t\to-\infty} \| U_1(t)\Psi - U_0(t)\Psi_- \| = 0 \tag{27}$$

A similar meaning can be given to the condition that Ω_+ exists. One could argue that the topology of strong convergence in $\mathcal{H}$ is perhaps not the most natural topology to use on physical grounds; this is a valid objection, and we shall come back on this point towards the end of this section; for the time being, we shall nevertheless conform to the widely accepted usage in the field, proceed with this topology, and see what it leads to.

Once Ω_- and Ω_+ are assumed to be defined by (24) for every vector in $E_\infty(H_0)$, the assumption that Ran $\Omega_- \subseteq$ Ran Ω_+ amounts to assuming that for each $\Psi = \Omega_-\Psi_-$ with $\Psi_- \in E_\infty(H_0)$ there exists $\Psi_+ \in E_\infty(H_0)$, the free evolution of which approximates, as $t \to +\infty$, the evolution of Ψ through $U_1(\mathbb{R})$ in the sense that

$$\lim_{t\to+\infty} \| U_1(t)\Omega_-\Psi_- - U_0(t)\Psi_+ \| = 0 \tag{28}$$

A similar interpretation, with the roles of $t \to -\infty$ and of $t \to +\infty$ interchanged, can be given for the condition that Ran $\Omega_+ \subseteq$ Ran Ω_-.

Finally, when (24a) [resp. (24b)] holds, the condition that Ran $\Omega_- = E_\infty(H_1)$ [resp. Ran $\Omega_+ = E_\infty(H_1)$] amounts to assuming that Ω_- [resp. Ω_+] is a map from $E_\infty(H_0)$ onto $E_\infty(H_1)$, thus establishing a map from the scattering states for $U_0(\mathbb{R})$ onto the scattering states for $U_1(\mathbb{R})$.

From these conditions, it follows that

$$\Omega_- : \Psi_- \in E_\infty(H_0) \mapsto s - \lim_{t\to-\infty} U_1(t)^* U_0(t)\Psi_- \in E_\infty(H_1) \tag{29a}$$

$$\Omega_+ : \Psi_+ \in E_\infty(H_0) \mapsto s - \lim_{t\to+\infty} U_1(t)^* U_0(t)\Psi_+ \in E_\infty(H_1) \tag{29b}$$

are isometric bijections, with adjoints

$$\Omega_-^* : \Psi \in E_\infty(H_1) \mapsto s - \lim_{t\to-\infty} U_0(t)^* U_1(t)\Psi \in E_\infty(H_0) \tag{30a}$$

$$\Omega_+^* : \Psi \in E_\infty(H_1) \mapsto s - \lim_{t\to+\infty} U_0(t)^* U_1(t)\Psi \in E_\infty(H_0) \tag{30b}$$

and that they satisfy, for all $t \in \mathbb{R}$, the *intertwining relations*

$$U_1(t)\Omega_\pm = \Omega_\pm U_0(t) \quad \text{on } E_\infty(H_0) \tag{31a}$$

$$U_0(t)\Omega_\pm{}^* = \Omega_\pm{}^* U_1(t) \quad \text{on } E_\infty(H_1) \tag{31b}$$

Mathematically, Ω_+ and Ω_- thus establish spatial equivalences between the unitary group obtained as the restriction of $U_0(\mathbb{R})$ to $E_\infty(H_0)$, and the unitary group obtained as the restriction of $U_1(\mathbb{R})$ to $E_\infty(H_1)$. Hence, the three conditions (24–26) together amount conceptually to the desired symmetric statement of the asymptotic equivalence between the two evolutions of $U_0(\mathbb{R})$ and $U_1(\mathbb{R})$ as viewed from their respective scattering states. At this abstract level, one could have exchanged the roles of $U_0(\mathbb{R})$ and $U_1(\mathbb{R})$, thus starting with the definition of the isometries $\Omega_\pm{}^*$ instead of $\Omega_\pm$. The reason for the traditional formulation presented here is therefore one of technical convenience: in the standard scattering situations, the proofs that the conditions (24–26) hold use explicitly the information immediately available (e.g. via Fourier transform) on the Hamiltonian H_0 of the free evolution (i.e. essentially on the differential operator $-\Delta/2m$) and its spectral family; no such information is available a priori on the total Hamiltonian H that includes the interaction V. In particular, a satisfactory scattering theory—in the sense that the limits (24) exist and define wave operators Ω_- and Ω_+ for which the asymptotic completeness condition (26) [and thus (25)] holds true—obtains when V belongs to the class of so-called Agmon potentials, of which (16) gives a most immediate example.

More work is nonetheless still necessary since, even in standard potential scattering theory, explicit situations can be described, that do not satisfy these conditions.

For instance, Pearson (1975) found a short range, spherically symmetric potential such that: (i) it is sufficiently well behaved (it is bounded on compact subsets of $\mathbb{R}^3 \setminus \{0\}$, and has compact support) for $H = -\Delta/2m + V$ to be well defined as a positive self-adjoint operator, and for $\Omega_\pm$ to exist; but (ii) it is nevertheless sufficiently wild (it is unbounded and oscillates violently as $r \to 0$) for Ran Ω_+ to be different from Ran Ω_-. Physically, "states exist which are asymptotically free as $t \to -\infty$, but have a non zero probability of absorption into the origin as $t \to +\infty$", i.e. there are incoming states which, in the distant future, break into two pieces, the first of which scatters away freely as expected in less singular situations, while the second piece gets trapped near the origin: the repulsive part of the oscillating potential prevents the incoming particle to reach the origin in a finite time, while the attractive part of the potential prevents the particle to be just reflected away.

Even more disturbingly, the Coulomb potential $V(r) = Cr^{-1}$ does not satisfy (16) for any $\epsilon > 0$. This is not a pathology of the method but, on the contrary, it is an intrinsic feature of the Coulomb potential itself: it decreases too slowly as $r \to \infty$ for $\Omega_\pm$ even to exist; see Dollard (1964). We shall come back to this point towards the end of this section.

For the time being, we nevertheless want to explore a few of the consequences of (24–26) since they can still serve as a guide for subsequent generalizations.

The existence of wave operators $\Omega_\pm$ satisfying the (weak) asymptotic completeness condition allows one to define the *scattering operator*

(32) $$S : \Psi_- \in E_\infty(H_0) \mapsto \Psi_+ \equiv \Omega_+{}^*\Omega_-\Psi_- \in E_\infty(H_0)\,.$$

The physical significance of this operator is that it ascribes to every asymptotically free, incoming state ψ_- (associated to Ψ_-), the asymptotically free, outgoing state ψ_+ (associated to Ψ_+); it thus describes the observable features of the scattering process without any explicit reference to the interpolating state ψ (associated to $\Omega_-\Psi_- = \Psi = \Omega_+\Psi_+$).

Note that S is an isometric bijection of $E_\infty(H_0)$ onto itself, and that the intertwining relations (31) imply

$$SU_0(t) = \Omega_+{}^* U_1(t)\Omega_- = U_0(t)S \text{ on } E_\infty(H_0). \tag{33}$$

In particular, for the standard scattering situation, where H_0 is defined from $-\Delta/2m$ acting on $\mathcal{L}^2(\mathbb{R}^3, d^3x)$, one has $E_\infty(H_0) = \mathcal{H}$; consequently S is then a unitary operator on $\mathcal{H}$, commuting with $U_0(\mathbb{R})$. If furthermore, V is spherically symmetric—and satisfies some assumptions of smoothness and of fast decrease at infinity, which we do not want to detail here beyond saying that they ensure, in particular, that the condition of asymptotic completeness is fulfilled; see Amrein, Jauch & Sinha (1977)—the rotational symmetry of the resulting scattering situation can be exploited to derive from S specific information on quantities of direct physical interest; we briefly indicate a few of them.

Recall first that $\mathcal{H} = \mathcal{L}^2(\mathbb{R}^3, d^3x)$ supports a unitary representation of the group $G = O^3_+$ of rotations in $\mathbb{R}^3$, continuous in the strong-operator topology, and defined by

$$[V(g)\Psi](x) = \Psi(g^{-1}[x]) \quad \forall\, g \in G \tag{34}$$

It is convenient to realize $\mathcal{H}$ as

$$\begin{aligned} &\mathcal{H} = \mathcal{F} \otimes \mathcal{G} \quad \text{with} \\ &\mathcal{F} \equiv \mathcal{L}^2(\mathbb{R}^+, r^2 dr) \text{ and } \mathcal{G} \equiv \mathcal{L}^2(S^2, d\mu) \end{aligned} \tag{35}$$

where $d\mu$ is the Haar measure on S^2 w.r.t. G. In this realization

$$V(g) = I_\mathcal{F} \otimes U(g) \quad \forall\, g \in G \tag{36}$$

where $U(G)$ decomposes $\mathcal{G}$ into a direct sum of mutually orthogonal subspaces $\{\mathcal{G}^l | l \in \mathbb{Z}^+\}$ such that: (i) dim $\mathcal{G}^l = (2l+1)$; (ii) each $\mathcal{G}^l$ is stable under the action of $U(G)$; (iii) the restriction $U^l(G)$ of $U(G)$ to $\mathcal{G}^l$ is irreducible and every irreducible unitary representation of G appears exactly once in this decomposition. One obtains in this manner

$$\mathcal{H} = \bigoplus_{l=0}^{\infty} \mathcal{H}^l \quad \text{with } \mathcal{H}^l \equiv \mathcal{F} \otimes \mathcal{G}^l \tag{37a}$$

and, for every $g \in G$

$$V(g) = \bigoplus_{l=0}^{\infty} V^l(g) \quad \text{with } V^l(g) \equiv I_\mathcal{F} \otimes U^l(g) \tag{37b}$$

The spherical symmetry of the potential V then implies that S, like $U_0(\mathbb{R})$, commutes with $V(G)$. From (36), (37) and the properties of $U(G)$ just mentioned, one concludes that, for every $t \in \mathbb{R}$

$$U_0(t) = \bigoplus_{l=0}^{\infty} U_0^l(t) \quad \text{with } U_0^l(t) \equiv U_{\mathcal{F}}^l(t) \otimes P^l \tag{38a}$$

$$S = \bigoplus_{l=0}^{\infty} S^l \quad \text{with } S^l(t) \equiv S_{\mathcal{F}}^l \otimes P^l \tag{38b}$$

where P^l is the projector from $\mathcal{G}$ onto $\mathcal{G}^l$. Moreover, since S commutes with $U_0(\mathbb{R})$ [see (33)], one has

$$S_{\mathcal{F}}^l\, U_{\mathcal{F}}^l(t) = U_{\mathcal{F}}^l(t)\, S_{\mathcal{F}}^l \quad \forall\, t \in \mathbb{R} \tag{39}$$

The principal feature of this decomposition is that, for each l, the von Neumann algebra $\mathcal{U}_{\mathcal{F}}^l$, defined on $\mathcal{F}$ by

$$\mathcal{U}_{\mathcal{F}}^l \equiv \{U_{\mathcal{F}}^l(t) | t \in \mathbb{R}\}'' \tag{40a}$$

satisfies

$$(\mathcal{U}_{\mathcal{F}}^l)' = \mathcal{U}_{\mathcal{F}}^l \tag{40b}$$

i.e. it is maximal abelian. Consequently, the spectrum of the generator $H_{\mathcal{F}}^l$ of $U_{\mathcal{F}}^l(\mathbb{R})$ is simple; moreover it is absolutely continuous and covers $\mathbb{R}^+$. This is the starting point of what physicists refer to as "partial wave analysis".

Let now $\{\mathcal{F}_\lambda^l | \lambda \in \mathbb{R}^+\}$ be the spectral family canonically associated to $H_{\mathcal{F}}^l$, and $\{P_\lambda^l \equiv \mathcal{F}_\lambda^l \otimes P^l\}$. As a consequence of (38)–(40) one has immediately

$$U_0^l(t) = \int \exp\{-i\lambda t\}\, dP_\lambda^l \quad \forall\, t \in \mathbb{R} \tag{41a}$$

$$S^l = \int S^l(\lambda)\, d\,P_\lambda^l \tag{41b}$$

Since, moreover, S^l is unitary, there exists a real-valued measurable function $\delta^l(\cdot)$ on $\mathbb{R}^+$ such that

$$S^l(\lambda) = \exp\{2\,i\,\delta^l(\lambda)\} \tag{41c}$$

$\delta^l(\lambda)$ is called the partial wave *phase shift* for angular momentum l and energy λ. These objects are used to compute a quantity of direct experimental significance: the quantum analog of the classical scattering *cross-section*, namely the energy dependent quantity

$$\sigma_{tot}(\lambda) \equiv \frac{4\pi}{\lambda} \sum_{l=0}^{\infty} (2l+1) \sin^2[\delta^l(\lambda)] \tag{42}$$

Amongst the applications of this type of analysis, we should at least mention here that resonance-energies have been identified with the values of λ where $\delta^l(\lambda) = \pi/2$ [and $d\delta^l(\lambda)/d\lambda$ positive and large]; also the behaviour of $\delta^l(\lambda)$ as $\lambda \to 0$ gives precise information on the number of bound states of H_1; see the texts by Newton (1966) and Amrein, Jauch & Sinha (1977).

Another quantity that can be linked to the scattering operator is the *time delay*. This is to be defined as the limit, when $r \to \infty$, of the difference between the actual and the free sojourn times—see (23)—in the ball S_r of radius r centered at the origin; specifically, let ψ_- be the scattering state given by some $\Psi_- \in E_\infty(H_0)$, and ψ be the corresponding interpolating state, given by $\Psi = \Omega_-\Psi_-$; one wants to define

$$\Delta T(\psi_-) = \lim_{r\to\infty} \Delta T_r(\psi_-) \tag{43a}$$

where

$$\begin{aligned} &\Delta T_r(\psi_-) = \tau_\psi(S_r) - \tau_{\psi_-}(S_r) = \\ &\int_{-\infty}^{+\infty} dt\{\| P_r U_1(t)\Psi \|^2 - \| P_r U_0(t)\Psi_- \|^2\} = \\ &\int_{-\infty}^{+\infty} dt(U_0(t)^*[\Omega_-^* P_r \Omega_- - P_r]U_0(t)\Psi_-, \Psi_-) \end{aligned} \tag{43b}$$

Conditions are known—in particular on the decrease at infinity of a spherically symmetric potential V; see Amrein, Jauch & Sinha (1977), and references quoted therein—which imply that there exists a dense family of $\Psi_- \in E_\infty(H_0) = \mathcal{H}$ with the following properties: (i) the limits involved in (43) do exist; (ii) one can rewrite (43) in the form

$$\Delta T(\psi_-) = \lim_{r\to\infty} \int_0^t dt(U_0(t)^*[S^* P_r S - P_r]U_0(t)\Psi_-, \Psi_-) \tag{44a}$$

and (iii) with the notation issued from (38)–(41)

$$\begin{aligned} &\Delta T(\psi_-) = (\Delta T\ \Psi_-, \Psi_-) \quad \text{where} \\ &\Delta T = \int \Delta T(\lambda)\, dP_\lambda \quad \text{and} \quad \Delta T(\lambda) = -i\, S(\lambda)^* \frac{dS}{d\lambda}(\lambda) \end{aligned} \tag{44b}$$

which one can rewrite

$$\Delta T(\lambda) = 2\frac{d\delta}{d\lambda}(\lambda) \quad \text{where } S(\lambda) = \exp\{2i\,\delta(\lambda)\} \tag{44c}$$

This formula thus expresses the time delay at energy λ as the derivative, with respect to λ, of the phase $2\delta(\lambda)$ of the unitary operator $S(\lambda)$, i.e. of the "on shell scattering operator"; (44c) is known as the *Eisenbud-Wigner* relation, and it can be given the following heuristic interpretation. Suppose an incoming spherical wave is described (for $t \ll 0$) by a radial wave packet of the form

$$\int_0^\infty dk\, \tilde{\Psi}(k) \exp\{-i(kr+\omega t)\} \quad \text{with } \omega \sim k^2/2m \tag{45a}$$

with $\tilde{\Psi}(k)$ strongly peaked around a given value of k, say k_0. The integral will thus differ appreciably from zero only when the phase $(kr+\omega t)$ is stationary at $k = k_0$, i.e. when

$$\frac{d}{dk}(kr+\omega t)|_{k=k_0} \simeq 0 \tag{45b}$$

a condition that can be rewritten in the suggestive form

$$r + v(k_0)t \simeq 0 \quad \text{with} \quad v(k_0) \equiv \frac{d\omega}{dk}(k_0) \tag{45c}$$

The corresponding outgoing spherical wave is described (for $t \gg 0$) by a radial wave packet of the form

$$\int_0^\infty dk\, \tilde{\Psi}(k) \exp\{i(kr+2\delta-\omega t)\} \quad \text{with } \omega \simeq k^2/2m \tag{46a}$$

where δ is the phase-shift for $\tilde{\Psi}(k)$—recall that the latter is strongly peaked at $k = k_0$. By the same argument as before one obtains

$$\frac{d}{dk}(kr+2\delta-\omega t)|_{k=k_0} \simeq 0 \quad \text{i.e.} \tag{46b}$$

$$r - v(k_0)[t - \Delta T(k_0)] \simeq 0 \quad \text{with } \Delta T(k_0) \equiv \frac{d\delta}{d\omega}(k_0) \tag{46c}$$

Upon comparing (45) and (46) one sees that $\Delta T(k_0)$ can indeed be interpreted as a time delay, providing thus the heuristic interpretation of (44c) that was looked for. Still another confirmation that the time delay ΔT does capture an essential aspect of the effect of the potential V on the motion of a quantum particle was provided by Lavine (1974) who proved, again under smoothness assumptions on V, that if $V + \frac{1}{2}x \cdot \nabla V$ is everywhere positive [resp. everywhere negative], then the time delay ΔT is positive [resp. negative].

We finally want to come back to a fact we did mention earlier, namely that the asymptotic condition (24) is too strong to be satisfied by some potentials of physical interest such as the Coulomb potential $V(r) = r^{-1}$. The question thus arises of whether it is possible to weaken this condition without losing its mathematical potency and its physical relevance. Several proposals have been made in answer to this question; see e.g. Dollard (1964); Jauch, Misra & Gibson (1968); Lavine (1970); Amrein, Martin & Misra (1970); Amrein, Georgescu & Martin (1974) and references quoted therein.

Physically, one should only require that a triple $\{S_\infty(\alpha_0), S_\infty(\alpha_1), \mathcal{A}\}$ be identified, where: $S_\infty(\alpha_0)$ [resp. $S_\infty(\alpha_1)$] is a distinguished class of states, called the scattering states with respect to the free evolution $\alpha_0(\mathbb{R})$ [resp. the evolution $\alpha_1(\mathbb{R})$ with interaction]; and $\mathcal{A}$ is a distinguished class of observables, with $S_\infty(\alpha_0)$, $S_\infty(\alpha_1)$ and $\mathcal{A}$ large enough to provide a useful description of the scattering process, while still allowing for the following three conditions to be satisfied. Firstly, one would require that

$$\begin{aligned} &\lim_{t\to\pm\infty} < \psi_\pm \circ \alpha_0(t)\, ;\, P_r >= 0 \\ &\lim_{t\to\pm\infty} < \psi \circ \alpha_1(t)\, ;\, P_r >= 0 \end{aligned} \tag{47}$$

for all $\psi_\pm \in S_\infty(\alpha_0)$, $\psi \in S_\infty(\alpha_1)$ and P_r defined by (17). Secondly, one would require that for every $\psi \in S_\infty(\alpha_1)$ there exist $\nu_\pm[\psi] \equiv \psi_\pm \in S_\infty(\alpha_0)$ such that $\forall\, A \in \mathcal{A}$

$$\lim_{t\to\pm\infty} \{< \psi \circ \alpha_1(t)\, ;\, A > - < \psi_\pm \circ \alpha_0(t)\, ;\, A >\} = 0 \tag{48}$$

Thirdly, in view of the way many scattering experiments are conducted, one may want to require that the observbles A in $\mathcal{A}$ satisfy

$$s - \lim_{t\to\pm\infty} \alpha_0(t)[A] \equiv A_\pm \quad \text{exist} \tag{49}$$

(here, as everywhere else in this section, $s-\lim$ denotes a limit that is taken in the strong-operator topology); one may even want to require that $A \in \mathcal{A}$ are constants of the free motion, i.e.

$$\alpha_0(t)[A] = A \quad \forall\, t \in \mathbb{R}\, , \forall\, A \in \mathcal{A}\, . \tag{50}$$

Mathematically, a condition that implements the above physical requirements on pure scattering states, and is still significantly weaker than (24), can be proven to be satisfied in situations of physical interest. Specifically, consider a potential scattering situation where $\alpha_0(\mathbb{R})$ is implemented by a unitary group $U_0(\mathbb{R})$, continuous for the strong operator topology, with generator $-\Delta/2m$; and where $\alpha_1(\mathbb{R})$ is implemented by a unitary group $U_1(\mathbb{R})$, continuous for the strong operator topology, with generator $-\Delta/2m + V$. Suppose further that V is spherically symmetric and can be written in the form $V = V_a + V_b$, where V_a is spherically symmetric, and is differentiable with respect to $r \equiv \| x \|$, with derivative dV_a/dr in $\mathcal{L}^1(\mathbb{R}^+, dr)$; and where V_b is such that—with H_a denoting $(-\Delta/2m + V_a)$—one has $\Omega_\pm(H_1, H_a)$ exist, with Ran $\Omega_\pm(H_1, H_a) \equiv \mathcal{H}_{ac}(H_1)$. In applications, the condition on V_a allows the potential to decrease to zero at infinity as slowly as $r^{-\epsilon}$ with $\epsilon > 0$, whereas the introduction of the term V_b allows to control some local singularities; in particular, the Coulomb potential $V(r) = Cr^{-1}$ is allowed by these assumptions. Under conditions that are satisfied for the situations just described, Lavine (1970) proved that the following limits exist

(51) $$s - \lim_{t \to \pm\infty} \alpha(t)[A] E_\infty(H_1) = \omega_\pm[A] \quad \forall\, A \in \mathcal{A}$$

when $\mathcal{A}$ is the C^*-algebra

(52a) $$\mathcal{A} \equiv \{ f(\mathbf{P}) | f : \mathbb{R}^3 \to \mathbb{C} \text{ bounded, continuous } \}$$

where the components P_1, P_2 and P_3 of $\mathbf{P}$ are the momentum operators defined by

(52b) $$U(\mathbf{a}) = \exp\{-i\mathbf{P} \cdot \mathbf{a}\} \quad \text{with } [U(\mathbf{a})\Psi](x) = \Psi(\mathbf{x} - \mathbf{a})$$

As a direct consequence of (51) and (52) one has $\forall\, A, B \in \mathcal{A}$, $\forall\, t \in \mathbb{R}$ and $\forall\, \lambda, \mu \in \mathbb{C}$:

(53a) $$\alpha_0(t)[A] = A \quad \text{and } \alpha(t) \circ \omega_\pm[A] = \omega_\pm[A]$$
(53b) $$E_\infty(H_1)\,\omega_\pm[A] = \omega_\pm[A] = \omega_\pm[A]\, E_\infty(H_1)$$
(53c) $$\omega_\pm[\lambda A + \mu B] = \lambda\omega_\pm[A] + \mu\omega_\pm[B]$$
(53d) $$\omega_\pm[AB] = \omega_\pm[A]\,\omega_\pm[B]$$
(53e) $$\omega_\pm[A^*] = \omega_\pm[A]^*$$
(53f) $$\| \omega_\pm[A] \| \leq \| A \|$$

This leaves open two questions. The first is whether $\omega_\pm$ are isometries, or equivalently, whether they are injective. For spherically symmetric potentials V, this question has been answered positively by Lavine (1970), provided one has, in addition to (51) on (52), that $V(H_0 + i)^{-1}$ is compact, a condition which is also satisfied for the Coulomb potential. The second question is: (a) to determine whether $\omega_\pm$ can be spatially implemented, i.e. whether there exist partial isometries $\Omega_\pm$ such that

(54) $$\omega_\pm[A] = \Omega_\pm A \Omega_\pm{}^* \,;$$

and (b), when (54) is satisfied, to interpret the meaning of the Hilbert space operators $\Omega_\pm$. This question is considered in Amrein, Jauch & Sinha (1977) who present there a synthesis of work done with several of their coworkers. The answer hinges on three technical questions: (i) whether $\omega_\pm$ are ultraweakly continuous, and thus naturally extend to the maximal abelian von Neumann algebra $\mathcal{A}''$; (ii) whether $\omega_\pm[\mathcal{A}'']$ admit vectors that are cyclic in $E_\infty(H_1)$, a condition that is satisfied if the algebras $\omega_\pm[\mathcal{A}'']$, when restricted to the stable subspace $E_\infty(H_1)$ [see (53b)], have abelian commutant in $E_\infty(H_1)$; (iii) whether there exist cyclic vectors Ψ° for $\mathcal{A}''$ and $\Psi^\circ_\pm$ for $\omega_\pm[\mathcal{A}'']$ such that

(55) $$(A\Psi^\circ, \Psi^\circ) = (\omega_\pm[A]\Psi^\circ_\pm, \Psi^\circ_\pm) \quad \forall\, A \in \mathcal{A}''.$$

Note that even if $\Omega_\pm$ exist in the sense of (54), and satisfy all the above conditions, it is still possible that they do *not* satisfy the original conditions (24). Nevertheless, conditions are known under which there exist two strongly continuous families $\{U_\pm(t)|t \in \mathbb{R}\}$ of unitary operators such that

$$\Omega_\pm = s - \lim_{t \to \pm\infty} U_1(t)^* U_\pm(t) \tag{56}$$

are satisfied, but not

$$U_\pm(t)U_\pm(s) = U_\pm(t+s) \tag{57}$$

The most interesting case for which: (i) $\Omega_\pm$ exist in the sense of (54), (ii) (56) is satisfied—with $U_\pm(t)$ functions of H_0—and (iii) (57) does *not* hold, is the scattering theory corresponding to the Coulomb potential $V(r) = Cr^{-1}$; this pathology was discovered by Dollard (1964). Notice further that even when $\Omega_\pm$ exist in the sense of (54), with both (56) and (57) satisfied, it is still possible that the generators $K_\pm$ of the groups $U_\pm(\mathbb{R})$ do not coincide with H_0; physicists refer to this situation as "energy renormalization"; the most trivial example for which this situation prevails is obtained when $H_1 - H_0 \equiv V = V_0 + cI$ where $c \in \mathbb{R} \setminus \{0\}$ and V_0 satisfies conditions ensuring that (54), or even (24), hold: in this case $K_\pm$ are evidently given by $H_0 + cI$.

Finally, we should mention again that we limited our scope in this section to 2-body scattering problems. Progress has recently been made on the theory of multi-channel scattering that is required to understand the quantum N-body problem; this work is still in progress, and the reader who desires to familiarize himself with the sophistication involved in treating this type of problems should consult Faddeev (1965); Part IV of Amrein, Jauch & Sinha (1977); Perry, Sigal & Simon (1981); Hagedorn (1980–83); Enss (1983); and references quoted therein.

CHAPTER 9.
THE ALGEBRAIC FORMULATION OF QUANTUM MECHANICS

SYNOPSIS

As seen in the preceding chapter, the Hilbert space formulation of quantum mechanics scored impressive successes when dealing with non-relativistic systems involving only a finite number of degrees of freedom. As the years passed, however, quantum statistical mechanics and relativistic quantum field theory were grudgingly recognized to lie somewhere beyond the reach of this formalism. Moreover, the somewhat *ad hoc*, or *a priori*, introduction of a Hilbert space on which to build the theory was leaving room for a conceptually tighter approach.

Following Jordan, von Neumann & Wigner (1934), Segal (1947) and Haag & Kastler (1964), it is argued in this chapter that a closer adherence to empirically verifiable structural relations between the fundamental objects of the theory—the observables and the states—suggests that the observables should be constructed from the self-adjoint elements of a C^*-algebra, characteristic of the system considered, and that the states should be identified as the elements of the convex set

$$S \equiv \{\varphi \in \mathcal{A}^* | < \varphi; A^*A > \geq 0 \ \forall \ A \in \mathcal{A}; \parallel \varphi \parallel = 1\}.$$

The Gelfand-Naimark-Segal theorem allows then to reconstruct a Hilbert space representation appropriate to each given physical situation, i.e. to a class of compatible preparations of the system. In statistical mechanics, for instance, such a class would include those preparations which differ from one another only by quasi-local perturbations of an equilibrium state at finite temperature; in quantum field theory such a class would include states which differ from one another only in finitely many particles.

To verify that "the baby has not been thrown out with the water of the bath", one checks that in those cases where the system considered has only finitely many degrees of freedom one recovers in this manner the formalism of Chapter Eight. Aside from the conceptual satisfaction one may derive from having thus established known things on firmer foundations, the main point is that the algebraic formulation is genuinely more general, and precisely so where more generality is needed. Drastically different (in technical terms "disjoint") representations do occur in cases of interest both to microphysics (leading to renewed hopes that one might handle elementary particle interactions in a mathematically consistent manner) and to macrophysics.

Some aspects of non-commutatiave ergodic theory are presented in Section Two, for use in Chapter Ten.

Section 1. THE FUNDAMENTAL POSTULATE

We described, in Section 8.3, von Neumann's mathematical synthesis of the ideas of Heisenberg, Schroedinger and Born on quantum mechanics. The main success of von Neumann's formulation is that it captures the essential aspects of these earlier theories, while providing quantum mechanics with a reliable and powerful tool to deal with situations, in microphysics, involving finitely many degrees of freedom. We now pursue this sanitation process a step further, with two aims in mind: (i) to increase the realm of the physical situations which can be mathematically handled; and (ii) to get closer to an epistemological critique of the physical premises of the theory.

THE C^*-ALGEBRAIC POSTULATE. *A physical system is characterized by a triple* $\{S^*, \mathcal{A}, < . \; ; \; . >\}$ *where:* $\mathcal{A}$*, the set of its observables (or measurable attributes), is the collection of all the self-adjoint elements* A *of a* C^**-algebra* $\mathcal{B}$ *with identity* I *satisfying* $\| I \| = 1$*;* S^**, the set of its states (or modes of preparation), is the collection of all real-valued, positive linear functionals* ϕ *on* $\mathcal{A}$*, normalized by the condition* $< \phi; I > = 1$*; and the prediction rule* $< . \; , \; . >$ *which attributes, to every pair* $\{\phi, A\} \in S^* \times \mathcal{A}$*, the value* $< \phi; A >$ *of* ϕ *at* A*, interpreted as the expectation of the observable* A *when the system is in the state* ϕ*.*

a. Basic mathematical structures

Recall from Subsection 8.3.b that a C^*-algebra $\mathcal{B}$ is an involutive Banach algebra over $\mathbb{C}$, satisfying the condition $\| B^*B \| = \| B \|^2$ for all $B \in \mathcal{B}$. An element $A \in \mathcal{B}$ is said to be self-adjoint whenever $A = A^*$. The set $\mathcal{A}$ appearing in the postulate is thus unambiguously defined as a mathematical object.

Clearly $\mathcal{A}$ inherits from $\mathcal{B}$ the structure of a linear space over $\mathbb{R}$. We, therefore, know what it means for a functional $\phi : A \in \mathcal{A} \mapsto < \phi; A > \in \mathbb{R}$ to be linear. We next have to say what we mean by the condition that ϕ be positive: we want to say that this condition is satisfied exactly when ϕ takes positive values on the positive elements of $\mathcal{A}$; for this condition to be more than a mere tautology, we need to know what we mean by a positive element of $\mathcal{A}$. There are several equivalent ways to define the latter concept; the way which is closer to its physical interpretation is to say that $A \in \mathcal{A}$ is *positive* exactly when its spectrum $\sigma(A)$ is a subset of $\mathbb{R}^+$. Recall that an element B

of an involutive Banach algebra $\mathcal{B}$ with unit I (satisfying $\| I \| = 1$) is said to be *regular* whenever there exists an element $X \in \mathcal{B}$, called the inverse of B and denoted by B^{-1}, such that $BX = I = XB$. An element $B \in \mathcal{B}$ which is not regular is said to be *singular*. The *spectrum* $\sigma(B)$ of an arbitrary element $B \in \mathcal{B}$ is the set

$$\sigma(B) \equiv \{z \in \mathbb{C} | (zI - B) \text{ is singular } \} \tag{1}$$

Recall further that $\sigma(B)$ is a non-void, closed subset of $\mathbb{C}$, satisfying

$$\sigma(B^*) = \sigma(B)^* \equiv \{z \in \mathbb{C} | z^* \in \sigma(B)\} \tag{2}$$

$$|\sigma(B)| \equiv \sup_{z\in\sigma(B)} |z| = \lim_{n\to\infty} \| B^n \|^{1/n} \tag{3}$$

$|\sigma(B)|$ is called the *spectral radius* of B. In particular, if $A \in \mathcal{A} \equiv \{A \in \mathcal{B} | A^* = A\}$, $\sigma(A)$ is a compact subset of $\mathbb{R}$; we could in fact have defined, for these self-adjoint elements A, $\sigma(A)$ as the subset of $\mathbb{R}$ consisting of the real numbers a such that $(aI - A)$ is singular in $\mathcal{A}$. In the particular case of interest to us here, namely when $\mathcal{B}$ is, in addition, a C^*-algebra, the defining relation $\| B^*B \| = \| B \|^2$ implies for every $A \in \mathcal{A}$: $\| A^2 \| = \| A \|^2$ and thus by induction $\| A^m \| = \| A \|^m$ for every m of the form $m = 2^n$ with $n \in \mathbb{Z}^+$. This implies (see 3):

$$|\sigma(A)| = \| A \| \quad \forall\ A \in \mathcal{A} \tag{4}$$

Note that while this relation still holds when A self-adjoint is replaced by N normal (i.e. $N^*N = NN^*$), it is not satisfied for an arbitrary element $B \in \mathcal{B}$; a counterexample can already be found in the C^*-algebra $\mathcal{M}(2, \mathbb{C})$. For $A \in \mathcal{A}$, the above discussion can be summarized by saying that the spectrum $\sigma(A)$ of any observable $A \in \mathcal{A}$ is a compact subset of $\mathbb{R}$, containing at least one of the two numbers $\pm \| A \|$.

All the mathematical terms entering the postulate are now defined, and we should comment on the resulting mathematical structures, with an eye to their physical import.

Scholium 1. *Let* $\mathcal{A}^+ \equiv \{A \in \mathcal{A} | \sigma(A) \subset \mathbb{R}^+\}$ *and* $\mathcal{A}^- \equiv \{A \in \mathcal{A} | -A \in \mathcal{A}^+\}$. *Then*

(5) $\lambda A \in \mathcal{A}^+ \quad \forall\, (\lambda, A) \in \mathbb{R}^+ \times \mathcal{A}^+$

(6) $\mathcal{A}^+ \cap \mathcal{A}^- = \{0\}$

(7) $A + C \in \mathcal{A}^+ \quad \forall\ (A, C) \in \mathcal{A}^+ \times \mathcal{A}^+$

(8) $\| A \| \, I \pm A \in \mathcal{A}^+ \quad \forall A \in \mathcal{A}$

For $A \in \mathcal{A}$ *the following two conditions are equivalent*

(9a) $A \in \mathcal{A}^+$

(9b) *for some* $a \in \mathbb{R}$ *with* $a \geq \| A \| : \| A - aI \| \leq a$

Proof: (5) follows immediately from the definition of the spectrum of an observable, and from the definition of $\mathcal{A}^+$. We prove (6) by noticing that $A \in \mathcal{A}^+ \cap \mathcal{A}^-$ implies $\sigma(A) \subset \mathbb{R}^+ \cap \mathbb{R}^- = \{0\}$; thus, by (4), this implies $\| A \| = 0$, i.e. $A = 0$. We keep (7) for the end of the proof, and proceed to prove (8). We have for any $A \in \mathcal{A} : \sigma(A \pm \| A \| I) = \sigma(A) \pm \| A \| = \sigma(A) \pm |\sigma(A)| \subset \mathbb{R}^\pm$, i.e. $\sigma(\| A \| I \pm A) \subset \mathbb{R}^+$ which indeed proves (8). Considering now $A \in \mathcal{A}$, we have $\| A - aI \| = |\sigma(A - aI)| = |\sigma(A) - a|$; with $a \in \mathbb{R}$ such that $a \geq \| A \| = |\sigma(A)|$, we have thus:

$$\| A - aI \| = \sup_{x \in \sigma(A)} (a - x) . \tag{10}$$

The condition (9b) therefore reads:

$$\sup_{x \in \sigma(A)} (a - x) \leq a \tag{11}$$

which is clearly equivalent to $\sigma(A) \subset \mathbb{R}^+$, i.e. to condition (9a). We now prove (7). From (9), we know $A \in \mathcal{A}^+$ and $C \in \mathcal{A}^+$ imply

$$\| (A - \| A \| I) \| \leq \| A \| \quad \text{and} \quad \| (C - \| C \| I) \| \leq \| C \| \tag{12}$$

Let now $a \equiv \| A \| + \| C \|$. Clearly $a \geq \| A + C \|$. With this a, we form

$$\begin{aligned} &\| A + C - aI \| = \| (A - \| A \| I) + (C - \| C \| I) \| \\ &\leq \| (A - \| A \| I) \| + \| (C - \| C \| I) \| \leq \| A \| + \| C \| = a \end{aligned} \tag{13}$$

where we have used (12). Since $a \geq \| A + C \|$, we can use the equivalence (9a–b) to conclude that (13) implies $(A + C) \in \mathcal{A}^+$. □

Corollary 2. *Every state ϕ belongs to the dual $\mathcal{A}^*$ of the real Banach space $\mathcal{A} = \{A \in B : A = A^*\}$; moreover*

$$\| \phi \| \equiv \sup_{\substack{A \in \mathcal{A} \\ \|A\| \leq 1}} | < \phi; A > | = 1 . \tag{14}$$

Proof: The assertion (8) of Scholium 1 implies for every positive linear functional ϕ on $\mathcal{A}$, and every $A \in \mathcal{A}$

$$- \| A \| < \phi; I > \leq < \phi; A > \leq \| A \| < \phi; I > \tag{15}$$

Consequently $| < \phi; A > |$ with $\| A \| \leq 1$ is bounded by $< \phi; I >$; this bound is reached since $\| I \| = 1$; hence $\| \phi \| = < \phi; I >$. The corollary then follows from the defining normalization of $\phi \in S^*$. □

The first three statements of Scholium 1 amount exactly to say that $\mathcal{A}^+$ is a *positive cone* in the space $\mathcal{B}$ (resp. $\mathcal{A}$) equipped with the partial ordering relation $B \leq C$ defined in $\mathcal{B}$ (resp. $\mathcal{A}$) by $(C - B) \in \mathcal{A}^+$.

The proof of (7) through the equivalence (9 a,b) as we gave it here, can be found in Kadison & Ringrose (1983) who also prove that our definition of the positivity of an observable (through the positivity of its spectrum) is equivalent to the two other definitions which are suggested by the following example.

Let K be a compact topological space, and $\mathcal{C}(K)$ be the C^*-algebra of all continuous, complex-valued functions on K; recall that the fundamental composition laws in $\mathcal{C}(K)$ are, with $f, g \in \mathcal{C}(K)$ and $\lambda \in \mathbb{C}$:

$$(\lambda f) : x \in K \mapsto \lambda f(x) \in \mathbb{C} \tag{16a}$$
$$(f + g) : x \in K \mapsto f(x) + g(x) \in \mathbb{C} \tag{16b}$$
$$(f \cdot g) : x \in K \mapsto f(x) \cdot g(x) \in \mathbb{C} \tag{16c}$$
$$f^* : x \in K \mapsto f(x)^* \in \mathbb{C} \tag{16d}$$
$$\| f \| \equiv \sup_{x \in K} |f(x)| \tag{16e}$$

To say that $f \in \mathcal{C}(K)$ is positive, in the sense defined above, clearly amounts to say that $f(K) \subset \mathbb{R}^+$; we shall see the abstract version of the latter condition later on; we however want now to point out that this condition is also equivalent to each of the following two conditions: (i) there exists $g \in \mathcal{C}(K)$ such that $f = g^*g$; and (ii) there exists $h \in \mathcal{C}(K)$ positive such that $f = h^2$; note that this h is unique: it is called the positive square root of f. These two equivalences generalize to the context we are concerned with in this section.

Let indeed $A = A^* \in \mathcal{A}$ and recall that $\sigma(A)$ is a compact subset of $\mathbb{R}$; write $K = \sigma(A)$ and define $\mathcal{C}(K)$ as in (16) above. Let $\mathcal{P}(K)$ the *-subalgebra of $\mathcal{C}(K)$, the elements of which are the polynomials from K to $\mathbb{C}$ (the same argument goes through if we replace $\mathbb{C}$ by $\mathbb{R}$). To every $p \in \mathcal{P}$ of the form

$$p : x \in K \mapsto \sum_{n=0}^{N} a_n x^n \tag{17}$$

(with $N \in \mathbb{Z}^+$, $a_n \in \mathbb{C}$, and $x^0 \equiv 1$), we associate

$$p(A) \equiv \sum_{n=0}^{N} a_n \ A^n \tag{18}$$

(with N and a_n as in (17), and $A^0 \equiv I$). Note now that the map $p \in \mathcal{P}(K) \mapsto p(A) \in \mathcal{B}$ is linear and satisfies $p^*(A) = p(A)^*$; furthermore, $\{p(A) | p \in \mathcal{P}(K)\}$ is an abelian subalgebra of $\mathcal{B}$. Moreover,

$$\begin{aligned} \| p(A) \| &= |\sigma(p(A))| \equiv \sup_{z \in \sigma(p(A))} |z| \\ &= \sup_{x \in \sigma(A)} |p(x)| \equiv \| p \| \end{aligned} \tag{19}$$

Proof: The idea for the construction of the space $\mathcal{H}_\phi$ lies in Proposition 4; see in particular (29)–(31). Since the map $< . \, , \, . >_\phi$ defined in (29) does satisfy all the properties of an inner product, *except* for the fact that $< B, B >_\phi = 0$ does not necessarily imply $B = 0$, we have to "quotient" this out. Specifically, let

$$K_\phi \equiv \{K \in \mathcal{B} | < \phi; K^*K >= 0\} \tag{45a}$$

From the Cauchy inequality (31), we see that

$$K_\phi = \{K \in \mathcal{B} | < \phi; B^*K >= 0 \quad \forall \; B \in \mathcal{B}\} \tag{45b}$$

Together with the continuity and linearity of ϕ, this implies that K_ϕ is a closed left-ideal of $\mathcal{B}$. We define then the equivalence relation $B' \sim B$ in $\mathcal{B}$ by the requirement $B' - B = K \in K_\phi$. For every $B \in \mathcal{B}$, we denote by $\Phi(B)$ the equivalence class of B, i.e. $\Phi(B) \in \mathcal{B}/K_\phi$. Notice now that, with $B' - B = K \in K_\phi$ and $C' - C = L \in K_\phi$, we have:

$$\begin{aligned} &< B', C' >_\phi - < B, C >_\phi = \\ &< \phi; C^*K > + < \phi; B^*L >^* + < \phi; L^*K >= 0 \end{aligned} \tag{46}$$

Consequently

$$(\Phi(B), \Phi(C)) \equiv < B, C >_\phi \tag{47}$$

defines a prehilbert space structure on $\mathcal{B}/K_\phi$. $\mathcal{H}_\phi$ is then defined as the Hilbert space completion of $\mathcal{B}/K_\phi$. Since K_ϕ is a left ideal of $\mathcal{B}$

$$\pi_\phi(A) : \Phi(B) \in \mathcal{B}/K_\phi \mapsto \Phi(AB) \in \mathcal{B}/K_\phi \tag{48}$$

defines, for every $A \in \mathcal{B}$, a linear map $\pi_\phi(A)$. Note that this map satisfies (see 32):

$$\| \pi_\phi(A)\Phi(B) \|^2 = \| \Phi(AB) \|^2 = < \phi; B^*A^*AB > \leq \| A^*A \| < \phi; B^*B > \tag{49}$$

which we can rewrite, since $\mathcal{B}$ is a C^*-algebra

$$\| \pi_\phi(A)\Phi(B) \| \leq \| A \| \cdot \| \Phi(B) \| \tag{50}$$

Hence $\pi_\phi(A)$ is bounded and can thus be extended uniquely, by continuity, to a bounded linear operator acting in $\mathcal{H}_\phi$; we denote again by $\pi_\phi(A)$ the operator so obtained. The resulting map

$$\pi_\phi : A \in \mathcal{B} \mapsto \pi_\phi(A) \in \mathcal{B}(\mathcal{H}_\phi) \tag{51}$$

is clearly linear (see in particular 48). Moreover, note that for all $A, B, C \in \mathcal{B}$

$$\pi_\phi(AB)\Phi(C) = \Phi((AB)C) = \Phi(A(BC)) = \pi_\phi(A)\Phi(BC) = \pi_\phi(A)\pi_\phi(B)\Phi(C) \tag{52}$$

$$(\pi_\phi(A)\Phi(B), \Phi(C)) = (\Phi(AB), \Phi(C)) = \; < \phi; C^*(AB) > = \; < \phi; (A^*C)^*B > = (\Phi(B), \Phi(A^*C)) = (\Phi(B), \pi_\phi(A^*)\Phi(C)) \tag{53}$$

Since $\{\Phi(B) | B \in \mathcal{B}\}$ is, by definition, dense in $\mathcal{H}_\phi$, and since $\{\pi_\phi(A) | A \in \mathcal{B}\} \subseteq \mathcal{B}(\mathcal{H}_\phi)$, (52) (resp. 53) implies that π_ϕ satisfies (42b) (resp. 42c). This completes the proof that π_ϕ is a representation of $\mathcal{B}$ on $\mathcal{H}_\phi$. Since we can assume, without loss of generality that $\mathcal{B}$ contains a unit I with $\| I \| = 1$, $\Phi \equiv \Phi(I)$ clearly satisfies (43) and (44).

The uniqueness (up to unitary equivalence) of the representation π_ϕ is established by the following construction. Suppose $\pi' : \mathcal{B} \to \mathcal{B}(\mathcal{H}')$ is another representation, satisfying (43) and (44) with respect to some $\Phi' \in \mathcal{H}'$. One then checks easily that

$$V : \pi'(A)\Phi' \mapsto \pi_\phi(A)\Phi \quad \forall \; A \in \mathcal{B} \tag{54}$$

extends to a linear isometry from $\mathcal{H}'$ onto $\mathcal{H}$, satisfying

$$V\pi'(A)V^{-1} = \pi_\phi(A) \quad \forall \; A \in \mathcal{B} \tag{55}$$ □

This construction is named after Gelfand & Naimark (1943) and Segal (1947). A variant of this construction, adapted to the needs of quantum field theory to deal with unbounded operators, is known as the Wightman (1956) reconstruction theorem (see e.g. Streater & Wightman, 1964; Jost, 1965; or Emch (1972); we shall come back on this aspect of the GNS construction in Chapter Ten).

The property (44) is referred to by saying that Φ is a *cyclic vector* for $\pi_\phi(\mathcal{B})$, or simply for the representation π_ϕ. Note that every vector $\Phi \in \mathcal{H}$, with $\Phi \neq 0$, is cyclic for $\mathcal{B}(\mathcal{H})$; hence for any pure *and* normal state ϕ on the von Neumann algebra $\mathcal{B}(\mathcal{H})$, the GNS construction gives us a representation π_ϕ unitarily equivalent to $\mathcal{B}(\mathcal{H})$ itself, i.e. we can identify $\mathcal{H}_\phi$ with $\mathcal{H}$, and $\pi_\phi(\mathcal{B})$ with $\mathcal{B}(\mathcal{H})$. The following result—Corollary 6—places this remark in proper perspective; we need for this two definitions, which we now proceed to give.

We first notice that for a representation $\pi : \mathcal{B} \to (\mathcal{H})$ the following conditions are equivalent: (i) $\{0\}$ and $\mathcal{H}$ are the only closed subspaces of $\mathcal{H}$ that are stable under the action of the collection of operators $\{\pi(B) | B \in \mathcal{B}\}$; (ii) the von Neumann algebra $\pi(\mathcal{B})''$, generated by $\pi(\mathcal{B})$, coincides with $\mathcal{B}(\mathcal{H})$; and (iii) the von Neumann algebra $\pi(\mathcal{B})' \equiv \{X \in \mathcal{B}(\mathcal{H}) | X\pi(B) = \pi(B)X \quad \forall B \in \mathcal{B}\}$ coincides with $\mathbb{C}I \equiv \{zI | z \in \mathbb{C}\}$. When any (and thus all) of the above three conditions is satisfied, we say that the representation π is *irreducible*.

Secondly, we can generalize from $\mathcal{B}(\mathcal{H})$ to an arbitrary C^*-algebra $\mathcal{B}$ (with unit I) the following features of S^* discussed in Subsection 8.3.c: S^* is a convex set, bounded in the norm topology of $\mathcal{B}^*$, and closed in the w^*-topology of $\mathcal{B}^*$. Consequently (again by the Krein-Milman theorem) S^* is the w^*-closed convex hull of the set $\mathcal{E}^*$ of its extreme points. Recall that $\phi \in \mathcal{E}^*$, and ϕ is then called a *pure state* on $\mathcal{B}$, whenever: (i) $\phi \in S^*$ and, in addition, (ii) $\psi, \chi \in S^*$ and $\lambda \in (0,1)$ with $\phi = \lambda\psi + (1-\lambda)\chi$ imply $\psi = \phi = \chi$. Clearly, $\phi \in \mathcal{E}^*$ if and only if its restriction to $\mathcal{A} \equiv \{A \in \mathcal{B} | A = A^*\}$ cannot be written as a non-trivial convex combination of states on the physical system considered.

We are now ready for the promised result on the role of the von Neumann algebra $\mathcal{B}(\mathcal{H})$ in the theory of representations of an arbitrary C^*-algebra $\mathcal{B}$ (with unit).

COROLLARY 6. *With the notations of Theorem 5, π_ϕ is irreducible if and only if ϕ is pure.*

Proof: We first prove that ϕ pure implies π_ϕ irreducible. Since $\pi_\phi(\mathcal{B})'$ is a von Neumann algebra, and hence is a $(C)^*$-algebra, we know that every $X \in \pi_\phi(\mathcal{B})'$ can be written as $X = Y + iZ$ with Y and Z self-adjoint in $\pi_\phi(\mathcal{B})'$. We also know that every self-adjoint element Y in a C^*-algebra, here $\pi_\phi(\mathcal{B})'$, can be written in the form $Y = Y^+ - Y^-$ with $Y^\pm$ positive in this C^*-algebra. It is, therefore, sufficient to prove that every positive operator $X \in \pi_\phi(\mathcal{B})'$ is a scalar multiple of the identity. Let thus $X \in \pi_\phi(\mathcal{B})'$ and X be positive. Let Φ be the cyclic vector corresponding to ϕ in Theorem 5. Note that $X\Phi = 0$, Φ cyclic for π_ϕ and $X \in \pi_\phi(\mathcal{B})'$ imply $X = 0$. We can, therefore, assume, without loss of generality, that $X\Phi \neq 0$; and then $\| X\Phi \| = 1$, and $\| X \|^2 < \lambda^{-1}$ with $\lambda \in (0,1)$. Consider now

(56) $$\psi : B \in \mathcal{B} \mapsto (\pi_\phi(B)X\Phi, X\Phi) \in \mathbb{C}$$

(57) $$\chi \equiv (1-\lambda)^{-1}(\phi - \lambda\psi)$$

Clearly

(58) $$\phi = \lambda\psi + (1-\lambda)\chi$$

Moreover

(59) $$< \psi; B^*B > = \| \pi_\phi(B)X\Phi \|^2 \geq 0 \quad \forall\ B \in \mathcal{B}$$

(60) $$< \psi; I > = \| X\Phi \|^2 = 1$$

Since ψ is obviously linear, (59) and (60) imply

(61) $$\psi \in S^*$$

Furthermore, since $X \in \pi_\phi(\mathcal{B})'$, (59) can be rewritten as

(62) $$\begin{aligned} &< \psi; B^*B > = \| X\pi_\phi(B)\Phi \|^2 \leq \\ &\| X \|^2 \cdot < \phi; B^*B > \ \leq\ \lambda^{-1} < \phi; B^*B > \end{aligned}$$

From (57) and (62), we conclude

$$< \chi; B^*B > \ \geq 0 \quad \forall \ B \in \mathcal{B} \tag{63}$$

while (58) and (60) give

$$< \chi; I >= 1 \tag{64}$$

Since $\chi : \mathcal{B} \mapsto \mathcal{B}(\mathcal{H})$ is obviously linear, (63) and (64) imply

$$\chi \in S^* \tag{65}$$

Since ϕ is pure, (58), (61) and (65) imply $\phi = \psi = \chi$. In particular, we have then for all $B, C \in \mathcal{B}$:

$$< \psi; B^*C >=< \phi; B^*C > \tag{66}$$

which we can rewrite (see 56) in the form

$$((X^2 - I)\pi_\phi(C)\Phi, \pi_\phi(B)\Phi) = 0 \tag{67}$$

Since Φ is cyclic, this implies $X^2 = I$; and since X is positive, that implies $X = I$ and thus $\pi_\phi(\mathcal{B})' = \mathbb{C}I$. We have, therefore, proven that ϕ pure implies π_ϕ irreducible. We now prove the converse, namely that $\pi_\phi(\mathcal{B})' = \mathbb{C}I$ implies ϕ pure. We proceed by showing that if there exist $\psi, \chi \in S^*$ and $\lambda \in (0, 1)$ such that

$$\phi = \lambda\psi + (1 - \lambda)\chi, \tag{68}$$

then $\psi = \phi = \chi$. From the Cauchy inequality (31), we have $\forall \, B, C \in \mathcal{B}$:

$$| < \psi; B^*C > | \leq \ < \psi; B^*B > \ < \psi; C^*C > \tag{69}$$

Since χ is a state, and, in particular, is positive, $(\lambda^{-1}\phi - \psi)$ is positive, and thus (69) implies

$$| < \psi; B^*C > | \leq \lambda^{-2} < \phi; B^*B >< \phi; C^*C > \tag{70}$$

Consequently, for every $B, C \in \mathcal{B}$ and $K, L \in \mathcal{K}_\phi$:

$$< \psi; (B + K)^*(C + L) >=< \psi; B^*C > \tag{71}$$

We can, therefore, define on $(\mathcal{B}/\mathcal{K}_\phi) \times (\mathcal{B}/\mathcal{K}_\phi)$:

$$(\Phi(B), \Phi(C))_\psi \equiv< \psi; C^*B > \tag{72}$$

which is a bounded, positive, sesquilinear form on $\mathcal{B}/K_\phi$; it can, therefore, be extended to a bounded, positive, sesquilinear form $(.\ ,\ .)_\psi$ on $\mathcal{H}_\phi$. By Riesz theorem, there exists a positive operator $X \in \mathcal{B}(\mathcal{H}_\phi)$ such that, $\forall\ \Psi_1, \Psi_2 \in \mathcal{H}_\phi$:

$$(\Psi_1, \Psi_2)_\psi = (X\Psi_1, \Psi_2) \tag{73}$$

In particular, we have $\forall\ A, B, C \in \mathcal{B}$:

$$\begin{aligned} &(X\pi_\phi(A)\pi_\phi(B)\Phi, \pi_\phi(C)\Phi) = (X\Phi(AB), \Phi(C)) = \\ &< \psi; C^*(AB) >=< \psi; (A^*C)^*B >= \\ &(X\Phi(B), \Phi(A^*C)) = (\pi_\phi(A)X\pi_\phi(B)\Phi, \pi_\phi(C)\Phi) \end{aligned} \tag{74}$$

Since Φ is cyclic for π_ϕ, (74) implies

$$X\pi_\phi(A) = \pi_\phi(A)X \quad \forall\ A \in \mathcal{B} \tag{75}$$

i.e. $X \in \pi_\phi(\mathcal{B})'$. Since we assumed $\pi_\phi(\mathcal{B})' = \mathbb{C}I$, this means $X = xI$ for some $x \in \mathbb{C}$. But ψ and ϕ are states, and thus $< \psi; I >= 1 =< \phi; I >$; with $A = B = C = I$ and $X = xI$, (74) implies $x = 1$, i.e. $\psi = \phi$, and thus $\chi = \phi$. Hence ϕ is pure. □

As we have already pointed out, the Krein-Milman theorem implies that $\mathcal{S}^*$ is the w^*-closed convex hull of the set $\mathcal{E}^*$ of its extreme points. In fact, the Krein-Milman theorem allows to refine this statement: the pure states *separate* the observables, i.e. $A, C \in \mathcal{A}$ and $< \phi; A >=< \phi; C > \quad \forall\ \phi \in \mathcal{E}^*$ imply $A = C$. An even stronger result can be proved, as we shall now see.

Scholium 7. *For every observable $A \in \mathcal{A}$ and every $a \in \sigma(A)$, there exists a pure state $\phi \in \mathcal{E}^*$ such that $< \phi; A >= a$ and $< \phi; (A - aI)^2 >= 0$.*

Proof: Recall from the proof of Scholium 3 that the smallest C^*-subalgebra $\mathcal{B}(A)$ of $\mathcal{B}$, containing A and I, is isometrically *-isomorphic to the C^*-algebra $\mathcal{C}(K)$ of continuous functions on the compact subset $K \equiv \sigma(A) \subset \mathbb{R}$. The Dirac measure δ_a, concentrated at $a \in \sigma(A)$, defines on $\mathcal{C}(K)$, and thus on $\mathcal{B}(A)$, a pure state

$$\delta_a : f \in \mathcal{C}(K) \mapsto f(a) \in \mathbb{C} \tag{76}$$

such that

$$< \delta_a; A >= a \ ; \ < \delta_a; (A - AI)^2 >= 0 \tag{77}$$

From the Hahn-Banach theorem, δ_a extends to at least one state on $\mathcal{B}$. Let $\mathcal{S}^*_a = \{\psi \in \mathcal{S}^* | < \psi; B >=< \delta_a; B > \quad \forall\ B \in \mathcal{B}(A)\}$; this again is a convex set, closed in the w^*-topology, and bounded in the metric topology of $\mathcal{B}^*$. Hence, by Krein-Milman theorem, $\mathcal{S}^*_a$ is the w^*-closed convex hull of the set $\mathcal{E}^*_a$ of its extreme points. Finally, we prove that $\mathcal{E}^*_a \subset \mathcal{E}^*$; let indeed $\phi \in \mathcal{E}^*_a$

and suppose that there exist ψ and $\chi \in S^*$ such that $\phi = \lambda\psi + (1-\lambda)\chi$. Upon restricting this equality to $\mathcal{B}(A)$ and using the fact that δ_a is a pure state on $\mathcal{B}(A)$, we see that $\psi, \chi \in S_a^*$. Since ϕ is extremal in S_a^*, this implies $\psi = \phi = \chi$; hence $\phi \in \mathcal{E}^*$. $\square$

Note that if $B \in \mathcal{B}$ and $< \phi; B >= 0$ for all $\phi \in \mathcal{E}^*$, we have $< \phi; A >= 0 =< \phi; C >$ for all $\phi \in \mathcal{E}^*$, where $A \equiv (B + B^*)/2$ and $C \equiv (B - B^*)/2i$. From Scholium 7, we conclude $\sigma(A) = \{0\} = \sigma(C)$, and thus $A = 0 = C$, i.e. $B = 0$. Hence $\mathcal{E}^*$ not only separates $\mathcal{A}$, but it separates $\mathcal{B}$ as well.

A particular consequence of Corollary 6 and Scholium 7 is that given an observable A, there always exists at least one irreducible representation π of $\mathcal{B}$ such that $\| \pi(A) \| = \| A \|$; construct indeed the GNS representation associated to a pure state on $\mathcal{B}$ that satisfies the conclusion of Scholium 7, with $a \in \sigma(A) \cap \{- \| A \|, + \| A \|\}$. Upon making the direct sum (see, e.g. Kadison & Ringrose, 1983) of all GNS representations π_ϕ, with ϕ running over any subset $\mathcal{X}$ satisfying $\mathcal{E}^* \subseteq \mathcal{X} \subseteq S^*$, one obtains a *faithful representation* π of $\mathcal{B}$, i.e. a representation $\pi : B \in \mathcal{B} \mapsto \pi(B) \in \mathcal{B}(\mathcal{H})$ such that $\pi(B) = 0$ exactly when $B = 0$. Consequently, Scholium 7 leads directly to a result of fundamental mathematical importance, known as the *Gelfand-Naimark theorem*, according to which every C^*-algebra is a "concrete" C^*-algebra, i.e. can be realized as a C^*-subalgebra of $\mathcal{B}(\mathcal{H})$ for some $\mathcal{H}$. When $\mathcal{X} = S^*$, π is called the universal representation of $\mathcal{B}$; every state ϕ on $\mathcal{B}$ is realized as a vector state in this representation; in a sense we do not need to make precise here, $\mathcal{B}^{**}$ can be identified with the von Neumann algebra $\pi(\mathcal{B})''$ generated by the universal representation π of $\mathcal{B}$. Fortunately, faithful representations do occur in a simpler context; for instance, π_ϕ is easily seen to be faithful when $\phi \in S^*$ is a *faithful state*, i.e. whenever $A = A^* \in \mathcal{A}$ and $A \neq 0$ imply $< \phi; A^2 > \neq 0$; such states play an important role in Statistical Mechanics.

c. Physical equivalence and quasi-equivalence.

Corollary 6 illustrates the fact that the GNS construction provides an informative way to associate a specific Hilbert space representation $\pi_\phi(\mathcal{A})$ of the observables on a physical system, given the physical situation prescribed by specifying a state ϕ—i.e. given a "mode of preparation" of the system.

We already pointed out, in the discussion following Theorem 5, that different "modes of preparation" may lead to the "same" representation; for instance, all states which are pure *and* normal on $\mathcal{A} \equiv \mathcal{B}_{sa}(\mathcal{H})$ lead to unitarily equivalent representations. We now want to formalize, and extend, this notion of equivalence; we shall see that this can be done in several ways.

Let $\pi : \mathcal{B} \to \mathcal{B}(\mathcal{H}_\pi)$ be an arbitrary representation of our C^*-algebra $\mathcal{B}$, satisfying $\pi(I) = I$. Since $\pi(\mathcal{B})$ is itself a C^*-algebra, we can define the collection S_π^* of states on $\pi(\mathcal{B})$, i.e.

$$S_\pi^* \equiv \{\phi \in \pi(\mathcal{B})^* | \phi \text{ is positive, and } \| \phi \| = 1\} \tag{78}$$

Since $\pi(\mathcal{B})$ is a C^*-subalgebra of $\mathcal{B}(\mathcal{H}_\pi)$, we can moreover consider the collection $\mathcal{S}_\pi$ of all normal states on the von Neumann algebra $\pi(\mathcal{B})''$, i.e.

(79) $$\mathcal{S}_\pi \equiv \{\phi \in [\pi(\mathcal{B})'']_* | \, \phi \text{ is positive, and } \parallel \phi \parallel = 1\}$$

where $[\pi(\mathcal{B})'']_*$ is the predual of $\pi(\mathcal{B})''$ (see Subsection 8.3.d). Finally, we can consider the collection $\mathcal{V}_\pi$ of all vector states on $\pi(\mathcal{B})$, i.e.

(80) $$\mathcal{V}_\pi \equiv \{\pi(B) \mapsto (\pi(B)\Phi, \Phi) \,|\, \Phi \in \mathcal{H}_\pi \text{ and } \parallel \Phi \parallel = 1\}$$

We already know (see expressions 8.3.126 and the discussion following them) that, to every state $\phi \in \mathcal{S}_\pi$ corresponds at least one density matrix ρ, i.e. a positive, trace-class operator, with $tr\ \rho = 1$, acting on $\mathcal{H}_\pi$, such that $< \phi; X >= tr(\rho X) \ \forall\ X \in \pi(\mathcal{B})''$. It is then clear that every state $\phi \in \mathcal{V}_\pi$ can be uniquely extended, by ultraweak continuity, to a state in $\mathcal{S}_\pi$, which we denote again by the same symbol ϕ. Under this identification, we, therefore, have $\mathcal{V}_\pi \subseteq \mathcal{S}_\pi$. Note that, given $\phi \in \mathcal{S}_\pi$, ρ is unique if and only if π is irreducible; when the latter condition is realized (see e.g. the situation covered by von Neumann's postulates, Section 8.3; and in a more general context, Corollary 6 above), one clearly has $\mathcal{V}_\pi \neq \mathcal{S}_\pi$ (unless evidently $\dim \mathcal{H}_\pi = 1$). However, $\mathcal{V}_\pi = \mathcal{S}_\pi$ can happen in the more general context of the present chapter (see e.g. Dixmier, 1957, p.233); for the applications to Quantum Statistical Mechanics (see Section One of Chapter Ten), this is in particular the case when $\pi(\mathcal{B})'$ admits a cyclic vector Ψ, or equivalently when Ψ is separating for $\pi(\mathcal{B})''$ (i.e. when there exists $\Psi \in \mathcal{H}_\pi$ such that $X\Psi = 0$, with $X \in \pi(\mathcal{B})''$, implies $X = 0$).

Moreover, upon restricting any $\phi \in \mathcal{S}_\pi$ to $\pi(\mathcal{B})$, one clearly obtains an element of $\mathcal{S}^*_\pi$, which we denote again by the same symbol ϕ. Under this identification, we, therefore, have $\mathcal{S}_\pi \subseteq \mathcal{S}^*_\pi$, and thus:

(81) $$\mathcal{V}_\pi \subseteq \mathcal{S}_\pi \subseteq \mathcal{S}^*_\pi$$

We can further lift (81) to an assertion on the states of $\mathcal{B}$. Indeed, since π is a representation, we can view every $\phi \in \mathcal{S}^*_\pi$ as a map

(82) $$\phi : B \in \mathcal{B} \mapsto < \phi; \pi(B) > \in \mathbb{C}$$

This map is clearly linear, positive, and takes the value 1 at $B = I$; hence (82) provides an embedding $\mathcal{S}^*_\pi \subseteq \mathcal{S}^*$. In fact:

(83a) $$\mathcal{S}^*_\pi = \mathcal{S}^* \cap (\text{ Ker } \pi)^\perp$$

where we used the notations

(83b) $$\text{Ker } \pi \equiv \{B \in \mathcal{B} | \pi(B) = 0\}$$

(83c) $$(\text{ Ker } \pi)^\perp \equiv \{\phi \in \mathcal{B}^* | < \phi; B >= 0 \ \forall\ B \in \text{ Ker } \pi\}$$

Note that Ker π is a closed, two-sided ideal of $\mathcal{B}$ and that $\pi(\mathcal{B})$ can be identified with $\mathcal{B}/$ Ker π; $\pi(\mathcal{B})$ inherits from $\mathcal{B}(\mathcal{H}_\pi)$ the properties that S_π^* (resp. S_π) is the w^*-closed (resp. norm-closed) convex hull of $\mathcal{V}_\pi$.

We can now address the question of what is meant by saying that two representations $\pi : \mathcal{B} \to \mathcal{B}(\mathcal{H}_\pi)$ and $\nu : \mathcal{B} \to \mathcal{B}(\mathcal{H}_\nu)$ are equivalent. A strict adherence to the spirit of the C^*-algebraic postulate (stated at the beginning of this section) requires us to say that two representations π and ν are *physically equivalent* whenever $S_\pi^* = S_\nu^*$, in which case $\pi(\mathcal{B})$ and $\nu(\mathcal{B})$ can be identified as C^*-algebras. Attention was drawn to the physical content of this concept of equivalence, by Haag & Kastler (1964); the mathematical origin of the concept can be found in Fell (1960).

Recall that if ϕ is a faithful state on $\mathcal{A}$, then π_ϕ is a faithful representation of $\mathcal{B}$, and thus Ker $\pi_\phi = \{0\}$. Hence all the representations of $\mathcal{B}$ that are obtained, by the GNS construction, from faithful states on $\mathcal{A}$ are physically equivalent. The condition that ϕ be faithful, while sufficient, is however not necessary. For instance recall, once more, that every state on $\mathcal{B}(\mathcal{H})$, that is pure and normal, gives, via the GNS construction, a representation which is faithful; still such a state is obviously not faithful (unless we have, trivially, dim $\mathcal{H} = 1$).

Consequently, the concept of physical equivalence needs to be refined. One can for instance require that $S_\pi = S_\nu$ or even that $\mathcal{V}_\pi = \mathcal{V}_\nu$. For a review of these conditions, see e.g. Emch (1972); a few remarks will suffice here.

Notice first that $\mathcal{V}_\pi = \mathcal{V}_\nu$ is a very stringent condition. For instance $\mathcal{V}$-equivalence singles out, among the normal states of $\mathcal{B}(\mathcal{H})$, those which are pure.

The condition that $S_\pi = S_\nu$ is more subtle, and its mathematical significance is discussed in Dixmier (1964) under the heading of "quasi-equivalence", a concept which is familiar in the theory of group representation (see, e.g. Mackey, 1955). What is involved can briefly be described as follows. A *subrepresentation* π_1 of a representation $\pi : \mathcal{B} \to \mathcal{B}(\mathcal{H}_\pi)$ of a C^*-algebra $\mathcal{B}$, is a representation of $\mathcal{B}$, obtained from π by restricting $\pi(\mathcal{B})$ to a non-zero, closed subspace $\mathcal{H}_1 \subseteq \mathcal{H}_\pi$ that is stable under the action of $\pi(B)$ for all $B \in \mathcal{B}$. Let $\mathcal{Q}(\pi)$ be the set of all subrepresentations of π. Two representations π and ν of $\mathcal{B}$ are said to be *disjoint* whenever $\pi_1 \in \mathcal{Q}(\pi)$ and $\nu_1 \in \mathcal{Q}(\nu)$ imply that π_1 and ν_1 are not unitarily equivalent. For any two representations π and ν of $\mathcal{B}$, let us denote by $\mathcal{Q}(\pi, \nu)$ the set of all subrepresentations of π that are disjoint from ν. The representations π and ν are said to be *quasi-equivalent* whenever $\mathcal{Q}(\pi, \nu)$ *and* $\mathcal{Q}(\nu, \pi)$ are empty. One can then prove that π and ν are quasi-equivalent exactly when there exists an isomorphism $\alpha : \pi(\mathcal{B})'' \to \nu(\mathcal{B})''$ such that $\alpha[\pi(B)] = \nu(B)$ for all $B \in \mathcal{B}$. This clearly happens if and only if $S_\pi = S_\nu$. Two representations that are quasi-equivalent are thus physically equivalent. The converse is not true. One even knows situations in Quantum Statistical Mechanics where two faithful states ϕ and ψ on $\mathcal{A}$ can generate, through the GNS construction, two representations $\pi \equiv \pi_\phi$ and $\nu \equiv \pi_\psi$ that are disjoint.

A remarkable example of this has been constructed by Takesaki (1970 b) where ϕ and ψ are equilibrium states (on the algebra of quasi-local observables) corresponding to different temperatures; this mathematical result makes more precise the physical intuition that such situations ought to be different at the level of description provided by the global (or macroscopic) observables; for more details, see Section Two below and Section One of Chapter Ten.

d. Physical meaning of the C^*-algebraic postulate.

We opened this section with a statement of the "C^*-algebraic postulate", and we then proceeded to describe some of the properties of the resulting structures. To close this introductory section, we must now address—at least briefly—the question of the status of this postulate in the realm of physical theories.

Firstly, there should be no argument on the fact that the postulate is both concise and mathematically legible.

Secondly, it is obtained by induction from the von Neumann synthesis of the quantum theories of Heisenberg, Schroedinger and Born (see Chapter Eight) which the C^*-algebraic postulate therefore encompasses.

Thirdly, the universe of discourse of the C^*-algebraic postulate is genuinely more general than allowed by the framework delineated by the von Neumann postulate: the algebras of observables which we can now consider are more general than $\mathcal{B}_{sa}(\mathcal{H})$; and the states which naturally appear in the theory are more general than the density-matrices of ordinary quantum mechanics. Moreover, the objects described by the C^*-algebraic postulate appear in all the physical situations so far encountered not only in ordinary quantum mechanics, but also in the quantum theories with superselection rules alluded to in Section 8.3, and even in classical mechanics. This is a synthesis which, therefore, goes beyond that achieved by the von Neumann postulate.

Fourthly, the GNS construction (Thm.5) allows to bring back into the formalism the technical powers of Hilbert space representations, on which depended many of the successes of the quantum mechanics described by the von Neumann postulate. Hence the generality added to the von Neumann framework of Section 8.3 is not crippling; quite to the contrary, in fact, since the Hilbert space representations, with which we are now prepared to work, correspond precisely to the physical situation in which one takes into account the modes of preparation of the systems considered. We shall see, in the remainder of this chapter and in Chapter Ten, how this flexibility can properly be taken advantage of.

Our *fifth* point has to do with the epistemological question of the empirical foundations of the C^*-algebraic postulate, beyond the inductive and pragmatic arguments advanced in the previous four remarks. To focus the problem we have in mind, the reader will notice that we could have reproached to the von Neumann postulate of Section 8.3 that it had left somewhat in the dark the empirical origin of the Hilbert space $\mathcal{H}$ on which the theory was built. A similarly remote role is played, in the C^*-algebraic postulate, by the C^*-algebra $\mathcal{B}$ of which only the self-adjoint elements are identified as observables. It is, therefore, a legitimate question to ask whether one can decide empirically when the observables of a physical system can be identified with the self-adjoint part of a C^*-algebra. This is a hard question, the roots of which can be traced back to the now 50-year old papers of Jordan (1933) and Jordan, von Neumann & Wigner (1934).

To organize this problem, let us first notice that the argument of Section 8.3 can be extended verbatim to $\mathcal{A} \equiv \mathcal{B}_{sa}$—where now $\mathcal{B}$ is any C^*-algebra with unit, and not just $\mathcal{B} = \mathcal{B}(\mathcal{H})$—thus leading to the conclusion that $\mathcal{A}$ is a Jordan-Lie algebra with unit, complete with respect to a norm for which $\| A^2 \| = \| A \|^2 \;\forall\; A \in \mathcal{A}$. Together with Proposition 4 above (see in particular the comments immediately preceding and following that result), this remark suggests to break our problem into two distinct parts. *Part (a)* will be to analyze the empirical content of the weaker postulate obtained by substituting, in the C^*-algebraic postulate, the questioned statement "$\mathcal{A}$ is the collection of all the self-adjoint elements of a C^*-algebra $\mathcal{B}$", with the weaker statement "$\mathcal{A}$ is a JB-algebra"; we will refer to the result of this substitution as the *JB-algebraic postulate*. Note that this postulate postpones the question of whether it is possible or not to equip the JB-algebra $\mathcal{A}$ with a consistent Lie structure making it a Jordan-Lie algebra. *Part (b)* of the problem will then be to find an *intrinsic* characterization of those JB-algebras which are the self-adjoint part of a C^*-algebra. This part of the problem has recently been solved by Alfsen, Hanche-Olsen & Shultz (1980). They proved that given a JB-algebra $\mathcal{A}$ there exists a C^*-algebra $\mathcal{B}$ such that $\mathcal{A}$ is isomorphic to $\mathcal{B}_{sa} \equiv \{B \in \mathcal{B} | B = B^*\}$ *if and only if* the state space S^* of $\mathcal{A}$ possesses the following two properties (which we shall describe below): (i) S^* has the 3-ball property; and (ii) S^* is orientable.

To understand the content of these two conditions, recall first that S^* is a convex set, compact in its weak*-topology, so that S^* is the w^*-closed convex hull of its extreme points. Recall also that a *face* F of a convex set K is a convex subset $F \subseteq K$ such that $\phi, \psi \in K$, $\lambda \in (0,1)$ and $\lambda\phi + (1-\lambda)\psi \in F$ together imply that ϕ and ψ belong to F. For instance, the faces of the 3-ball $B^3 \equiv \{x \in \mathbb{R}^3 | \; \| x \|^2 \leq 1\}$ are ϕ, B^3 and each $\phi \in \mathcal{E}(B^3) = S^2 \equiv \{x \in \mathbb{R}^3 | \; \| x \|^2 = 1\}$. The relevance of this 3-ball in quantum theory stems from the fact that

$$x \in B^3 \mapsto \rho = \frac{1}{2}\begin{pmatrix} 1 + x^3 & x^1 - ix^2 \\ x^1 + ix^2 & 1 - x^3 \end{pmatrix} \tag{84}$$

is an affine isomorphism from B^3 onto the state space of the JB-algebra $\mathcal{M}(2, \mathbb{C})_{sa}$. Note that, through this isomorphism, the change of orientation $(x^1, x^2, x^3) \in B^3 \mapsto (x^1, -x^2, x^3) \in B^3$ corresponds to the transposition map $t_2 : \rho \to \rho^t$ with $(\rho^t)_{jk} = \rho_{kj} = (\rho_{jk})^*$ (where * denotes the complex conjugation in $\mathbb{C}$).

Recall further, from Subsection 8.3.e, that the symmetries α of $\mathcal{B}(\mathcal{H})$ fall in two classes: the *-automorphism $\alpha : B \in \mathcal{B}(\mathcal{H}) \mapsto UBU^* \in \mathcal{B}(\mathcal{H})$ with U unitary on $\mathcal{H}$, and the *-antiautomorphisms $\alpha : B \in \mathcal{B}(\mathcal{H}) \mapsto UB^*U^* \in \mathcal{B}(\mathcal{H})$ with U antiunitary on $\mathcal{H}$; in the first case $\alpha[AB] = \alpha[A]\alpha[B]$, while in the second case $\alpha[AB] = \alpha[B]\alpha[A]$. Hence α always preserves the Jordan structure of $\mathcal{B}_{sa}(\mathcal{H})$, while $\alpha[\{A, B\}] = \pm\{\alpha[A], \alpha[B]\}$ depending on whether α is a *-automorphism or a *-antiautomorphism. This remark illustrates the fact that even when a Lie structure can be superimposed to a Jordan structure, one still has a sign ambiguity. To see how this ties up with the above question of the orientation of B^3, note that a symmetry α of $\mathcal{M}(2, \mathbb{C})$ is a *-automorphism (resp. a *-antiautomorphism) if and only if $\alpha \otimes id_2$ (resp. $\alpha \otimes t_2$) is positive, and this occurs exactly when α^*—the dual action of α on the state space of $\mathcal{M}(2, \mathbb{C})$—induces, through the affine isomorphism (84), an orientation preserving (resp. an orientation reversing) affine isomorphism of B^3.

As a further preliminary, notice that if ϕ and ψ are two distinct, pure, normal states on $\mathcal{B}(\mathcal{H})$ (or equivalently on $\mathcal{B}_{sa}(\mathcal{H})$), then the smallest face $B(\phi, \psi)$ of the state space S^* of $\mathcal{B}(\mathcal{H})$ that contains both ϕ and ψ is isomorphic to B^3, i.e. $\dim B(\phi, \psi) = 3$. To avoid misunderstandings, note, however, that if $\mathcal{B}$ is the C^*-algebra $\mathbb{C} \oplus \mathbb{C}$, and ϕ and ψ are distinct pure states on $\mathcal{B}$ (or equivalently on $\mathcal{B}_{sa}$) then $B(\phi, \psi)$ is isomorphic to $[0, 1]$, i.e. $\dim B(\phi, \psi) = 1$.

We are now ready for the definition of the two characteristic properties of the state space of a JB-algebra which is the self-adjoint part of a C^*-algebra.

A JB-algebra $\mathcal{A}$ is said to have the *3-ball property* if for each pair $\{\phi, \psi\}$ of distinct pure states on $\mathcal{A}$, the minimal face $B(\phi, \psi)$ of S^*, that contains both ϕ and ψ, is of dimension 3 or 1.

Alfsen *et al.* (1980) showed that the state space of every C^*-algebra has this property; hence the condition is necessary. It is, however, not sufficient, and they exhibited a counterexample (see below) of a JB-algebra having the 3-ball property while it is not the self-adjoint part of a C^*-algebra.

The second condition of Alfsen *et al.* (1980)—namely the orientability of S^*—is arrived at as follows.

A JB-algebra $\mathcal{A}$ with the 3-ball property is supposed to be given. With S^* denoting the state space of $\mathcal{A}$, the topological space $B(S^*)$ is defined as follows. Its elements are the maps $a : B^3 \to S^*$ such that a is an affine isomorphism from B^3 onto a face $a(B^3)$ of S^*. The topology on $B(S^*)$ is that of the point-wise convergence. Note that the following conditions are equivalent for $a, b \in B(S^*)$: (a) $a(B^3) = b(B^3)$; (b) $b^{-1} \circ a \in O^3$, the group of all orthogonal transformations of $\mathbb{R}^3$; and (c) $a = b \circ g$ for some $g \in O^3$. Note that O^3 acts continuously (from the right) on $B(S^*)$. When the above three conditions are satisfied we write $a \sim b \,(\mathrm{mod}\, O^3)$; this is clearly an equivalence relation. We further write $a \approx b \,(\mathrm{mod}\, O^3_+)$ whenever $a \sim b \,(\mathrm{mod}\, O^3)$ *and* $\det(b^{-1} \circ a) = +1$; this, too, is an equivalence relation.

It is now natural to call $B(S^*)/O^3$ the space of facial 3-balls of S^*, and to call $B(S^*)/O^3_+$ the space of oriented facial 3-balls of S^*. Note that $B(S^*)/O^3_+ \to$

$B(S^*)/O^3$ is a locally trivial Z_2-bundle. When this bundle is (globally) trivial, we say that S^* is *orientable*.

Alfsen *et al* (1980) showed that the state space of every C^*-algebra has this property; hence the orientability condition is necessary to the conclusion of their main theorem.

The reader must notice, however, that the 3-ball property and the orientability are *intrinsic* conditions, properly pertaining to the geometry of the state space of a JB-algebra: to check whether these conditions are satisfied for a given JB-algebra $\mathcal{A}$ no extra structure is needed. For instance Alfsen *et al.* (1980) offer the example of the JB-algebra $\mathcal{A}$, the elements of which are the continuous functions f from $S^1 \equiv \{x \in \mathbb{R}^2 | \parallel x \parallel^2 = 1\}$ to $\mathcal{M}(2, \mathbb{C})_{sa}$ that satisfy $f(-x) = f(x)^t \;\; \forall\, x \in S^1$; the algebraic operations of $\mathcal{A}$ are defined point-wise, and the norm is the sup-norm. While these are compatible with the natural embedding of $\mathcal{A}$ in the C^*-algebra $\mathcal{B}$ of the continuous functions $F : S^1 \rightarrow M(2, \mathbb{C})$, $\mathcal{A}$ is not the self-adjoint part of $\mathcal{B}$; its state space satisfies the 3-ball condition, but is *not* orientable; hence $\mathcal{A}$ cannot be the self-adjoint part of *any* C^*-algebra.

The affine isomorphism (84) gives an intuitive indication on how the 3-ball property and the orientability can be linked to the possibility of superimposing, to a JB-structure, a consistent Lie structure. The beauty of the paper of Alfsen, Hanche-Olsen & Shultz (1980) is that it shows how such a primitive intuition can, in fact, be raised to the level of a full characterization of those JB-algebras which are the self-adjoint part of a C^*-algebra.

This result is an answer to part (b) of the question raised above, concerning the empirical status of the C^*-algebraic postulate. The question is then reduced to its part (a): to understand the empirical content of the JB-algebraic postulate.

In an earlier paper, Alfsen & Schultz (1978) had already succeeded to obtain an intrinsic characterization of the compact convex sets that are the state space of a JB-algebra. Their characterization, however, is rather involved and it did not seem to lend itself to a physical interpretation that would be both simple and compelling; this line of attack, trying to characterize the algebra of observables through the properties of its state space, has been pursued in Araki (1980).

In view of this situation, the reader may be willing to settle, at least tentatively, with an operational description of the composition laws and idealizations involved in the JB-algebraic postulate. A proposal in this direction, involving a step-by-step axiomatization of the map

$$< ; >: S^* \times \mathcal{A} \rightarrow \mathbb{R} \tag{85}$$

has been worked out in Emch (1972). We only outline the argument below.

The first ingredient is to assume that, through the map (85), $\mathcal{A}$ and S^* are separating for one another. With this in mind, S^* can naturally be equipped, via the maps

$$< ;A >: \phi \in S^* \to < \phi; A > \in \mathbb{R} \tag{86}$$

with the structure of a *convex set*: the convex sum $(\sum_k \lambda_k \phi_k)$ is interpreted as the mode of preparation consisting in preparing, separately for each k, a fraction λ_k of the samples following the mode of preparation prescribed by ϕ_k. Similarly $\mathcal{A}$ can naturally be equipped, via the maps

$$< \phi; >: A \in \mathcal{A} \mapsto < \phi; A > \in \mathbb{R} \tag{87}$$

with the structure of a *real vector space*: the linear combination $(\sum_k a_k A_k)$ is interpreted as the observable, the expectation of which is given, for every mode of preparation ϕ, by $\Sigma_k a_k < \phi; A_k >$. By the same token, every $\phi \in S^*$ is thus a linear functional on $\mathcal{A}$. Several authors, see e.g. Jauch (1968), seem to have been reluctant to assume, without further ado, that the observables form a real vector space $\mathcal{A}$—or, equivalently, that the states are linear functionals on $\mathcal{A}$; instead, Jauch first summons a theorem, due to Gleason (1957), according to which every normalized, σ-additive positive functional on the lattice $\mathcal{P}(\mathcal{H})$ of the projectors on a Hilbert space is of the form $< \phi; P >= tr(\rho P)$ for some $\rho \in \mathcal{T}(\mathcal{H})$ with ρ positive and $tr\ \rho = 1$; that characterization, however, is too restrictive for our purpose: we want more general states on more general JB-algebras. We, therefore, keep (87) and its consequence as stated above.

The second ingredient, in our operational approach to JB-algebraic structures, is the requirement that the collection S_A^* of all dispersion-free states on any given observable A is rich enough so that giving S_A^* and the map

$$\phi \in S_A^* \mapsto < \phi; A > \in \mathbb{R} \tag{88}$$

determines A uniquely. One then posits that the square of an observable A is the unique observable (A^2) such that

$$< \phi; A^2 >=< \phi; A >^2 \quad \forall\, \phi \in S_A^* \tag{89}$$

The *Jordan product* is then introduced as

$$A \circ B = \frac{1}{2}((A+B)^2 - A^2 - B^2) \tag{90}$$

Note that (90) does not involve any operation not previously defined.

The third ingredient is the spectrum $\sigma(A)$ of an observable A; it is defined by

$$\sigma(A) \equiv \{< \phi; A > | \phi \in S_A^*\} \tag{91}$$

The spectral radius $|\sigma(A)|$ of an observable A is then defined by

$$|\sigma(A)| \equiv \sup_{a \in \sigma(A)} |a| \tag{92}$$

and it is shown that this can serve as a *norm* on $\mathcal{A}$.

The aforementioned reference then spells out in detail the chain of operational axioms involved when one postulates that $\mathcal{A}$ has the structure of a JB-algebra. Here again, one could object that the resulting axiomatic is not compelling; this is, in fact, not our claim; and we accept that there could possibly exist physical systems that the JB-algebraic postulate does not capture. The claim is, rather, that the formulation presented there is intrinsic and explicit enough so that one can empirically decide whether each one of the axioms is a reasonably idealized description of a given physical system.

Together with the result of Alfsen *et al.* (1980) reported above, and the four remarks presented at the beginning of Subsection d, this concludes what we wanted to say in this section in support of the case for the C^*-algebraic postulate.

We should, nevertheless, still mention that the emphasis placed, in this section, on bounded observables was mostly for axiomatic purposes; it should certainly not be construed as a denial of the physical relevance of unbounded observables. What the C^*-algebraic postulate subsumes in this connection is that the physical systems which it covers are already recognizable from their bounded observables. The postulate also serves as a prototype: one hopes that the techniques developed for the study of algebras of bounded observables will be adaptable to the ulterior study of more general algebras. For stimulating accounts of the possibilities open towards a systematic algebraic theory, where unbounded objects play a central role rather than being brought in as some kind of afterthought, see e.g. Borchers (1972, 1975), Lassner (1975), Yngvason (1976), Araki & Jurzak (1981) and Powers (1982). We also will have occasions, later in this book, to address the question of the proper handling and role of unbounded observables and fields.

Section 2. NON-COMMUTATIVE ERGODIC THEORY

An important branch of mathematical research has been devoted to the ergodic problems of *classical* mechanics; see e.g. Arnold & Avez (1968). The main purpose of non-commutative ergodic theory is to adapt this framework to the needs of *quantum* statistical mechanics; some of its results are moreover relevant to certain aspects of quantum field theory.

The general context is as follows. One supposes given the *action* $\alpha : G \to \text{Aut}(\mathcal{B})$ of a topological group G on a C^*-algebra $\mathcal{B}$, satisfying the following three conditions: (i) for every $g \in G$, $\alpha(g)$ is an automorphism of $\mathcal{B}$; (ii) α is a group homomorphism, i.e. $\alpha(g_1)\alpha(g_2) = \alpha(g_1 g_2)$ for all $g_1, g_2 \in G$; and (iii) α is continuous in the weak*-topology of $\mathcal{B}$, i.e. one assumes that the maps $g \in G \mapsto \langle \phi; \alpha(g)[A] \rangle \in \mathbb{R}$ are continuous for every observable $A = A^* \in \mathcal{B}$ and every state $\phi \in S^*$. One is then typically interested in the following questions: characterize the properties of the set $S_G^* \equiv \{\phi \in S^* | \phi \circ \alpha(g) = \phi \; \forall g \in G\}$ and of its elements; discuss the asymptotic properties of $\alpha(g)[A]$ as g "tends to infinity" (when this makes sense); and obtain a description of what it means to average an observable, or a state, with respect to the group action α. Several of these questions were already discussed by Eberlein (1949) in the general context where $\mathcal{B}$ is an abstract topological vector space.

In this section we explore the general answers one can hope to give to these questions in the framework of the theory of C^*-algebraic dynamical systems, to prepare for the specific applications that will be dealt with in Section One of Chapter Ten. The reader should, nevertheless, be warned, already here, that the explicit construction of the action $\alpha : G \to \text{Aut}(\mathcal{B})$ relative to a given physical situation is often far from being a trivial matter. We shall, therefore, only deal in this section, with the simplest cases, so that the essential features of the theory can be outlined.

a. C^*-inductive limits.

Recall that a *directed set* $\{\mathcal{F}, \subseteq\}$ is a partially ordered set $\mathcal{F}$, with partial-order relation $\subseteq$, such that for every pair $\{\Omega_1, \Omega_2\}$ of elements of $\mathcal{F}$ there exists at least one element $\Omega \in \mathcal{F}$ for which $\Omega_1 \subseteq \Omega$ and $\Omega_2 \subseteq \Omega$.

The examples we have in mind in this section are: the collection $\mathcal{F}$ of all finite subsets of $\mathbb{Z}^n$ (with n fixed), and the collection $\mathcal{F}$ of all bounded open subsets of $\mathbb{R}^n$ (with n fixed); in both cases the partial-order relation is the set-theoretical inclusion. Occasionally in this section, but mostly in Chapter Ten, we will also consider the case where $\mathcal{F}$ is the collection of all finite-dimensional subspaces of an infinite-dimensional Hilbert space $\mathcal{T}$, with again the set-theoretical inclusion providing the partial ordering. We will also consider, in Chapter Ten, the case where the Euclidean space $\mathbb{R}^n$, in the second of the above examples, is replaced by the Minkowskian space $M^{n,1}$.

A collection $\{\mathcal{B}_\Omega | \Omega \in \mathcal{F}\}$ of C^*-algebras $\mathcal{B}_\Omega$ (with unit I_Ω), indexed by a directed set $\mathcal{F}$, is said to be *isotonic* if for every ordered pair $\Omega_1 \subseteq \Omega_2$ of elements of $\mathcal{F}$, an injective *-homomorphism $i_{21} : \mathcal{B}_{\Omega_1} \to \mathcal{B}_{\Omega_2}$ is given, satisfying:

(1a) $\Omega_1, \Omega_2, \Omega_3 \in \mathcal{F}$ with $\Omega_1 \subseteq \Omega_2 \subseteq \Omega_3$ implies $i_{31} = i_{32} \circ i_{21}$

(1b) $\Omega_1, \Omega_2 \in \mathcal{F}$ with $\Omega_1 \subseteq \Omega_2$ implies $i_{\Omega_1}(I_{\Omega_1}) = I_{\Omega_2}$

A C^*-algebra $\mathcal{B}$ (with unit I) is said to be the *C^*-inductive limit* of a collection $\{\mathcal{B}_\Omega | \Omega \in \mathcal{F}\}$ of C^*-algebras $\mathcal{B}_\Omega$ (with unit I_Ω), indexed by a directed set $\mathcal{F}$, if for every $\Omega \in \mathcal{F}$ there exists an injective *-homomorphism $i_\Omega : \mathcal{B}_\Omega \to \mathcal{B}$ such that

(2a) $\Omega_1, \Omega_2 \in \mathcal{F}$ with $\Omega_1 \subseteq \Omega_2$ implies $i_{\Omega_1}(\mathcal{B}_{\Omega_1}) \subseteq i_{\Omega_2}(\mathcal{B}_{\Omega_2})$

(2b) $\mathcal{B}_0 \equiv \bigcup_{\Omega \in \mathcal{F}} i_\Omega(\mathcal{B}_\Omega)$ is norm-dense in $\mathcal{B}$

(2c) $i_\Omega(I_\Omega) = I \quad \forall \; \Omega \in \mathcal{F}$

Takeda (1955) proved that every isotonic collection $\{\mathcal{B}_\Omega | \Omega \in \mathcal{F}\}$ of C^*-algebras $\mathcal{B}_\Omega$ (with unit I_Ω), indexed by a directed set $\mathcal{F}$, admits a unique C^*-inductive limit $\mathcal{B}$. This is the object we will work with in this subsection.

To illustrate this structure, we will describe the C^*-algebra of a *quantum lattice*. As we proceed, the reader who thinks naturally in the terms of classical probability theory will realize that the object we will be constructing is a quantum generalization of a setting in which one could describe the Bernoulli schemes associated to infinite sequences of coin tossings, or dice throwings. Other readers, who are used to view Bernoulli schemes as some sort of Ising models, will see that the concept of a quantum lattice encompasses the situation encountered, for instance, in the study of the Heisenberg models in quantum statistical mechanics.

We consider a lattice $\mathbb{Z}^n$ and the directed set $\{\mathcal{F}, \subseteq\}$ of its finite subsets, with partial-ordering provided by the set-theoretical inclusion. To every site $\omega \in \mathbb{Z}^n$ we associate a copy $\mathcal{B}_\omega$ of the von Neumann algebra $\mathcal{B}(\mathcal{H})$ of all bounded operators on a fixed, separable Hilbert space $\mathcal{H}$. For instance, if $\mathcal{H} = \mathcal{L}^2(\mathbb{R}^\nu, d^\nu x)$ $\mathcal{B}_\omega$ can be interpreted as the algebra of observables describing a ν-dimensional harmonic oscillator attached to the site ω. If, however, dim $\mathcal{H} = m < \infty$, i.e. $\mathcal{H} = \mathbb{C}^m$, $\mathcal{B}_\omega$ is isomorphic to the algebra $\mathcal{M}(m, \mathbb{C})$ of the $m \times m$ matrices with complex entries, describing a "spin $s = (m-1)/2$ particle" localized at the site ω; the particular case $s = 1/2$ corresponds to the original Heisenberg model for ferromagnetism. Classical Bernoulli schemes (of finite entropy, or Ising models) would obtain if we had associated, instead, to each site $\omega \in \mathbb{Z}$, a copy of the diagonal part of $\mathcal{M}(m, \mathbb{C})$, i.e. of the abelian algebra $\mathcal{C}(K)$ with $K = \{0, 1, \ldots, m-1\}$.

To every *finite* subset $\Omega \subset \mathbb{Z}^n$ we associate the von Neumann algebra

$$\mathcal{B}_\Omega \equiv \bigotimes_{\omega \in \Omega} \mathcal{B}_\omega \equiv \mathcal{B}(\bigotimes_{\omega \in \Omega} \mathcal{H}_\omega) \equiv \mathcal{B}(\mathcal{H}_\Omega) \tag{3}$$

where, for each $\omega \in \mathbb{Z}^n$, $\mathcal{H}_\omega$ is a copy of $\mathcal{H}$. Note in particular, that if dim $\mathcal{H} = m < \infty$

$$\mathcal{B}_\Omega \cong \mathcal{M}(m^{|\Omega|}, \mathbb{C}) \tag{4}$$

where $|\Omega|$ denotes the number of sites in Ω. For every ordered pair $\Omega_1 \subseteq \Omega_2$ of elements of $\mathcal{F}$, we write Ω_2 as the disjoint union $\Omega_1 \cup (\Omega_2 \setminus \Omega_1)$ where $\Omega_2 \setminus \Omega_1 \equiv \{\omega \in \Omega_2 | \omega \neq \Omega_1)$. Clearly then

$$\mathcal{B}_{\Omega_2} = \mathcal{B}_{\Omega_1} \otimes \mathcal{B}_{\Omega_2 \setminus \Omega_1} \tag{5}$$

and the maps

$$i_{21} : B \in \mathcal{B}_{\Omega_1} \mapsto B \otimes I_{\Omega_2 \setminus \Omega_1} \in \mathcal{B}_{\Omega_2} \tag{6}$$

equip $\{\mathcal{B}_\Omega | \Omega \in \mathcal{F}\}$ with the structure of an isotonic collection of C^*-algebras. Its C^*-inductive limit $\mathcal{B}$ gives a precise meaning to the a priori ambiguous notation

$$B = \bigotimes_{\omega \in \mathbb{Z}^n} B_\omega \tag{7}$$

Note that, while each $\mathcal{B}_\Omega$ is a concrete C^*-algebra acting on a definite Hilbert space $\mathcal{H}_\Omega$ (see 3), $\mathcal{B}$ comes to us as an abstract C^*-algebra, free of any underlying Hilbert space (beyond, evidently, the Hilbert space given by its universal representation—see Subsection 9.1.b—which is irrelevant to our purposes). We shall use this freedom, later on in this section, to define the global observables (e.g. the magnetization) of our quantum lattice; as we shall see there, these global observables do not belong to $\mathcal{B}$ itself; for this reason, we refer to the self-adjoint elements of $\mathcal{B}$ as the *quasi-local* observables, reserving the name of *local* observables to the self-adjoint elements of $\mathcal{B}_0$ (see 2b). By extension, $\mathcal{B}$ is referred to the C^*-algebra of quasi-local observables, and $\mathcal{B}_0$ as the *-algebra of local observables.

b. Norm-asymptotic abelianness and observables at infinity.

We now define an action $\alpha : \mathbb{Z}^n \to \mathrm{Aut}(\mathcal{B})$ of the group $\mathbb{Z}^n$ of translations, on the quasi-local C^*-algebra $\mathcal{B}$ of the quantum lattice defined in Subsection a.

We first define $\alpha(a)[B]$ for every $a \in \mathbb{Z}^n$, $B \in \mathcal{B}_\Omega$ and $\Omega \in \mathcal{F}$. Let $a[\Omega] \equiv \{\omega + a \,|\, \omega \in \Omega\}$; clearly $a[\Omega] \in \mathcal{F}$ and $\Lambda \equiv \Omega \cup a[\Omega] \in \mathcal{F}$. Let

$$i : \mathcal{B}_\Omega \to \mathcal{B}_\Lambda \tag{8a}$$

$$i_a : \mathcal{B}_{a[\Omega]} \to \mathcal{B}_\Lambda \tag{8b}$$

denote the injective *-homomorphisms prescribed by the isotony relation (6). With $B \in \mathcal{B}_\Omega$, we have thus

$$i[B] = B \otimes I_{\Lambda\backslash\Omega} \tag{9a}$$

and we can then define $\alpha(a)[B]$ as the unique element of $\mathcal{B}_{a[\Omega]}$ such that

$$i_a[\alpha(a)[B]] = I_{\Lambda\backslash a[\Omega]} \otimes B \tag{9b}$$

Note that $\alpha(a)$ establishes a *-isomorphism from $\mathcal{B}_\Omega$ onto $\mathcal{B}_{a[\Omega]}$; in particular $\| \alpha(a)[B] \| = \| B \|$. We use now the injective *-homomorphisms (see 2)

$$i_\Omega : \mathcal{B}_\Omega \to \mathcal{B}_0 \subset \mathcal{B} \tag{10a}$$

$$i_{a[\Omega]} : \mathcal{B}_{a[\Omega]} \to \mathcal{B}_0 \subset \mathcal{B} \tag{10b}$$

to lift $\alpha(a)$ to a bijective *-homomorphism from $\mathcal{B}_0$ onto itself, which we denote again by the same symbol $\alpha(a)$. Since $\| \alpha(a) \| = 1$, and since by definition $\mathcal{B}_0$ is norm-dense in $\mathcal{B}$, $\alpha(a)$ extends uniquely by continuity to an automorphism of $\mathcal{B}$. We denote again this automorphism by the symbol $\alpha(a)$. It is then easy to verify that

$$\alpha : a \in \mathbb{Z}^n \to \alpha(a) \in \mathrm{Aut}(\mathcal{B}) \tag{11}$$

defines an action, called the *canonical action*, of $\mathbb{Z}^n$ on the quasi-local algebra $\mathcal{B}$.

SCHOLIUM 1. *Let $\mathcal{B}$ be the C^*-algebra of the quasi-local obervables of a quantum lattice on $\mathbb{Z}^n$; and let α be the canonical action of $\mathbb{Z}^n$ on $\mathcal{B}$. Then, for every B and C in $\mathcal{B}$:*

$$n - \lim_{|a|\to\infty} [B, \alpha(a)[C]] = 0 \tag{12}$$

(where $n - \lim$ means limit in the norm-topology of $\mathcal{B}$)

Proof: With B and C in $\mathcal{B}$, and $\epsilon > 0$ all fixed, let $\eta > 0$ be such that:

$$2\eta\,(3\eta + \| B \| + \| C \|) = \epsilon \tag{13}$$

Since $\mathcal{B}_0$ is norm-dense in $\mathcal{B}$, there exist B_0 and C_0 in $\mathcal{B}_0$ such that

$$\| B - B_0 \| < \eta \quad \text{and} \quad \| C - C_0 \| < \eta \tag{14}$$

Since B_0 and C_0 belong to $\mathcal{B}_0$, there exist Ω in $\mathcal{F}$ such that $\mathcal{B}_\Omega$ contains both B_0 and C_0. Since Ω is finite there exists K such that

$$\Omega \cap a[\Omega] = \emptyset \quad \forall\ a \text{ with } |a| > K \tag{15}$$

and thus

$$[B_0, \alpha(a)[C_0]] = 0 \quad \forall\ a \text{ with } |a| > K \tag{16}$$

We have thus, for all $a \in \mathbb{Z}^n$ with $|a| > K$:

$\| [B, \alpha(a)[C]] \| \leq$

$\| [B - B_0, \alpha(a)[C - C_0]] \| + \| [B_0, \alpha(a)[C - C_0]] \| + \| [B - B_0, \alpha(a)[C_0]] \| \leq$

$2 \| B - B_0 \| \cdot \| C - C_0 \| + 2 \| B_0 \| \cdot \| C - C_0 \| + 2 \| B - B_0 \| \cdot \| C_0 \| \leq$

$2\eta(\eta + \| B_0 \| + \| C_0 \|) \ \leq$

$2\eta(\eta + \| B - B_0 \| + \| B \| + \| C - C_0 \| + \| C \|) \leq$

$2\eta(3\eta + \| B \| + \| C \|) = \epsilon\,.$

Clearly K depends only on ϵ, B and C, thus proving (12). □

Remarks.

1) This result implies that $\alpha(b)$ with $b \in \mathbb{Z}^n$ and $b \neq 0$ cannot be an inner automorphism of $\mathcal{B}$, i.e. there cannot be any unitary $U(b) \in \mathcal{B}$ such that $\alpha(b)[B] = U(b)BU(b)^* \ \ \forall B \in \mathcal{B}$. Suppose indeed that such an element existed;

we would then have:

$$\begin{aligned}&\| [U(b), \alpha(a)[B]] \| = \| (U(b)\alpha(a)[B]U(b)^* - \alpha(a)[B])U(b) \| = \\ &\| \alpha(b) \circ \alpha(a)[B] - \alpha(a)[B] \| = \| \alpha(a)(\alpha(b)[B] - B) \| = \\ &\| \alpha(b)[B] - B \| \ .\end{aligned}$$

From the scholium, we know that we can choose a large enough so that the LHS of this equality becomes as small as one wishes; since the RHS does not depend on a, this means that $\| \alpha(b)[B] - B \| = 0$, i.e. $\alpha(b)[B] = B$ for all $B \in \mathcal{B}$, and thus in particular for all $B \in \mathcal{B}_0$ which is clearly absurd when $b \neq 0$. This shows, among other things that the C^*-algebra $\mathcal{B}$ cannot be realized as the algebra of *all* bounded linear operators on any Hilbert space.

2) The above scholium can trivially be extended to the case where $\mathcal{F}$ is replaced by the set of all bounded open region in $\mathbb{R}^n$, where the group $\mathbb{Z}^n$ is replaced by the group $\mathbb{R}^n$, and where the isotonic family $\{\mathcal{B}_\Omega | \Omega \in \mathcal{F}\}$ defined from (3) is replaced by any isotonic collection $\{\mathcal{B}_\Omega | \Omega \in \mathcal{F}\}$ of C^*-algebras $\mathcal{B}_\Omega$ (with unit I_Ω), indexed by our new directed set $\mathcal{F}$, and satisfying the two conditions: (i) $\Omega_1 \cap \Omega_2 = \emptyset$ implies $[i_{\Omega_{31}}(B_1), i_{\Omega_{32}}(B_2)] = 0$ for $\Omega_3 = \Omega_1 \cup \Omega_2$ and all $B_1 \in \mathcal{B}_1$, $B_2 \in \mathcal{B}_2$; and (ii) for every $a \in \mathbb{R}^n$ and every $\Omega \in \mathcal{F}$ there exists an isomorphism $\alpha(a) : \mathcal{B}_\Omega \rightarrow \mathcal{B}_{a[\Omega]}$. Our remark 1 extends to this case as well, and should thus serve as a warning against any attempt to try identifying the total momentum as a quasi-local observable. As we shall see in Chapter Ten, this warning carries over to the relativistic case where $\mathbb{R}^n$ is further replaced by $M^{n,1}$ and Einstein causality is assumed.

3) The scholium and the above remark 2 suggest the following definition. With G denoting $\mathbb{Z}^n$ or $\mathbb{R}^n$, an action $\alpha : G \rightarrow \mathrm{Aut}(\mathcal{B})$ on a C^*-algebra $\mathcal{B}$ (with unit I) is said to be *norm-asymptotic abelian* if for every pair $\{B, C\}$ of elements of $\mathcal{B}$ and every $\epsilon > 0$, there exists a finite number $K(B, C, \epsilon)$ such that

$$\| [B, \alpha(a)[C]] \| < \epsilon \tag{17}$$

for all $a \in G$ with $|a| > K(B, C, \epsilon)$. The scholium thus asserts that this property is satisfied by the group of translations of any quantum lattice.

While norm-asymptotic abelianness does say that

$$n - \lim_{|a| \rightarrow \infty} [B, \alpha(a)[C]] = 0, \tag{18}$$

it does not make, by itself, any statement on the existence of the limit of $\alpha(a)[C]$ as $|a| \rightarrow \infty$. The next result (Theorem 2 below) goes in that direction. It is our first "ergodic theorem" in the sense that it characterizes a class of extremal G-invariant states and shows that these states satisfy a certain mixing property. Before we state this theorem, however, we need a few definitions which we now give.

For any action $\alpha : G \to \text{Aut}(\mathcal{B})$ of a topological group G on a C^*-algebra $\mathcal{B}$, we denote by $\mathcal{S}_G^*$ the set of all G-invariant states on $\mathcal{B}$, i.e.:

$$\mathcal{S}_G^* \equiv \{\phi \in \mathcal{S}^* | \phi \circ \alpha(g) = \phi \quad \forall\ g \in G\} \tag{19}$$

Under very mild conditions—see for instance, Subsection c below—one can prove that $\mathcal{S}_G^*$ is not empty. The reader should, nevertheless, realize that some condition, however mild, must be imposed. Let indeed $\mathcal{C}(\mathcal{H})$ be the C^*-algebra of all compact operators acting on a separable Hilbert space $\mathcal{H}$. Since $\mathcal{C}(\mathcal{H})$ is a two-sided ideal in $\mathcal{B}(\mathcal{H})$, we have that

$$\alpha(a)[C] \equiv U(a)CU(a)^* \tag{20}$$

belongs to $\mathcal{C}(\mathcal{H})$ for all $C \in \mathcal{C}(\mathcal{H})$ and all $U(a) \in \mathcal{U}(\mathcal{H})$. Let in particular $\mathcal{H} = \mathcal{L}^2(\mathbb{R}, dx)$ and $U(a)$ be defined, for every $a \in \mathbb{R}$, by

$$(U(a)\Psi)(x) \equiv \Psi(x - a) \tag{21}$$

for all Ψ in $\mathcal{H}$. It is then easy to verify that (20) and (21) define an action

$$\alpha : a \in \mathbb{R} \mapsto \alpha(a) \in \text{Aut}\,(\mathcal{C}(\mathcal{H})) \tag{22}$$

Recall also that *every* state on the C^*-algebra $\mathcal{C}(\mathcal{H})$ can be written in the form

$$\phi : C \in \mathcal{C}(\mathcal{H}) \mapsto tr\, \rho C \in \mathbb{C} \tag{23}$$

with ρ a positive, trace-class operator such that $tr\,\rho = 1$, i.e.

$$\rho = \sum_i \lambda_i\ P_i \quad \text{with} \tag{24a}$$

$$\lambda_i \geq 0 \quad \text{and} \quad \sum_i \lambda_i \dim(P_i) = 1 \tag{24b}$$

If there were any $\alpha(\mathbb{R})$-invariant state ϕ on $\mathcal{C}(\mathcal{H})$ we would have, for the corresponding density matrix ρ and its spectral projectors P_i:

$$U(a)^*\rho\ U(a) = \rho \quad \forall\ a \in \mathbb{R} \tag{25}$$

$$P_i \in \{U(a) | a \in \mathbb{R}\}' = \{U(a) | a \in \mathbb{R}\}'' \tag{26}$$

The von Neumann algebra $\{U(a)|a \in \mathbb{R}\}''$ admits, however, no finite dimensional projector. Together with (24) this implies $\rho = 0$, in clear contradiction with $tr\,\rho = 1$. Hence, the action (22) admits no G-invariant state, i.e. $\mathcal{S}_G^* = \emptyset$.

For a quantum lattice on $\mathbb{Z}^n$, the following construction produces, in an instructive manner, a class of $\mathbb{Z}^n$-invariant states, thus establishing, for this case, that $\mathcal{S}_G^* \neq \emptyset$ (more, and in fact all, $\mathbb{Z}^n$-invariant states will be constructed in Subsection c below).

Recall that a quantum lattice is characterized by associating, to every $\omega \in \mathbb{Z}^n$, a copy $\mathcal{B}_\omega$ of the von Neumann algebra $\mathcal{B}(\mathcal{H})$. Let ψ be a normal state on $\mathcal{B}(\mathcal{H})$, and for each $\omega \in \mathbb{Z}^n$, let ϕ_ω be a copy of ψ on $\mathcal{B}_\omega$. For each $\Omega \in \mathcal{F}$ let

$$\phi_\Omega \equiv \bigotimes_{\omega\in\Omega} \phi_\omega \tag{27}$$

be the unique normal state on $\mathcal{B}_\Omega$ such that for every choice $\{B_\omega \in \mathcal{B}_\omega | \omega \in \Omega\}$:

$$< \phi_\Omega; \bigotimes_{\omega\in\Omega} B_\omega >= \prod_{\omega\in\Omega} < \phi_\omega ; B_\omega > \tag{28}$$

The family $\{\phi_\Omega | \Omega \in \mathcal{F}\}$ is consistent, in the sense that for every $\Omega_1, \Omega_2 \in \mathcal{F}$ with $\Omega_1 \subseteq \Omega_2$ and every $B \in \mathcal{B}_{\Omega_1}$:

$$< \phi_{\Omega_2}; i_{21}(B) >=< \phi_{\Omega_1}; B > \tag{29}$$

This allows to define, without ambiguities, a map

$$\phi : B \in \mathcal{B}_0 \mapsto \lim_{\Omega\to\mathbb{Z}^n} < \phi_\Omega; B >\in \mathbb{C} \tag{30}$$

which is linear, positive and of norm 1. This map extends therefore uniquely, by continuity, to a state on $\mathcal{B}$, which we denote

$$\phi = \bigotimes_{\omega\in\mathbb{Z}^n} \phi_\omega \tag{31}$$

This state is clearly $\mathbb{Z}^n$-invariant.

Let us now return to the general case of an action $\alpha : G \mapsto \text{Aut}(\mathcal{B})$ of a topological group G on a C^*-algebra $\mathcal{B}$ with unit. We assume henceforth that $\mathcal{S}_G^*$ is not empty.

One verifies easily that $\mathcal{S}_G^*$ is convex and is closed in the weak*-topology of $\mathcal{B}^*$; since it is bounded in the metric topology of $\mathcal{B}^*$, we thus have that $\mathcal{S}_G^*$ is a w^*-compact convex set. We can, therefore, use again the Krein-Milman theorem to assert that $\mathcal{S}_G^*$ is the w^*-closed convex hull of the set $\mathcal{E}_G^*$ of its extreme points. Hence, there exist extremal G-invariant states whenever $\mathcal{S}_G^*$ is not empty.

THEOREM 2. *Let $\alpha : G \to \text{Aut}(\mathcal{B})$ be a norm-asymptotic abelian action of G (e.g. $\mathbb{Z}^n$ or $\mathbb{R}^n$) on a C^*-algebra $\mathcal{B}$ (with unit I). Let ϕ be a G-invariant state on $\mathcal{B}$, such that the GNS representation π_ϕ of $\mathcal{B}$ associated to ϕ satisfy $\pi_\phi(\mathcal{B})'' \cap \pi_\phi(\mathcal{B})' = \mathbb{C}I$. Then:*

(32) (i) *for every* $B \in \mathcal{B}$
$$w - \lim_{|a|\to\infty} \pi_\phi(\alpha(a)[B]) = <\phi; B> I$$

(33) (ii) *for every* $A, B, C \in \mathcal{B}$
$$\lim_{|a|\to\infty} <\phi; B^*\alpha(a)[A]C> = <\phi; A><\phi; B^*C>$$

(iii) ϕ *is extremal* G *-invariant.*

Proof: For an arbitrary representation $\pi : \mathcal{B} \to \mathcal{B}(\mathcal{H}_\pi)$, let $Z_\pi(\mathcal{B}) \equiv \pi(\mathcal{B})'' \cap \pi(\mathcal{B})'$. The commutant $Z_\pi(\mathcal{B})'$ of $Z_\pi(\mathcal{B})$ is the von Neumann algebra $\{\pi(\mathcal{B}), \pi(\mathcal{B})'\}''$ generated by $\pi(\mathcal{B})$ and $\pi(\mathcal{B})'$. Hence $Z_\pi(\mathcal{B})'$ is the closure, in the strong operator topology on $\mathcal{B}(\mathcal{H}_\pi)$, of the set $\mathcal{Y}_0$ of all *finite* sums of the form

$$Y_0 = \sum_i \pi(B_i)X_i \tag{34}$$

with $B_i \in \mathcal{B}$ and $X_i \in \pi(\mathcal{B})'$. Hence, with $Y = Y^* \in Z_\pi(\mathcal{B})'$, Φ and $\Psi \in \mathcal{H}$, and $\epsilon > 0$ all fixed, there exists $Y_0 = Y_0^* \in \mathcal{Y}_0$, such that

$$\| (Y - Y_0)\Phi \| < \epsilon \| \Phi \| \text{ and } \| (Y - Y_0)\Psi \| < \epsilon \| \Psi \| \tag{35}$$

Write Y_0 in the form (34); since $X_i \in \pi(\mathcal{B})'$, the norm-asymptotic abelianness condition implies that, at fixed $B \in \mathcal{B}$, there exists a finite number $K(Y_0, B, \epsilon)$ such that

$$\| [Y_0, \pi(\alpha(a)[B])] \| < \epsilon \tag{36a}$$

for all $a \in G$ with $|a| > K(Y_0, B, \epsilon)$. We have thus:

$$\begin{aligned} &|([Y, \pi\{\alpha(a)[B]\}]\Phi, \Psi)| \leq \\ &|([Y_0, \pi\{\alpha(a)[B]\}]\Phi, \Psi)| + |([Y - Y_0, \pi\{\alpha(a)[B]\}]\Phi, \Psi)| \leq \\ &\| [Y_0, \pi\{\alpha(a)[B]\}] \| \cdot \| \Phi \| \cdot \| \Psi \| \\ &\quad + \| (Y - Y_0)\Psi \| \cdot \| B \| \cdot \| \Phi \| \\ &\quad\quad + \| B \| \cdot \| \Psi \| \cdot \| (Y - Y_0)\Phi \| < \\ &\epsilon(1 + 2 \| B \|) \cdot \| \Phi \| \cdot \| \Psi \| \end{aligned} \tag{36b}$$

for all $a \in G$ with $|a| > K(Y_0, B, \epsilon)$. We conclude from this:

$$\lim_{|a|\to\infty} |([Y, \pi\{\alpha(a)[B]\}]\Phi, \Psi)| = 0 \tag{37a}$$

i.e. since Φ and Ψ are arbitrary:

$$w - \lim_{|a|\to\infty} [Y, \pi(\alpha(a)[B])] = 0 \tag{37b}$$

for all $B \in \mathcal{B}$ and all $Y \in Z_\pi(\mathcal{B})'$, the restriction $Y = Y^*$ being easily disposed of here.

We now impose the condition that π is primary, i.e. that $\pi(\mathcal{B})'' \cap \pi(\mathcal{B})' \equiv Z_\pi(\mathcal{B}) = \mathbb{C}I$. In this case, (37) holds for every $Y \in Z_\pi(\mathcal{B})' = \mathcal{B}(\mathcal{H}_\pi)$. Let Φ_1 and Φ_2 in $\mathcal{H}_\pi$, with $\| \Phi_1 \|= 1 =\| \Phi_2 \|$. There exists then at least one unitary operator U, acting in $\mathcal{H}_\pi$, such that $U\Phi_1 = \Phi_2$. Let us now define, for $k = 1, 2$:

$$\phi_k : B \in \mathcal{B} \mapsto (\pi(B)\Phi_k, \Phi_k) \in \mathbb{C} \tag{38}$$

We have then:

$$\begin{aligned} &< \phi_1; \alpha(a)[B] > - < \phi_2; \alpha(a)[B] >= \\ &([U, \pi\{\alpha(a)[B]\}]\Phi_1, \Phi_2) \end{aligned} \tag{39}$$

Upon inserting (37b) into (39) we obtain:

$$\lim_{|a|\to\infty} \{< \phi_1; \alpha(a)[B] > - < \phi_2; \alpha(a)[B] >\} = 0 \tag{40}$$

We finally impose that $\pi = \pi_\phi$ be the GNS representation associated to a G-invariant, primary state ϕ on $\mathcal{B}$. Let $\Phi_1 = \Phi$ be the cyclic vector canonically associated to ϕ and $\Phi_2 = \Psi$ be any normalized vector in the representation space $\mathcal{H}_\phi$. (40) now reads:

$$\lim_{|a|\to\infty} (\pi_\phi\{\alpha(a)[B]\}\Psi, \Psi) =< \phi; B > \tag{41}$$

By polarization, (41) implies, for every Ψ_1, Ψ_2 in $\mathcal{H}_\phi$

$$\lim_{|a|\to\infty} (\pi_\phi\{\alpha(a)[B]\}\Psi_1, \Psi_2) =< \phi; B > (\Psi_1, \Psi_2) \tag{42}$$

which is exactly the assertion (32) of part (i) of the theorem. Notice further that the assertion (33) of part (ii) of the theorem is only a particular case of (42). Indeed

$$\begin{aligned} &< \phi; A >< \phi; B^*C >=< \phi; A > (\pi_\phi(C)\Phi, \pi_\phi(B)\Phi) = \\ &\lim_{|a|\to\infty} (\pi_\phi\{\alpha(a)[A]\}\pi_\phi(C)\Phi, \pi_\phi(B)\Phi) = \\ &\lim_{|a|\to\infty} < \phi; B^*\alpha(a)[A]C > . \end{aligned} \tag{43}$$

Therefore, only part (iii) of the theorem remains to be proven. To do this, suppose on the contrary that ϕ is *not* extremal G-invariant, i.e. that there exist: $\lambda \in (0,1)$, ψ and $\chi \in S_G^*$ such that

$$\phi = \lambda\psi + (1-\lambda)\chi \tag{44}$$

By a reasoning akin to that carried out in the proof of Corollary 9.1.6, we conclude that there exists $X \in \pi_\phi(\mathcal{B})'$ such that

$$< \psi; B >= (\pi_\phi(B)X\Phi, \Phi) \quad \forall \ B \in \mathcal{B} \tag{45a}$$

Since ψ is G-invariant, we can rewrite (45a) as:

$$< \psi; B >= (\pi_\phi\{\alpha(a)[B]\}X\Phi, \Phi) \quad \forall \ (a, B) \in G \times \mathcal{B} \tag{45b}$$

and thus:

$$< \psi; B >= \lim_{|a|\to\infty} (\pi_\phi\{\alpha(a)[B]\}X\Phi, \Phi) \tag{45c}$$

We now use (42) to rewrite (45c) as:

$$< \psi; B >=< \phi; B > \ (X\Phi, \Phi) \quad \forall \ B \in \mathcal{B} \tag{46}$$

Computed at $B = I$, this equality gives $(X\Phi, \Phi) = 1$, and thus (46) reads $\psi = \phi$. From this definition, we conclude, therefore, that $\chi = \phi$ as well. So there is no, non-trivial, convex decomposition of ϕ into G-invariant states. This is exactly the conclusion (iii) of the theorem. □

The conclusions of the theorem are reminiscent of situations encountered in classical ergodic theory. Compare for instance the classical notion of metric indecomposability, and the present notion that ϕ be extremal G-invariant. The latter property was obtained here (see part iii of the theorem) on the strength of a stronger property, namely (33), akin to the classical mixing condition. In the physics literature, properties of the type (33), are referred to as 'clustering' or 'vanishing of long-range correlations'; we shall see several conditions of this type in this subsection and the next. We already commented—immediately after the definition (17) of norm asymptotic abelianness—on the thrust of conclusion (i) of the theorem: compare (18) and (32). We still, however, have to understand the background for one assumption of the theorem, namely that ϕ be primary, i.e. $\pi_\phi(\mathcal{B})'' \cap \pi_\phi(\mathcal{B})' = \mathbb{C}I$. While the full physical meaning of this condition in quantum equilibrium statistical mechanics will be explained in Section One of Chapter Ten, we can already here make a few comments on its mathematical meaning. First of all, it is obviously a weakening of the condition that the representation be irreducible. Hence, this condition is in particular satisfied when ϕ is pure. We are, however, in position to say significantly more, as we shall presently show.

For the remarks that immediately follow, the essential ingredients are that $\mathcal{B}$ is the C^*-inductive limit of an isotonic collection $\{\mathcal{B}_\Omega | \Omega \in \mathcal{F}\}$ of C^*-algebras $\mathcal{B}_\Omega$

(with unit I_Ω), indexed by a directed set $\mathcal{F}$ satisfying the following conditions. A 'local disjunction' $\perp$ is defined on $\mathcal{F}$ such that, with Ω, Ω_1, Ω_2, Λ in $\mathcal{F}$: (i) $\Omega_1 \subseteq \Omega_2$ and $\Omega_2 \perp \Lambda$ imply $\Omega_1 \perp \Lambda$; (ii) $\Omega_1 \perp \Lambda$ and $\Omega_2 \perp \Lambda$ imply that there exists $\Omega \in \mathcal{F}$ with $\Omega_1 \subseteq \Omega$, $\Omega_2 \subseteq \Omega$, and $\Omega \perp \Lambda$; and (iii) $\Omega \perp \Lambda$ implies $[i_\Omega(B), i_\Lambda(C)] = 0$ for all B in $\mathcal{B}_\Omega$ and all C in $\mathcal{B}_\Lambda$. We summarize this situation by saying that $\mathcal{B}$ is the C^*-algebra of the quasi-local observables for a system where *local commutativity* holds.

These conditions are clearly satisfied for a quantum lattice where $\Omega \perp \Lambda$ is defined to mean $\Omega \cap \Lambda = \emptyset$. They are also satisfied in the continuous-space situation described in Remark 2 to Scholium 1, where again $\Omega \perp \Lambda$ means $\Omega \cap \Lambda = \emptyset$. Furthermore, a relativistic version of the latter can be worked out where $\mathbb{R}^n$ is replaced by $M^{n,1}$, $\mathcal{F}$ is the directed set of all subsets of $M^{n,1}$ which are bounded and open in the Euclidean topology of $\mathbb{R}^{n+1}$, the partial ordering in $\mathcal{F}$ is the set-theoretical inclusion, *and* the disjointness relation $\Omega \perp \Lambda$ in $\mathcal{F}$ is defined to mean that Ω and Λ are causally disjoint, i.e. that no light ray can join any point in Ω to any point in Λ or vice versa, i.e. $g_{\mu\nu}(x-y)^\mu(x-y)^\nu > 0$ for all $x \in \Omega$ and $y \in \Lambda$.

Let $\mathcal{B}$ be the C^*-algebra of the quasi-local observables for a system where local commutativity holds. For notational convenience, let us drop again the explicit mention of the injective *-homomorphisms $i_\Omega : \mathcal{B}_\Omega \to \mathcal{B}$. For every $\Omega \in \mathcal{F}$, let

$$(47)\qquad \begin{array}{l} \mathcal{F}^c_\Omega \equiv \{\Lambda \in \mathcal{F} \mid \Omega \perp \Lambda\} \\ \mathcal{B}^c_\Omega \subseteq \mathcal{B} \text{ be the norm closure of } \bigcup\limits_{\Lambda \in \mathcal{F}^c_\Omega} \mathcal{B}_\Lambda \end{array}$$

$\mathcal{B}^c_\Omega$ is thus the C^*-subalgebra of $\mathcal{B}$ generated by the local observables $\{C \in \mathcal{B}_\Lambda | \Lambda \in \mathcal{F}^c_\Omega\}$. In the same way that $\mathcal{B}_\Omega$ is referred to as the C^*-algebra of local observables relative to Ω, $\mathcal{B}^c_\Omega$ is referred to as the C^*-algebra of quasi-local observables disjoint from Ω.

A state ϕ on $\mathcal{B}$ is said to be *uniformly clustering* if, for every $B \in \mathcal{B}$ and every $\epsilon > 0$, there exists $\Omega(B,\epsilon) \in \mathcal{F}$ such that:

$$(48)\qquad \begin{array}{l} |<\phi; BC> - <\Phi; B><\phi; C>| \leq \epsilon \, \| C \| \\ \text{for every } C \in \mathcal{B}^c_{\Omega(B,\epsilon)} \end{array}$$

Note that this definition does not require any group action to be defined on $\mathcal{B}$; so, even in the case of a quantum lattice on $\mathbb{Z}^n$, where an action of $\mathbb{Z}^n$ is defined, we do not require, yet, that ϕ be $\mathbb{Z}^n$-invariant in (48).

We now want to give a characterization of uniformly clustering states ϕ in terms of a property of the GNS representation $\pi_\phi : \mathcal{B} \to \mathcal{B}(\mathcal{H}_\phi)$ canonically associated to ϕ.

Let first $\pi : \mathcal{B} \to \mathcal{B}(\mathcal{H}_\pi)$ be an arbitrary representation of the C^*-algebra of quasi-local observables for a system where local commutativity holds. We define for π the von Neumann algebras

$$B_\pi(\Omega^c) \equiv \pi(\mathcal{B}_\Omega^c)'' \tag{49}$$

$$B_\pi \equiv \bigcap_{\Omega \in \mathcal{F}} B_\pi(\Omega^c) \tag{50}$$

B_π is referred to as the von Neumann algebra of *observables at infinity* relative to the representation π. When π is the GNS representation associated to a state ϕ on $\mathcal{B}$, we write B_ϕ for B_π and we refer to it as the algebra of observables at infinity relative to ϕ.

Scholium 3. *For every representation $\pi : \mathcal{B} \to \mathcal{B}(\mathcal{H}_\pi)$ of the C^*-algebra of quasi-local observables for a system where local commutativity holds, $B_\pi \subseteq Z_\pi \equiv \pi(\mathcal{B})'' \cap \pi(\mathcal{B})'$.*

Proof: By definition, we have on the one hand

$$B_\pi \subseteq \pi(\mathcal{B}_\Omega^c)'' \quad \forall\ \Omega \in \mathcal{F} \tag{51}$$

and thus, since $\mathcal{B}_\Omega^c \subseteq \mathcal{B}$:

$$B_\pi \subseteq \pi(\mathcal{B})'' \tag{52}$$

On the other hand, we have for every $\Omega \in \mathcal{F}$ and every $\Lambda \in \mathcal{F}_\Omega^c$:

$$\pi(\mathcal{B}_\Lambda) \subseteq \pi(\mathcal{B}_\Omega)' \tag{53}$$

and thus

$$\pi(\bigcup_{\Lambda \in \mathcal{F}_\Omega^c} \mathcal{B}_\Lambda) \subseteq \pi(\mathcal{B}_\Omega)' \tag{54}$$

Since π is continuous for the norm topologies in $\mathcal{B}$ and $\mathcal{B}(\mathcal{H}_\pi)$, and since a von Neumann algebra, here $\pi(\mathcal{B}_\Omega)'$, is closed in the norm topology of $\mathcal{B}(\mathcal{H}_\pi)$, we obtain from (54) and (47):

$$\pi(\mathcal{B}_\Omega^c) \subseteq \pi(\mathcal{B}_\Omega)' \tag{55}$$

and thus

$$B_\pi(\Omega^c) \equiv \pi(\mathcal{B}_\Omega^c)'' \subseteq \pi(\mathcal{B}_\Omega)' \tag{56}$$

Consequently

$$B_\pi \equiv \bigcap_{\Omega \in \mathcal{F}} B_\pi(\Omega^c) \subseteq \bigcap_{\Omega \in \mathcal{F}} \pi(\mathcal{B}_\Omega)' \tag{57}$$

and thus since the RHS of (57) is $\{\pi(\mathcal{B}_\Omega) | \Omega \in \mathcal{F}\}'$:

$$B_\pi \subseteq \pi(\mathcal{B})'. \tag{58}$$

The relations (52) and (58) together imply $B_\pi \subseteq Z_\pi$. □

THEOREM 4. *Let $\mathcal{B}$ be the C^*-algebra of quasi-local observables for a system where local commutativity holds. Then the following conditions on a state ϕ of $\mathcal{B}$ are equivalent: (i) ϕ is uniformly clustering; (ii) the von Neumann algebra $\mathcal{B}_\phi$ of observables at infinity relative to ϕ is trivial (i.e. $\mathcal{B}_\phi = \mathbb{C}I$).*

Proof: We first prove that (i) implies (ii). Let $X \in \mathcal{B}_\phi$ and $A, D \in \mathcal{B}$. Form $B \equiv D^*A$. Since ϕ is supposed to be uniformly clustering, we have, at arbitrary $\epsilon > 0$, that there exists $\Omega \equiv \Omega(B,\epsilon) \in \mathcal{F}$ such that

$$(59) \qquad |<\phi; BC> - <\phi; B><\phi; C>| \leq \epsilon \| C \| \quad \forall\ C \in \mathcal{B}_\Omega^c$$

Since $X \in \mathcal{B}_\phi$, we have in particular $X \in \mathcal{B}_\phi(\Omega^c) \equiv \pi_\phi(\mathcal{B}_\Omega^c)''$. Since $\pi_\phi(\mathcal{B}_\Omega^c)$ is a self-adjoint algebra of operators on $\mathcal{H}_\phi$, dense in $\pi_\phi(\mathcal{B}_\Omega^c)''$ with respect to the strong-operator topology, we know from Kaplanski density theorem (see e.g. Kadison & Ringrose, 1983), that there exists $C \in \mathcal{B}_\Omega^c$ such that

$$(60a) \qquad \| C \| \leq \| X \|$$
$$(60b) \qquad |(\{X - \pi_\phi(C)\}\Phi, \pi_\phi(B)^*\Phi)| \leq \epsilon$$
$$(60c) \qquad |(\{X - \pi_\phi(C)\}\Phi, \Phi)| \cdot |(\pi_\phi(B)\Phi, \Phi)| \leq \epsilon$$

where $\epsilon > 0$ was chosen in (59), and Φ is the cyclic vector associated to ϕ by the GNS construction. From (59) and (60) we obtain

$$|(\pi_\phi(B)X\Phi, \Phi) - (\pi_\phi(B)\Phi, \Phi)(X\Phi, \Phi)| \leq$$

$$|(\{X - \pi_\phi(C)\}\Phi, \pi_\phi(B)^*\Phi) + (\pi_\phi(BC)\Phi, \Phi)$$
$$- (\pi_\phi(B)\Phi, \Phi)(\{X - \pi_\phi(C)\}\Phi, \Phi) - (\pi_\phi(B)\Phi, \Phi)(\pi_\phi(C)\Phi, \Phi)| \leq$$

$$\epsilon(2 + \| C \|) \leq \epsilon(2 + \| X \|).$$

Since the LHS of this inequality is independent of ϵ, we can make the RHS arbitrarily small, and thus conclude:

$$(61) \qquad (\pi_\phi(B)X\Phi, \Phi) - (\pi_\phi(B)\Phi, \Phi)(X\Phi, \Phi) = 0$$

Recall now that $X \in \mathcal{B}_\phi$ and thus (see Scholium 3) $X \in \pi_\phi(\mathcal{B})'$, and that $B \equiv D^*A$ is arbitrary in $\mathcal{B}$. We can, therefore, rewrite (61) in the form

$$(62) \qquad (\{X - (X\Phi, \Phi)I\}\pi_\phi(A)\Phi, \pi_\phi(D)\Phi) = 0 \quad \forall\ A, D \in \mathcal{B}$$

Since Φ is cyclic in $\mathcal{H}_\phi$ for $\pi_\phi(\mathcal{B})$, (62) implies

$$(63) \qquad X = (X\Phi, \Phi)I$$

Since X is arbitrary in $\mathcal{B}_\phi$, we conclude indeed that $\mathcal{B}_\phi = \mathbb{C}I$.

We next prove that (ii) implies (i) by contradiction. Let $\mathcal{B}_\phi = \mathbb{C}I$ and suppose that ϕ were not uniformly clustering. This would imply that there exist: (a) $B \in \mathcal{B}$; (b) an increasing, absorbing, directed subset $\{\alpha\}$ of $\mathcal{F}$; (c) a collection $\{C_\alpha \in \mathcal{B}^c_\alpha \mid \ \| C_\alpha \| = 1\}$; and (d) a positive number $\eta > 0$, such that

$$(64) \qquad | < \phi; BC_\alpha > - < \phi; B > < \phi; C_\alpha > | \geq \eta \quad \forall \ \alpha$$

Since the unit ball in $\mathcal{B}(\mathcal{H}_\phi)$ is compact in the weak-operator topology (see e.g. Kadison & Ringrose, 1983), we can assume without loss of generality that there exists a subnet $\{C_\beta\}$ of $\{C_\alpha\}$ such that $\{\pi_\phi(C_\beta)\}$ converges in the weak-operator topology. Since the net is absorbing, this limit must belong to $\mathcal{B}_\phi$. Since $\mathcal{B}_\phi = \mathbb{C}I$, we must therefore have:

$$(65) \qquad w - \lim_\beta \ \pi_\phi(C_\beta) = \lambda I \quad \text{for some} \quad \lambda \in \mathbb{C}$$

This implies

$$(66) \qquad \begin{aligned} &0 \equiv (\pi_\phi(B)\lambda\Phi, \Phi) - (\pi_\phi(B)\Phi, \Phi)\lambda = \\ &\lim_\beta \{(\pi_\phi(B)\pi_\phi(C_\beta)\Phi, \Phi) - (\pi_\phi(B)\Phi, \Phi)(\pi_\phi(C_\beta)\Phi, \Phi)\} = \\ &\lim_\beta \{< \phi; BC_\beta > - < \phi; B > < \phi; C_\beta >\} \end{aligned}$$

which contradicts (64). Hence, when $\mathcal{B}_\phi = \mathbb{C}I$, our tentative hypothesis, that ϕ be not uniformly clustering, is untenable. This proves that (ii) implies (i). □

To compare this result with that of Theorem 2, notice that Scholium 3 and Theorem 4 together imply that every primary state on $\mathcal{B}$ is uniformly clustering, whenever $\mathcal{B}$ is the C^*-algebra of the quasi-local observables on a system where local commutativity holds. It is, therefore, a natural question to ask whether one could strengthen the conditions on $\mathcal{B}$ in such a manner that the converse implication holds as well. The following result shows that this is indeed possible.

Corollary 5. *Let $\mathcal{B}$ be the C^*-algebra of quasi-local observables of a quantum lattice on $\mathbb{Z}^n$, and assume that for every $\omega \in \mathbb{Z}^n$, $\mathcal{B}_\omega \cong \mathcal{M}(m, \mathbb{C})$ the algebra of all $m \times m$ matrices with complex entries ($m < \infty$!). Then the following conditions on a state ϕ on $\mathcal{B}$ are equivalent: (i) ϕ is uniformly clustering; and (ii) ϕ is primary* (i.e. $Z_\phi \equiv \pi_\phi(\mathcal{B})'' \cap \pi_\phi(\mathcal{B})' = \mathbb{C}I$).

Proof: $\mathcal{B}$ is the C^*-algebra of the quasi-local observables on a system where local commutativity holds; it is therefore sufficient, by Theorem 4, to show that $Z_\phi = \mathcal{B}_\phi$. We already know from Scholium 3 that $\mathcal{B}_\phi \subseteq Z_\phi$; it is therefore sufficient to show that $Z_\phi \subseteq \mathcal{B}_\phi$, which we now prove. Let $\pi : \mathcal{B} \to \mathcal{B}(\mathcal{H})$ be an arbitrary representation of $\mathcal{B}$, and let $Z \in Z_\pi \equiv \pi(\mathcal{B})'' \cap \pi(\mathcal{B})'$.

Since $Z \in \pi(\mathcal{B})''$ there exists a net $\{B_\alpha\}$ of elements $B_\alpha \in \mathcal{B}$ such that $\{\pi(B_\alpha)\}$ converges to Z in the strong-operator topology; since $\mathcal{B}$ is the closure, in the norm topology, of the union over $\mathcal{F}$ of the local algebras $\mathcal{B}_\Omega$, we can require that each B_α is a strictly local observable. Hence, we can assume, without loss of generality that, at fixed $\Omega \in \mathcal{F}$, there exist: (a) an increasing, absorbing, directed subset $\{\alpha\}$ of $\mathcal{F}$; and (b) a collection $\{B_\alpha \in \mathcal{B}_\alpha\}$ such that:

$$\Omega \subseteq \alpha \quad \forall\ \alpha \tag{67a}$$

$$s - \lim_\alpha \ \pi(B_\alpha) = Z \tag{67b}$$

We now use the fact that $\mathcal{B}_\Omega \cong \mathcal{M}(m^{|\Omega|}, \mathbb{C})$ where $|\Omega|$ is the number of sites $\omega \in \mathbb{Z}^n$ belonging to the finite subset Ω. There exists then a *finite* collection $\{e_{jk} | 1 \leq j, k \leq m^{|\Omega|}\}$ of matrices $e_{jk} \in \mathcal{B}_\Omega$ such that

$$e_{jk} \cdot e_{ln} = \delta_{kl} e_{jn} \quad \text{and} \quad e^*_{jk} = e_{kj}, \tag{68a}$$

$$\sum_j e_{jj} = I \quad \text{and} \quad \mathcal{B}_\Omega = \{\sum_{jk} b_{jk} e_{jk} \mid b_{jk} \in \mathbb{C}\} \tag{68b}$$

We have then

$$Z = ZI = Z\pi(\sum_j e_{jj}) = Z \sum_j \pi(e_{jk})\pi(e_{kj}) \tag{69}$$

Since $Z \in \pi(\mathcal{B})'$, we obtain from (67b) and (69):

$$Z = s - \lim_\alpha \ \pi(\sum_j e_{jk} B_\alpha e_{kj}) \tag{70}$$

For every α, we have because of (67a):

$$\sum_j e_{jk} B_\alpha e_{kj} \ \in \ \mathcal{B}_\alpha \tag{71}$$

Moreover, a straightforward computation using (68) shows that for every $B \in \mathcal{B}_\Omega$

$$[B, \sum_j e_{jk} B_\alpha e_{kj}] = 0 \tag{72}$$

Upon combining (71) and (72), we obtain

$$\sum_j e_{jk} B_\alpha e_{kj} \ \in \ \mathcal{B}_{\alpha \backslash \Omega} \subset \mathcal{B}^c_\Omega \tag{73}$$

and thus

$$\pi(\sum_j e_{jk} B_\alpha e_{kj}) \ \in \ \mathcal{B}_\pi(\Omega^c) \tag{74}$$

Since $\{\alpha\} \subset \mathcal{F}$ is absorbing, we conclude from (70) and (74) that:

$$Z \in \bigcap_{\Omega \in \mathcal{F}} \mathcal{B}_\pi(\Omega^c) \equiv \mathcal{B}_\pi \ ,$$

i.e. since Z was arbitrary in $\mathcal{Z}_\pi : \mathcal{Z}_\pi \subseteq \mathcal{B}_\pi$. As we saw in the beginning of the proof, this suffices to prove the corollary. □

A few bibliographical notes will help situate these results in a larger picture.

The concept of the algebra of observables at infinity $\mathcal{B}_\pi$, as in (50), was introduced by Lanford & Ruelle (1969); see also Ruelle (1970). The proof of $\mathcal{Z}_\pi = \mathcal{B}_\pi$, which is at the heart of Scholium 5 is essentially theirs; see also Powers (1967). So is Theorem 4.

For other quasi-local algebras which admit a reasonably general class of states for which $\mathcal{Z}_\pi = \mathcal{B}_\phi$, see e.g. Dell'Antonio, Doplicher & Ruelle (1966) and Ruelle (1967).

A sharpening of the cluster property (48) appearing in Theorem 4, valid for every primary state ϕ on a C^*-algebra $\mathcal{B}$ which is the uniform limit of an increasing net $\{\mathcal{B}_\alpha\}$ of *-subalgebras $\mathcal{B}_\alpha$ of $\mathcal{B}$, has been obtained by Araki & Kishimoto (1976), namely for every $B \in \mathcal{B}$ and every $\epsilon > 0$, there exists α such that

$$\begin{aligned} &| < \phi; BC > - < \phi; B >< \phi; C > | \leq \\ &\epsilon\{< \phi; C^*C > + < \phi; CC^* >\}^{1/2} \\ &\text{for all } C \in \mathcal{B} \text{ such that } [C, A] = 0 \ \forall \ A \in \mathcal{B}_\alpha \end{aligned} \tag{75}$$

The notion of uniform clustering itself is the generalization to the realm of non-commutative C^*-algebras of an idea that originated in study of the K-cascades of classical ergodic theory; see e.g. Sinai (1962, 1963). We shall say more on the non-commutative generalizations of Kolmogorov flows in Section One of Chapter Ten.

Note that every state of the form (31) is uniformly clustering. Let indeed $B \in \mathcal{B}$ and $\epsilon > 0$; there exist then $\Omega \in \mathcal{F}$ and $B_0 \in \mathcal{B}_\Omega$ such that $\| B - B_0 \| \leq \epsilon/2$; for every $C \in \mathcal{B}_\Omega^c$, we have $< \phi; B_0 C > = < \phi; B_0 >< \phi; C >$ and thus: $| < \phi; BC > - < \phi; B >< \phi; C > | = | < \phi; (B - B_0)C > - < \phi; B - B_0 >< \phi; C > | \leq 2 \| B - B_0 \| \cdot \| C \| \leq \epsilon \| C \|$, which is (48). When $\mathcal{B}_\omega \cong \mathcal{M}(m, \mathbb{C})$, Scholium 5 ensures that these states are primary. They provide, therefore, very special, but very explicit, examples of states that satisfy the assumptions of Theorem 2. These states, in fact, can be used to produce an interesting variety of "matricial factors", which we now briefly describe.

Recall from Subsections 8.3.c and d, that a von Neumann algebra $\mathcal{N}$ is generated by the collection $\mathcal{P} \equiv \{P \in \mathcal{N} | P^* = P = P^2\}$ of its projectors; and that $\mathcal{P}$ comes naturally equipped with the structure of an orthocomplemented lattice. We now say that two elements P and Q in $\mathcal{P}$ are *equivalent*, a relation which we denote $P \sim Q$, if there exists an element U in $\mathcal{N}$ such that $UU^* = P$ and $U^*U = Q$. An element $P \in \mathcal{P}$ is said to be *finite* if $Q \in \mathcal{P}$, $Q \subseteq P$ and

$Q \sim P$ imply $Q = P$. An element $P \in \mathcal{P}$ is said to be *infinite* if it is not finite. With Q and P in $\mathcal{P}$, we write $Q \leq P$ if there exists $R \in \mathcal{P}$ such that $R \sim Q$ and $R \subseteq P$. We further write $Q < P$ when $Q \leq P$ and Q is not equivalent to P. The relation $\leq$ has the following remarkable property (see e.g. Dixmier, 1957: III.1.2, Theorem 1): Given P and $Q \in \mathcal{P}$, there exists $R \in \mathcal{P} \cap \mathcal{N}'$ such that $PR \leq QR$ and $Q(I - R) \leq P(I - R)$. Hence, if $\mathcal{N}$ is a *factor*, i.e. if $\mathcal{N}$ is a von Neumann algebra satisfying $\mathcal{N} \cap \mathcal{N}' = \mathbb{C}I$, any two projectors P and Q in $\mathcal{N}$ satisfy either $P \leq Q$ or $Q \leq P$. This is the key for the generalization, to the lattice $\mathcal{P}$ of the projectors of a factor $\mathcal{N}$, of the notion of dimension familiar for the projectors in $\mathcal{B}(\mathcal{H})$. For this, we need a generalization of the notion of trace; compare with formulas 8.3.4–6 and 8.3.102.

Let $\mathcal{N}$ be an arbitrary von Neumann algebra, and $\mathcal{N}^+$ be the cone of its positive elements (see Scholium 9.1.1). A map $\psi : \mathcal{N}^+ \to [0, +\infty]$ is said to be a *trace* for $\mathcal{N}$ if for every $A, B \in \mathcal{N}^+$, every $C \in \mathcal{N}$ and every $\lambda \in \mathbb{R}^+$:

(76a) $$\psi(A + B) = \psi(A) + \psi(B)$$
(76b) $$\psi(\lambda A) = \lambda \psi(A)$$
(76c) $$\psi(CC^*) = \psi(C^*C)$$

A trace ψ is said to be *faithful* if $A \in \mathcal{N}^+$ and $\psi(A) = 0$ imply $A = 0$. It is said to be *normal* if, for every increasing net $\{A_\alpha\} \subset \mathcal{N}^+$ with upper bound $A \in \mathcal{N}^+$, $\psi(A)$ is the upper bound of $\{\psi(A_\alpha)\}$. A trace ψ is said to be *semi-finite* if, for every $A \in \mathcal{N}^+$ with $A \neq 0$, there exists $B \in \mathcal{N}^+$ with $A - B \in \mathcal{N}^+$, $B \neq 0$ and $\psi(B) < \infty$. Another remarkable property of factors which will bear on the forthcoming definition of a dimension-function is (see Dixmier, 1957: I.6.4, Cor. to Thm. 3): any two semi-finite, normal faithful traces on a factor, if they exist, are necessarily proportional.

A von Neumann algebra $\mathcal{N}$ is said to be *semi-finite* if, for every $A \in \mathcal{N}^+$ with $A \neq 0$, there exists a semi-finite normal trace ψ for $\mathcal{N}$ such that $\psi(A) \neq 0$; it is then possible (see Dixmier, 1957: I.6.7, Prop. 9) to find a trace ψ for such an algebra $\mathcal{N}$, with ψ semi-finite, normal and faithful. For instance, $\mathcal{B}(\mathcal{H})$ (with $\mathcal{H}$ separable) is semi-finite, and $\psi : A \in \mathcal{B}(\mathcal{H})^+ \mapsto tr\,A \in [0, +\infty]$ is a semi-finite, normal, faithful trace. Clearly the restriction of this trace to $\mathcal{P}(\mathcal{H})$ coincides with the usual concept of dimension, which we now are ready to generalize to an arbitrary factor $\mathcal{N}$.

A *dimension-function* for a factor $\mathcal{N}$ is the restriction d to $\mathcal{P} \equiv \{P \in \mathcal{N} | P^* = P = P^2\}$ of a faithful normal trace ψ for $\mathcal{N}$, with ψ supposed to be semi-finite whenever $\mathcal{N}$ is semi-finite. The last restriction in this definition ensures that a dimension-function is uniquely defined up to a multiplicative constant $0 < \lambda < \infty$; indeed either: $\mathcal{N}$ is a semi-finite factor, and we already mentioned that two faithful normal traces which are semi-finite must then be proportional; or: $< \psi; A >= \infty \;\; \forall \;\; A \in \mathcal{N}^+$ with $A \neq 0$ (see e.g. Dixmier, 1957: I.6.7, Prop. 8). The reader will verify that d satisfies all the properties one could reasonably expect for a measure of the "dimension" of projectors.

We have in particular:

(77a) $d(P) \geq 0 \;\; \forall\; P \in \mathcal{P}$, and $d(P) = 0$ iff $P = 0$
(77b) $P, Q \in \mathcal{P}$ with $P \perp Q$ implies $d(P+Q) = d(P) + d(Q)$
(77c) $P, Q \in \mathcal{P}$ with $P \sim Q$ implies $d(P) = d(Q)$
(77d) $P \in \mathcal{P}$ then $d(P) = \infty$ iff P is infinite
(77e) $P, Q \in \mathcal{P}$ with $P < Q$ and P finite imply $d(P) < d(Q)$

Murray & von Neumann (1936), who introduced this notion of dimension, showed that factors can be classified according to the range of their dimension-functions d. Indeed, the following five, disjoint, cases do occur. A factor is said to be of *type* I_n (resp. I_∞, or II_1) if one can, by a suitable choice of a multiplicative constant, make $d(\mathcal{P})$ coincide with $\{0, 1, 2, \ldots, n\}$ (resp. $\mathbb{Z}^+$, or $[0, 1]$). It is said to be of *type* II_∞ (resp. III) if $d(\mathcal{P}) = [0, +\infty]$ (resp. $\{0, +\infty\}$). The following nomenclature is used to characterize the factors: *discrete* (I_n, I_∞) vs. *continuous* $(II_1, II_\infty$, and $III)$; *finite* (I_n, II_1) vs. *infinite* $(I_\infty, II_\infty$, and $III)$; *semi-finite* $(I_n, I_\infty, II_1, II_\infty)$ vs. *purely infinite* (III).

Murray & von Neumann (1943) introduced further the concept of an *approximately finite* factor, as one which is obtained as the strong-operator closure of an increasing sequence of finite, type I factors; the words *hyperfinite*, or *matricial* are used interchangeably today to refer to these factors. In the same paper, Murray & von Neumann proved that all approximately finite factors of type II_1 are isomorphic.

In sharp contrast with this result, Powers (1967) constructed—from his version of Scholium 5 above—a continuum(!) of non-isomorphic matricial factors of type III. Consider indeed the most simple case where $\mathcal{B}$ is the C^*-algebra of quasi-local observables for a quantum lattice on $\mathbb{Z}$, with $\mathcal{B}_\omega \cong \mathcal{M}(2, \mathbb{C})$ and ϕ is of the form (31) with $\phi_\omega \cong \psi$. We already saw that ϕ is uniformly clustering and thus, by Scholium 5, ϕ is primary, i.e. $\pi_\phi(\mathcal{B})''$ is a factor. This factor is matricial since it is the strong-operator closure of a representation $\pi_\phi(\mathcal{B})$ of a C^*-algebra $\mathcal{B}$ which is the uniform closure of an increasing sequence of finite, type I factors, namely $\{\mathcal{B}_{[-n,n]} \cong \mathcal{M}(2^{(2n+1)}, \mathbb{C}) | n \in \mathbb{Z}\}$. Note in passing, that $\mathcal{B}$ is thus an example of what Glimm (1960) called UHF-algebras, namely of a C^*-algebra which is the uniform closure of an increasing sequence $\{\mathcal{N}_{m(n)} | n \in \mathbb{Z}^+\}$ of factors $\mathcal{N}_{m(n)}$ of type $I_{m(n)}$ with $m(n)$ finite and $m(n) \to \infty$ as $n \to \infty$. The representations of UHF-algebras were studied by Glimm (1961) and by Powers (1967); see also Pukanszky (1956) for some useful preliminary results. As a result of these papers, one can be very precise on the type of the matricial factors $\pi_\phi(\mathcal{B})''$ as ψ varies over the possible states of $\mathcal{M}(2, \mathbb{C})$. If ψ is pure, so is ϕ and $\pi_\phi(\mathcal{B})$ is irreducible; since $\mathcal{H}_\phi$ is infinite-dimensional, $\pi_\phi(\mathcal{B})''$ is a type I_∞ factor. If ψ is the normalized trace on $\mathcal{M}(2, \mathbb{C})$, $\pi_\phi(\mathcal{B})''$ is the unique matricial type II_1 factor of Murray & von Neumann. In all other cases, one can assume without loss of generality that $\psi : B_\omega \in \mathcal{M}(2, \mathbb{C}) \mapsto tr\, \rho B_\omega \in \mathbb{C}$ with $\rho = \exp(-\beta\sigma)/tr\{\exp(-\beta\sigma)\}$ with $\beta \in (0, \infty)$ and σ is any Hermitian non-degenerate matrix, e.g. assume $tr\, \sigma = 0$ and $\sigma^2 = I$ so that σ is a Pauli matrix

(8.3.156). Glimm (1961) had used a result of Pukanski (1956) to show that, for every $\beta \in (0, \infty)$, the matricial factor $\pi_\phi(\mathcal{B})''$ is of type III. Powers (1967) then showed that any two factors obtained in this way are non-isomorphic as soon as they come from two different β, thus establishing indeed the existence of uncountably many non-isomorphic matricial factors of type III.

As we shall see in Section One of Chapter Ten, type III factors appear, in quantum statistical mechanics, under rather general circumstances from the GNS representations associated to canonical equilibrium states at finite temperatures. The above Glimm-Powers results will thus turn out to be the most simple, but genuine, examples of this phenomenon.

c. Averages over group actions.

The purpose of this subsection is to define the average of states and of observables with respect to group actions, and to study their properties. For this purpose, we need the notion of an amenable group; there are several equivalent definitions of this condition in the literature, and we present here the definition which is most directly relevant to our purpose. Moreover, throughout this subsection, we will restrict our attention to locally compact groups; this restriction, although mathematically non-trivial, is natural in the context of this book, since many of the groups encountered in physics either are discrete, or are Lie groups, so that they are locally compact. Locally compact amenable groups have been extensively studied; see e.g. Day (1957), Hewitt & Ross (1963), Pier (1965), Effros & Hahn (1968), or Greenleaf (1969); for a brief review, see also Emch (1972).

Let G be a locally compact group, and $\mathcal{C}(G)$ be the C^*-algebra of all bounded, continuous functions $f : G \to \mathbb{C}$, equipped with its usual point-wise composition laws and its sup-norm. A left action $g[.] : f \in \mathcal{C}(G) \mapsto g[f] \in \mathcal{C}(G)$, and a right action $[.]g : f \in \mathcal{C}(G) \mapsto [f]g \in \mathcal{C}(G)$ of G on $\mathcal{C}(G)$ are defined by:

$$(78) \qquad g[f] : h \in G \mapsto f(g^{-1}h) \in \mathbb{C} \ ; \ [f]g : h \in G \mapsto f(hg) \in \mathbb{C}$$

Now, G is said to be *amenable* whenever $\mathcal{C}(G)$ admits at least one state η which is invariant under the left and right actions of G on $\mathcal{C}(G)$ just defined. When such a G-invariant state on $\mathcal{C}(G)$ exists, it is said to be an *invariant mean* on G.

Clearly, every compact group is amenable: the normalized Haar measure provides the desired mean. One can also prove (see Dixmier, 1950; Greenleaf, 1959; or for an elementary presentation of the argument: Emch, 1972) that every abelian group is amenable. Note also that the class $\mathcal{G}$ of all locally compact amenable groups enjoys the following properties: if $G \in \mathcal{G}$, then every closed subgroup $H \subset G$ is also amenable; if moreover H is normal in G, then G/H is again amenable. Conversely if G is locally compact, and if H is a closed normal subgroup of G such that both H and G/H are amenable, then so is G. Hence, $\mathbb{R}^n$, O^3_+, E^3 are all amenable.

There are, however, locally compact groups of relevance to physics which are *not* amenable: no non-compact, semi-simple Lie group is amenable; in particular, the (homogeneous) Lorentz group, and the (inhomogeneous) Poincaré group of special relativity are not amenable.

Scholium 6. *Let $\alpha : G \to \mathrm{Aut}(\mathcal{B})$ be an action of an amenable group G on a C^*-algebra $\mathcal{B}$ with unit. Then, for every state ϕ on $\mathcal{B}$, and every invariant mean η on G,*

$$\eta\phi : B \in \mathcal{B} \mapsto \eta < \phi; \alpha(\cdot)[B] > \in \mathbb{C} \tag{79}$$

is a G-invariant state on $\mathcal{B}$.

Proof: By definition of an action α of G on $\mathcal{B}$:

$$g \in G \mapsto < \phi; \alpha(g)[B] > \in \mathbb{C} \tag{80}$$

is a continuous function on G, i.e. belongs to $\mathcal{C}(G)$. Hence, the RHS of (79) is well-defined. Clearly $\eta\phi$ is linear. If $B \in \mathcal{B}$ is positive, so is $\alpha(g)[B]$ for all $g \in G$; hence, in this case, the function (80) is positive; since η is a state on $\mathcal{C}(G)$, we have then that the RHS of (79) is a positive number for every positive $B \in \mathcal{B}$, i.e. $\eta\phi$ is positive. Moreover, if $B = I$, the function (80) is the constant function 1, and thus $< \eta\phi; I >= 1$. Hence $\eta\phi$ is a state on $\mathcal{B}$. Finally, since η is an invariant mean on $G, \eta\phi$ is G-invariant. □

Note in particular that the assumptions of the scholium ensures that $\mathcal{S}_G^*$ is not empty. Clearly every G-invariant state on $\mathcal{B}$ can then be obtained in this way, i.e. $\eta(\mathcal{S}^*) \equiv \mathcal{S}_G^*$. The G-invariant state $\eta\phi$ is said to be the *average of the state* ϕ with respect to the group action α, and the invariant mean η.

Proposition 7. *Let $\alpha : G \to \mathrm{Aut}(\mathcal{B})$ be an action of an amenable group G on a C^*-algebra $\mathcal{B}$ (with unit). Let ϕ be a G-invariant state on $\mathcal{B}$, and $\pi_\phi : \mathcal{B} \to \mathcal{B}(\mathcal{H}_\phi)$ be the GNS representation of $\mathcal{B}$ associated to ϕ, with cyclic vector $\Phi \equiv \Phi(I)$. Then*

$$\pi_\phi(B)\Phi \mapsto \pi_\phi\{\alpha(g)[B]\}\Phi \tag{81}$$

extends to a continuous unitary representation $U_\phi : G \to \mathcal{U}(\mathcal{H}_\phi)$ such that, $\forall\ g \in G$ and $\forall\ B \in \mathcal{B}$:

$$\pi_\phi\{\alpha(g)[B]\} = U_\phi(g)\pi_\phi(B)U_\phi(g)^*. \tag{82}$$

For every invariant mean η on G, there exists a linear, *-homomorphism:

$$\eta_\phi : B \in \mathcal{B} \mapsto \eta_\phi(B) \in \pi_\phi(\mathcal{B})'' \cap U_\phi(G)' \tag{83}$$

such that, $\forall\ \Psi_1$ and $\Psi_2 \in \mathcal{H}_\phi$, and $\forall\ B \in \mathcal{B}$:

$$(\eta_\phi(B)\Psi_1, \Psi_2) = \eta\,(\pi_\phi\{\alpha(\cdot)[B]\}\Psi_1, \Psi_2) \tag{84}$$

$$\eta_\phi(B)E_\phi = E_\phi\eta_\phi(B) = E_\phi\pi_\phi(B)E_\phi \tag{85}$$

where E_ϕ is the projector on the subspace

(86) $$\mathcal{E}_\phi \equiv \{\Psi \in \mathcal{H}_\phi | U_\phi(g)\Psi = \Psi \ \forall g \in G\}.$$

Proof: The notation established in the proof of Theorem 9.1.5 is used systematically. Since now ϕ is G-invariant, $\mathcal{K}_\phi$ is stable under the action of $\alpha(g)$ for every g in G. Consequently

(87) $$\pi_\phi(B)\Phi \equiv \Phi(B) \mapsto \Phi(\alpha(g)[B]) \equiv \pi_\phi\{\alpha(g)[B]\}\Phi ,$$

which is (81), is well-defined as a linear map of $\mathcal{B}/\mathcal{K}_\phi$ onto itself. Moreover, this map is isometric; indeed:

(88) $$\| \Phi(B) \|^2 = < \phi; B^*B > = < \phi; \alpha(g)[B^*B] > = \| \Phi(\alpha(g)[B]) \|^2$$

and, therefore, extends uniquely, by continuity, to a unitary operator $U_\phi(g)$ acting on $\mathcal{H}_\phi$. This operator satisfies

(89) $$\begin{aligned} &U_\phi(g)\pi_\phi(B)\Phi(C) = U_\phi(g)\Phi(BC) = \\ &\Phi(\alpha(g)[BC]) = \pi_\phi\{\alpha(g)[B]\}\Phi(\alpha(g)[C]) = \\ &\pi_\phi\{\alpha(g)[B]\}U_\phi(g)\Phi(C) \end{aligned}$$

Since $\{\Phi(C)|C \in \mathcal{B}\} \equiv \mathcal{B}/\mathcal{K}_\phi$ is dense in $\mathcal{H}_\phi$, this establishes (82). Clearly, $U_\phi(g_1)U_\phi(g_2) = U_\phi(g_1 g_2)$ for all $g_1, g_1 \in G$; and $U_\phi(g)$ is continuous in g for the weak-operator topology. Hence $\{U_\phi(g)|g \in G\}$ is indeed a unitary representation of G on $\mathcal{H}_\phi$. This establishes the first part of the Proposition. Notice now, by polarization, that for every $B \in \mathcal{B}$, and every $\Psi_1, \Psi_2 \in \mathcal{H}_\phi$:

(90) $$g \in G \mapsto (\pi_\phi\{\alpha(g)[B]\}\Psi_1, \Psi_2) \in \mathbb{C}$$

is continuous; it is also bounded, namely by $\| B \| \cdot \| \Psi_1 \| \cdot \| \Psi_1 \|$. Hence the RHS of (84) is well-defined, and extends uniquely by continuity to a bounded sesquilinear form on $\mathcal{H}_\phi$. By the Riesz representation theorem, (84) defines then $\eta_\phi(B) \in \mathcal{B}(\mathcal{H}_\phi)$. Straightforward computations show that $B \mapsto \eta_\phi(B)$ is a linear *-homomorphism. Since η is an invariant mean, we conclude from (84) and (82) that

(91) $$U_\phi(g)\eta_\phi(B)U_\phi(g)^* = \eta_\phi(B) \quad \forall \ g \in G$$

and thus

(92) $$\eta_\phi(B) \in U_\phi(G)'$$

Furthermore, for every $X \in \pi_\phi(\mathcal{B})'$, and every $\Psi_1, \Psi_2 \in \mathcal{H}_\phi$, we have

(93) $$\begin{aligned} &(\eta_\phi(B)X\Psi_1, \Psi_2) = \eta\,(\pi_\phi\{\alpha(\cdot)[B]\}X\Psi_1, \Psi_2) = \\ &\eta\,(\pi_\phi\{\alpha(\cdot)[B]\}\Psi_1, X^*\Psi_2) = (X\eta_\phi(B)\Psi_1, \Psi_2) \end{aligned}$$

and thus

$$\eta_\phi(B)X = X\eta_\phi(B) \quad \forall\ X \in \pi_\phi(\mathcal{B})' \tag{94}$$

i.e.

$$\eta_\phi(B) \in \pi_\phi(\mathcal{B})'' \tag{95}$$

The conclusions (92) and (95) establish (83). Finally, from the definition (86) of E_ϕ as an eigenprojector of $U_\phi(G)$, we have $E_\phi \in U_\phi(G)''$. Together with (92) this implies that

$$\eta_\phi(B)E_\phi = E_\phi\eta_\phi(B) = E_\phi\eta_\phi(B)E_\phi \tag{96}$$

We now compute, for arbitrary $\Psi_1, \Psi_2 \in \mathcal{H}_\phi$:

$$\begin{aligned} &(E_\phi\eta_\phi(B)E_\phi\Psi_1, \Psi_2) = \\ &\eta\left(\pi_\phi\{\alpha(\cdot)[B]\}E_\phi\Psi_1, E_\phi\Psi_2\right) = \\ &\eta\left(\pi_\phi(B)U_\phi(\cdot)^*E_\phi\Psi_1, U_\phi(\cdot)^*E_\phi\Psi_2\right) = \\ &\eta\left(\pi_\phi(B)E_\phi\Psi_1, E_\phi\Psi_2\right) = (E_\phi\pi_\phi(B)E_\phi\Psi_1, \Psi_2) \end{aligned} \tag{97}$$

and thus

$$E_\phi\eta_\phi(B)E_\phi = E_\phi\pi_\phi(B)E_\phi \tag{98}$$

From (96) and (98), we conclude that (85) holds true. □

We say that the self-adjoint operator $\eta_\phi(A) \in \pi_\phi(\mathcal{B})'' \cap U_\phi(G)'$ is the *average of the observable* $A = A^* \in \mathcal{B}$, with respect to: the group action α, the G-invariant state ϕ, and the invariant mean η on G. While it is true, in general, that $\eta_\phi(A)$ may depend on the choice of the mean η used to define it, there are nevertheless cases, of importance for quantum statistical mechanics, when $\eta_\phi : B \in \mathcal{B} \to \eta_\phi(B) \in \pi_\phi(\mathcal{B})'' \cap U_\phi(G)'$ is in fact independent of η. The following result provides such an example, and we shall see some more, later on in this subsection.

Corollary 8. *With the assumption of Proposition 7, suppose further that* Φ *is cyclic in* $\mathcal{H}_\phi$ *with respect to* $\pi_\phi(\mathcal{B})'$. *Then, for every invariant mean* ς *on* $G : \varsigma_\phi(B) = \eta_\phi(B)$ *for every* B *in* $\mathcal{B}$.

Proof: From (81) and (86) we have $E_\phi\Phi = \Phi$. Together with (85) this gives

$$\{\eta_\phi(B) - \varsigma_\phi(B)\}\Phi = E_\phi\{\pi_\phi(B) - \pi_\phi(B)\}E_\phi\Phi = 0 \tag{99}$$

Now, Φ cyclic with respect to $\pi_\phi(\mathcal{B})'$ implies Φ separating for $\pi_\phi(\mathcal{B})''$, i.e. with $X, Y \in \pi_\phi(\mathcal{B})''$: $(X - Y)\Phi = 0$ if and only if $X = Y$. Since $\eta_\phi(B)$ and $\varsigma_\phi(B)$ belong to $\pi_\phi(\mathcal{B})''$ (see 83), the above relation (99) thus implies $\eta_\phi(B) = \varsigma_\phi(B)$. □

The first part of Proposition 7 suggests the introduction of the notion of a *covariant representation* of a group action $\alpha : G \to \mathrm{Aut}(\mathcal{B})$ to be defined as a couple $\{\pi, U\}$ formed by a representation $\pi : \mathcal{B} \to \mathcal{B}(\mathcal{H})$ of the C^*-algebra $\mathcal{B}$, and a representation $U : G \to \mathcal{U}(\mathcal{H})$ of the topological group G satisfying

$$\pi\{\alpha(g)[B]\} = U(g)\pi(B)U(g)^* \;\; \forall\, (g, B) \in G \times \mathcal{B} \tag{100}$$

The reader will verify that the proof of the second part of Proposition 7 only depends of the structure that $\{\pi_\phi, U_\phi\}$ has as a covariant representation of the action of an amenable group on a C^*-algebra.

The structure of covariant representation $\{\pi, U\}$ of a group action $\alpha : G \to \mathrm{Aut}(\mathcal{B})$ is also all one needs to extend α to a group action $\alpha_\pi : G \to \mathrm{Aut}\{\pi(\mathcal{B})''\}$. Indeed since the von Neumann algebra $\pi(\mathcal{B})''$ is the closure of $\pi(\mathcal{B})$ in topologies for which the product of operators in $\mathcal{B}(\mathcal{H})$ is separately continuous, we have that, for every g in G:

$$\alpha_\pi(g) : X \in \pi(\mathcal{B})'' \mapsto U(g)XU(g)^* \in \pi(\mathcal{B})''. \tag{101a}$$

Similarly, every state ϕ on $\mathcal{B}$ that can be written in the form

$$\phi : B \in \mathcal{B} \mapsto \sum_k (\pi(B)\Psi_k, \Psi_k) \text{ with } \sum_k \| \Psi_k \|^2 = 1 \tag{101b}$$

can be uniquely extended to a normal state $\tilde{\phi}$ on $\pi(\mathcal{B})''$; and every normal state $\tilde{\phi}$ on $\pi(\mathcal{B})''$ can be obtained in this way, namely as the unique normal extension of its restriction ϕ to $\pi(\mathcal{B})$. A normal state $\tilde{\phi}$ on $\pi(\mathcal{B})''$ is then said to be G-invariant, with respect to the action $\alpha : G \to \mathrm{Aut}\,(\mathcal{B})$, whenever $\tilde{\phi} \circ \alpha_\pi(g) = \tilde{\phi} \;\forall\; g \in G$.

The following theorem and its corollaries give, under successively sharper conditions, some characterizations of extremal G-invariant states and of the associated averaged obervables.

Theorem 9. *Let $\alpha : G \to \mathrm{Aut}\,(\mathcal{B})$ be an action of an amenable group G on a C^*-algebra $\mathcal{B}$ (with unit). Let ϕ be a G-invariant state on $\mathcal{B}$; η be an invariant mean on G; π_ϕ be the GNS representation of $\mathcal{B}$ associated to ϕ; U_ϕ, E_ϕ and η_ϕ be as in Proposition 7. Then the implication-scheme*

$$\begin{array}{ccccccccc} & & (ix) & \Leftrightarrow & (ii) & \Leftarrow & (i) & \Rightarrow & (iv) \\ & & & & \Downarrow & & & & \Downarrow \\ (viii) & \Leftarrow & (vii) & \Leftarrow & (vi) & \Leftrightarrow & (iii) & \Leftarrow & (v) \end{array}$$

gives the logical relations between the following nine conditions

(*i*) $\eta_\phi(B) = <\phi; B> I \quad \forall\ B \in \mathcal{B}$
(*ii*) $\eta <\phi; \alpha(\cdot)[B]C> = <\phi; B><\phi; C> \quad \forall\ B, C \in \mathcal{B}$
(*iii*) ϕ *is extremal G-invariant*
(*iv*) $\tilde{\phi}$ *is the only G-invariant normal state on* $\pi_\phi(\mathcal{B})''$
(*v*) ϕ *is the only G-invariant vector state on* $\pi_\phi(\mathcal{B})$
(*vi*) $\pi_\phi(\mathcal{B})' \cap U_\phi(G)' = \mathbb{C}I$
(*vii*) $\{\pi_\phi(\mathcal{B}), U_\phi(G)\}'' \cap \pi_\phi(\mathcal{B})' \cap U_\phi(G)' = \mathbb{C}I$
(*viii*) $\pi_\phi(\mathcal{B})'' \cap \pi_\phi(\mathcal{B})' \cap U_\phi(G)' = \mathbb{C}I$
(*ix*) $\dim\ E_\phi = 1$.

Proof: From set-theoretical considerations alone, we see immediately that (vii) implies (viii), and that (vi) implies (vii). From the definition (84) of η_ϕ, it is clear that (i) implies (ii). Since a G-invariant vector state ψ on $\pi_\phi(\mathcal{B})$ extends uniquely to a G-invariant normal state $\tilde{\psi}$ on $\pi_\phi(\mathcal{B})''$, we obtain that (iv) implies (v).

To prove that (i) implies (iv), recall that every normal state $\tilde{\psi}$ on the von Neumann algebra $\pi_\phi(\mathcal{B})''$ can be written in the form

(102) $$\tilde{\psi} : X \in \pi_\phi(\mathcal{B})'' \mapsto \sum_i (X\Psi_i, \Psi_i) \text{ with } \sum_i \| \Psi_i \|^2 = 1$$

In particular, we have for the restriction ψ to $\pi_\phi(\mathcal{B})$ of any G-invariant normal state $\tilde{\psi}$ on $\pi_\phi(\mathcal{B})''$:

(103) $$<\psi; B> = \eta <\psi; B> = \sum_i (\eta_\phi(B)\Psi_i, \Psi_i).$$

Hence, when (i) holds (103) reduces to $<\psi; B> = <\phi; B> \quad \forall\ B \in \mathcal{B}$, and thus by continuity $\tilde{\psi} = \tilde{\phi}$. Hence (i) indeed implies (iv).

To see that (ii) is equivalent to (ix), we use Proposition 7 to compute the LHS of condition (ii):

(104) $$\begin{aligned} &\eta <\phi; \alpha(\cdot)[B]C> = \eta\, (\pi_\phi\{\alpha(\cdot)[B]\}\pi_\phi(C)\Phi, \Phi) = \\ &(\eta_\phi(B)\pi_\phi(C)\Phi, E_\phi\Phi) = (\pi_\phi(B)E_\phi\pi_\phi(C)\Phi, \Phi) \end{aligned}$$

The RHS of condition (ii) can be rewritten as

(105a) $$<\phi; B><\phi; C> = (\pi_\phi(B)P_\Phi\pi_\phi(C)\Phi, \Phi)$$

where P_Φ is the projector

(105b) $$P_\Phi : \Psi \in \mathcal{H}_\phi \mapsto (\Psi, \Phi)\Phi \in \mathcal{H}_\phi$$

Upon comparing (104) and (105), we see that condition (ii) reads simply:

(106) $$P_\Phi = E_\phi$$

Upon comparing (81) and (86) wih (105b), we see that in general

$$P_\Phi \subseteq E_\phi \tag{107}$$

Hence condition (ix) is equivalent to (106), i.e. to (ii).

We next show that (ii) implies (vi). Let $S \in \pi_\phi(\mathcal{B})' \cap U_\phi(G)'$. Since Φ is cyclic for $\pi_\phi(\mathcal{B})$, it is separating for $\pi_\phi(\mathcal{B})'$. Hence to show that $S = \lambda I$ for some $\lambda \in \mathbb{C}$, it is sufficient to show that

$$(S - \lambda I)\Phi = 0 \tag{108}$$

since $(S - \lambda I) \in \pi_\phi(\mathcal{B})'$. Since moreover $S \in U_\phi(G)'$, we have

$$U_\phi(g)S\Phi = SU_\phi(g)\Phi = S\Phi \quad \forall\ g \in G \tag{109}$$

i.e. $S\Phi \in E_\phi$. Under condition (ii), see (106) above, this is equivalent to $S\Phi \in P_\Phi$, i.e. to (108). Hence (ii) implies (vi) indeed.

To prove that (v) and (vi) separately imply (iii), let us assume that there exists $\lambda \in (0,1)$ and $\psi, \chi \in \mathcal{S}_G^*$ such that

$$\phi = \lambda\psi + (1-\lambda)\chi \tag{110}$$

From the second part of the proof of Corollary 9.1.6, we know that there exists $X \in \pi_\phi(\mathcal{B})'$ such that

$$< \psi; B >= (X\pi_\phi(B)\Phi, \Phi) \quad \forall\ B \in \mathcal{B} \tag{111}$$

The G-invariance of ψ thus reads:

$$\begin{aligned} &(X\pi_\phi(B^*)\pi_\phi(C)\Phi, \Phi) =< \psi; B^*C >= \\ &< \psi; \alpha(g)[B^*C] >= (X\pi_\phi\{\alpha(g)[B^*]\}\pi_\phi\{\alpha(g)[C]\}\Phi, \Phi) \end{aligned} \tag{112}$$

and thus, upon using $X \in \pi_\phi(\mathcal{B})'$:

$$(X\pi_\phi(C)\Phi, \pi_\phi(B)\Phi) = (U(g)^*XU(g)\pi_\phi(C)\Phi, \pi_\phi(B)\Phi) \tag{113}$$

which is to say $X \in U_\phi(G)'$. Note further that $< \psi; B^*B >\geq 0$ implies $X \geq 0$. Upon denoting by Y the positive square root of X, we thus obtain

$$< \psi; B >= (\pi_\phi(B)Y\Phi, Y\Phi) \quad \forall\ B \in \mathcal{B} \quad \text{where } Y \in \pi_\phi(\mathcal{B})' \cap U_\phi(G)' \tag{114}$$

This relation shows that (v) or (vi) separately imply that $\psi = \phi$ and thus that the decomposition (110) is trivial, i.e. that ϕ is extremal G-invariant, which is condition (iii).

We only have left to prove that (iii) implies (vi). This can be done by a straightforward modification of the first part of the proof of Corollary 9.1.6, consisting in replacing $X \in \pi_\phi(\mathcal{B})'$ by $X \in \pi_\phi(\mathcal{B})' \cap U_\phi(G)'$. As a result of this substitution, (9.1.56–57) provide a decomposition of ϕ into a convex sum of G-invariant states, which contradicts (iii) unless $X = xI$. Hence (iii) implies (vi). □

The ergodic flavour carried by the various conditions of Theorem 9 invites comparison with the results of classical ergodic theory. One notices then that the non-abelianness of $\mathcal{B}$ introduces new distinctions between some of the notions that do coincide in the classical theory; compare, for instance, the relations we found between our conditions (i), (iii) and (ix) and Definition 7.1, Corollary 7.6 and Theorem 9.7 in Arnold & Avez (1968).

A further comparison with the results of Subsection b above does suggest, however, to investigate what happens to the relations between the nine conditions of Theorem 9 when one assumes, in addition, some kind of asymptotic abelianness in the action $\alpha : G \to \text{Aut}(\mathcal{B})$. The next three corollaries are results in this direction.

Corollary 10. *With the general assumptions of Theorem 9, the following three supplementary conditions are mutually equivalent:*

$$\text{(115a)} \qquad \text{With } \mathcal{N}_\phi \equiv \{\pi_\phi(\mathcal{B}), U_\phi(G)\}'' : E_\phi \mathcal{N}_\phi E_\phi \text{ is abelian}$$

$$\text{(115b)} \qquad [E_\phi \pi_\phi(B) E_\phi, E_\phi \pi_\phi(C) E_\phi] = 0 \quad \forall\ B, C \in \mathcal{B}$$

$$\text{(115c)} \qquad \eta < \psi; [\alpha(\cdot)[B], C] >= 0 \quad \forall\ B, C \in \mathcal{B}$$
$$\text{whenever } \psi : B \in \mathcal{B} \mapsto (\pi_\phi(B)\Psi, \Psi) \text{ with } \Psi \in \mathcal{E}_\phi$$

When any, and thus all, of the conditions (115) is satisfied, the conditions (ii), (iii), (vi), (vii) and (ix) of Theorem 9 are equivalent.

Proof: Clearly (115a) implies (115b). To see that the converse is also true, we notice first that $E_\phi \pi_\phi(\mathcal{B}) E_\phi$ is abelian exactly when $E_\phi \pi_\phi(\mathcal{B})'' E_\phi$ is abelian, and that $E_\phi \in U_\phi(G)''$. From (82), we conclude that $\mathcal{N}_\phi$ is the closure, in the strong operator topology, of the set of all finite linear combinations of operators of the form $\pi_\phi(B_i) U_\phi(g_i)$ with $B_i \in \mathcal{B}$ and $g_i \in G$. Moreover, from (86), we have $U_\phi(g) E_\phi = E_\phi$ for all $g \in G$. Consequently

$$\text{(116)} \qquad \mathcal{N}_\phi E_\phi = \pi_\phi(\mathcal{B})'' E_\phi \text{ and } E_\phi \mathcal{N}_\phi E_\phi = E_\phi \pi_\phi(\mathcal{B})'' E_\phi$$

Hence (115b) implies (115a).

To see that (115b) and (115c) are equivalent, we compute for every $\Psi \in \mathcal{H}_\phi$ and

$$\text{(117)} \qquad \psi : B \in \mathcal{B} \mapsto (\pi_\phi(B) E_\phi \Psi, E_\phi \Psi)$$

$$\text{(118)} \qquad \begin{array}{l} \eta < \psi; [\alpha(\cdot)[B], C] >= \eta\left([\pi_\phi\{\alpha(\cdot)[B]\}, \pi_\phi(C)] E_\phi \Psi, E_\phi \Psi\right) = \\ ([\eta_\phi(B), \pi_\phi(C)] E_\phi \Psi, E_\phi \Psi) = ([E_\phi \pi_\phi(B) E_\phi, E_\phi \pi_\phi(C) E\phi] \Psi, \Psi) \end{array}$$

The equivalence of (115b) and (115c) follows then readily from (118). This completes the proof of the equivalence of the three conditions (115).

To prove the second part of the Corollary, we see from Theorem 9 that it is sufficient to show, under condition (115) that (vi) implies (ix) and that (vii) implies (vi).

To show that (vi) implies (ix), notice that (vi) can be rewritten equivalently as $\mathcal{N}_\phi = \mathcal{B}(\mathcal{H}_\phi)$. So (115) and (vi) together imply that $E_\phi \mathcal{B}(\mathcal{H}_\phi) E_\phi$ is abelian; this is only possible when E_ϕ is one-dimensional. Hence, under (115), (vi) implies indeed (ix).

Condition (vii) reads $\mathcal{N}_\phi \cap (\mathcal{N}_\phi)' = \mathbb{C}I$, while (vi) reads $(\mathcal{N}_\phi)' = \mathbb{C}I$. It is, therefore, clear that (vii) implies (vi) whenever $(\mathcal{N}_\phi)' \subseteq \mathcal{N}_\phi$, i.e. since $\mathcal{N}_\phi$ is a von Neumann algebra, whenever

(119) $\qquad (\mathcal{N}_\phi)'$ is an abelian von Neumann algebra

The proof of the Corollary will, therefore, be completed when we have proven that (115) implies (119). To prove this, let $\mathcal{M}_\phi$ denote $E_\phi \mathcal{N}_\phi E_\phi$ when considered as a von Neumann algebra acting within the closed subspace $\mathcal{E}_\phi \subseteq \mathcal{H}_\phi$. Clearly Φ cyclic for $\pi_\phi(\mathcal{B})$ in $\mathcal{H}_\phi$ implies Φ cyclic for $\mathcal{M}_\phi$ in $\mathcal{E}_\phi$. Hence, when (115) is satisfied, $\mathcal{M}_\phi$ is a von Neumann algebra which is abelian *and* admits a cyclic vector. Consequently (see e.g. Dixmier, 1957, I. 6.3, Cor. 2) $\mathcal{M}_\phi$ is maximal abelian, i.e. $\mathcal{M}_\phi = (\mathcal{M}_\phi)'$; and thus $(\mathcal{M}_\phi)'$ is an abelian von Neumann algebra. Since $E_\phi \in \mathcal{N}_\phi$ we have (see, e.g. Dixmier, 1957, I.2.1, Prop. 1): $E_\phi(\mathcal{N}_\phi)'E_\phi = (\mathcal{M}_\phi)'$. Hence:

(120) $\qquad E_\phi(\mathcal{N}_\phi)'E_\phi \equiv \{E_\phi X E_\phi | X \in (\mathcal{N}_\phi)'\}$ is an abelian von Neumann algebra

To prove (119) it is, therefore, sufficient to construct a *-algebraic isomorphism between $(\mathcal{N}_\phi)'$ and $E_\phi(\mathcal{N}_\phi)'E_\phi$. The map

(121) $$\nu : X \in (\mathcal{N}_\phi)' \mapsto E_\phi X E_\phi \in E_\phi(\mathcal{N}_\phi)'E_\phi$$

is clearly a *-algebraic homomorphism, since $E_\phi \in \mathcal{N}_\phi$, and it is surjective. It is, therefore, sufficient to show that ν is injective. Suppose $X \in (\mathcal{N}_\phi)'$ and $E_\phi X E_\phi = 0$; we have then, for every $Y \in \mathcal{N}_\phi$:

(122) $$XY\Phi = YX\Phi = YXE_\phi\Phi = YE_\phi XE_\phi\Phi = 0$$

Since Φ is cyclic in $\mathcal{H}_\phi$ for $\pi_\phi(\mathcal{B}) \subseteq \mathcal{N}_\phi$, (122) implies $X = 0$; and thus ν is injective. □

Condition (115a), when assumed to hold for all $\phi \in S_G^*$, is referred to by saying that the action α is *G-abelian*. It was introduced in this form by Lanford & Ruelle (1967). Note that (115a) does not require, for its formulation, any invariant mean η on G; in fact it does not even require that G be amenable. When G is amenable, however, the equivalence between (115a) and (115c)

namely that Φ is also cyclic, in the representation space $\mathcal{L}(\mathcal{H})$, for the von Neumann algebra $\pi(\mathcal{B}(\mathcal{H}))'$. Notice further that

$$J\Delta J = \Delta^{-1} \tag{66}$$

so that, in particular:

$$JU(t)J = U(t) \quad \forall\ t \in \mathbb{R} \tag{67}$$

Hence J, although it is antiunitary, does not reverse the direction of time; this is a reflection of the symmetry, around $O \in \mathbb{R}$, of the spectrum of the generator of $U(\mathbb{R})$, a symmetry which is evident from (57).

The representation (50–51) was singled out only because it allowed a straightforward computation of the operators Δ, J and S and of their relations. Other realizations of this representation may, however, be used for other purposes. For instance, the GNS representation of $\mathcal{B}(\mathcal{H})$, with $\mathcal{H} = \mathcal{L}^2(\mathbb{R}, dx)$, associated to any canonical equilibrium state on $\mathcal{B}(\mathcal{H})$, can also be described in $\mathcal{L}(\mathcal{H}) = \mathcal{L}^2(\mathbb{R}^2, dpdq)$ as the weak-operator closure of the representation of the canonical commutation relations (see Subsection 8.3.f) defined by:

$$(E(z)\Psi)(p,q) = \exp\{-i\hat{a}(q - \frac{a}{2})\}\ \Psi(p + \hat{a}, q - a) \tag{68}$$

To see this, it is sufficient to choose arbitrarily $\Theta > 1$ and to verify, by straightforward computations, that the function

$$\Phi(p,q) = \{(\Theta^2 - 1)/\pi^2\}^{1/4} \exp\{-\frac{1}{2}(\frac{\Theta^2 - 1}{\Theta}p^2 + \Theta q^2)\} \tag{69}$$

is cyclic in $\mathcal{L}^2(\mathbb{R}^2, dpdq)$ under the action of Span $\{E(z) | z \in \mathbb{C}\}$, and that

$$(E(z)\Phi, \Phi) = \exp\{-\frac{\Theta}{4}(a^2 + b^2)\} = \langle \phi_\beta; E(z) \rangle \tag{70}$$

where ϕ_β—see (8.3.221), with $\lambda = \omega = 1$—is the canonical equilibrium state of the harmonic oscillator at natural temperature β given by $\Theta = \coth(\beta/2)$.

The interest of the representation (68) is that its infinitesimal form gives

$$P = -i\partial_q \quad \text{and} \quad Q = q + i\partial_p \tag{71}$$

which is exactly the prequantization representation (8.1.117). The isomorphism (65) between $\pi(\mathcal{B}(\mathcal{H}))$ and $\pi(\mathcal{B}(\mathcal{H}))'$ is reflected here in the fact that $\mathcal{L}^2(\mathbb{R}^2, dpdq)$ hosts another representation of the CCR, commuting with (68), and giving, in its infinitesimal form

$$\hat{P} = -i\partial_p \quad \text{and} \quad \hat{Q} = p + i\partial_q \tag{72}$$

which is precisely (8.1.118). The generalization to n degrees of freedom is straightforward, and some of the implications of this remark for the "geometric quantization" scheme have been explored in Emch (1981).

Returning now to a more conventional view of the canonical equilibrium states on $\mathcal{B}(\mathcal{H})$, we review a version of the results of Kubo (1957) and Martin & Schwinger (1959), the importance of which, for a mathematically rigorous discussion of infinite thermodynamical systems, was recognized by Haag, Hugenholtz & Winnink (1967).

THEOREM 6. *Let ϕ be a faithful normal state on $\mathcal{B}(\mathcal{H})$. With $\beta \in (0,\infty)$ fixed, let $\alpha : t \in \mathbb{R} \mapsto \alpha(t) \in \mathrm{Aut}(\mathcal{B}(\mathcal{H}))$ be the evolution (54) corresponding to the Hamiltonian canonically associated to ϕ by Scholium 4. Then, for every pair $\{A, B\}$ of observables in $\mathcal{B}(\mathcal{H})$ there exists a bounded continuous function f on the strip*

(73) $$\Omega_\beta \equiv \{z \in \mathbb{C} \,|\, 0 \leq Im\, z \leq \beta\}$$

such that f is holomorphic in the interior $\Omega_\beta \setminus \partial\Omega_\beta$ of the strip, and satisfies the boundary conditions:

(74) $$\left.\begin{aligned} f(t) &= < \phi; A\,\alpha(t)[B] > \\ f(t+i\beta) &= < \phi; \alpha(t)[B]\,A > \end{aligned}\right\} \forall\, t \in \mathbb{R}$$

Proof: *Part (a).* Notice that for every $A \in \mathcal{B}(\mathcal{H})$ and every operator E_{mn} of the form (48), the identity (49) and the identification of the Hamiltonian H provided by Scholium 4 imply:

(75) $$< \phi; E_{mn}A > = \exp\{-\beta(\epsilon_m - \epsilon_n)\} < \phi; AE_{mn} >$$

Let further B a finite sum of the form

(76) $$B = \sum_{mn} b_{mn} E_{mn}$$

Notice now that the entire function $F : z \in \mathbb{C} \to \mathbb{C}$ defined by

(77) $$F(z) = \sum_{mn} b_{mn} \exp\{iz(\epsilon_m - \epsilon_n)\} < \phi; AE_{mn} >$$

satisfies, for all t in $\mathbb{R}$:

(78a) $$F(t) = < \phi; A\alpha(t)[B] >$$
(78b) $$F(t+i\beta) = < \phi; \alpha(t)[B]A >$$

With f denoting the restriction of F to Ω_β, and $\partial\Omega_\beta$ denoting the boundary of Ω_β, we have from the Phragmen-Lindelöf theorem:

$$\begin{aligned}\sup_{z\in\Omega_\beta}|f(z)| = \sup_{z\in\partial\Omega_\beta}|f(z)| \equiv \\ \sup_{t\in\mathbb{R}}\{|f(t)|\vee|f(t+i\beta)|\} \leq \| A \| \cdot \| B \|\end{aligned} \tag{79}$$

Hence the theorem is proven in the particular case where B is restricted to belong to $\mathcal{E}$, the collection of all elements of $\mathcal{B}(\mathcal{H})$ satisfying (76).

Part (b). $\mathcal{E}$ is a *-subalgebra of $\mathcal{B}(\mathcal{H})$ and $\mathcal{E}'' = \mathcal{B}(\mathcal{H})$. Consequently $\mathcal{E}$ is dense in $\mathcal{B}(\mathcal{H})$ for the ultrastrong topology. Since π is the GNS representation associated to a normal state, namely ϕ, we already know that π is ultrastrongly continuous, so that $\pi(\mathcal{E})$ is dense in $\pi(\mathcal{B}(\mathcal{H}))$ for the ultrastrong, and hence the strong-operator, topologies in $\mathcal{B}(\mathcal{L}(\mathcal{H}))$; note, incidentally, that this conclusion is obvious from the explicit form (51) of this representation. Let now $B = B^* \in \mathcal{B}(\mathcal{H})$. By Kaplanski's density theorem (see e.g. Kadison & Ringrose, 1983, Thm. 5.3.5), B belongs to the closure, in the strong-operator topology, of the set of all self-adjoint operators in the ball $\{E \in \mathcal{E} |\ \| E \| \leq \| B \|\}$. Since $\mathcal{H}$ is assumed to be separable throughout this subsection, $\mathcal{L}(\mathcal{H})$ is also separable, and thus the ball $\{C \in \mathcal{B}(\mathcal{L}(\mathcal{H}))|\ \| C \| \leq \| B \|\}$ is metrizable in the strong-operator topology. Upon bringing together the above two facts, we conclude that there exists a sequence $\{B_n = B_n^*\} \subset \mathcal{E}$ such that $\| B_n \| \leq \| B \|\ \ \forall\, n$, and $\pi(B_n)\Phi$ converges strongly to $\pi(B)\Phi$. With $A = A^* \in \mathcal{B}(\mathcal{H})$, we thus have

$$\begin{aligned}&| < \phi; A\alpha(t)[B_n] > - < \phi; A\alpha(t)[B] > | = \\ &|(U(t)\pi(B_n)\Phi - U(t)\pi(B)\Phi, \pi(A)\Phi)| \leq \\ &\| \pi(B_n)\Phi - \pi(B)\Phi \| \cdot \| A \|\end{aligned} \tag{80a}$$

Similarly

$$\begin{aligned}&| < \phi; \alpha(t)[B_n]A > - < \phi; \alpha(t)[B]A > | = \\ &|(\pi(A)\Phi, U(t)\pi(B_n)\Phi - U(t)\pi(B)\Phi)| \leq \\ &\| \pi(B_n)\Phi - \pi(B)\Phi \| \cdot \| A \|\end{aligned} \tag{80b}$$

Recall now that we proved that, for each B_n there exists a bounded continuous function f_n on the strip Ω_β, such that f_n is holomorphic in the interior of the strip and satisfies, for all $t \in \mathbb{R}$:

$$f_n(t) = < \phi; A\alpha(t)[B_n] > \tag{81a}$$
$$f_n(t+i\beta) = < \phi; \alpha(t)[B_n]A > \tag{81b}$$

Note further that the functions f_n are uniformly bounded, namely

$$|f_n(z)| \leq \| A \| \cdot \| B_n \| \leq \| A \| \cdot \| B \| \tag{82}$$

Moreover, again from the Phragmen-Lindelöf theorem

$$
\begin{aligned}
&|f_n(z) - f_m(z)| \leq \sup_{z\in\Omega_\beta} |f_n(z) - f_m(z)| = \sup_{z\in\partial\Omega_\beta} |f_n(z) - f_m(z)| = \\
&\sup_{t\in\mathbb{R}}\{| < \phi; A\alpha(t)[B_n - B_m] > | \vee | < \phi; \alpha(t)[B_n - B_m]A > |\} = \\
&\sup_{t\in\mathbb{R}}\{|(U(t)\pi(B_n - B_m)\Phi, \pi(A)\Phi)| \vee |(\pi(A)\Phi, U(t)\pi(B_n - B_m)\Phi)|\} \leq \\
&\| \pi(B_n)\Phi - \pi(B_m)\Phi \| \cdot \| A \|
\end{aligned}
\tag{83}
$$

i.e. $\{f_n\}$ is uniformly convergent. From (80)–(83), we conclude that $\{f_n\}$ converges to a function f that is holomorphic in the interior of the strip Ω_β, continuous and bounded on Ω_β, and takes on $\partial\Omega_\beta$ the values required by (74). □

In closing this subsection, we want to make three brief comments on the role played by the discreteness of the spectrum of the Hamiltonian H appearing in Scholium 4. *First*: an unescapable consequence of this discreteness is the undesirable implication stated in part (b) of Scholium 5. *Second*: the discreteness of the spectrum of an operator is very unstable against perturbations and limiting procedures. *Third*: while it is true that the discreteness of the spectrum of H was used in most of the proofs presented in this subsection, it is also true that this condition does not appear in the definition (61) of the operator S and in the statement of the conclusion in Theorem 6; one should, therefore, not rule out the possibility that some of the positive aspects of the structure uncovered in the present particular case, e.g. (60), (64), (65) and (66) extend to more general circumstances where part (b) of Scholium 5 can be circumvented. One of the major advances in the theory was the recognition that this potentiality can indeed be realized; as we shall see in the next subsection.

b. KMS condition and modular actions

The results of the previous subsection indicate that the framework provided by the von Neumann postulates is too restrictive for the formulation of a quantum statistical mechanics with good ergodic properties. Our purpose now is to present the simplest cogwheels of a mathematical apparatus, building selectively on the strengths of the von Neumann formalism, to bypass some of the shortcomings signalled in the previous subsection. The physical aspects of the situation will be discussed in the next three subsections.

Theorem 6 above suggests the following definition. Let $\alpha : \mathbb{R} \to \text{Aut}(\mathcal{B})$ be an action of $\mathbb{R}$ on a C^*-algebra $\mathcal{B}$ (with unit I), and let $\beta \in (0, \infty)$. A state ϕ on $\mathcal{B}$ is said to satisfy the *KMS condition* at natural temperature β if, for every pair $\{A, B\}$ of observables in $\mathcal{B}$, there exists a complex-valued, bounded, continuous function f on the strip

$$\Omega_\beta \equiv \{z \in \mathbb{C} \mid 0 \leq \text{Im}\, z \leq \beta\} \tag{84}$$

such that f is holomorphic in the interior $\Omega_\beta \setminus \partial\Omega_\beta$ of the strip, and satisfies the boundary conditions

$$\left.\begin{aligned} f(t) &= <\phi; A\,\alpha(t)[B]> \\ f(t+i\beta) &= <\phi; \alpha(t)[B]\,A> \end{aligned}\right\} \quad \forall\, t \in \mathbb{R} \tag{85}$$

In applications, it is often useful to have a slightly different formulation of this condition, which we briefly describe.

An element $B \in \mathcal{B}$ is said to be *analytic* for α if there exists a function

$$\alpha(\cdot)[B] : z \in \mathbb{C} \mapsto \alpha(z)[B] \in \mathcal{B} \tag{86a}$$

such that: (a) it extends to $\mathbb{C}$ the function

$$\alpha(\cdot)[B] : t \in \mathbb{R} \mapsto \alpha(t)[B] \in \mathcal{B} \tag{86b}$$

and (b) for every $\phi \in \mathcal{S}^*$

$$z \in \mathbb{C} \mapsto <\phi; \alpha(z)[B]> \in \mathbb{C} \tag{86c}$$

is analytic. We can clearly replace, in (86c), $\mathcal{S}^*$ by $\mathcal{B}^*$. Recall (see, e.g. Dunford & Schwartz, 1958, Part I, Chap. VII; or Hille, 1972, Chap. 8) that B analytic implies that $\alpha(\cdot)[B]$ is strongly continuous, and strongly differentiable, uniformly on compacta in $\mathbb{C}$.

A *strongly continuous action* $\alpha : \mathbb{R} \to \mathrm{Aut}(\mathcal{B})$ is a group homomorphism— i.e. $\alpha(s)\alpha(t) = \alpha(s+t)\ \forall\, s,t \in \mathbb{R}$—such that for every $B \in \mathcal{B}$, the map $t \in \mathbb{R} \mapsto \alpha(t)[B]$ is strongly continuous. Clearly, then, $t \in \mathbb{R} \mapsto <\phi;\ \alpha(t)[B]> \in \mathbb{C}$ is continuous for every $(\phi, B) \in \mathcal{S}^* \times \mathcal{B}$, and α is an action in the ordinary sense defined in Section 9.2; the converse however, is not true in general, unless $\mathcal{B}$ satisfies some separability assumption (see, e.g. Hille & Phillips, 1957, Section 10.2, Corollary, p.306).

Proposition 7. *Let $\alpha : \mathbb{R} \to \mathrm{Aut}(\mathcal{B})$ be a strongly continuous action. (a) $\mathcal{B}_\alpha \equiv \{B \in \mathcal{B} | B$ analytic for $\alpha\}$ is then a strongly dense *-subalgebra of $\mathcal{B}$. (b) Let $\mathcal{D} \subset \mathcal{B}$ be a strongly dense set of analytic elements for α; then ϕ satisfies the KMS condition at natural temperature $\beta[\in (0,\infty)]$ with respect to α, if and only if, for every $(A, B, z) \in \mathcal{B} \times \mathcal{D} \times \mathbb{C}$:*

$$<\phi; A\alpha(z+i\beta)[B]> = <\phi; \alpha(z)[B]A> \tag{87}$$

(c) If ϕ satisfies (87), then ϕ is $\alpha(\mathbb{R})$-invariant.

Proof: With $(B, n, z) \in \mathcal{B} \times \mathbb{Z}^+ \times \mathbb{C}$ fixed,

$$t \in \mathbb{R} \mapsto B_n(z,t) \equiv (n/\pi)^{1/2}\alpha(t)[B]\exp\{-n(t-z)^2\} \in \mathcal{B} \tag{88}$$

condition on ϕ implies that there exists, for each a in the directed set I, a bounded continuous function $f_a : \Omega_\beta \to \mathbb{C}$, holomorphic in the interior of this strip, and such that:

$$\left.\begin{aligned} f_a(t) &= < \phi; A_a\, \alpha(t)[B] > \\ f_a(t+i\beta) &= < \phi; \alpha(t)[B]\, A_a > \end{aligned}\right\} \ \forall\ t \in \mathbb{R} \tag{102}$$

i.e. since ϕ is $\alpha(\mathbb{R})$-invariant:

$$\left.\begin{aligned} f_a(t) &= (U_\phi(t)\pi_\phi(B)\Phi, \pi_\phi(A_a)\Phi) \\ f_a(t+i\beta) &= (\pi_\phi(A_a)\Phi, U_\phi(t)\pi_\phi(B)\Phi) \end{aligned}\right\} \ \forall\ t \in \mathbb{R} \tag{103}$$

We now use again the Phragmen-Lindelöf theorem on Ω_β, to conclude that the net $\{f_a | a \in I\}$ of functions converges uniformly, on Ω_β, to a bounded continuous function f_B on Ω_β, holomorphic in the interior of this strip, and satisfying the boundary conditions:

$$\left.\begin{aligned} f_B(t) &= (U_\phi(t)\pi_\phi(B)\Phi, \Psi) \\ f_B(t+i\beta) &= (\Psi, U_\phi(t)\pi_\phi(B)\Phi) \end{aligned}\right\} \ \forall\ t \in \mathbb{R}\,. \tag{104}$$

Let Ψ' be another element of K; there exists then a net $\{B_b | b \in J\}$ of elements of $\mathcal{B}_{sa}$, such that $\{\pi_\phi(B_b)\Phi | b \in J\}$ converges to Ψ'. Just as before, except that we start now with (104), we can prove that there exists a bounded continuous function $f : \Omega_\beta \to \mathbb{C}$, holomorphic in the interior of this strip, and such that

$$\left.\begin{aligned} f(t) &= (U_\phi(t)\Psi', \Psi) \\ f(t+i\beta) &= (\Psi, U_\phi(t)\Psi') \end{aligned}\right\} \ \forall\ t \in \mathbb{R} \tag{105}$$

Let now $\Psi \in K \cap (iK)$ and $\Psi' \in K$. From the above argument, we know that there exists two bounded continuous functions f and g on Ω_β, holomorphic in the interior of this strip, and satisfying the boundary conditions

$$\left.\begin{aligned} f(t) &= (U_\phi(t)\Psi', \Psi) \\ f(t+i\beta) &= (\Psi, U_\phi(t)\Psi') \\ g(t) &= (U_\phi(t)\Psi', i\Psi) \\ g(t+i\beta) &= (i\Psi, U_\phi(t)\Psi') \end{aligned}\right\} \ \forall\ t \in \mathbb{R} \tag{106}$$

Note, however, that these equations imply

$$g(t) = -if(t) \quad \text{and} \quad g(t+i\beta) = +if(t+i\beta)\ \forall\ t \in \mathbb{R} \tag{107}$$

and thus

$$g(z) = -if(z) \quad \text{and} \quad g(z) = +if(z)\ \forall\ z \in \Omega_\beta \tag{108}$$

Consequently $f \equiv 0 \equiv g$, and in particular:

$$(109) \qquad (\Psi, \Psi') = 0 \quad \forall\, \Psi \in K \cap (iK) \quad \text{and} \quad \forall\, \Psi' \in K$$

Since Φ is cyclic for $\pi_\phi(\mathcal{B})$, this implies $\Psi = 0$, which indeed proves (101).

Notice now that $Z = Z^* \in \pi_\phi(\mathcal{B})''$ implies $Z\Phi \in K$. For any $X = X^* \in \pi_\phi(\mathcal{B})''$ and any $Y \in \pi_\phi(\mathcal{B}_{sa})$, both $(XY + YX)$ and $\{(XY - YX)/i\}$ are self-adjoint elements of $\pi_\phi(\mathcal{B})''$. Hence

$$(110) \qquad (XY + YX)\Phi \in K \text{ and } (XY - YX)\Phi \in iK\,.$$

If now $X\Phi = 0$, then $XY\Phi \in K \cap (iK)$, and thus we obtain from (101) that $XY\Phi = 0$. Again from the cyclicity of Φ with respect to $\pi_\phi(\mathcal{B})$, this implies $X = 0$. Finally, for any $X \in \pi_\phi(\mathcal{B})''$ with $X\Phi = 0$, we have $X^*X\Phi = 0$ and thus, from the result just proven, $X^*X = 0$ i.e. $\| X \|^2 = 0$, i.e. $X = 0$. This proves part (a) of the theorem.

To prove part (b), let $B \in \mathcal{B}$ such that $< \phi; B^*B >= 0$. This implies $\pi_\phi(B)\Phi = 0$, and hence, by part (a): $\pi_\phi(B) = 0$. Hence

$$(111) \qquad \{B \in \mathcal{B} \mid < \phi; B^*B >= 0\} \subseteq \operatorname{Ker} \pi_\phi$$

The opposite inclusion is obvious, and thus part (b) is proven. Finally, part (c) of the theorem is a trivial consequence of part (a), since

$$(112) \qquad \tilde{\phi} : X \in \pi_\phi(\mathcal{B})'' \mapsto (X\Phi, \Phi) \in \mathbb{C}$$

and thus $< \tilde{\phi}; X^*X >= 0$ if and only if $X\Phi = 0$. □

This result establishes an important bridge *from* the study of KMS-states relative to (strongly continuous) actions α of $\mathbb{R}$ on a C^*-algebra $\mathcal{B}$, *to* the study of faithful normal states on von Neumann algebras. This bridge can, in fact, be established under technically less stringent conditions. We shall, however, not enter into these technicalities here, although they are sometimes required by certain applications to quantum statistical mechanics—see e.g. Winnink (1972) and references quoted therein.

Since this book is concerned with the conceptual structures of the theory, viewed from a mathematical vantage point, it is a legitimate preliminary question to ask whether the bridge just established can also be traveled in the opposite direction. A major advance in the theory of modular Hilbert algebras made this possible, as we shall now indicate.

Let $\mathcal{M}$ be a von Neumann algebra acting on a Hilbert space $\mathcal{H}$, and ϕ be a faithful normal state on $\mathcal{M}$. Since ϕ is normal and faithful, $\pi_\phi : \mathcal{M} \to \pi_\phi(\mathcal{M})$ is an ultraweakly, and ultrastrongly, bicontinuous isomorphism. We can, therefore, assume without loss of generality, that $\mathcal{M}$ is presented to us as $\pi_\phi(\mathcal{M})$, and that there exists a vector Φ in $\mathcal{M}$, cyclic and separating for

Theorem 11. *Let ϕ be a faithful normal state on a von Neumann algebra $\mathcal{M}$, and let $\mathcal{M}_\sigma$ denote the fixed point algebra of $\mathcal{M}$ under the modular action σ associated to ϕ for the natural temperature β. If there exists a convex decomposition:*

$$\phi = \lambda\psi + (1-\lambda)\chi \tag{128}$$

into $\sigma(\mathbb{R})$-invariant states ψ and χ, then there exists a unique $Z \in \mathcal{M}_{sa}$ such that, for all $M \in \mathcal{M}$

$$< \psi; M >= \lambda^{-1} < \phi; ZM > \tag{129a}$$

and this Z is a positive element in the unit ball of $\mathcal{M}_\sigma$, such that for all $M \in \mathcal{M}$

$$< \chi; M >= (1-\lambda)^{-1} < \phi; (I-Z)M > \tag{129b}$$

If, furthermore, the decomposition (128) is into states that satisfy the KMS condition at the natural temperature β, with respect to σ, then $Z \in \mathcal{M} \cap \mathcal{M}' \equiv \mathcal{Z}$. Conversely, for every positive element Z in the unit ball of $\mathcal{M}_\sigma$ (resp. $\mathcal{Z}$) with $O \neq Z \neq I$, (129) defines a convex decomposition (128) of ϕ into $\sigma(\mathbb{R})$-invariant (resp. β-KMS w.r.t. σ) normal states on $\mathcal{M}$.

Proof: We first prove the "converse" part of the theorem. Let Z be a positive element in the unit ball of $\mathcal{M}_\sigma$, with $O \neq Z \neq I$. Since ϕ is faithful $\lambda \equiv\ < \phi; Z > \in (0,1)$. Since $\mathcal{M}_\sigma$ is a von Neumann algebra $Z^{1/2} \in \mathcal{M}_\sigma$; and since $\mathcal{M}_\sigma = \mathcal{M}_\phi$, (129a) reads

$$\psi : M \in \mathcal{M} \mapsto \lambda^{-1} < \phi; Z^{1/2} M Z^{1/2} > \tag{130}$$

Hence ψ is a vector state on $\mathcal{M}$, and is therefore normal. Moreover for all $t \in \mathbb{R}$:

$$\begin{aligned} \lambda < \psi; \sigma(t)[M] > &=< \phi; Z\sigma(t)[M] >= \\ < \phi; \sigma(t)[ZM] > &=< \phi; ZM >= \lambda < \psi; M > \end{aligned} \tag{131}$$

Hence ψ is $\sigma(\mathbb{R})$-invariant. The same argument applies to χ. Note that (129a) determines $Z \in \mathcal{M}$ uniquely. Indeed, if

$$< \phi; YM >=< \phi; ZM > \quad \forall\ M \in \mathcal{M} \tag{132}$$

we have

$$(M\Phi, \{Y^* - Z^*\}\Phi) = 0 \quad \forall\ M \in \mathcal{M} \tag{133}$$

Since Φ is cyclic and separating for $\mathcal{M}$, this implies indeed $Z = Y$.

If now $Z \in \mathcal{Z}$, we prove that ψ satisfies the KMS condition at temperature β, with respect to σ, as follows. Since ϕ satisfies this condition, there exists for every pair $\{M, N\}$ of elements of $\mathcal{M}$ a complex-valued function f, bounded and continuous on the strip Ω_β, and holomorphic in its interior, such that, for all $t \in \mathbb{R}$:

$$f(t) = \;<\phi;(ZM)\sigma(t)[N]> = \lambda <\psi; M\sigma(t)[N]> \tag{134a}$$

$$\begin{aligned} f(t+i\beta) = \;& <\phi;\sigma(t)[N](ZM)> = \\ & <\phi; Z(\sigma(t)[N]M)> = \lambda <\psi;\sigma(t)[N]M> \end{aligned} \tag{134b}$$

which proves indeed that ψ is β-KMS w.r.t. σ. Again the same argument applies to χ. This proves the "converse" part of the theorem.

We now prove the "direct" part of the theorem. If ψ satisfies (128), we already know from the proof of Corollary 9.1.6 that ψ is a vector state, and hence, a normal state on $\mathcal{M}$. From the first Radon-Nikodym theorem of Sakai (1971)—see his Prop. 1.24.4—there exists a positive element Z in the unit ball of $\mathcal{M}$ such that

$$2\lambda <\psi; M> = <\phi; ZM + MZ> \quad \forall\, M \in \mathcal{M} \tag{135}$$

Suppose that there exists an element $Z' \in \mathcal{M}$ for which (135) holds with Z' substituted for Z. With $X = Z - Z'$ we would then have

$$0 = <\phi; XM> + <\phi; MX> \quad \forall\; M \in \mathcal{M} \tag{136}$$

In particular, for $M = X^*$, this gives

$$0 = <\phi; XX^*> + <\phi; X^*X> \tag{137}$$

Since ϕ is positive, this is only possible if $<\phi; X^*X> = 0$; since ϕ is faithful, this implies $X^*X = 0$ and thus $X = 0$. Hence Z satisfying (135) is unique. We now introduce the condition that ψ is $\sigma(\mathbb{R})$-invariant. Eq.(135) then reads, with t arbitrary in $\mathbb{R}$:

$$\begin{aligned} & 2\lambda <\psi; M> = 2\lambda <\psi;\sigma(t)[M]> = \\ & <\phi; Z\sigma(t)[M] + \sigma(t)[M]Z> = \\ & <\phi;\sigma(-t)[Z]M + M\sigma(-t)[Z]> \end{aligned} \tag{138}$$

where we used the fact–see Lemma 10(b)—that ϕ is $\sigma(\mathbb{R})$-invariant. The uniqueness of Z then gives, upon comparing (135) and (138), that $Z \in \mathcal{M}_\sigma$. From Lemma 10(a), we see that (135) can be rewritten in the form (129a), and we already saw that this determines Z uniquely. (129b) follows then immediately, thus establishing the first part of the theorem.

To prove the second half of the direct part of the theorem, we first notice that if the decomposition (128) is into β-KMS states w.r.t. σ, it is also—see Lemma 10(b)—into $\sigma(\mathbb{R})$-invariant states. We can, therefore, use the above result, and it is sufficient, then, to show that Z belongs to $\mathcal{Z}$. The KMS condition on ϕ reads

$$f(t) = \lambda < \psi; N\sigma(t)[M] >=< \phi; ZN\sigma(t)[M] > \tag{139a}$$

$$f(t+i\beta) =< \phi; \sigma(t)[M]ZN > \tag{139b}$$

while the KMS condition on ψ gives

$$f(t+i\beta) = \lambda < \psi; \sigma(t)[M]N >=< \phi; Z\sigma(t)[M]N > \tag{139c}$$

Upon comparing (139b) and (139c) we obtain, for $t = 0$,

$$< \phi; [M, Z]N >= 0 \quad \forall\ N, M \in \mathcal{M} \tag{140}$$

Since Φ is cyclic and separating for $\mathcal{M}$, we obtain from (140) that $Z \in \mathcal{M}'$; since $Z \in \mathcal{M}$, this concludes the proof of the theorem. □

We say that a state is *extremal β-KMS* for an action $\alpha : \mathbb{R} \to \mathrm{Aut}(\mathcal{M})$ when it satisfies the KMS condition, at natural temperature $\beta[\in (0,\infty)]$, with respect to α, *and* it cannot be decomposed into a non-trivial convex combination of states satisfying the same condition.

Corollary 12. *Let ϕ be a faithful normal state on a von Neumann algebra $\mathcal{M}$, and σ be the modular action of $\mathbb{R}$ on $\mathcal{M}$ canonically associated to ϕ at the natural temperature $\beta[\in (0,\infty)]$. Then: (a) ϕ is extremal β-KMS for σ, exactly when $\mathcal{M}$ is a factor; and it is extremal $\sigma(\mathbb{R})$-invariant, exactly when the fixed point algebra $\mathcal{M}_\sigma$ of $\mathcal{M}$, under the action σ, is $\mathbb{C}I$; (b) in general ϕ extremal β-KMS for σ does not imply ϕ extremal $\sigma(\mathbb{R})$-invariant, unless the action σ is η-abelian on ϕ; (c) the action σ is necessarily η-abelian on ϕ when ϕ is extremal $\sigma(\mathbb{R})$-invariant, and then conditions (i)–(x) in Theorem 9.2.9 and Corollary 9.2.11 are all satisfied.*

Proof: Part (a) is an immediate consequence of Theorem 11. To prove, next, that there exist extremal β-KMS states, w.r.t. σ, which are not extremal $\sigma(\mathbb{R})$-invariant, it is sufficient to produce one counterexample; Scholium 5 and Theorem 6 of Subsection a above give such a counterexample. Suppose now that $\sigma(\mathbb{R})$ is a η-abelian action on ϕ. Corollary 9.2.11 then asserts that conditions (i)–(x) are equivalent. We now have an additional ingredient, namely Lemma 10(d), asserting that $\sigma(\mathbb{R})$ leaves the center $\mathcal{Z} \equiv \mathcal{M} \cap \mathcal{M}'$ of $\mathcal{M}$ pointwise invariant. Consequently, condition (viii) is satisfied, with $U_\phi(G) \equiv U_\sigma(\mathbb{R})$, exactly when $\mathcal{Z}$ is trivial, which we just saw is the case when ϕ is extremal β-KMS. In this case, we thus have, from Corollary 9.2.11, that (x) is satisfied, i.e. that $\mathcal{M}_\sigma = \mathbb{C}I$; we just saw that this is equivalent to the condition that ϕ be extremal $\sigma(\mathbb{R})$-invariant. Hence, part (b) of the corollary is proven. To prove part (c), it is sufficient to show that $\sigma(\mathbb{R})$ is an η-abelian action on ϕ. This follows from Corollary 9.2.12, since ϕ is faithful on $\mathcal{M}$ implies that Φ is cyclic and separating for $\mathcal{M}$, and thus that Φ is cyclic for $\mathcal{M}'$. □

Applications of the above theorem—and its corollary—to quantum statistical mechanics will be discussed in the next subsections, where we will need two more results of a general character that makes them belong, rather, to the present subsection.

SCHOLIUM 13. *Let σ be the modular action associated to a faithful normal state ϕ on a von Neumann algebra $\mathcal{M}$, and $\alpha \in \mathrm{Aut}(\mathcal{M})$ with $\phi \circ \alpha = \phi$. Then $\alpha \circ \sigma(t) = \sigma(t) \circ \alpha \ \forall t \in \mathbb{R}$.*

Proof: For every t in $\mathbb{R}$, let

$$\alpha(t) \equiv \alpha^{-1} \circ \sigma(t) \circ \alpha \tag{141}$$

Given $\beta \in (0, \infty)$, we know that ϕ satisfies the KMS condition w.r.t. σ; hence, for every pair $\{M, N\}$ of elements of $\mathcal{M}$, there exists a complex-valued function f, bounded and continuous on the strip Ω_β, and holomorphic inside this strip, such that:

$$\begin{aligned} f(t) &= <\phi; \alpha[M]\sigma(t)[\alpha[N]]> \\ &= <\phi \circ \alpha; M\alpha^{-1} \circ \sigma(t) \circ \alpha[N]> \\ &= <\phi; M\alpha(t)[N]> \\ f(t+i\beta) &= <\phi; \sigma(t)[\alpha[N]]\alpha[M]> \\ &= <\phi \circ \alpha; \alpha^{-1} \circ \sigma(t) \circ \alpha[N]M> \\ &= <\phi; \alpha(t)[N]M> \end{aligned} \tag{142}$$

The uniqueness of the modular action, with respect to the KMS condition, implies: $\alpha(t) = \sigma(t)$ for all $t \in \mathbb{R}$; with this information, (141) reads $\sigma(t) = \alpha^{-1} \circ \sigma(t) \circ \alpha$, which is precisely the conclusion of the scholium. □

The argument, based on a repeated use of the Pragmen-Lindelöf theorem—which we first presented in the simple case of Theorem 6, and then in various versions later on in the present subsection, see e.g. Theorem 9—can be lifted up to the following general result, the detailed proof of which can be found in Kadison & Ringrose (1983).

PROPOSITION 14. *Let ϕ be a faithful normal state on a von Neumann algebra $\mathcal{M}$, $\alpha : \mathbb{R} \to \mathrm{Aut}(\mathcal{M})$ be a one-parameter group of automorphisms of $\mathcal{M}$, and $\mathcal{B}$ be a *-subalgebra of $\mathcal{M}$, dense in $\mathcal{M}$ for the strong-operator topology. If, for every pair $\{A, B\}$ of elements of $\mathcal{B}$, there exists a complex-valued function f, bounded and continuous on the strip $\Omega_\beta = \{z \in \mathbb{C} \mid 0 \leq Im\ z \leq \beta\}$, with β fixed in $(0, \infty)$, such that f is analytic on the interior of Ω_β, and:*

$$\left.\begin{aligned} f(t) &= <\phi; A\alpha(t)[B]> \\ f(t+i\beta) &= <\phi; \alpha(t)[B]A> \end{aligned}\right\} \ \forall\ t \in \mathbb{R} \tag{143}$$

then α coincides with the modular action σ canonically associated to ϕ at the natural temperature β.

Together with results of the type of Theorem 9, this Proposition gives the theory the resilience necessary for the applications we want to consider.

c. Canonical Equilibrium for infinite quantum lattices.

The concepts of KMS conditions and modular actions, presented in the preceding subsection, offer a mathematical paradigm that fits the canonical equilibrium theory of finite quantum systems, as described in Subsection a. The question we now want to address is whether the mathematical generality thus added allows to account for situations which are genuinely more general from a physical point of view.

Two methodological remarks should be made from the outset. The first is that we shall proceed from the particular to the general; and the second is that we will deal with infinitely extended systems.

Concerning the first of these qualifications, we should sound a warning: while it is true that there exist models, some of which we shall present, that fit the KMS scheme most naturally, it is also true that there are several situations, of physical interest, for which modular actions only appear after some rather delicate preliminary analysis has been mastered—as we shall also indicate.

The second of the above two qualifications is even more pragmatic, and it has some of its roots in the last remarks made in Subsection a. The reader may very well object, against the proposal to consider infinitely extended systems, that his ordinary cup of tea is *not* an infinite system, while a truly physical description of it would require a microscopic account for the thermodynamics of the liquid-gas phase transition. To this objection, we answer by arguing that even a finite cup of tea is an extremely large system when looked at from a microscopic point of view—it contains more than 10^{24} molecules H_2O– and that bulk properties of matter, such as the possible existence of more than one thermodynamical phase—for which surface effects are expected to play only a triggering, but secondary role—are better modelled with infinitely extended systems, i.e. more effectively than they would be with models of finite extension. This strategy is thus akin to the familiar one consisting in searching for the leading term of an asymptotic expansion. The recognition that an explicit "thermodynamical" limit, involving the passage from large but finite systems to infinite systems, was the proper thing to do did not come without pangs; Pais (1979) reports indeed the following anecdote: " ... The point is that a sharp phase transition can only occur in the so-called thermodynamic limit ... This view emerged in a morning-long debate which took place during the van der Waals Centenary Conference in November 1937. The issue was: does the partition function contain the information necessary to describe a sharp phase transition? The transition implies the existence of analytically distinct parts of isotherms. It was not clear how this would come about. The debate was inconclusive and Kramers, the chairman, put the question to a vote. Uhlenbeck recalls that the ayes and nays were about evenly divided.

However, Kramers' suggestion to go to the thermodynamic limit eventually was realized to be the correct answer."

The Onsager (1944) solution of the Lenz (1920)-Ising (1925) model in two-dimensions, and zero external magnetic field—which, incidentally, appeared for some time to be quite a baffling tour de force; for reviews, see Schultz, Mattis & Lieb (1964) and Brush (1983)—was the first treatment of a truly infinite system. This treatment, however, focussed exclusively on the one macroscopic observable of immediate physical interest: the magnetization.

The first model of an infinite system in thermodynamical equilibrium, for which a representation of the algebra of quasi-local observables has been explicitly constructed is the free Bose gas. This construction, due to Araki & Woods (1963) is so explicit that, with hindsight, it is easy to recognize there the essence of the structures discussed in the preceding subsection.

With time, and the growing recognition that the KMS condition was at the core of the subject, simpler models were exhibited, and we shall start with these.

We consider a quantum lattice of the type discussed in Corollary 9.2.5, with now $n = 1$; i.e. to every $k \in \mathbb{Z}$, we associate a copy $\mathcal{B}_k$ of the algebra $\mathcal{M}(m, \mathbb{C})$, with $m < \infty$, of all $m \times m$ matrices with complex entries. The algebra of quasi-local observables for this sytem is defined, as in (9.2.7) to be

$$\mathcal{B} \equiv \bigotimes_{k \in \mathbb{Z}} \mathcal{B}_k \tag{144}$$

Let further

$$\alpha : a \in \mathbb{Z} \mapsto \alpha(a) \in \operatorname{Aut}(\mathcal{B}) \tag{145}$$

be the action of the translation group $\mathbb{Z}$ on $\mathcal{B}$, defined by (9.2.11).

We now add a new element to this structure, namely the dynamics. We first formalize the notion of an *interaction*, defining it as a map

$$V : \Omega \in \mathcal{F} \mapsto V_\Omega \in \mathcal{B} \tag{146a}$$

subject to the two conditions:

$$V_\emptyset = 0 \quad \text{and} \quad V_\Omega^* = V_\Omega \in \mathcal{B}_\Omega \;\; \forall \; \Omega \in \mathcal{F} \tag{146b}$$

The interaction is said to be *translation invariant* if

$$\alpha(a)[V_\Omega] = V_{a[\Omega]} \quad \forall \, (a, \Omega) \in \mathbb{Z} \times \mathcal{F} \tag{147}$$

and it is said to be of *finite range* r if

$$V_\Omega = 0 \quad \text{unless} \quad \exists \, k \in \mathbb{Z} \quad \text{such that} \quad \Omega \subseteq [k, k + r]. \tag{148}$$

Under the composition laws naturally suggested by (146), the collection of all translation-invariant, finite-range interactions becomes a real vector space, which we will denote $\mathcal{V}_0$.

As an illustration, consider the case of nearest neighbour interactions, i.e. $r = 1$, and assume further that $\mathcal{B}_k \cong \mathcal{M}(2, \mathbb{C})$. Let $\{\tau^j | j = x, y, z\}$ be the Pauli matrices introduced in (8.3.156b), namely:

$$\tau^x \equiv \begin{pmatrix} 0 & 1 \\ 1 & 0 \end{pmatrix}; \tau^y \equiv \begin{pmatrix} 0 & -i \\ i & 0 \end{pmatrix}; \tau^z \equiv \begin{pmatrix} 1 & 0 \\ 0 & -1 \end{pmatrix} \tag{149}$$

With B, J^z, $J^{x,y}$, ς fixed in $\mathbb{R}$, and k running over $\mathbb{Z}$, let

$$\begin{aligned} V_{[k]} &\equiv -B\tau_k^z \\ V_{[k,k+1]} &\equiv V_{[k,k+1]}^z + V_{[k,k+1]}^{x,y} \\ V_{[k,k+1]}^z &\equiv -J^z \tau_k^z \otimes \tau_{k+1}^z \\ V_{[k,k+1]}^{x,y} &\equiv -J^{x,y}\{(1+\varsigma)\tau_k^x \otimes \tau_{k+1}^x + (1-\varsigma)\tau_k^y \otimes \tau_{k+1}^y\} \end{aligned} \tag{150}$$

The particular cases of historical interest are: the *Lenz-Ising model* ($J^{x,y} = 0, J^z > 0$), the isotropic *Heisenberg model* ($J^{x,y} = J^z > 0, \varsigma = 0$), and the *$x-y$ model* ($J^z = 0, J^{x,y} > 0$, $0 \leq \varsigma < 1$).

We can now return to the general case described by (146–8), i.e. $V \in \mathcal{V}_0$, and we define, for every finite interval $[j,k]$ in $\mathbb{Z}$, the local *Hamiltonian*

$$H_{[j,k]} \equiv \sum_{\Omega \subseteq [j,k]} V_\Omega \,. \tag{151}$$

The formalism of Subsection a applies to each of these finite intervals $[j,k]$ separately when it is cut-off from the rest of the chain. In this manner, we obtain the following local objects.

The local evolution is:

$$\sigma_{[j,k]}(t) : B \in \mathcal{B}_{[j,k]} \mapsto \exp\{iH_{[j,k]}t\} \cdot B \cdot \exp\{-iH_{[j,k]}t\}; \tag{152a}$$

the local partition function, at natural temperature $\beta [\in (0,\infty)]$ is:

$$Z_{[j,k]}(\beta, V) \equiv tr \exp\{-\beta H_{[j,k]}\} \tag{152b}$$

the local Helmoltz free energy is:

$$A_{[j,k]}(\beta, V) = -\frac{1}{\beta} \ln Z_{[j,k]}(\beta, V); \tag{152c}$$

the local canonical equilibrium density matrix is:

$$\rho_{[j,k]}(\beta, V) = Z_{[j,k]}(\beta, V)^{-1} \exp\{-\beta H_{[j,k]}\}; \tag{152d}$$

and the local canonical equilibrium state is:

$$\phi_{[j,k]} : B \in \mathcal{B}_{[j,k]} \mapsto tr\{\rho_{[j,k]}(\beta, V) B\}. \tag{152e}$$

At $[j,k]$ fixed, the expressions (151–2) all belong, by definition, to the formalism of Subsection a: they describe a finite system. This, however, is clearly not the case for the collection of all these objects, as $[j,k]$ runs over unboundedly large, although finite, intervals of $\mathbb{Z}$. We now want to indicate how the new situation differs from the previous one.

For this class of models, specifically for every $V \in \mathcal{V}_0$, i.e. for every translation-invariant, finite-range interaction V, Araki (1969) proved the following results.
(i) The limit

$$\beta P(\beta, V) = \lim_{\substack{j\to-\infty \\ k\to+\infty}} (k-j)^{-1} \ell n\, Z_{[j,k]}(\beta, V) \tag{153}$$

exists, is finite, and defines a real analytic function in β on $(0,\infty)$, and in V on finite-dimensional subspaces of $\mathcal{V}_0$.
(ii) For every local observable $B \in \mathcal{B}_0$, the limit

$$< \phi; B > = \lim_{\substack{j\to-\infty \\ k\to+\infty}} < \phi_{[j,k]}; B > \tag{154}$$

exists, and extends uniquely to a state ϕ on $\mathcal{B}$, that is uniformly clustering and $\alpha(\mathbb{Z})$-invariant.
(iii) For every local observable $B \in \mathcal{B}_0$, the limit

$$\sigma(t)[B] = \lim_{\substack{j\to-\infty \\ k\to+\infty}} \sigma_{[j,k]}(t)[B] \tag{155}$$

exists, and extends uniquely to a strongly continuous action $\sigma : t \in \mathbb{R} \mapsto \sigma(t) \in \mathrm{Aut}(\mathcal{B})$; moreover, ϕ satisfies the KMS condition, at natural temperature β, with respect to σ.

In this precise mathematical sense, the infinite extension limit does indeed exist and is smooth. For reasons which will become clear as we proceed, we refer to this as the thermodynamical limit for the quantum lattices considered.

The above three properties will now be used to illustrate the general results we obtained earlier in this chapter and in Section Two of Chapter Nine; we shall indeed conclude, from these results, that the quantum lattices considered so far—one dimensional, infinite extension, interaction $V \in \mathcal{V}_0$—possess the following properties, already recognized by Araki.
(iv) The algebra $\mathcal{B}_\phi$ of observables at infinity, defined by (10.1.50) with respect to the state ϕ, defined in (ii) above, is trivial since ϕ is uniformly clustering—see Theorem 9.2.4.
(v) The von Neumann algebra $\pi_\phi(\mathcal{B})''$, generated by the GNS representation canonically associated to ϕ, is a factor—see Corollary 9.2.5.
(vi) ϕ is extremal $\mathbb{Z}$-invariant—see Theorem 9.2.2—and for all $B \in \mathcal{B}$

$$w - \lim_{|a|\to\infty} \pi_\phi\{\alpha(a)[B]\} = < \phi; B > I\,. \tag{156}$$

Consequently, the space average of every $B \in \mathcal{B}$ is given by

$$\eta_\phi(B) = <\phi; B> I \; ; \tag{157}$$

and thus ϕ is ergodic, with respect to the action α of the translation group $\mathbb{Z}$, in the sense of every one of the conditions (i)–(ix) of Theorem 9.2.9, and condition (x) of Corollary 9.2.11; in fact in each of these results space-averages can be replaced by point-wise convergence since (156) and not only (157) holds.
(vii) ϕ extends uniquely to a faithful normal state $\tilde{\phi}$ on $\pi_\phi(\mathcal{B})''$—see Theorem 10.1.9.
(viii) The natural extension to $\pi_\phi(\mathcal{B})''$ of the time evolution σ, defined in (iii) above, coincides with the modular action σ on $\pi_\phi(\mathcal{B})''$—see Proposition 10.1.14; and the evolution σ is translation invariant in the sense of Scholium 10.1.13.
(ix) ϕ is extremal β-KMS for σ—see Corollary 10.1.12b and property (v) above.
(x) ϕ is stationary under the evolution σ—see Lemma 10.1.10b or Proposition 10.1.7c.

Looking back at the construction of the Powers' type III factors at the end of Subsection 9.2.b, the reader will see that these factors are obtained from KMS states that result from an interaction of the form (150) with $J^z = 0 = J^{x,y}$ and $B \neq 0$. The fact that factors of type III, acting on separable Hilbert spaces, occur under much more general circumstances in quantum statistical mechanics, was realized soon after Powers' pioneering paper appeared in 1967. This is the combination of the works of several authors, among whom we mention only: Dell'Antonio (1967), Porta & Schwartz (1967), Ruelle (1967), Størmer (1967, 1969), Hugenholtz (1967), Hugenholtz & Wieringa (1969), Haag, Kadison & Kastler (1970), Takesaki (1970). The main steps in the argument are as follows—for more details, see Emch (1972), pp.209–211 and pp.281–283.

Let $\{\mathcal{M}_\Omega | \Omega \in \mathcal{F}\}$ be an isotonic family of von Neumann algebras, indexed by a directed set $\mathcal{F}$,

$$\mathcal{B}_0 = \bigcup_{\Omega \in \mathcal{F}} \mathcal{M}_\Omega \tag{158}$$

and $\mathcal{B}$ be the C^*-inductive limit of this isotonic family. Let $\{\Omega(n) | n \in \mathbb{Z}^+\}$ be an increasing *sequence*, that we assume to be absorbing, i.e. we assume that for every $\Omega \in \mathcal{F}$, there exists a positive integer $N(\Omega) < \infty$, such that $\Omega \subseteq \Omega(n) \;\; \forall\, n \geq N(\Omega)$. Suppose further that for each $n \in \mathbb{Z}^+$, a *normal* state ϕ_n is given on $\mathcal{M}_n \equiv \mathcal{M}_{\Omega(n)}$ in such a manner that for every $B \in \mathcal{B}_0$ the limit

$$<\phi; B> = \lim_{n \to \infty} <\phi_n; B> \tag{159}$$

exists. This defines a state ϕ on $\mathcal{B}$. For each $\Omega \in \mathcal{F}$, let ϕ_Ω denote the restriction of ϕ to $\mathcal{M}_\Omega$. Since the sequence $\{\Omega(n)\}$ is absorbing, ϕ_Ω can be seen as the point-wise limit, as $n \to \infty$, of a sequence of normal states ψ_n on $\mathcal{M}_\Omega$, namely the restrictions of ϕ_n to $\mathcal{M}_\Omega$ for all n such that $\Omega \subseteq \Omega(n)$. This is sufficient—see Sakai (1957) or Dell'Antonio (1967)—to conclude that ϕ_Ω is a normal state on $\mathcal{M}_\Omega$. The state ϕ is then said to be *locally normal.*

Let us now introduce the additional assumption that, for each $n \in \mathbb{Z}^+$, the von Neumann algebra $\mathcal{M}_n$ acts on a separable Hilbert space $\mathcal{X}_n$. One can then show—upon using a result of Porta & Schwartz (1967)—that the GNS representation π_n of $\mathcal{M}_n$, canonically associated to the normal state (see above) obtained as the restriction of ϕ to $\mathcal{M}_n$, acts on a separable Hilbert space $\mathcal{H}_n$. Upon using, next, the fact that $\{\Omega(n)\}$ is absorbing, we notice that the GNS representation π_ϕ of $\mathcal{B}$, associated to ϕ, acts on a Hilbert space $\mathcal{H}_\phi$ that is the closure of a countable union of separable Hilbert spaces $\approx \mathcal{H}_n$. Hence $\mathcal{H}_\phi$ is *separable.*

Suppose finally—as it is the case, for instance, under the general assumptions of Theorem 10.1.9—that the normal extension $\tilde{\phi}$ of ϕ to $\mathcal{M} \equiv \pi_\phi(\mathcal{B})''$ is *faithful.* We know then, from the general theory of modular actions described in Subsection b, that for $\beta \in (0,\infty)$, $\tilde{\phi}$ satisfies the KMS condition for its modular action σ; that this condition determines σ uniquely; and that $\tilde{\phi}$ is extremal β-KMS if and only if $\mathcal{M}$ is a *factor.*

From Takesaki (1970) we know that the modular action σ canonically associated with a faithful normal state $\tilde{\phi}$ of a von Neumann algebra $\mathcal{M}$ is inner if and only if $\mathcal{M}$ is semi-finite. Hence, if $\mathcal{M}$ is a factor, *either* $\mathcal{M}$ is of type III and σ is not inner, *or* $\mathcal{M}$ is not of type III (i.e. is of type I or II) and σ is inner, i.e. for each $t \in \mathbb{R}$, there exists a unitary element $U(t) \in \mathcal{M}$ such that $\sigma(t)[M] = U(t)^* M U(t) \;\; \forall\, M \in \mathcal{M}$. Clearly, the canonical equilibrium states of the finite systems discussed in Subsection a belong to the second term of the above alternative: $\mathcal{M} \cong \mathcal{B}(\mathcal{H})$ and σ is given by 10.1.54b. We now want to indicate that, *on the contrary,* it is the first term of the alternative that is realized, under very general circumstances, for systems of infinite extension.

Suppose indeed that an action α of an amenable group G is given on $\mathcal{M}$—we have principally in mind here the group G of space translations—such that ϕ is extremal G-invariant, and α is η-abelian on ϕ. If $\mathcal{M}$ were of type I or II, there would exist a semi-finite, faithful normal trace ψ on $\mathcal{M}$, unique up to a multiplicative constant. We now show that

$$\psi \circ \alpha(g) = \psi \;\; \forall\; g \in G\,. \tag{160}$$

Since ϕ is G-invariant, α is unitarily implemented (see Proposition 9.2.7), and $\psi \circ \alpha(g)$ is again a semi-finite, faithful normal trace on $\mathcal{M}$. Hence there exists a group homomorphism $\lambda : G \to \mathbb{R}^+$ such that $\psi \circ \alpha(g) = \lambda(g)\psi \;\forall\, g \in G$. We thus have to show $\lambda(g) = 1 \;\; \forall\;\; g \in G$. Given $g \in G$, we have either $\lambda(g) \leq 1$ or $\lambda(g)^{-1} = \lambda(g^{-1}) \leq 1$. The proof will thus be completed if we can show that $\lambda(g) < 1$ for some $g \in G$ leads to a contradiction. Since $\mathcal{M}$ is semi-finite, there exists a finite projector $E \in \mathcal{M}$ with $E \neq 0$. We would thus have $< \psi; E >< \infty$ and

$$\lim_{n\to\infty} < \psi \circ \alpha(g)^n; E >=< \psi; E > \lim_{n\to\infty} \lambda(g)^n = 0 \tag{161}$$

Since ψ is faithful and normal, this would imply

$$s - \lim_{n\to\infty} \ \alpha(g)^n[E] = 0 \tag{162}$$

and thus in particular

$$\lim_{n\to\infty} < \phi \circ \alpha(g)^n; E >= 0 \tag{163}$$

Since ϕ is G-invariant, this would imply $< \phi; E >= 0$, and since ϕ is faithful $E = 0$, which is the desired contradiction.

We can thus use (160), in conjunction with Corollary 9.2.11 and our assumption that ϕ is extremal G-invariant, to show that, for any finite projector $E \in \mathcal{M}$ with $E \neq 0$, and thus $< \phi; E > \neq 0$:

$$\begin{aligned} \infty > \ < \psi; E >=< \psi; \alpha(g)[E] >= \\ \eta < \psi; \alpha(\cdot)[E] >=< \phi; E >< \psi; I > \end{aligned} \tag{164}$$

Hence $< \psi; I > \equiv \lambda < \infty$, i.e. ψ is finite, and thus $\mathcal{M}$ is of type II_1 or I_n. Note further that $\lambda^{-1}\psi$ is a G-invariant normal state on $\mathcal{M}$; by Theorem 9.2.9 (iv) this implies $\lambda^{-1}\psi = \phi$ and thus ϕ is a faithful normal trace on $\mathcal{M}$. Consequently, the uniqueness of the modular action implies $\sigma(t) = id \ \forall \ t \in \mathbb{R}$. This means that the dynamics, given by the modular action σ associated to ϕ is trivial. This case is of very little interest for a quantum statistical mechanics at finite temperature. We are, therefore, left with the other term of the alternative only: $\mathcal{M}$ is a type III factor *and* the modular action σ is not inner.

Note that the assumptions, under which we obtained the above general results, are all satisfied for the states ϕ defined, via (154), for any one-dimensional, infinite quantum lattice with translation-invariant, finite-range interactions (see 146–148); they thus apply in particular to the situation where Powers (1967) constructed his type III-factors. The point of the above excursion was two-fold: (a) it indicates that, rather than being an accident, the occurence of type III-factors, acting on separate Hilbert spaces, is to be expected under general circumstances that one understands; (b) it shows that the mathematical structure of the quantum statistical mechanics of infinitely extended systems, while still compatible with the KMS condition, differs in an interesting manner from that of finite systems.

Having thus established that we now have a mathematically genuine generalization of the formalism discussed in Subsection a, we have to address the question of whether this larger framework could possibly help with the physical problems we encountered when we tried—see in particular Scholium 5, part b—to investigate the ergodic behaviour of the canonical equilibrium states of finite systems.

We already saw that the state ϕ, defined by (154), has good ergodic properties with respect to space translations. However, the question remains open as to whether the dynamics $\sigma : \mathbb{R} \to \text{Aut}(\mathcal{B})$, defined by (155), can possibly have similar properties. The interaction (150), with $J^z = 0 = J^{x,y}$ and $B \neq 0$, on which the Powers factors are built, shows that this cannot be the case in general: some conditions—not satisfied by that model—must be imposed on the interaction, if one is to expect that ϕ will be ergodic with respect to the evolution σ. The following discussion is an indication that infinite systems do exist, for which this is not an unreasonable hope.

Consider indeed the *$x - y$ model*, i.e. the interaction (150) with $J^z = 0$, $J^{x,y} > 0$, $0 \leq \varsigma < 1$; assume further that $B = 0$.

Notice, first, that for each $\mathcal{B}_k \cong \mathcal{M}(2, \mathbb{C})$, the linear map $\gamma_k : \mathcal{B}_k \to \mathcal{B}_k$ defined by:

$$\begin{array}{lll} \gamma_k[I_k] = +I_k & ; & \gamma_k[\tau_k^z] = +[\tau_k^z] \\ \gamma_k[\tau_k^x] = -\tau_k^x & ; & \gamma_k[\tau_k^y] = -[\tau_k^y] \end{array} \tag{165}$$

is an idempotent automorphism. For every $\Omega \in \mathcal{F}$, let

$$\gamma_\Omega \equiv \bigotimes_{k\in\Omega} \gamma_k \tag{166a}$$

and note that $\{\gamma_\Omega | \Omega \in \mathcal{F}\}$ defines an idempotent automorphism

$$\gamma \equiv \bigotimes_{k\in\mathbb{Z}} \gamma_k \tag{166b}$$

of $\mathcal{B}$. We introduce now the *even algebra*

$$\mathcal{B}_\gamma \equiv \{B \in \mathcal{B} \mid \gamma[B] = B\} \tag{167}$$

Note, incidentally, that this C^*-algebra is isomorphic to $\mathcal{B}$ itself (Størmer, 1970). Indeed γ_Ω is unitarily implemented by

$$U_\Omega \equiv \bigotimes_{k\in\Omega} \tau_k^z \in \mathcal{B}_\Omega \tag{168}$$

This unitary matrix has exactly two eigenvalues: $+1$ and -1, with the same multiplicity, namely $2^{|\Omega|-1}$. Let then E_Ω and F_Ω denote the corresponding eigenprojectors, and write $U_\Omega = E_\Omega - F_\Omega$. We have then $\mathcal{B}_\Omega \cap \mathcal{B}_\gamma = \{B \in \mathcal{B}_\Omega | B = E_\Omega B E_\Omega + F_\Omega B F_\Omega\} \cong \mathcal{B}(E_\Omega \mathcal{H}_\Omega) \oplus \mathcal{B}(F_\Omega \mathcal{H}_\Omega)$. Upon recalling that $\mathcal{B}_\Omega$ is the algebra of the $2^{|\Omega|} \times 2^{|\Omega|}$ matrices, it is then easy to check that $\mathcal{B}_\gamma$ is isomorphic to $\mathcal{B}$, thus proving for this particular case a result obtained by

Glimm (1960) for general UHF-algebras. Hence, $\mathcal{B}_\gamma$ is quite large, certainly large enough to make the following considerations appear as more than a minute curiosity.

From the fact that $\gamma^2 = id$, we have:

$$(169) \qquad \mathcal{B}_\gamma = \{(B + \gamma[B])/2 \mid B \in \mathcal{B}\}$$

We say that a state ψ on $\mathcal{B}$ is *even* if $\psi \circ \gamma = \psi$, and notice, from (169), that an even state ψ on $\mathcal{B}$ is determined by its restriction to $\mathcal{B}_\gamma$.

Notice now that γ leaves invariant the local Hamiltonians

$$(170) \qquad H_{[j,k]} \equiv -J^{x,y}\{\sum_{i=j}^{k-1}(1+\varsigma)\,\tau_i^x\tau_{i+1}^x + (1-\varsigma)\,\tau_i^y\tau_{i+1}^y\}$$

Consequently, we obtain from (152), (154) and (155):

$$(171) \qquad \phi \circ \gamma = \phi \quad \text{and} \quad \gamma \circ \sigma(t) = \sigma(t) \circ \gamma$$

Hence ϕ is uniquely determined by its restriction ϕ_γ to $\mathcal{B}_\gamma$, and $\mathcal{B}_\gamma$ is stable under $\sigma(t)$ for each t in $\mathbb{R}$; let $\sigma_\gamma : t \in \mathbb{R} \mapsto \sigma_\gamma(t) \in \mathrm{Aut}(\mathcal{B}_\gamma)$ be the action obtained by restricting σ to $\mathcal{B}_\gamma$. Similarly $\mathcal{B}_\gamma$ is stable under $\alpha(a)$ for each $a \in \mathbb{Z}$, and we denote by $\alpha_\gamma : a \in \mathbb{Z} \mapsto \alpha_\gamma(a) \in \mathrm{Aut}(\mathcal{B}_\gamma)$ the action obtained by restricting α to $\mathcal{B}_\gamma$.

One then verifies that $\{\phi_\gamma, \mathcal{B}_\gamma, \sigma_\gamma, \alpha_\gamma\}$ inherits from $\{\phi, \mathcal{B}, \sigma, \alpha\}$ the properties that the state ϕ_γ is extremal $\mathbb{Z}$-invariant and is extremal β-KMS with respect to σ_γ.

The interest of the $x - y$ model, in this general framework, is that the dynamics σ_γ of this infinite system escapes—for details, see Emch & Radin (1971)—the deadlock imposed by Scholium 10.1.5. Indeed, one can transcribe for $\{\mathcal{B}_\gamma, \sigma_\gamma\}$ an argument, originally due to Narnhofer (1970), and prove—see also Abraham, Baruch, Gallavotti & Martin-Löf (1970)—that $\sigma_\gamma : t \in \mathbb{R} \mapsto \sigma_\gamma(t) \in \mathrm{Aut}(\mathcal{B}_\gamma)$ is a norm-asymptotic abelian action, i.e.

$$(172) \qquad \lim_{|t|\to\infty} \| [B, \sigma_\gamma(t)[C]] \| = 0 \quad \forall\ B, C \in \mathcal{B}_\gamma$$

This implies, in particular, that for every invariant mean ξ on $\mathbb{R}$, this action is ξ-abelian on ϕ_γ, i.e.

$$(173) \qquad \xi < \phi_\gamma; A^*[\sigma_\gamma(\cdot)[B], C]A > = 0 \quad \forall\ A, B, C \in \mathcal{B}_\gamma$$

Consequently—see Corollary 10.1.12—ϕ_γ is extremal invariant for the dynamics σ_γ, and thus satisfies with respect to this group action all the ergodic conditions of Theorem 9.2.9 and Corollary 9.2.11. In particular, with η denoting an invariant mean over the translation group $\mathbb{Z}$, and ξ an invariant mean on the evolution group $\mathbb{R}$, we have for every $B \in \mathcal{B}_\gamma$:

$$\xi_{\phi_\gamma}(B) = < \phi_\gamma : B > I = \eta_{\phi_\gamma}(B) \tag{174}$$

i.e. the space- and time-averages of B coincide, and they are given by the expectation of B over the state ϕ_γ. This certainly is a good ergodic result, but one can actually say more, since this result only uses the weakened form (173) of (172). Indeed, since ϕ_γ is extremal β-KMS, we know—Corollary 10.1.12—that the GNS representation of $\mathcal{B}_\gamma$ associated to ϕ_γ is primary; we conclude then, from Theorem 9.2.2, that

$$w - \lim_{|t|\to\infty} \pi_{\phi_\gamma}\{\sigma_\gamma(t)[B]\} = < \phi_\gamma; B > I \quad \forall \quad B \in \mathcal{B}_\gamma \tag{175}$$

so that, in particular, ϕ_γ satisfies, for every $A, B, C \in \mathcal{B}_\gamma$, the clustering property:

$$\lim_{|t|\to\infty} < \phi_\gamma; A^*\sigma_\gamma(t)[B]C > = < \phi_\gamma; A^*C > < \phi_\gamma; B > \tag{176}$$

This, in turn, implies that for every $A \in \mathcal{B}_\gamma$, the state on $\mathcal{B}$ [not $\mathcal{B}_\gamma$ only!] defined by

$$\psi : B \in \mathcal{B} \mapsto \frac{< \phi; A^*BA >}{< \phi; A^*A >} \in \mathbb{C} \tag{177}$$

satisfies—see (169) and (176):

$$\lim_{|t|\to\infty} < \psi; \sigma(t)[B] > = < \phi; B > \tag{178}$$

i.e. every state ψ on the full $x - y$ model, provided it is of the form (177), approaches the equilibrium state ϕ as time tends to infinity. This is in particular the case for any state ψ obtained from ϕ by the following procedure. Cut the chain at two points j and k (with $1 < k - j < \infty$); change the temperature of the piece $[j, k]$, and bring it back in place, switching on again the interaction between $[j, k]$ and the rest of the chain: the situation so prepared relaxes back to ϕ. In this sense, we can say that the infinite dynamical system, described by the $x - y$ model, serves as a *thermal reservoir* for any of its finite parts.

The $x - y$ model is admittedly "rigged just the right way". Nevertheless, it does provide a proof that certain things are now possible which were rigourously impossible when one only had a formalism limited to the study of finite systems. Recall in particular that, for these systems, any evolution compatible with the existence of a canonical equilibrium state was generated by a Hamiltonian with discrete spectrum; such evolutions are, at best, almost periodic, thus ruling out taking limits as time tends to infinity. We now have learned that taking the thermodynamical limit *first* is a well-defined process, and that it not only makes it possible for such limits to exist, but also to converge to the result one would expect from thermodynamics.

The first technical problem, that one has to master next, is to control the thermodynamical limit in more general circumstances.

This is necessary at least on three accounts: (a) one should be able to deal with interactions that decrease to zero at infinity, but do not identically vanish outside a finite range; (b) quantum lattices on $\mathbb{Z}^n$ (especially with $n = 3$, but also with $n = 2$ and $n = 4$) are of physical interest; (c) continuous systems—i.e. systems in which particles are allowed to move on $\mathbb{R}^n$, or even on more general manifolds—should also be accounted for.

Still for quantum lattices on $\mathbb{Z}^n$ with $n = 1$, and $\mathcal{B}_k$ finite factors of type I, Araki (1975) proved that the thermodynamical limit of the evolution again has exactly one β-KMS state if the interaction (146) is restricted only by the following two conditions. Firstly, one assumes that there exists $\lambda > 0$ such that:

(179)
$$\| V \|_\lambda \equiv \sup_{k \in \mathbb{Z}} \{ \sum_{\Omega \in \mathcal{F}_k} \| V_\Omega \| e^{\lambda |\Omega|} \} < \infty$$
$$\text{where } \mathcal{F}_k \equiv \{\Omega \in \mathcal{F} | k \in \Omega\}$$

Secondly, for every $\Omega \in \mathcal{F}$, define

(180)
$$W_\Omega \equiv \sum_{\Lambda \in \mathcal{F}_{\partial\Omega}} V_\Lambda \quad \text{where}$$
$$\mathcal{F}_{\partial\Omega} \equiv \{\Lambda \in \mathcal{F} \mid \Lambda \cap \Omega \neq \emptyset \text{ and } \Lambda \cap \Omega^c \neq \emptyset\}$$

The second condition one imposes on the interaction V is that there exists an increasing, absorbing sequence $\{\Omega(n) | n \in \mathbb{Z}^+\} \subset \mathcal{F}$ such that

(181)
$$\sup_n \| W_{\Omega(n)} \| < \infty$$

The physical meaning of this condition, and in fact the key to the abstract formulation of Araki's assumptions, stems from the following comparison. On the one hand, the time evolution is defined by its restriction to $\mathcal{B}_0$ on which it satisfies

(182a)
$$\sigma(t)[B] = \lim_{n \to \infty} \sigma_n(t)[B]$$

with

(182b)
$$\sigma_n(t)[B] = \exp\{iH_n t\} \cdot B \cdot \exp\{-iH_n t\}$$

where n is chosen to be large enough so that $B \in \mathcal{B}_{\Omega(n)}$, and H_n is defined for every $n \in \mathbb{Z}^+$ by:

(182c)
$$H_n \equiv \sum_{\Lambda \in \mathcal{F}(n)} V_\Lambda \quad \text{with}$$
$$\mathcal{F}(n) \equiv \{\Lambda \in \mathcal{F} \mid \Lambda \subseteq \Omega(n)\}$$

On the other hand, for each $n \in \mathbb{Z}^+$ and every $B \in \mathcal{B}_{\Omega(n)}$,

$$\frac{d}{dt}\sigma(t)[B]|_{t=0} = i[H'_n, B] \tag{183a}$$

where

$$H'_n \equiv \sum_{\Lambda \in \mathcal{F}'(n)} V_\Lambda \quad \text{with} \tag{183b}$$
$$\mathcal{F}'(n) \equiv \{\Lambda \in \mathcal{F} \mid \Lambda \cap \Omega(n) \neq \emptyset\}$$

Upon comparing (180), (182) and (183), we see that

$$W_{\Omega(n)} = H'_n - H_n \tag{184}$$

Hence $W_{\Omega(n)}$ appears here as the contribution of interactions across the boundary to the effective Hamiltonian H'_n—physicists refer to $W_{\Omega(n)}$, irrespectively of dimension, as a "surface energy" term—for the system $\mathcal{B}_{\Omega(n)}$ embedded in $\mathcal{B}$. Condition (181) requires, thus, that this physical quantity be bounded in norm, uniformly in $n \in \mathbb{Z}^+$; this evidently is in sharp contrast with what happens for H_n.

The result of Araki (1975) on one-dimensional quantum lattices is thus strongly reminiscent—both in its assumptions (179) and (181), and in its conclusion, namely that there exists exactly one β-KMS state—of the theorems on the *absence* of phase transitions in one-dimensional classical systems, both discrete and continuous; see van Hove (1950), Dobrushin (1968), Ruelle (1968), Gallavotti & Miracle-Sole (1970); for surveys see Lieb & Mattis (1966), Ruelle (1969), Domb & Green (1972).

As for quantum lattices with $n > 1$ and translation invariant interaction, condition (179) had been known, even before Araki (1969) to be sufficient to ensure the existence of the limiting evolution σ as a strongly continuous action on $\mathcal{B}$—see Robinson (1968), Lanford & Robinson (1968), Manuceau & Trottin (1969); and for surveys Ruelle (1969) or Emch (1972). Under some mild geometric conditions, these authors further show that the limiting equilibrium state ϕ exists and satisfies the KMS condition, at natural temperature β, with respect to σ.

In conformity with the expected occurence of phase transitions for systems with even short range interaction, but infinite spacial extension in more than one dimension—see already Peierls (1936)—one should not expect conditions of the type of (179) to be strong enough to ensure that ϕ be the unique β-KMS state for σ, nor that $\beta P(\beta, V)$—see (153)—be real analytic for all β. One can nevertheless expect, and in fact prove—see Robinson (1967), and again the texts of Ruelle (1969) or Emch (1972)—under the very mild condition

$$\| V \|_0 \equiv \sum_{\Omega \in \mathcal{F}_k} \| V_\Omega \| \cdot |\Omega|^{-1} < \infty \ , \tag{185}$$

that $P(\beta, V)$ exists, is finite, and is continuous in β, and in V with respect to the norm $\|\cdot\|_0$.

Hence, for infinite quantum lattices, the thermodynamical limit seems to be under control, in the sense that the basic mathematical structures have been proven to exist; this is evidently not to say that there are no problems left—paramount among these the question of a genuine qualitative and quantitative understanding of phase transitions.

d. Ideal Fermi and Bose gases

This subsection surveys the thermodynamical limit for two free gases, each satisfying one of the two fundamental types of quantum statistics: the ideal Fermi gas, which we shall discuss first; and the ideal Bose gas, the first exactly solvable model exhibiting a phase transition.

For the description of the Fermi gas, we need first to define the C^*-algebra of the canonical anticommutation relations, or *CAR-algebra*. This is an object we already know, albeit under another name. Consider indeed $\{\mathcal{B}_\omega | \omega \in \mathbb{Z}^+\}$ with $\mathcal{B}_\omega \simeq \mathcal{M}(2, \mathbb{C})$ for every $\omega \in \mathbb{Z}^+$; except for the convenience of indexing with $\mathbb{Z}^+$ instead of $\mathbb{Z}$, this is a familiar object indeed—see Subsection c—and we write:

$$\mathcal{B} = \bigotimes_{\omega \in \mathbb{Z}^+} \mathcal{B}_\omega \ . \tag{186}$$

With the Pauli matrices $\{\tau^j | j = x, y, z\}$ as in (149), let

$$a_\omega \equiv \{ \bigotimes_{1 \leq \nu < \omega} \tau_\omega{}^z \} \bigotimes \{ \frac{1}{2}(\tau_\omega{}^x + i\tau_\omega{}^y) \} \tag{187a}$$

and its adjoint

$$a_\omega{}^* \equiv \{ \bigotimes_{1 \leq \nu < \omega} \tau_\omega{}^z \} \bigotimes \{ \frac{1}{2}(\tau_\omega{}^x - i\tau_\omega{}^y) \} \ . \tag{187b}$$

The elements $\{a_\omega, a_\omega{}^* | \omega \in \mathbb{Z}^+\}$ defined in this manner satisfy the *canonical anticommutation relations*

$$\begin{aligned} \{a_\nu, a_\omega\} = 0 &= \{a_\nu{}^*, a_\omega{}^*\} \\ \{a_\nu, a_\omega{}^*\} &= \delta_{\nu\omega} I \end{aligned} \tag{188}$$

where $\{A, B\} \equiv AB + BA$.

These are the mathematical objects we will be dealing with. These objects are approached in a different way by most physicists, although presentations of the CAR akin to (187) go back as far as Jordan & Wigner (1928); for a textbook presentation see e.g. Kramers (1957).

Firstly, instead of dealing with an abstract algebra, physicists usually prefer to use a concrete representation by operators acting on some Hilbert space, and they most often start with a specific irreducible representation, namely the *Fock space representation*, characterized by the existence of a (cyclic!) vector Φ_0 satisfying $a_\omega \Phi_0 = 0 \ \forall \ \omega \in \mathbb{Z}^+$. This in itself doesn't change the basic mathematical object under consideration since—as a consequence of (186), and of the fact that $\mathcal{M}(2, \mathbb{C})$ is simple—every non-zero morphism π from $\mathcal{B}$ into a C^*-algebra $\mathcal{A}$ is isometric, i.e. $\mathcal{B}$ is simple and all its non-zero representations are isometric and thus faithful. However, for reasons which will become clear, the Fock representation is ill-suited for the situation of an infinite Fermi gas at non-zero temperature.

Secondly, physicists usually interpret the index set $\mathbb{Z}^+$ as that set which is used to enumerate the elements of a specific orthonormal basis in a separable Hilbert space $\mathcal{T}$ identified with the "one particle space".

In particular, for the description of an ideal Fermi gas enclosed in a cubic box $\Lambda \subset \mathbb{R}^3$, of volume $V = L^3$, $\mathcal{T}_\Lambda$ is taken to be $\mathcal{L}^2(\Lambda, d^3x)$, in which the following orthonormal basis $\{\psi_k | k \in \mathbb{Z}^3\}$ is chosen:

$$\psi_k(x) = V^{-1/2} \exp\{-2\pi i k \cdot x / L\} \tag{189}$$

The main property of this basis is to satisfy

$$H_0 \psi_k = \epsilon_k \psi_k \tag{190}$$

where the free Hamiltonian is

$$H_0 = -\frac{1}{2m}(\frac{h}{2\pi})^2 \Delta \ ; \tag{191}$$

where h is the Planck constant, m is the mass of the particles one wants to consider, Δ is the Laplacian in $\mathbb{R}^3$, and ϵ_k, the eigenvalues of H_0 are

$$\epsilon_k = \frac{h^2}{2m} \cdot |\frac{k}{L}|^2. \tag{192}$$

This reinterpretation of the index set $\mathbb{Z}^+ \simeq \mathbb{Z}^3$ suggests to consider the maps

$$\begin{aligned} a^* : f \in \mathcal{T}_\Lambda &\mapsto \sum_{k \in \mathbb{Z}^3} (f, \psi_k) a_k{}^* \in \mathcal{B} \\ a : f \in \mathcal{T}_\Lambda &\mapsto \sum_{k \in \mathbb{Z}^3} (f, \psi_k)^* a_k \in \mathcal{B} \end{aligned} \tag{193}$$

satisfying the canonical anticommutation relations

$$\begin{aligned} \{a(f), a(g)\} = 0 = \{a^*(f), a^*(g)\} \\ \{a(f), a^*(g)\} = (g, f) I \end{aligned} \tag{194}$$

Note in particular that in the Fock space representation, the distinguished vector Φ_0, satisfies

$$a(f)\Phi_0 = 0 \quad \forall \ f \in \mathcal{T}_\Lambda \tag{195a}$$

while

$$\{a^*(f)\Phi_0 | f \in \mathcal{T}_\Lambda\} \tag{195b}$$

spans a Hilbert space isomorphic to the one-particle space $\mathcal{T}_\Lambda$ we started with. Each $a(g)$ separately sends back the vectors of the form (195b) in the space spanned by Φ_0. Similarly

$$\{a^*(f)a^*(g)\Phi_0 | f, g \in \mathcal{T}_\Lambda\}$$

spans a Hilbert space isomorphic to the antisymmetric part of $\mathcal{T}_\Lambda \otimes \mathcal{T}_\Lambda$, etc. The reader will verify that the following words give a picturesque—and consistent—description of what is going on in Fock space: Φ_0 is called the *vacuum* of the theory; $a^*(f)$ the *creation operator* for a particle of wave-function f; $a(f)$ the *annihilation operator* for that particle; $a^*(f) \cdot a(f)$ the *number operator*, counting how many particles are in wave function f; and

$$N = \sum_{k \in \mathbb{Z}^3} a_k^* a_k \quad \text{resp.} \quad H_F = \sum_{k \in \mathbb{Z}^3} \epsilon_k \, a_k^* a_k \tag{196}$$

are called the *total number operator*, resp. the *total energy operator*. The following two elementary facts will be important for what follows. First, N and H_F commute with one another. Second, the particles created by the a^*-operators satisfy the *Pauli exclusion* rule; specifically, for each $f \in \mathcal{T}_\Lambda$, the number operator $a^*(f) \cdot a(f)$ has exactly two eigenvalues: zero and one, which means that no two of these particles can be found in the same wave function f.

For every $\mu \in \mathbb{R}$, consider the action σ_μ of $\mathbb{R}$ on $\mathcal{B}$ prescribed by defining, for t (resp. f) running over $\mathbb{R}$ (resp. $\mathcal{T}_\Lambda$):

$$\sigma_\mu(t)[a^*(f)] = a^*(\exp\{iH_\mu t\}f) \quad \text{with} \quad H_\mu = H_0 - \mu I \tag{197}$$

where H_0 is the one particle free Hamiltonian entering (189)–(192).

The β-KMS state $\phi(\beta,\mu)$ for σ_μ can be computed as follows. Notice first that a_k is an analytic element for σ_μ. For the two-point correlation $< \phi(\beta,\mu); a^*(f)\, a_k >$, the KMS condition reads then:

$$< \phi(\beta,\mu); a^*(f)\sigma_\mu(t+i\beta)[a_k] > = < \phi(\beta,\mu); \sigma_\mu(t)[a_k]a^*(f) > \tag{198}$$

and thus:

$$\begin{aligned} &\exp\{\beta(\epsilon_k - \mu)\} < \phi(\beta,\mu); a^*(f)\, a_k > = \\ &< \phi(\beta,\mu); a_k\, a^*(f) > = (f, \psi_k) - < \phi(\beta,\mu); a^*(f)\, a_k > \end{aligned} \tag{199}$$

from which we conclude:

$$\begin{aligned} &< \phi(\beta,\mu); a^*(f)a(g) > = (K(\beta,\mu)f, g) \\ &\text{with } K(\beta,\mu) = [\exp\{\beta H_\mu\} + 1]^{-1} \end{aligned} \tag{200}$$

For higher correlation-functionals, a similar computation gives by recursion:

$$\begin{aligned}(201)\qquad & < \phi(\beta,\mu); a^*(f_1)a^*(f_2)\ldots a^*(f_n)a(g_m)\ldots a(g_2)a(g_1) > \\ & = \delta_{nm} \det\{(K(\beta,\mu)f_i, g_j)\}.\end{aligned}$$

By a succession of doublings of the linear dimensions of the cube Λ, one obtains an absorbing sequence $\{\Lambda_n\}$ of cubes in $\mathbb{R}^3$, which allows to control easily the limit: $\mathcal{T}$ becomes $\mathcal{L}^2(\mathbb{R}^3, d^3x)$, and the correlation functionals keep the form they have in (201), except that

$$(202)\qquad (K(\beta,\mu)f, g) \equiv \frac{1}{(2\pi)^3} \int d^3k\, \tilde{f}(k)\, \tilde{g}(k)^* \rho(\beta,\mu;k)$$

where $\tilde{f}$ denotes the Fourier transform of $f \in \mathcal{T}$, and

$$\begin{aligned}(203)\qquad & \rho(\beta,\mu;k) \equiv [\exp\{\beta[\epsilon(k)-\mu]\}+1]^{-1} \\ & \text{with } \epsilon(k) \equiv h^2|k|^2/2m\,.\end{aligned}$$

These correlation functionals determine, on $\mathcal{B}$, a state $\phi(\beta,\mu)$ that satisfies the KMS condition, at natural temperature β, for the action σ_μ determined by (197) where now:

$$(204)\qquad (\exp\{iH_\mu t\}f)^\sim(k) = \exp\{i[\epsilon(k)-\mu]t\}\tilde{f}(k).$$

The above expressions define, mathematically, the *ideal Fermi gas*, at *natural temperature* β, with *chemical potential* μ—or *fugacity* $z \equiv \exp(\beta\mu)$.

The meaning of μ is obtained by retracing our steps, and noticing that, at finite volume V, we did not really compute the canonical equilibrium state corresponding to the total energy H_F. Recall that the canonical equilibrium state was obtained, in Subsection a, by maximalizing the entropy, with the constraint that the expectation value of the energy, here $< H_F >$, be fixed; see Proposition 10.1.3. Mathematically, this brought in the Lagrange multiplier β. Physically, this procedure was interpreted by stating that the system considered was in contact with a thermal reservoir at natural temperature β.

Instead, we computed here what is called the *grand canonical equilibrium state*. This is obtained by maximalizing the entropy subject to two constraints, namely that the expectation values of both the total energy H_F and the total number of particle N be fixed. Mathematically, this brings about a second Lagrange multiplier that is conventionally written $(\beta\mu)$. Physically, this procedure is interpreted by stating that the system considered is in contact with a thermal reservoir, with which it can also exchange particles.

Since H_F and N commute—see (196)—the grand canonical equilibrium state, at finite volume, is given by

$$(205a)\qquad \rho \equiv \exp\{-\beta(H_F-\mu N)\}/Z(\beta,\mu)$$

where

(205b) $$Z(\beta,\mu) \equiv tr[\exp\{-\beta(H_F - \mu N)\}]$$

Note that $H_F - \mu N$, restricted to the one-particle subspace of the Fock space, reduces to the operator H_μ appearing in (197). Notice also that the *grand canonical partition* function $Z(\beta,\mu)$ is given by:

(206) $$\ell n\ Z(\beta,\mu) = \sum_{k\in\mathbb{Z}^3} \ell n\ [1 + \exp\{-\beta(\epsilon_k - \mu)\}]$$

From this, one defines

(207) $$\begin{aligned} \frac{1}{v} &\equiv z\partial_z[\frac{1}{V}\ell n\ \ Z^*(T,z)] \\ \frac{P}{kT} &\equiv \frac{1}{V}\ell n\ \ Z^*(T,z) \end{aligned}$$

where we wrote $Z^*(T,z) = Z(\beta,\mu)$ with $\beta = 1/kT$ and $z = \exp(\beta\mu)$.

Note that $v^{-1} = < N > /V$ can be interpreted as the expected value of the total number of particles per unit volume; stated differently $v^{1/3}$ can be interpreted as the average interparticle distance. P will be interpreted as the pressure.

In the limit $V \to \infty$, (206)–(207) give:

(208) $$\begin{aligned} \frac{1}{v} &= 4\pi\int_0^\infty dp\ \ p^2 z[z + \exp\{h^2p^2/2mkT\}]^{-1} \\ \frac{P}{kT} &= 4\pi\int_0^\infty dp\ \ p^2\ \ \ell n\ [1 + z\exp\{-h^2p^2/2mkT\}] \end{aligned}$$

i.e.

(209a) $$\frac{1}{v} = \lambda^{-3} f(\frac{3}{2}, z)$$

(209b) $$\frac{P}{kT} = \lambda^{-3} f(\frac{5}{2}, z)$$

where

(210) $$\lambda = h\ (2\pi mkT)^{-1/2}$$

(211a) $$f(s,z) = -g(s,-z)$$

(211b) $$g(s,z) = \frac{z}{\Gamma(s)}\int_0^\infty dt \frac{t^{s-1}}{e^t - z}$$

Drawing on an analogy with the de Broglie wavelength—see 7.2.35a—physicists refer to λ as the *thermal wavelength.* The classical analyst will recognize in (211b) the *Appell integral*—see, e.g. Whittaker & Watson (1927), p. 280, ex. 7. For every s with $\mathbb{R}e(s) > 0$, this integral defines a function of z that is analytic in the cut complex plane $\mathbb{C} \setminus [1, \infty)$. For z with $|z| < 1$, and s with $\mathbb{R}e(s) > 0$, $g(s, z) = z\varsigma(s, z)$ where $\varsigma(s, z)$ is the *Lerch zeta function*

$$\varsigma(s, z) = \sum_{n=0}^{\infty} z^n \, (n+1)^{-s} \tag{211c}$$

—see, e.g. Erdélyi (1953), Vol. I, pp. 27–31, or C. Truesdell (1945), where the reader will also find historical notes. If $z = 1$ and $\mathbb{R}e(s) > 1$, the series (211c) is still convergent, and converges to the *Riemann zeta function* $\varsigma(s)$.

For our study of the ideal Fermi gas, we need only the functions $g(s, \cdot) : z \in (-\infty, 0] \mapsto g(s, z) \in \mathbb{R}$ for the special values $s = \frac{3}{2}$ and $s = \frac{5}{2}$. Of direct interest to us are the following three facts: (i) $f(\frac{3}{2}, \cdot) : z \in \mathbb{R}^+ \mapsto f(\frac{3}{2}, z) \in \mathbb{R}^+$ is a monotonically increasing, real analytic function; in particular, (209a) can be inverted to give z as a uniquely defined function of (λ^3/v); (ii) $z^{-1} f(\frac{3}{2}, z) \to 1$ as $z \to 0$; and (iii) the leading term in the asymptotic expansion of $f(\frac{3}{2}, z)$ as $z \to \infty$ is $[(4/3)\pi^{-1/2}(\ell n \; z)^{3/2}]$. The following two limiting cases are indeed of special physical interest.

Firstly, the case $(\lambda^3/v) \ll 1$ is the so-called high temperature/low density limit of the physicists. Note that, formally, this limit can be viewed as the limit $h \to 0$, and we thus expect to find back the classical limit. This is indeed the case: from (209) and (211) one finds in the limit $(\lambda^3/v) \to 0$

$$\frac{Pv}{kT} = 1 + 2^{-5/2}(\lambda^3/v) + : \ldots \tag{212}$$

The zeroth order approximation in (212) reproduces the formula for the classical ideal gas. The main feature of the first order correction is that it is positive; upon comparing this result with the classical virial expansion, one obtains a classical picture of the quantum effects in the ideal Fermi gas at high temperature and low density: the Pauli exclusion rule acts in the same way as would a small effective repulsion between classical particles.

Secondly, the case $(\lambda^3/v) \gg 1$ is the so-called low temperature/high density limit. Note that in this limit the thermal wavelength λ is much larger than the average interparticle distance $v^{1/3}$; one should, therefore, expect quantum effects to dominate, and thus the Fermi gas to behave, under these circumstances, quite differently from a classical gas. From the leading term of the asymptotic expansion of $f(\frac{3}{2}, z)$ as $z \to \infty$, one finds

$$\lim_{(\lambda^3/v)\to\infty} [z \; \exp(-\beta\epsilon_F)] = 1 \qquad \text{where } \epsilon_F \equiv \frac{\hbar^2}{2m}\left(\frac{6\pi^2}{v}\right)^{2/3} \tag{213}$$

The most drastic illustration of what happens in this limit is obtained by computing from (203), at v fixed.

$$\lim_{T\to 0} \rho^*(T,z;k) = \begin{cases} 1 & \text{if } \epsilon(k) < \epsilon_F \\ 0 & \text{if } \epsilon(k) > \epsilon_F \end{cases} \tag{214}$$

Loosely speaking, this situation can be characterized by saying that all energy levels, up to the *Fermi energy* ϵ_F, are occupied while all energy levels above ϵ_F are empty. From the macroscopic point of view, the Pauli exclusion rule has still another consequence. Recall that it prevents any two particles to be in the same state; in particular, again speaking loosely, only one particle can be in the state of zero momentum; the other particles, with non-zero momentum, should thus be expected to prevent the pressure to vanish, even at zero absolute temperature. Specifically, upon using for the infinite Fermi gas, our asymptotic expansions as $z \to \infty$, we compute from (209):

$$\lim_{(\lambda^3/v)\to\infty} P v^{5/3} = \frac{(6\pi^2)^{2/3}}{5} \cdot \frac{\hbar^2}{m} \tag{215}$$

Hence, at v fixed, the pressure P approaches indeed a finite, non-zero value as $T \to 0$, in contrast with what happens for a classical gas.

A gas of He^3 particles at very low temperature ($<\sim 1°K$) would satisfy the condition $(\lambda^3/v) \gg 1$; and so would electrons in metals, under ordinary laboratory conditions—say room temperature (note, however, that such an electron gas is not quite an ideal Fermi gas). The fact that it is the limit $(\lambda^3/v) \to \infty$ that counts, and not the limit $T \to 0$ alone, is emphasized by the fact that the condition $(\lambda^3/v) \gg 1$ is also realized in white dwarf stars, where the central temperature is of the order of $10^7\ °K$, whereas the electron density v^{-1} is so enormous that $T_F \equiv k^{-1}\epsilon_F$ is of the order of $10^{11}\ °K$, so high in fact that a relativistic treatment of this gas is called for.

Note finally that the specific heat of the ideal Fermi gas can be proven to be a smooth, monotonically increasing function of the temperature for $T \in (0,\infty)$; it approaches the classical limit $(3k/2)$ as $T \to \infty$, and goes linearly to zero as $T \to 0$.

This remark concludes what we wanted to say here about the ideal Fermi gas; for more details, see e.g. Landau & Lifshitz (1958), Huang (1963), or from a point of view more in tune with modern mathematical physics, Thirring (1983).

We now turn our attention to the ideal Bose gas. This system is obtained, formally, by replacing the canonical anticommutation relations (194) by the *canonical commutation relations*

$$\begin{gathered} [a(f), a(g)] = 0 = [a^*(f), a^*(g)] \\ [a(f), a^*(g)] = (g,f)I \end{gathered} \tag{216}$$

where we have $[A, B] = AB - BA$, instead of $\{A, B\} = AB + BA$ as in (194). There are, however, three difficulties to be attended to when one attempts to generalize to (216) the analysis previously carried on from (194). These difficulties are: (i) one cannot represent $a(f)$ and $a^*(f)$ by bounded operators acting on a Hilbert space; (ii) a phase transition does occur, first evidenced by a breakdown in analyticity; (iii) the time evolution for the infinite system cannot be obtained by the usual limiting procedure.

We first take care of some of the domain questions that are associated to the fact that we must consider, for our basic objects, unbounded $a(f)$ and $a^*(f)$.

With $\mathcal{T} \equiv \mathcal{L}^2(\mathbb{R}^3, d^3x)$, we consider $\mathcal{T}^{(N)} \equiv \mathcal{L}^2(\mathbb{R}^{3N}, d^{3N}x)$ and the operator S acting on $\mathcal{T}^N$ defined by

$$(S\Psi)(x_1, \ldots, x_N) = \frac{1}{N!} \sum_{p \in \mathcal{P}_N} \Psi(x_{p(1)}, \ldots, x_{p(n)}) \tag{217}$$

where $\mathcal{P}_N$ is the group of all permutations of the N indices $\{1, \ldots, N\}$. We now introduce:

$$\mathcal{F}^{(N)} \equiv \{\Psi \in \mathcal{T}^N \mid S\Psi = \Psi\} \tag{218}$$

Hence $\mathcal{F}^{(N)}$ is spanned by the square integrable functions on $(\mathbb{R}^3)^N$ that are symmetric in all their arguments $x_1, \ldots, x_N \in \mathbb{R}^3$. With $\mathcal{F}^{(O)} \equiv \mathbb{C}$, we define the *Fock space of the CCR on* $\mathbb{R}^3$ as the completed direct sum:

$$\mathcal{F} \equiv \bigoplus_{N=0}^{\infty} \mathcal{F}^{(N)} \tag{219}$$

Let further $\mathcal{F}_o$ be the incomplete direct sum

$$\mathcal{F}_o \equiv \bigcup_{M=1}^{\infty} \bigoplus_{N=0}^{M} \{\mathcal{F}^{(N)} \cap \mathcal{D}\} \tag{220}$$

where $\mathcal{D}$ denotes the space of all infinitely differentiable functions with compact support. On the dense subspace $\mathcal{F}_o \subset \mathcal{F}$, we can now define the creation operator $a^*(f)$ and annihilation operator $a(f)$ with f arbitrary in $\mathcal{T}$ by:

$$\begin{aligned}(a^*(f)\Psi)^{(N)}(x_1, \ldots, x_N) = \\ N^{-1/2} \sum_{n=1}^{N} f(x_n)\, \Psi^{(N-1)}(x_1, \ldots, \tilde{x}_n, \ldots, x_N)\end{aligned} \tag{221a}$$

where $\sim$ over a variable x_n means that this argument has been deleted—so that $\Psi^{(N-1)}$ is indeed a function of $(N-1)$ variables; we also define on $\mathcal{F}_o$, for every $f \in \mathcal{T}$:

$$\begin{aligned}(a(f)\Psi)^{(N)}(x_1, \ldots, x_N) = \\ (N+1)^{1/2} \int dx \;\; f(x)^* \, \Psi^{(N+1)}(x, x_1, \ldots, x_N)\end{aligned} \tag{221b}$$

Upon noticing that the operator

$$\Phi(f) = 2^{-1/2}\{a^*(f) + a(f)\} \tag{222}$$

is essentially self-adjoint on $\mathcal{F}_o$, i.e. admits exactly one self-adjoint extension, one can introduce the unitary operator

$$W(f) = \exp\{i\Phi(f)\} \tag{223}$$

We further remark that the operators (221) satisfy the CCR relations (216) on $\mathcal{F}_o$, or equivalently:

$$W(f)W(g) = W(f+g)\exp\{-i\ \mathrm{Im}(f,g)/2\} \tag{224}$$

Compare with (8.3.189); note further that, for every $f \in \mathcal{T}$,

$$\lambda \in \mathbb{R} \mapsto W(\lambda f) \in \mathcal{U}(\mathcal{F}) \tag{225}$$

is continuous for the weak operator topology; moreover

$$\mathcal{B}_0 \equiv \{\sum_n c_n\, W(f_n) \mid c_n \in \mathbb{C},\ f_n \in \mathcal{D}(\mathbb{R}^3)\} \tag{226}$$

is an irreducible *-subalgebra of $\mathcal{B}(\mathcal{F})$. We denote by $\mathcal{B}$ the C^*-algebra obtained as the norm closure of $\mathcal{B}_0$, and we refer to $\mathcal{B}$ (resp. $\mathcal{B}_0$) as the *algebra of quasi local* (resp. *strictly local*) *observables* of our system. Note, in particular that a functional

$$\hat{\phi} : f \in \mathcal{D}(\mathbb{R}^3) \mapsto \hat{\phi}(f) \in \mathbb{C} \tag{227a}$$

defines uniquely a state ϕ on $\mathcal{B}$, such that

$$< \phi; W(f) > = \hat{\phi}(f) \quad \forall\ f \in \mathcal{D}(\mathbb{R}^3), \tag{227b}$$

if and only if $\hat{\phi}$ satisfies the following three conditions:

$$\hat{\phi}(0) = 1\ ; \tag{228a}$$

for all f and $g \in \mathcal{D}(\mathbb{R}^3)$:

$$\lambda \in \mathbb{R} \mapsto \hat{\phi}(\lambda f + g) \in \mathbb{C} \tag{228b}$$

is continuous; for all $n \subset \mathbb{Z}^+$, all $\{\lambda_k | k = 1, \ldots, n\} \subset \mathbb{C}$, and all $\{f_k | k = 1, \ldots, n\} \subset \mathcal{D}(\mathbb{R}^3)$:

$$\sum_{j,k=1}^{n} \lambda_j\, \lambda_k^*\, \hat{\phi}(f_j - f_k)\exp\{-i\,\mathrm{Im}(f_j, f_k)/2\} \geq 0 \tag{228c}$$

Compare with (8.3.211b). For instance:

(229) $$\hat{\phi}_0(f) = \exp\{- \| f \|^2 /4\}$$

corresponds to the pure state

(230) $$< \phi_0; B > = (B\Phi_0, \Phi_0)$$

where $\Phi_0 \in \mathcal{F}^{(0)}$ with $\| \Phi_0 \| = 1$.

We complete the algebraic description of our system by noticing that $\mathcal{B}$ is the C^*-inductive limit of an increasing sequence $\{\mathcal{B}_{\Lambda(n)} | n \in \mathbb{Z}^+\}$ of local von Neumann algebras namely:

(231) $$\mathcal{B}_{\Lambda(n)} \equiv \{\sum_m \lambda_m W(f_m) \mid c_m \in \mathbb{C},\ f_m \in \mathcal{D}(\Lambda(n))\}''$$

with, L being fixed, and n running over $\mathbb{Z}^+$, and:

(232) $$\Lambda(n) \equiv \{x \in \mathbb{R}^3 | - n\frac{L}{2} \leq x^k \leq n\frac{L}{2}\, ;\, k = 1, 2, 3\}.$$

With $\Lambda = \Lambda(n)$ fixed, we define $\mathcal{T}_\Lambda$, $\{\psi_k, \epsilon_k | k \in \mathbb{Z}^3\}$, H_0 as we did for the Fermi gas—see (189) to (192)—and then the actions $\sigma_\mu : \mathbb{R} \to \text{Aut}(\mathcal{B}_\Lambda)$:

(233) $$\begin{aligned} &\sigma_\mu(t)[W(f)] = W(\exp\{iH_\mu t\} f) \\ &\text{with} \quad H_\mu = H_0 - \mu I \quad \text{and} \quad \mu \in (-\infty, 0). \end{aligned}$$

Compare with (197); the present restriction of the range of μ will be made physically transparent in the immediate sequel. A simple computation, akin to that performed in the case of the Fermi gas, gives that each action σ_μ has exactly one β-KMS state $\phi(\beta, \mu)$, determined by the functional

(234) $$\begin{aligned} &\hat{\phi}(\beta, \mu;\ \cdot\ ) : f \in \mathcal{D}(\Lambda) \mapsto \exp\{-(\Theta(\beta, \mu) f, f)/4\} \\ &\text{with } \Theta(\beta, \mu) \equiv \ \coth\ \{\beta H_\mu /2\} \end{aligned}$$

Compare with (8.3.221), and note in particular that, in contrast with (200), we now have:

(235) $$< n_k > \equiv < \phi(\beta, \mu); a_k^* a_k > = [\exp\{\beta(\epsilon_k - \mu)\} - 1]^{-1}$$

Hence the restriction $\mu \in (-\infty, 0)$ is directly interpretable as the condition that, in the state $\phi(\beta, \mu)$ on the CCR-algebra $\mathcal{B}_\Lambda$, the mean value $< n_k >$ of the number of particles in the wave function ψ_k be strictly positive and finite for all $k \in \mathbb{Z}^3$.

The thermodynamical quantities v^{-1} and (P/kT) are again computed as the limit, over the absorbing sequence $\{\Lambda(n) | n \in \mathbb{Z}^+\}$ of the expressions (207) where now (206) is replaced by:

$$\ell n\, Z(\beta,\mu) = -\sum_{k\in\mathbb{Z}^3} \ell n[1-\exp\{-\beta(\epsilon_k-\mu)\}]. \tag{236}$$

Note that (236) satisfies

$$< n_k > = -\frac{1}{\beta}\cdot\frac{\partial}{\partial\epsilon_k}\ell n\, Z(\beta,\mu) \tag{237}$$

with $< n_k >$ given by (235), as it should. In the above limit, we obtain again the thermodynamical quantities v^{-1} and (P/kT) in the form of Appell's integrals, namely now

$$\begin{aligned} \frac{1}{v} &= 4\pi\int_0^\infty dp\ p^2\ z[\exp\{h^2p^2/2mkT\}-z]^{-1} \\ \frac{P}{kT} &= -4\pi\int_0^\infty dp\ p^2\ \ell n[1-z\exp\{-h^2p^2/2mkT\}] \end{aligned} \tag{238}$$

One can rewrite these expressions in the form

$$\begin{aligned} \frac{1}{v} &= \lambda^{-3} g(\frac{3}{2},z) \\ \frac{P}{kT} &= \lambda^{-3} g(\frac{5}{2},z) \end{aligned} \tag{239a}$$

where $g(s,z)$ can be expressed—compare with the Lerch ς-functions defined in (211c)—as:

$$g(s,z) = \sum_{n=1}^{\infty} z^n/n^s\,. \tag{239b}$$

The limiting case $(\lambda^3/v) \ll 1$ of (239) is without surprise, and we receive

$$\frac{Pv}{kT} = 1 - 2^{-5/2}(\lambda^3/v) + \ldots \tag{240}$$

From the defining relation (210) of λ, we again interpret this limit as the high temperature/low density limit; the zeroth order approximation of (240) coincides again with the result obtained for the ideal classical gas, as we should expect from the fact that, at fixed temperature T, λ is proportional to the Planck constant h. Note that the first order correction in (240) is now negative, in contrast with the result obtained for the ideal Fermi gas—compare with (212) and the discussion following it: the classical picture of the first order correction in (240) is that of a small effective attraction between the particles.

Upon returning to the general expression (239), we must note that the function $g(\frac{3}{2},\cdot) : (0,1) \to \mathbb{R}^+$ is a smooth, strictly increasing function which approaches a finite limit as $z \to 1$, namely $\varsigma(\frac{3}{2}) = 2.612\ldots$ From this we conclude that (239) holds only as long as:

$$(241a) \qquad (\lambda^3/v) < \varsigma(\frac{3}{2})$$

which we can equivalently rewrite as the condition:

$$(241b) \qquad T\, v^{2/3} > C_1 \equiv h^2(2\pi m k)^{-1}\varsigma(\frac{3}{2})^{-2/3}$$

or, still in another equivalent form, as the condition

$$(241c) \qquad P\, v^{5/3} > C_2 \equiv h^2(2\pi m)^{-1}\varsigma(\frac{5}{2})\,\varsigma(\frac{3}{2})^{-5/3}$$

These three equivalent inequalities characterize what is called the *gaseous phase* of the *ideal Bose gas*.

This leaves open the question of whether the boundary $(\lambda^3/v) = \varsigma(\frac{3}{2})$, or $Tv^{2/3} = C_1$, or $Pv^{5/3} = C_2$ can be physically reached, or even crossed through. Mathematically the answer involves studying whether, and how, the limit $z = 1$ can be reached. To understand what is happening, we return briefly to the finite volume expression of v^{-1}. From (235) we obtain

$$(242) \qquad \frac{1}{V}\Big(\sum_{k\in\mathbb{Z}^3} < n_k >\Big) = \frac{1}{V}\Big(\sum_{k\in\mathbb{Z}^3}{}' < n_k >\Big) + \frac{1}{V}\cdot\frac{z}{1-z}$$

where: $< n_k >$ is given explicitly in (235), and $\sum'$ means that the term corresponding to $k = 0$ is omitted from the summation. Note that each one of the $< n_k >$ with $k \neq 0$, is well-behaved as $z \equiv \exp(\beta\mu) \to 1$. This suggests taking simultaneously the two limits $V \to \infty$ and $z \to 1$ in such a manner that the second term in (242) approaches a finite limit, say v_0^{-1}. If we do this, we obtain instead of (239):

$$(243) \qquad \begin{aligned} \frac{1}{v} &= \lambda^{-3}\varsigma(\frac{3}{2}) + \frac{1}{v_0} \\ \frac{P}{kT} &= \lambda^{-3}\varsigma(\frac{5}{2}) \end{aligned}$$

These are the characteristic equations that link the density v^{-1}, the pressure P and the temperature T—recall that the latter enters in λ through (210)—of what is called the *condensed phase* of the *ideal Bose gas*; the word "condensed" refers to the fact that, in the course of the combined limit $(V, z) \to (\infty, 1)$ that we have taken, we have imposed that the ground state $k = 0$ be occupied with a finite density $v_0^{-1} =< n_0 > /V$. We shall come back to the mathematical meaning of this limiting procedure later in this subsection.

Before doing so, however, we want to complete the thermostatic description of the model. A typical isotherm through both phases is pictured in Fig. 1; the analytic form of this curve is given by (239) and (243), while the dashed line represents the coexistence curve $Tv^{2/3} = C_1$—see (241b). Fig. 2 graphs the specific heat at constant v.

The reader will verify that $c_v \equiv (\partial u/\partial T)_v$ is given analytically by:

$$c_v = k\left\{\frac{15}{4}\left(\frac{v}{\lambda^3}\right)\cdot g\left(\frac{5}{2}, z\right) - \frac{9}{4}\cdot\frac{g(\frac{3}{2}, z)}{g(\frac{1}{2}, z)}\right\} \tag{244a}$$

for the gaseous phase; and

$$c_v = k\left\{\frac{15}{4}\left(\frac{v}{\lambda^3}\right)\cdot\varsigma\left(\frac{5}{2}\right)\right\} \tag{244b}$$

for the condensed phase. Three qualitative comments are called for here. First, c_v approaches the classical value $(3k/2)$ as $T \to \infty$; thus confirming our earlier remark that $(\lambda^3/v) \to 0$ corresponds to the classical limit. Second, c_v behaves like $T^{3/2}$ as $T \to 0$; this is to be contrasted to the Debye behaviour (7.2.34b), namely T^3; this difference is due to the difference in the behaviour, as $|k| \to 0$ of the one-particle energy spectrum of the ideal gas considered here—where $\epsilon_k \sim |k|^2$—and that of the phonon gas—where $\epsilon_k \sim |k|$. At the critical temperature T_c, the derivative of c_v presents a discontinuity, but c_v itself is finite. In the laboratory, He^4 is a candidate for the theoretical model discussed here. When the numbers are brought in, the critical temperature falls in the right ballpark: $\sim 2.2°K$ (experiment) instead of $\sim 3.2°K$ (theory). However, c_v diverges as T approaches T_c from either side. The ideal Bose gas model is not quite sufficient to predict the details of this phase transition: inter-particle interactions have to be brought to bear on the problem.

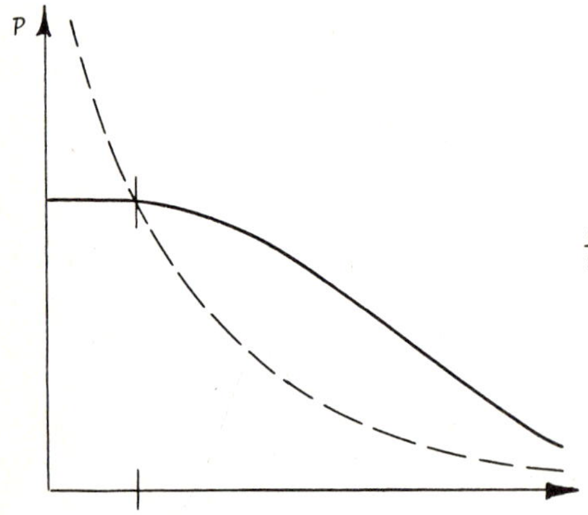

Fig.1

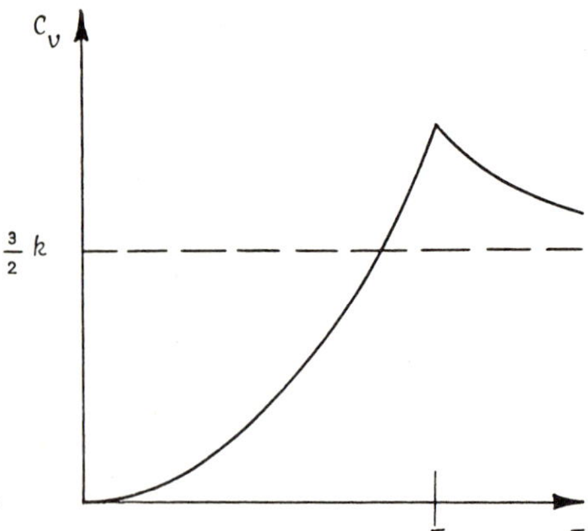

Fig.2

From the point of view of the conceptual foundations of condensed matter physics, the principal lesson to be learned from the study of the ideal Bose gas is that a description of phase transitions—across which certain thermodynamical functions show a non analytic behaviour—is within the purview of quantum statistical mechanics, provided the latter is formulated broadly enough to include the description of infinite systems: the mathematical investments involved in controlling the thermodynamical limit are warranted by the physics of the problem.

The particular history of the investigations of the ideal Bose gas is recounted in Brush (1983); see also Pais (1982). Let it suffice here to mention a few salient points to fix the time scale of events. The original theoretical discovery of the existence of a condensed phase—the phenomenon known today as the *Bose-Einstein condensation*—is due to Einstein (1924–5); it seems to be true that, at the time, neither Einstein nor his contemporaries did recognize the importance of this contribution for the general theory of phase transitions. For the conceptual difficulties involved in the need to use the thermodynamical limit for a proper description of phase transitions within the framework of statistical mechanics, see Uhlenbeck (1927) and Kahn & Uhlenbeck (1938). The interpretation of the observed Helium phase transition as a Bose-Einstein condensation was proposed by London (1938). The robustness of the Bose-Einstein condensation against the inclusion, in the theory, of interparticle interactions was recognized by Feynman (1953). To these milestones in the discussion of the physical aspects of the phenomenon at hand, we should add the following references to its mathematical description: Araki & Woods (1963), Dubin & Sewell (1970) and Cannon (1973).

As a result of the work of Cannon (1973)—see also the references quoted therein—we know that, in the thermodynamical limit, the canonical and the grand canonical equilibrium states converge nicely to states on the quasi local algebra $\mathcal{B}$, and that their explicit form is as follows.

We choose units such that $(h/2\pi) = 1 = m$. The limits are taken at fixed, but arbitrary, values of the following two thermodynamical parameters: the natural temperature $\beta \in (0, \infty)$ and the density $\rho = v^{-1} \in (0, \infty)$. We denote by $\rho_c(\beta)$ the value of ρ, as a function of β, on the phase separation curve—see (241b), where the inequality is replaced by an equality. At fixed β, $\rho \leq \rho_c(\beta)$ corresponds therefore to the gaseous phase, where z is determined, as a function of ρ and β, through (239a). For the condensed phase, we write $\rho_0 = \rho - \rho_c(\beta)$. For both phases we denote by $\hat{\phi}_c(\beta, \rho; \cdot)$ [resp. $\hat{\phi}_g(\beta, \rho; \cdot)$] the functional—see (227b)—corresponding to the canonical [resp. grand canonical] equilibrium state on $\mathcal{B}$ in the thermodynamical limit.

For the gaseous phase, the canonical and the grand canonical equilibrium states are identical—in the thermodynamical limit!—and one has:

$$\hat{\phi}_g(\beta, \rho; f) = \hat{\phi}_c(\beta, \rho; f) = \hat{\phi}(\beta, z; f) \tag{245}$$

where

(246)
$$\hat{\phi}(\beta, z; f) \equiv \exp\{-\frac{1}{4} \| f \|^2 - \frac{1}{2} \cdot \frac{1}{(2\pi)^3} \int d^3k \;\; \rho(\beta, z; k) \, |\tilde{f}(k)|^2\}$$
$$\rho(\beta, z; k) \equiv z[\exp\{\beta\epsilon(k)\} - z]^{-1} \text{ with } \epsilon(k) \equiv |k|^2/2$$

and z is determined by ρ and β—see (238) or (239)—by

(247)
$$\rho = \frac{1}{(2\pi)^3} \int d^3k \;\; \rho(\beta, z; k).$$

In the condensed phase one has:

(248)
$$\hat{\phi}_g(\beta, \rho; f) = \hat{\phi}(\beta, 1; f) \; \exp\{-\frac{1}{2} \rho_0 \, |\tilde{f}(0)|^2\}$$

(249)
$$\hat{\phi}_c(\beta, \rho; f) = \hat{\phi}(\beta, 1; f) \; J_0([2\rho_0]^{1/2} \, |\tilde{f}(0)|)$$

where $\hat{\phi}(\beta, 1; f)$ is given by (246) at $z = 1$, and where J_0 is the Bessel function

(250)
$$J_0(\xi) = \frac{1}{2\pi} \int_0^{2\pi} d\theta \exp\{i\xi \cos\theta\}.$$

Araki & Woods (1963) obtained the explicit form of the GNS representations $\pi(\beta, \rho; \cdot)$ of the quasi local algebra $\mathcal{B}$ associated to the canonical equilibrium states $\phi_c(\beta, \rho)$ determined by the functionals (245, 246)—for the gaseous phase—and (249)—for the condensed phase. These representations are given by exhibiting the explicit form of $W(\beta,, \rho; f) \equiv \pi(\beta, \rho; W(f))$ for f arbitrary in $\mathcal{D}(\mathbb{R}^3)$. This is done as follows.

We first describe the representations corresponding to the gaseous phase. In the Hilbert space

(251)
$$\mathcal{H} \equiv \mathcal{F} \otimes \mathcal{F}$$

where $\mathcal{F}$ is the Fock space (219), let

(252)
$$\Phi \equiv \Phi_0 \otimes \Phi_0$$

where Φ_0 is the vacuum vector (229) in $\mathcal{F}^{(0)}$. Consider then the unitary operators, defined for all $f \in \mathcal{D}(\mathbb{R}^3)$, by:

(253)
$$W(\beta, \rho; f) \equiv W(\varsigma_+ f) \otimes W(K\varsigma_- f)$$

where $W(\cdot)$ is defined in (223), and

(254)
$$(\varsigma_+ f)^\sim(k) \equiv [1 + \rho(\beta, z; k)]^{1/2} \tilde{f}(k)$$
$$(\varsigma_- f)^\sim(k) \equiv [\rho(\beta, z; k)]^{1/2} \tilde{f}(k)$$
$$(Kf)^\sim(k) \equiv \tilde{f}(k)^*$$

where $\rho(\beta, z; k)$ as in (246). It is then easy to check that $W(\beta, \rho; \cdot)$ satisfies (224), and determines, therefore, a representation of the quasi local algebra $\mathcal{B}$. Similarly, one checks that, for every $f \in \mathcal{D}(\mathbb{R}^3)$

$$(W(\beta, \rho; f)\Phi, \Phi) = \hat{\phi}_c(\beta, \rho; f) \tag{255}$$

with $\hat{\phi}_c(\beta, \rho; \cdot)$ as in (245). Because the linear span of $\{W(\beta, \rho; f)\Phi \mid f \in \mathcal{D}(\mathbb{R}^3)\}$ is dense in $\mathcal{H}$, we have obtained the GNS representation of $\mathcal{B}$ associated to the canonical equilibrium states of the gaseous phase. Since the properties of the von Neumann algebra

$$\mathcal{M} \equiv \{W(\beta, \rho; f) \mid f \in \mathcal{D}(\mathbb{R}^3)\}'', \tag{256}$$

generated by any one of these representations, played a seminal role in the algebraic approach to quantum statistical mechanics, it seems proper to outline them here.

One notices immediately that $\mathcal{H}$ harbors another representation of $\mathcal{B}$, namely:

$$\overline{W}(\beta, \rho; f) \equiv W(K\varsigma_- f) \otimes W(\varsigma_+ f) \tag{257}$$

With $\overline{\mathcal{M}}$ denoting the von Neumann algebra generated by this representation, one notices that $\overline{\mathcal{M}} \subseteq \mathcal{M}'$ and that $\{\mathcal{M}, \overline{\mathcal{M}}\}'' = \mathcal{B}(\mathcal{H})$, and thus that $\mathcal{M}$ is a factor. Since Φ is cyclic for $\overline{\mathcal{M}}$ (for the same reason it was for $\mathcal{M}$), it is cyclic for $\mathcal{M}'$ and thus separating for $\mathcal{M}$. Consequently, the canonical equilibrium state $\phi_c(\beta, \rho)$ extends to a faithful normal state on $\mathcal{M}$. The corresponding modular action—see Subsection b—is determined by:

$$\sigma(t)[W(\beta, \rho; f)] = W(\beta, \rho; \exp\{iH_\mu t\}f) \tag{258a}$$

with

$$(\exp\{iH_\mu t\}f)^\sim(k) = \exp\{i[\epsilon(k) - \mu]t\}\tilde{f}(k) \tag{258b}$$

where μ is a function of β and ρ, determined by $z = \exp(\beta\mu)$ and (247); see further discussion of (258) later in this subsection.

Since $\mathcal{M}$ is a factor, $\phi_c(\beta, \rho)$ is extremal β-KMS for σ—see Corollary 10.1.12.

Moreover, an action α of $\mathbb{R}^3$ is determined by

$$\alpha(a)[W(\beta, \rho; f)] = W(\beta, \rho; \exp\{iP \cdot a\}f) \tag{259a}$$

with

$$(\exp\{iP \cdot a\}f)(x) = f(x + a) \tag{259b}$$

This action is clearly norm asymptotic abelian, and satifies

$$\phi_c(\beta, \rho) \circ \alpha(a) = \phi_c(\beta, \rho) \tag{260}$$

Since $\mathcal{M}$ is a factor, $\phi_c(\beta,\rho)$ is extremal $\mathbb{R}^3$-invariant and $\mathbb{R}^3$—uniformly clustering—see Theorem 9.2.2, Scholium 9.2.3 and Theorem 9.2.4.

From these facts we conclude—as we did in Subsection c—that the factors $\mathcal{M}$ are of type III. Finally, since $\alpha(a)$ and $\sigma(t)$ leave $\phi_c(\beta,\rho)$ invariant, they are unitarily implemented, and for $a \neq 0 \neq t$, the construction described in Proposition 9.2.7 produces operators $U(a)$ and $V(t)$ that do not commute with $\overline{\mathcal{M}}$, and therefore do not belong to $\mathcal{M}$; this explicitly illustrates results we already knew, on general grounds, from Subsections a and c, to the effect that α and σ cannot be inner.

For the condensed phase, the situation is somewhat more complicated. Upon using the integral representation (250) of the Bessel function appearing in (249), one can rewrite $\phi_c(\beta,\rho)$ as an integral over a family—parametrized by $\theta \in S^1 \equiv [0,2\pi]$—of primary states, described by the functionals:

$$\hat{\phi}_\theta(\beta,\rho;f) = \hat{\phi}(\beta,1;f)\ \exp\{i\,\mathbb{R}e\,[\tilde{f}(0)e^{i\theta}]\} \tag{261}$$

The representation of $\mathcal{B}$ corresponding to $\phi_c(\beta,\rho)$ is thus a direct integral of primary representations; in that sense, the condensed phase of the ideal Bose gas appears, at the microscopic level of description, as a continuum of pure phases. An understanding of this ambiguity can be gained from the following remark.

Let us consider the gauge action $\gamma : S^1 \to \mathrm{Aut}(\mathcal{B})$ defined by:

$$\gamma(\theta)[W(f)] = W(e^{i\theta}f) \tag{262}$$

and its fixed point algebra

$$\mathcal{B}_\gamma \equiv \{B \in \mathcal{B} \,|\, \gamma(\theta)[B] = B \ \ \forall\, \theta \in S^1\} \tag{263}$$

Compare this to the effect of (165–166) on (193) through (187). Physically, $\gamma(\theta)$ is the automorphism implemented by $U(\theta) = \exp\{i\theta N\}$ where N is the number operator in the Fock representation on which $\mathcal{B}$ is originally defined; hence, in this representation, $\mathcal{B}_\gamma$ is the algebra of the elements of $\mathcal{B}$ that have zero matrix elements between vectors of $\mathcal{F}$ belonging to different $\mathcal{F}^{(N)}$. Notice now, from (249) and (246), that

$$\phi_c(\beta,\rho) = \phi_c(\beta,\rho)\circ\gamma(\theta) \quad \forall\,\theta \in S^1 \tag{264a}$$

while we see, from (261) and (246), that

$$\phi_\theta(\beta,\rho) = \phi_{\theta_\circ=0}(\beta,\rho)\circ\gamma(\theta) \quad \forall\,\theta \in S^1. \tag{264b}$$

Hence, whereas the states $\phi_\theta(\rho,\beta)$ are different, and

$$\phi_c(\beta,\rho) = \frac{1}{2\pi}\int d\theta\ \phi_\theta(\beta,\rho) \tag{265}$$

is thus a genuine decomposition on $\mathcal{B}$, all these states do agree on $\mathcal{B}_\gamma$. This is one of the earliest examples of the kind of spontaneous symmetry breaking that can accompany a phase transition. Other examples of this phenomenon will be presented in the following subsection.

One of the interests of exactly solvable models in the exploration of the conceptual foundations of a theory is that one can test on them the reasonableness of one's axioms. From the results obtained in Subsection c for quantum lattice systems—see in particular (155)—one might have been led to expect that the dynamics $\sigma : \mathbb{R} \to \mathrm{Aut}(\mathcal{B})$ of an infinite system, say in $\mathbb{R}^3$, could be obtained as a point-wise norm limit of the dynamics $\sigma^{(n)}(t)$ defined over an increasing absorbing sequence of bounded regions $\Lambda(n) \subset \mathbb{R}^3$, in the sense that for every $t \in \mathbb{R}$ and every $B \in \mathcal{B}$ one would have

$$\lim_{n\to\infty} \| \sigma^{(n)}(t)[B] - \sigma(t)[B] \| = 0 \tag{266}$$

Dublin & Sewell (1970) showed that this is not the case for the ideal Bose gas. Their argument runs as follows. Note first that the Weyl form (224) of the CCR can be extended to f and g in the Schwartz space $\mathcal{S}(\mathbb{R}^3)$ of infinitely differentiable functions such that they and all their derivatives decrease at infinity faster than all inverse powers of x. One then verifies that for every $f \in \mathcal{S}(\mathbb{R}^3)$ the generator $\Phi(f)$ of $\{W(\lambda f) | \lambda \in \mathbb{R}\}$ is an unbounded, self-adjoint operator with continuous spectrum. Consequently, for $f \in \mathcal{S}(\mathbb{R}^3)$:

$$\| W(f) - I \| = 2 \quad \text{unless } f \equiv 0 \tag{267}$$

With $\epsilon(k) = |k|^2/2$, $f \in \mathcal{D}(\mathbb{R}^3)$, $\Lambda(n)$ as in (232), n large enough so that $\mathrm{supp}(f) \subseteq \Lambda(n)$, and $t \in \mathbb{R}$, let

$$[U(t)f](x) \equiv \frac{1}{(2\pi)^3} \int d^3k\, \tilde{f}(k) \exp\{i[k \cdot x + \epsilon(k)t]\} \tag{268}$$

$$[U^{(n)}(t)f](x) = \chi_{\Lambda(n)}(x) \sum_{k \in \frac{2\pi}{L(n)} \mathbb{Z}^3} \tilde{f}(k) \exp\{i[k \cdot x + \epsilon(k)t]\} \tag{269}$$

Then the dynamics σ_μ prescribed by (258) is given by the action of the unitary group $\{\exp(-i\mu t)U(t) \mid t \in \mathbb{R}\}$ on the one particle space $\mathcal{T}$; similarly $\{\exp(-i\mu t)U^{(n)}(t) \mid t \in \mathbb{R}\}$ gives the dynamics $\sigma_\mu^{(n)}$ in $\Lambda(n)$. Moreover,

$$\begin{aligned} & s - \lim_{n\to\infty} [U^{(n)}(t) - U(t)]f = 0 \\ & U^{(n)}(t)f \quad \text{has compact support} \\ & U(t)f \quad \text{belongs to } \mathcal{S}(\mathbb{R}^3), \text{ but} \\ & \qquad \text{it does not have compact support unless } f \equiv 0 \end{aligned} \tag{270a}$$

Consequently, for $f \neq 0$:

$$\| [U^{(n)}(t) - U(t)]f \| \neq 0 \quad \forall\ n < \infty \tag{270b}$$

$$\lim_{n\to\infty} (U^{(n)}(t)f, U(t)f) = \| f \|^2 \tag{270c}$$

Suppose now that (266) were to hold. We would then have in particular, for $0 \neq f \in \mathcal{D}(\mathbb{R}^3)$:

$$\lim_{n\to\infty} \| \sigma_\mu^{(n)}(t)[W(f)] - \sigma_\mu(t)[W(f)] \| = 0 \tag{271a}$$

which we can rewrite—upon using (268), (269), the unitarity of $W(\cdot)$, the Weyl form of the CCR, and (270c)—in the form:

$$\lim_{n\to\infty} \| W(e^{-i\mu t}[U^{(n)}(t) - U(t)]f) - I \| = 0 \tag{271b}$$

From (267) and (271b), we conclude that there exists $N < \infty$ such that $\| [U^{(n)}(t) - U(t)]f \| = 0 \ \forall \ n > N$ which contradicts (270b). Hence (266) is not satisfied. As we saw, this does not prevent σ_μ to be well defined, as the modular group of the normal extension of the state $\phi_c(\beta, \mu)$ to the von Neumann algebra generated by the GNS representation of $\mathcal{B}$ associated to this state. The interpretation of σ_μ as the physically meaningful dynamics of the infinite ideal Bose gas is also contained in Dubin & Sewell (1970) who proposed a weakened form of (266), involving only the physically relevant objects, namely the thermodynamical limit of the multiple times correlation functions. They then proved that their conditions were satisfied for the ideal Bose gas, and that these conditions are in general sufficient to recover the essentials of the KMS theory discussed in Subsection b. The situation is, therefore, now completely under control.

e. Spontaneous symmetry breaking and stability.

Let $H(\psi) = 0$ denote, symbolically, the setting of a problem where H is invariant under a group of symmetries G; and let ψ_0 stands for any one of the fundamental, or stable, solutions of the problem. One speaks of a spontaneous symmetry breaking when ψ_0 is invariant only under a lower symmetry group than G.

One of the earliest examples of this phenomenon is provided by the bucking of a rod under a load; this problem was already addressed, in the first half of the Eighteenth Century by Jakob Bernoulli and Leonhard Euler; $H(\psi) = 0$ stands there for a variational principle, and one of the successes of the theory was Euler's classification of the equilibrium *elastic curves* ψ_0; for a brief, but insightful overview, see Truesdell (1983).

Closer to us in the history of mathematical physics, another example can be found in Poincaré (1913); see in particular pp. LX–LXI, 53–58, and 184–189; see also Bertin & Radicati (1976). The question discussed there is a tentative explanation for the origin of the Moon/Earth system, or possibly the double stars, from the cooling of an homogeneous, rotating, self-gravitating nebulous mass. In the course of time, radiation losses cause the temperature to decrease, and thus the fluid to contract; as this happens, the angular velocity increases and the following scenario unfolds. At small angular velocities ω, one has the expected result: the stable equilibrium configuration is a rotationally symmetric ellipsoid, flattened along the direction of the rotation axis; these are the so-called *MacLaurin ellipsoids*, one for each value of ω. As the angular velocity increases and reaches a first critical value ω_C, these configurations become unstable and new solutions develop that are stable but rotationally assymmetric: elongated ellipsoids, with three unequal axes; these are the so-called *Jacobi ellipsoids*. As a still higher critical value ω_M is reached, the Jacobi ellipsoids themselves become unstable, and new solutions do develop that Poincaré refers to as pearlike configurations; their stability was discussed by Darwin and by Lyapunov. It is these pearlike configurations that would ultimately break up into two distinct bodies, thus giving birth to the Moon/Earth system, or to a double star. From the point of view of a discussion of spontaneously broken symmetries, the sharp transition, at precisely $\omega = \omega_C$, from the MacLaurin ellipsoids to the Jacobi ellipsoids, does break the S^1-symmetry of the former around the axis of rotation. Note, nevertheless, that this is only a property of individual solutions. For $\omega > \omega_C$, the original S^1-symmetry of the problem still manifests itself in the fact that any rotation $\varphi \in S^1$ brings any non-rotationally invariant solution onto another solution with the same control parameters—compare with (264) valid for the condensed phase of the ideal Bose gas.

The purpose of this subsection is to explore some of the general reasons that might monitor the occurence of similar phenomena in the realm of quantum statistical mechanics, where the compatibility of the following two facets of the theory have to be understood. On the one hand, the fundamental microscopic interactions, that enter in the Hamiltonian, are invariant under a large group of symmetries, e.g. the group E^3 of Euclidean transformations in $\mathbb{R}^3$. On the other hand there exists, in the macroscopic world, forms of matter with a much lower residual symmetry, e.g. magnets and crystals.

At least three conceptually different avenues to a solution of this problem have been proposed, each with its own merits.

The first avenue is to approach the situation as if it were a kind of structural instability. One indeed first breaks the symmetry by fiat; in the symbolic notation used in the opening sentences of this subsection, one adds explicitly a small, low symmetry perturbation ϵH_1 to H, and looks for the fundamental solution ψ_ϵ of $(H + \epsilon H_1)\psi = 0$; one then studies the limit of ψ_ϵ as ϵ tends to zero; if this limit exists and has a lower symmetry than that of H, one can claim to have a model with spontaneous symmetry breaking. This programme is difficult to implement, but it corresponds, e.g. to the physicist's intuition on the expected behaviour of a magnetizable substance, where H_1 would represent the effect of a magnetic field, constant in both magnitude and direction.

The second avenue is to impose, at finite volume Λ, boundary conditions (BC) that explicitly violate the original symmetry of H; one then gains enough control on the solution ψ_Λ^{BC} of $H_\Lambda^{BC}(\psi) = 0$ (recall that we use this only as a symbolic notation) to be able to compute the thermodynamical limit ψ of ψ_Λ^{BC};

one finally examines whether ψ depends on the boundary conditions chosen to reach it. This programme has been implemented successfully with lattice models for ferromagnetism, starting from the seminal paper of Peierls (1936); and it has been given a mathematically firm footing in the work of Dobrushin (1968), Minlos & Sinai (1967–8); see also references quoted therein, on both this approach to the problem and on the connection of their method with the programme sketched in the above "first avenue".

The third avenue starts where we left the ideal Bose gas—see in particular (264)—and leads to the study of the decomposition of β-KMS states into their extremal β-KMS components. The intuitive idea behind this approach is this: since the KMS condition seems to be an essential ingredient in the make-up of equilibrium states, extremal KMS states should be expected to have something to do with pure thermodynamical phases. In this approach, one thus assumes that a thermodynamical limiting procedure, that preserves the full invariance G of the theory, has produced a G-invariant state ϕ, satisfying the KMS condition at natural temperature β for some (unique!) evolution $\sigma(\mathbb{R})$; we also assume this limiting procedure to be such that the GNS representation space $\mathcal{H}_\phi$ associated to ϕ is separable—see e.g. the discussion following (158) and (159). From Corollary 12, we know that ϕ is extremal β-KMS on the von Neumann algebra $\mathcal{M} \equiv \pi_\phi(\mathcal{B})''$ if and only if $\mathcal{Z} \equiv \mathcal{M} \cap \mathcal{M}' = \mathbb{C}I$; we, however, do not want to assume that ϕ is extremal β-KMS. To explore what can happen, suppose for an instant that there exists a finite, non trivial, partition $\{F_k\}$ of the identity in $\mathcal{H}_\phi$ such that $\mathcal{Z} = \{\sum_k z_k F_k \mid z_k \in \mathbb{C}\}$. From Theorem 11, we read that

$$
(272) \qquad \begin{aligned} &\phi = \sum_k \lambda_k \phi_k \quad \text{with} \\ &\lambda_k = < \phi; F_k > \quad \text{and} \\ &\phi_k : M \in \mathcal{M} \mapsto \lambda_k^{-1} < \phi; F_k M > \in \mathbb{C} \end{aligned}
$$

is a decomposition of ϕ into a convex combination of β-KMS states ϕ_k w.r.t $\sigma(\mathbb{R})$. Moreover, since F_k is a minimal projector in $\mathcal{Z}$, the restriction $\mathcal{M}_k$ of $\mathcal{M}$ to $\mathcal{H}_k = F_k \mathcal{H}_\phi$ is a factor; with Φ denoting the cyclic vector canonically associated to ϕ, note that $F_k \Phi$ is cyclic for $\mathcal{M}_k$ in $\mathcal{H}_k$; from this we conclude that ϕ_k are extremal β-KMS states on $\mathcal{M}$, and—see again Theorem 11—that (272) is the unique decomposition of ϕ into extremal β-KMS states w.r.t $\sigma(\mathbb{R})$. Note further that for every $M \in \mathcal{M}$

$$
(273) \qquad < \phi; ZM > = \sum_k \lambda_k \, z_k < \phi_k; M > \quad \forall \quad Z = \sum_k z_k F_k \in \mathcal{Z}
$$

From our assumption that ϕ is G-invariant, i.e. that we are given an action $\alpha : g \in G \mapsto \text{Aut}(\mathcal{B})$ such that $\phi \circ \alpha(g) = \phi$ for all $g \in G$, we know—see Proposition 9.2.7—that there exists a continuous unitary representation $U_\phi : G \to \mathcal{U}(\mathcal{H}_\phi)$ defining, through $\alpha(g)[M] \equiv U_\phi(g) M U_\phi(g)^*$, a continuous group of automorphism of $\mathcal{M}$ such that $\alpha(g)[\pi_\phi(B)] = \pi_\phi\{\alpha(g)[B]\} \;\; \forall \, g \in G$

and $\forall\ B \in \mathcal{B}$. While Z is clearly stable under $\alpha(g)$, we have no a priori reason to assume that it is point-wise invariant under each $\alpha(g)$. Since F_k is a minimal projector in Z, so is $\alpha(g)[F_k]$; consequently $\phi_k \circ \alpha(g)$ is again one of the extremal β-KMS states occuring in the decompositon (272) of ϕ. If G is a discrete group, and if for some g and k $\alpha(g)[F_k] \neq F_k$, this scheme could produce a mechanism for spontaneous symmetry breaking. If, however, G is a connected topological group, the continuity of the action α implies that $\alpha(g)[F_k] = F_k$; this suggests that the simplifying assumption that Z is atomic is overly restrictive. Fortunately, the above scheme can be generalized along the following lines, which do not involve this assumption.

One learns from Sakai (1965, 1971) and Wils (1968, 1971)—for a didactic presentation, see also Pedersen (1979)—that, to every state ϕ on a C^*-algebra $\mathcal{B}$ (with unit I) is associated a unique measure μ_ϕ on the state space S of $\mathcal{B}$, such that there exists a σ-continuous isomorphism

$$\gamma : Z \in \pi_\phi(\mathcal{B})'' \cap \pi_\phi(\mathcal{B})' \mapsto \gamma(Z, \cdot) \in \mathcal{L}^\infty(S, \mu_\phi) \tag{274}$$

with the property that, for all $Z \in \pi_\phi(\mathcal{B})'' \cap \pi_\phi(\mathcal{B})'$ and all $B \in \mathcal{B}$:

$$(Z\, \pi_\phi(B)\, \Phi, \Phi) = \int d\mu_\phi(\psi)\, \gamma(Z, \psi) < \psi; B > . \tag{275}$$

The measure μ_ϕ is moreover concentrated, in the Baire sense, on the set $\mathcal{F} \equiv \{\psi \in S | \psi \text{ primary }\}$; for this reason μ_ϕ is called the *central measure* of ϕ.

With $\alpha : G \to \text{Aut}(\mathcal{B})$, $S_G \equiv \{\psi \in S | \psi \circ \alpha(g) = \psi \ \ \forall\ g \in G\}$ and $\phi \in S_G$, we speak of *spontaneous symmetry breaking* when there is no subset $\mathcal{T} \subseteq S_G$ such that $\mu_\phi(\mathcal{T}) = 1$. Note that (275) is indeed the proper generalization of (273) in the following sense. If $\phi \in S_\beta \cap S_N$ where $S_\beta \equiv \{\psi \in S | \psi$ is β-KMS w.r.t. $\sigma(\mathbb{R})\}$ and $S_N \equiv \{\psi \in S | \psi \text{ is locally normal }\}$, then μ_ϕ is concentrated, in the Borel sense, on $\mathcal{E}_\beta \cap S_N$ where $\mathcal{E}_\beta$ is the set of all extremal β-KMS states w.r.t. $\sigma(\mathbb{R})$. For details, see Ruelle (1970), Emch, Knops & Verboven (1970), Takesaki & Winnink (1973), and Andersson (1976); much of this study was stimulated by the concept of Choquet simplex—see Phelps (1966)—which already appeared in this context in the fundamental work of Takesaki (1970 a).

The spontaneous symmetry breaking mechanism (275) may seem tailored especially to fit the situation (264) encountered in the condensed phase of the ideal Bose gas, where the symmetry that is spontaneously broken is the invariance, under the action of the gauge group S^1 defined by (262), of the Hamiltonian from which the canonical equilibrium state was computed.

For Fermi systems, one of the best studied examples of a spontaneous breaking of the gauge invariance of the theory is the BCS model for superconductivity; see Bardeen, Cooper & Schrieffer (1957), Haag (1962), Schrieffer (1964), Ezawa (1964), Emch & Guenin (1966), Thirring & Wehrl (1967), Thirring (1968,

1969), Jelinek (1968), Dubin & Sewell (1970). This model being of the mean free field variety, the convergence of the dynamics is a delicate problem. What happens in this type of situations is best illustrated with the quintessential mean free field system, namely the Weiss model for ferromagnetism.

The Weiss model is an infinite quantum lattice—see Subsection c—where the algebra of quasilocal observables is

$$\mathcal{B} = \bigotimes_{k\in\mathbb{Z}} \mathcal{B}_k \quad \text{with} \quad \mathcal{B}_k \simeq \mathcal{M}(2,\mathbb{C}) \tag{276}$$

The local Hamiltonians are

$$H_\Lambda = -|\Lambda|^{-1} \sum_{j,k\in\Lambda} (\tau_j^x \otimes \tau_k^x + \tau_j^y \otimes \tau_k^y + \tau_j^z \otimes \tau_k^z) \tag{277}$$

where $\{\tau_k^\alpha | \alpha = x, y, z\}$ are the Pauli matrices (149). Let $S^2 = \{n \in \mathbb{R}^3 | \; \| n \| = 1\}$, $S^1 = [0, 2\pi)$ and $\gamma : 0^3 \to \mathrm{Aut}(\mathcal{B})$ be the action defined, for every $(n,\theta) \in S^2 \times S^1$ and τ_k^α by

$$\begin{aligned} &\gamma(n,\theta)[\tau_k^\alpha] \equiv \sum_\beta \hat{\gamma}(n,\theta)^\alpha_\beta \, \tau_k^\beta \text{ where} \\ &\hat{\gamma}(n,\theta) \equiv \exp\{\hat{R}(n)\theta\} \quad , \quad \hat{R}(n) \equiv \begin{pmatrix} 0 & n^z & -n^y \\ -n^z & 0 & n^x \\ n^y & -n^x & 0 \end{pmatrix} \end{aligned} \tag{278}$$

Clearly each H_Λ is invariant under that action. An even simpler version of this model, that still exhibit the phenomenon we are interested in, is the Weiss-Ising model, where (277) is replaced by

$$H_\Lambda = -|\Lambda|^{-1} \sum_{j,k\in\Lambda} \tau_j^z \otimes \tau_k^z \tag{279}$$

For this model, restricting our attention, e.g., to the invariance group $Z_2 \simeq \{\gamma(n,\theta) | n = (0,1,0); \theta = 0, \pi\}$ will serve our purpose.

Both models, and variation thereof, have been studied extensively; see e.g. Kac (1968), Siegert & Vezetti (1968), Thompson, Siegert & Vezetti (1970), Emch & Knops (1970), Thirring (1983). The thermodynamical limit of the canonical partition function can be computed explicitly—for instance by the method of steepest descent in the classical case (279)—and one obtains that both models have a phase transition at the critical temperature $\beta_c = 1/2$ where the equation

$$\varsigma = 2 \; \tanh\{\beta\varsigma\} \tag{280}$$

develops a solution $\varsigma > 0$. For $\beta < \beta_c$ the canonical equilibrium state is the trivial tracial state

$$\begin{aligned} &\phi = \bigotimes_{k\in\mathbb{Z}} \phi_k \\ &\phi_k : B_k \in \mathcal{B}_k \mapsto \frac{1}{2} tr\, B_k \in \mathbb{C} \end{aligned} \tag{281}$$

For $\beta > \beta_c$, the canonical equilibrium state becomes

$$\phi(\beta) = \int_{S^2} d\mu(n)\, \phi(\beta, n) \tag{282a}$$

where, for the Weiss model, $d\mu$ is the normalized invariant measure on S^2; and, for the Weiss-Ising model, $d\mu = \{\delta(0,0,1) + \delta(0,0,-1)\}/2$; the primary states appearing in the decompostion (282a) are given by:

$$\begin{aligned} \phi(\beta, n) &= \bigotimes_{k \in \mathbb{Z}} \phi_k(\beta, n) \\ \phi_k : B_k \in \mathcal{B}_k &\mapsto tr\{\rho_k(\beta, n)\, B_k\} \in \mathbb{C} \\ \rho_k(\beta, n) &= Z(\beta, n)^{-1} \exp(-\beta\varsigma < n, \tau >) \\ Z(\beta, n) &= tr\{\exp(-\beta\varsigma < n, \tau >)\} \\ < n, \tau > &= \sum_\alpha n^\alpha \tau^\alpha \end{aligned} \tag{282b}$$

The corresponding modular automorphisms are

$$\begin{aligned} \sigma(\beta, n; t) &= \bigotimes_{k \in \mathbb{Z}} \sigma_k(\beta, n; t) \\ \sigma_k(\beta, n; t) &\simeq \gamma(n, 2\varsigma t) \end{aligned} \tag{283}$$

with $\gamma(n, \theta)$ given by (278). Note that this result refers separately to the primary representation $\pi(\beta, n; \cdot)$ associated to each $\phi(\beta, n)$; in particular, $\sigma(\beta, n; \cdot)$ depends explicitly on the direction n, and on β through the value of ς given by (280). There is, therefore, no way to obtain each of these dynamics as a point-wise norm limit: the state has to enter the limiting process in which these infinite volume dynamics are computed.

The spontaneous symmetry breaking (282) again fits the proposed symmetry breaking mechanicsm (275). We should, moreover, mention that these models can still be solved exactly in the thermodynamical limit when one adds to the original Hamiltonians (277) or (279) a symmetry breaking term of the form

$$V_\Lambda = -\epsilon \sum_{k \in \Lambda} < n, \tau_k > \quad \epsilon > 0 \tag{284}$$

with n arbitrary in S^2 for (277) or n restricted to $(0, 0, \pm 1)$ in the case of (279). After the thermodynamical limit has been taken, one can then take the limit $\epsilon \to +0$; one obtains back the tracial state (281) if $\beta < \beta_c$, and the state $\phi(\beta, n)$ if $\beta > \beta_c$, thus establishing the desired link with the first avenue to spontaneous symmetry breaking described in this subsection.

The spontaneous symmetry breaking mechanism (275) having been tested on some explicitly solvable models, the next steps are to try using it to explore more speculative situations. Among these is the question of whether crystals can be made to occur in a theory where the fundamental interactions are invariant under the Euclidean group E^3; some progresses in this direction have been presented in Emch, Knops & Verboven (1970), and in Kastler, Mebkhout, Loupias & Michel (1972); for a comparative review of these two approaches and of their antecedents, see Emch (1980). A complete, group theoretical, classification of the various degrees in which the Euclidean symmetry can be spontaneously broken has been obtained, as well as a list of the spectral and clustering criteria characterizing the various states of matter that appear in the classification. This investigation has also led to a confirmation of the Landau argument on the absence of critical point in the fluid/solid phase transition. Still missing, however, is a full treatment, based on specific (quantum) Hamiltonians, that would determine which interactions are responsible, under specified circumstances, for the actual occurrence of the various states of matter present in the classification one has obtained; nevertheless, see Gates (1971), and the systematic programme of Radin *et al.* (1979–) to get control, mathematically and conceptually, over the crystalline states of matter.

Moving from the analysis of specific models to the consideration of more abstract general questions, such as the classification of all possible states of matter in equilibrium, gives one occasions to pause and reflect on the formalism one uses. We accordingly close this subsection with a brief review of the main ingredients that entered the theory.

Firstly, we assumed that the algebra $\mathcal{B}$ of quasilocal observables was the C^*-inductive limit of an isotonic family of von Neumann algebras $\mathcal{B}_\Omega$—e.g. $\mathcal{B}(\mathcal{H}_\Omega)$—indexed by an absorbing directed set $\mathcal{F}$ of bounded open regions Ω in the configuration space Γ—typically $\mathbb{R}^3$. Secondly, we assumed that an action $\alpha : G \to \mathrm{Aut}(\mathcal{B})$ was given, with G containing at least the group of isometries of Γ—typically E^3. Thirdly, we assumed that the states ϕ we wanted to consider as equilibrium states were locally normal, i.e. that the restriction of ϕ to $\mathcal{B}_\Omega$ is a density matrix ρ_Ω acting in $\mathcal{H}_\Omega$. Fourthly, we assumed that these states ϕ satisfied the KMS condition, at natural temperature β, with respect to an action $\sigma : \mathbb{R} \to \mathrm{Aut}(\pi_\phi(\mathcal{B})'')$, identified with the dynamics.

The fourth assumption is the one that most needs discussing. Our use of this assumption was first prompted by the fact that it was satisfied for finite systems, where we knew exactly what it meant in terms of first principles. We further noticed that the KMS condition was satisfied for the equilibrium states of all the particular models of infinite systems that we considered. We also remarked that, in these models, ϕ was a faithful normal state on $\pi_\phi(\mathcal{B})''$, and that the modular action $\sigma : \mathbb{R} \to \mathrm{Aut}(\pi_\phi(\mathcal{B})'')$, determined uniquely by ϕ, coincided with the proper limit of the dynamics we obtained from local Hamiltonians. Finally, we took note that the measure μ_ϕ, appearing in the central decomposition (275) was unique, and was concentrated on the primary states of $\mathcal{B}$; stated differently, the set S_β of β-KMS states on $\pi_\phi(\mathcal{B})''$ was a Choquet simplex.

Nonetheless, our fourth assumption, that ϕ be β-KMS, still belongs to the realm of plausible analogies, and one would like to have a few more physical reasons marshalled to its support; see Sewell (1980) where the following properties are reviewed, with an emphasis on their physical meaning as equilibrium criteria and their relation to the KMS condition; see also Thirring (1983) for a critical evaluation of these, and related, proposals.

The first of these properties is the following correlation inequality, valid for all B in the domain $\mathcal{B}_\delta$ of the generator δ of the dynamics $\sigma(\mathbb{R})$:

$$-i\beta < \phi; B^*\delta[B] > \geq \Phi(< \phi; B^*B >, < \phi; BB^* >) \tag{285a}$$

where:

$$\Phi(u,v) = \begin{cases} u\,\ell n\, u - u\,\ell n\, v & u,v \geq 0, u+v > 0 \\ 0 & u = 0 = v \end{cases} \tag{285b}$$

see Sewell (1977), Fannes & Verbeure (1977), and Araki (1978).

The next two properties belong to the thermostatics of infinite systems, and their formulation requires a few preliminary definitions. Two locally normal states ϕ and ψ are said to be locally equivalent—which we write $\phi \sim \psi$—if there exists $\Omega \in \mathcal{F}$ such that $< \phi; B >=< \psi; B >$ for all $B \in \mathcal{B}_\Lambda$ with $\Lambda \cap \Omega = \emptyset$. Let now $\rho_\Omega(\phi)$ be the density matrix given by the restriction to $\mathcal{B}_\Omega$ of the locally normal state ϕ, and H_Ω be the local Hamiltonian associated to Ω. In line with (33) we define the *free energy* $A_\Omega(\phi)$ of the state ϕ for the region Ω, as

$$A_\Omega(\phi) \equiv tr\{\rho_\Omega(\phi)[H_\Omega + kT\,\ell n\, \rho_\Omega(\phi)]\} \tag{286}$$

The locally normal state ϕ is said to be *locally thermostatically stable* if for every $\psi \sim \phi$, the increment in the free energy density due to the transition from ϕ to ψ is positive, i.e.

$$0 \leq \Delta(\phi,\psi) \equiv \lim_{\Omega \in \mathcal{F}} |\Omega|^{-1}\{A_\Omega(\psi) - A_\Omega(\phi)\} \tag{287a}$$

Note that this condition extends to the infinite case the fact that, for finite systems, the canonical equilibrium states are those that minimize the free energy. A more stringent extension of this condition is to require that ϕ be G-invariant and minimize the free energy density

$$A(\phi) \equiv \lim_{\Omega \in \mathcal{F}} |\Omega|^{-1} A_\Omega(\phi) \tag{287b}$$

When this condition is realized, ϕ is said to be *globally thermostatically stable*. Note, however, that systems with long-range forces are known to support KMS states that are not globally thermostatically stable. The equivalence between the KMS condition and the condition of local thermostatic stability was studied by Araki & Sewell (1977) and by Sewell (1977); its classical equivalent is the DLR condition of Dobrushin (1968D) and Lanford & Ruelle (1969 LR); for the condition of global thermostatic stability, see Ruelle (1969), Araki (1974) and

Araki & Sewell (1977). For a comparison between quantum KMS condition and the classical equilibrium conditions, see Gallavotti & Verboven (1975), Aizenman, Goldstein, Gruber, Lebowitz & Martin (1977), Pulvirenti (1977), Andersson (1976); see also Demoen, Vanheuverzwijn & Verbeure (1978).

The next property is an attempt to formalize, at the microscopic level, the second principle of thermodynamics. Let $C_o^{1+}(\mathbb{R}, \mathcal{B}_{sa})$ be the set of all continuously differentiable functions, with compact support in $\mathbb{R}^+$, from $\mathbb{R}$ to the self-adjoint part $\mathcal{B}_{sa}$ of $\mathcal{B}$. For every $h \in C_o^{1+}(\mathbb{R}, \mathcal{B}_{sa})$, and the given dynamics σ (with generator δ), consider the family $\alpha_h(\mathbb{R})$ of automorphisms of $\mathcal{B}$ defined by:

(288)
$$\left.\begin{aligned} \frac{d}{dt}\alpha_h(t)[B] &= \alpha_h(t)[\delta[B] + i[h(t), B]] \\ \alpha_h(0)[B] &= B \end{aligned}\right\} \ \forall\, B \in \mathcal{B}_\delta$$

The state ϕ is said to be *passive* if for every $h \in C_o^{1+}(\mathbb{R}, \mathcal{B}_{sa})$ and every $T > \sup\{t \in \mathbb{R} \mid h(t) \neq 0\}$

(289)
$$\int_0^T dt \ < \phi; \alpha_h(t)[\frac{dh}{dt}(t)] > \ \ \geq 0$$

This concept was proposed by Pusz & Woronowicz (1978); they proved that KMS states are passive, and that the converse is also true provided ϕ is η-clustering—see Subsection 9.2.c—with respect to the action $\alpha : G \to \mathrm{Aut}(\mathcal{B})$ of an amenable group G such that $\alpha(g) \circ \sigma(t) = \sigma(t) \circ \alpha(g) \ \ \forall\, (g,t) \in G \times \mathbb{R}$; the translation group is a natural candidate for G.

The next condition is boldly stringent, but is nevertheless indicative of the kind of dynamical stability one may want to expect. Suppose a dynamics $\sigma(\mathbb{R})$ is given, with generator δ. For every $h \in \mathcal{B}_{sa}$, let $\alpha_h(\mathbb{R})$ be the action the generator of which is $(\delta + i[h, \cdot])$. A state ϕ with $\phi \circ \sigma(t) = \phi \ \ \forall\, t \in \mathbb{R}$ is said to be *locally dynamically stable* if for every $h \in \mathcal{B}_{sa}$, there exists a neighbourhood $V(h)$ of 0 in $\mathbb{R}$ such that for every $\lambda \in V(h)$ there exists a state $\phi_{h,\lambda}$ with the following properties

(290)
$$\begin{aligned} &\phi_{h,\lambda} \circ \alpha_{\lambda h}(t) = \phi_{h,\lambda} \quad \forall\, t \in \mathbb{R} \\ &w^* - \lim_{\lambda \to 0} \phi_{h,\lambda} = \phi \\ &w^* - \lim_{|t| \to \infty} \phi_{h,\lambda} = \phi \end{aligned}$$

To prove that such a state satisfies the KMS condition, one seems to require some additional assumptions, among which: (i) a condition to the effect that ϕ is primary, and (ii) a condition of asymptotic abelianness on $\sigma(\mathbb{R})$, namely that there exists a norm-dense, $\sigma(\mathbb{R})$-stable, *-subalgebra $\mathcal{B}_0$ of $\mathcal{B}$ such that

(291)
$$\int_{-\infty}^{+\infty} dt \ \| \, [A, \sigma(t)[B]] \, \| < \infty \quad \forall\, A, B \in \mathcal{B}_0 \, .$$

see Haag, Kastler & Trych-Pohlmeyer (1974), Haag & Trych-Pohlmeyer (1977), Brattelli & Kastler (1976), Brattelli, Kishimoto & Robinson (1978) and, Araki (1973).

Two other types of properties have been examined in connection with the search for the conceptual content of the KMS condition for infinite systems, namely: (i) the systems considered can serve as thermal baths for finite systems—see Sewell (1974) and Frigerio, Gorini & Verri (1978); and (ii) these systems are stable under localized interactions with systems satisfying property (i), see Sewell (1980); in addition, this paper also explores the concepts of *meta-stability* in statistical mechanics.

f. Non equilibrium quantum statistical mechanics

In many situations where matter in bulk is studied, one observes an irreversible evolution that can be described by deterministic equations—e.g. the diffusion equation—the form of which is essentially different from the form of the reversible equations of classical or quantum Hamiltonian mechanics. While the statistical mechanics of Boltzmann was designed to derive such irreversible equations by matching a mechanistic treatment of the microscopic evolution equations with probabilistic considerations based on the macroscopic character of what is observed, serious conceptual difficulties appeared very early; see for instance the critical review of P. & T. Ehrenfest (1911). These difficulties are epitomized in the so-called paradoxes of Thomson (1874)—Loschmidt (1876) and of Zermelo (1896); see e.g. Brush (1983). Work done in the last thirty years, starting with the investigations of van Hove (1955) on the quantum many-body problem, has shown that the mathematical control of various physical limits is necessary to sort out what is essential.

We first summon here a simple, exactly solvable model to help visualize the problems at hand; this model has also the interest of having been rigged for the purpose of interpreting an actual experiment, the so-called free induction relaxation; see Lowe & Norberg (1957) and S. P. Heims (1965). A CaF_2 crystal was placed in a magnetic field ς, the direction of which we shall call z, and first allowed to reach thermal equilibrium. A radio frequency pulse was then applied with the effect of turning the net magnetic moment in a direction x orthogonal to z. The evolution of the magnetic moment was then observed: it exhibited an oscillatory decay towards an equilibrium value zero. This is interpreted as the result of the dipolar interactions between the spins $\{\tau_k\}$ of the fluorine nuclei; no relaxation via lattice vibrations seems involved: the spins system can be considered as a rigid quantum lattice in the sense of Subsection c. The idealizations involved in the model—see Emch (1966), and also Thirring (1983)—are: (i) the lattice is taken to be one dimensional; (ii) the spin-spin interaction is supposed to be so anisotropic that it can be modelled by an Ising type interaction; (iii) the interaction is supposed to decrease exponentially with the distance; the latter assumption is not essential, but it allows to express the final results in a simple form that is particularly well-suited to our illustrative purpose here.

Mathematically the model is defined as follows. We consider a chain with $(2N+1)$ sites, and to each of these sites we associate a copy $\mathcal{B}_k$ of $\mathcal{M}(2,\mathbb{C})$. With $\{\tau_k^j\}$ denoting the copy, at the site k, of the Pauli matrices (149), ς_k and $\epsilon(n)$ in $\mathbb{R}$, we define the Hamiltonian

(292)
$$H_N = \sum_{k=-N}^{N} H_k \quad \text{with}$$
$$H_k = \varsigma_k \tau_k^z + \sum_{n=1}^{2N} \epsilon(n) \tau_k^z \tau_{k+n}^z$$

where $k = -N, -N+1, \ldots, N-1, N$; $k+n$ is similarly defined by identifying $N+m$ with $-N+m-1$; in this way, the variable n appearing in $\epsilon(n)$ is the distance between the sites k and $k+n$; hence each site interacts symmetrically with the sites on its right and on its left. The features that make the model immediately tractable are: (i) all the terms in H_N commute with one another; and (ii), with $\tau^\pm \equiv (\tau^x \pm i\tau^y)/2$, we have

(293)
$$\tau_j^z \tau_k^\pm = \begin{cases} \tau_k^\pm \tau_j^z & \text{if } k \neq j \\ \tau_k^\pm(\tau_k^z \pm 2I) & \text{if } k = j \end{cases}$$

In particular, the evolution $\alpha_N : t \in \mathbb{R} \mapsto \mathrm{Aut}(\mathcal{B}_N)$, with $\mathcal{B}_N \simeq \mathcal{M}(2^{(2N+1)}, \mathbb{C})$, defined by:

(294)
$$\alpha_N(t)[B] \equiv \exp(iH_N t)\, B\, \exp(-iH_N t)$$

is easily computed to give:

(295)
$$\alpha_N(t)[\tau_k^z] = \tau_k^z$$
$$\alpha_N(t)[\tau_k^\pm] = V_k^\pm(t)\tau_k^\pm \quad \text{with}$$
$$V_k^\pm(t) \equiv \exp\{\pm 2it[\varsigma_k I + \sum_{n=1}^{N} \epsilon(n)(\tau_{k+n}^z + \tau_{k-n}^z)]\}$$

With the conditions of the experiment in mind, we compute the expectation of these observables on any state of the form

(296a)
$$\phi_N = \bigotimes_{k=-N}^{N} \phi_k$$

where ϕ_k is a pure state, on $\mathcal{B}_k$, of the form:

(296b)
$$< \phi_k; \tau_k^z > \; = 0$$
$$< \phi_k; \tau_k^\pm > \; = \frac{1}{2}\exp(\pm i\eta_k) \;;$$

we find:

$$(297a)\qquad \begin{aligned} < \phi_k \circ \alpha_N(t); \tau_k^z > &= 0 \\ < \phi_N \circ \alpha_N(t); \tau_k^\pm > &= \frac{1}{2}\exp\{\pm i(\eta_k + 2\varsigma_k t)\} f_N(t) \end{aligned}$$

where

$$(297b)\qquad f_N(t) = \prod_{n=1}^{N} \cos^2\{2\epsilon(n)t\}$$

Up to this point, the special form of ϵ has not been used. We now impose:

$$(298)\qquad \epsilon(n) = \epsilon_0 2^{-n}$$

which allows to compute (297b) explicitly, namely

$$(299a)\qquad \begin{aligned} f_N(t) &= [f(2^{-N}t)]^{-1} f(t) \\ f(t) &= \left[\frac{\sin(2\epsilon_0 t)}{2\epsilon_0 t}\right]^2 \end{aligned}$$

Clearly, at fixed t:

$$(299b)\qquad \lim_{N\to\infty} f_N(t) = f(t)$$

and, therefore, for every $B_k \in \mathcal{B}_k$ and every $k \in \mathbb{Z}$:

$$(300)\qquad \lim_{N\to\infty} < \phi_N \circ \alpha_N(t); \tau_k^j >= f(t) < \phi \circ \alpha_0(t); \tau_k^j >$$

where α_0 is the free evolution obtained from (292) with $\epsilon_0 = 0$; f thus gives the contribution to the evolution brought in by the interparticle interactions. From its explicit form (299), we see that we obtain the desired oscillatory approach to the equilibrium value given, for every $B_k \in \mathcal{B}_k$, by:

$$(301)\qquad \lim_{t\to\infty} \lim_{N\to\infty} < \phi_N \circ \alpha_N(t); B_k >= \frac{1}{2} tr\ B_k$$

Note that, although $\phi_N \circ \alpha_N(t)$ is a pure state, its restriction to any single-spin algebra $\mathcal{B}_k$, approaches the perfectly mixed tracial state when the limits (301) are taken (in the order indicated!). More general initial states can in fact be allowed without changing this result. Firstly, we can take, evidently, for ϕ_k any mixture of the pure states we considered to far; this is, in essence, the situation corresponding to the experimental set-up we had described. If one further wants to introduce more general pure states ϕ_k, namely:

$$< \phi_k; \tau_k^z >= \xi; \; < \phi_k; \tau_k^{\pm} >= \frac{1}{2}(1-\xi^2)^{1/2}\exp(\pm i\eta_k) \tag{302}$$

the only change required in (301) is to replace the tracial state by the canonical equilibrium state corresponding to a magnetic field, in the direction z, the magnitude ς of which satisfies $\tanh(\beta\varsigma) = \xi$.

In connection with the Loschmidt paradox, note that $f_N(t) = f_N(-t)$ and $f(t) = f(-t) \;\forall\, t \in \mathbb{R}$: reversibility shows up in the fact that the security of a postdiction is the same as that of a prediction. As for the Zermelo paradox, it manifests itself only in the finite volume case; in fact, we do have in (299) the finite volume correction to thermodynamical behaviour; $f_N(t)$ comes close again to its initial value only for $t \simeq T_N = 2^{N+1}\pi/\epsilon_0$; hence, the dreaded recurrence time T_N grows exponentially with N, while the time needed to come close to the equilibrium value is independent of N; compare, for instance, with the Ehrenfest "dogs-and-fleas" probabilistic model, as discussed by Kac (1959).

Notice finally that, in a translation-invariant theory, where the magnetic field is homogeneous—i.e. ς_k in (292) is independent of k—and where the initial state ϕ is translation invariant—i.e. $\phi_k \simeq \psi \;\; \forall\, k$—the analysis presented here carries over directly to the study of macroscopic observables such as the magnetization

$$M_N^j \equiv (2N+1)^{-1} \sum_{k=-N}^{N} \tau_k^j \, . \tag{303}$$

Thirring (1983) gives also a semi-qualitative discussion of (mean square) fluctuations in this model, comparing microscopic and macroscopic observables.

While a single model does not prove much in terms of what happens in the general case, it does nonetheless illustrate why the Loschmidt and Zermelo paradoxes are not anymore the pressing issues they once were; together with the $x-y$ model discussed in Subsection c, the above model does serve as an indication of where to look for the way out of the old deadlock. One essential step is to control the thermodynamical limit; a second step is to restrict one's attention to a privileged set of observables, e.g. one-particle observables, or the local observables associated to a finite subsystem, or even an algebra of macroscopic observables; we shall argue later, on the basis of further models, that a third step is involved in solving the remaining outstanding problem, namely that the approach to equilibrium be governed by a dissipative equation of the desired type.

In general mathematical terms, the second step involves the generalization, to the quantum realm, of the concept of conditional expectation familiar in the classical theory of stochastic processes; see the characterization given by Moy (1954). One natural candidate for this notion is obtained by considering the following situation; see Nakamura & Turumaru (1954), Umegaki (1954–1962),

Tomiyama (1957–1959), and Arveson (1967). Let $\mathcal{M}$ be a von Neumann algebra, ϕ be a faithful normal state on $\mathcal{M}$, $\mathcal{N}$ be a von Neumann subalgebra of $\mathcal{M}$; let ψ denote the restriction of ϕ to $\mathcal{N}$. A *conditional expectation* $\mathcal{E} : \mathcal{M} \to \mathcal{N}$ *with respect to* ϕ is defined as a projection onto $\mathcal{N}$, that is of norm 1, is faithful and normal, and satisfies $\psi \circ \mathcal{E} = \phi$. Takesaki (1972)—see also Golodez (1972)—proved that $\mathcal{E}$ exists, and is unique, if and only if $\mathcal{N}$ is stable with respect to the action of the modular group $\sigma(\mathbb{R})$ associated to ϕ. For instance, if $\mathcal{M} = \mathcal{N}_1 \otimes \mathcal{N}_2$, $\phi = \phi_1 \otimes \phi_2$, and $\mathcal{N} \equiv \mathcal{N}_2$, $\mathcal{E}$ is uniquely defined by $\mathcal{E}(N_1 \otimes N_2) = < \phi_1; N_1 > N_2 \ \forall\, N_1 \otimes N_2 \in \mathcal{N}_1 \otimes \mathcal{N}_2$. For various reasons, see e.g. Davies (1970), Accardi (1976) or Gudder & Marchand (1977), it has been advocated that the insistence on the requirement that $\mathcal{E}$ be a projection might be wrong-headed. Accardi & Cecchini (1982) proposed an alternate characterization of classical expectations, generalizing to the non-commutative case, and giving a unique completely positive (see definition immediately below), identity preserving, faithful and normal map $\mathcal{E} : \mathcal{M} \to \mathcal{N}$ with $\psi \circ \mathcal{E} = \phi$, irrespective of whether Takesaki's condition on $\mathcal{N}$ is satisfied or not; in fact, they remark that $\{M \in \mathcal{M} | \mathcal{E}(M) = M\}$ being the largest subalgebra of $\mathcal{N}$ that satisfies Takesaki's condition, implies that when $\mathcal{E}$ is a projection, the two definitions do coincide. We shall not explicitly need this generalization here; it is nevertheless useful to know that it exists, when one wants to consider a general theory of quantum stochastic processes: see Accardi, Frigerio & Lewis (1982) and, for a survey, Accardi & Frigerio (to appear).

We mentioned earlier the necessity to consider dissipative equations. The simplest non-classical example is the *Bloch equation* for the relaxation of a spin:

$$\begin{aligned} &\frac{d}{dt}\{\tau^z(t) - \epsilon\} = -\mu\,\{\tau^z(t) - \epsilon\} \\ &\frac{d}{dt}\tau^{\pm}(t) = (\pm i\omega - \lambda)\,\tau^{\pm}(t) \end{aligned} \tag{304}$$

where $\tau^z, \tau^{\pm} : \mathbb{R}^+ \to \mathcal{M}(2, \mathbb{C}); \tau^z(0),\ \tau^{\pm}(0) = \{\tau^x(0) \pm i\tau^y(0)\}/2$ are defined from the Pauli matrices (149); $-1 < \epsilon < 1$; $\omega, \lambda, \mu \in \mathbb{R}$ with $0 \le \mu \le 2\lambda$. It is only when $\lambda = 0 = \mu$ that the solution of this equation has the form $\tau^j(t) = \alpha(t)[\tau^j(0)]$ with $\alpha(t) \in \mathrm{Aut}[\mathcal{M}(2, \mathbb{C})]$ for $t \neq 0$. We must, therefore, generalize our concept of dynamics; for this we need the notion of completely positive maps.

Recall—see Scholium 9.1.3—that an element X of a C^*-algebra $\mathcal{B}$ is said to be positive if there exists $B \in \mathcal{B}$ such that $X = B^*B$; let thus $\mathcal{B}^+ = \{B^*B | B \in \mathcal{B}\}$. With $\mathcal{B}$ and $\mathcal{C}$ two C^*-algebras, a linear map $\gamma : \mathcal{B} \to \mathcal{C}$ is said to be *positive* whenever $\gamma(\mathcal{B}^+) \subseteq \mathcal{C}^+$. For $n \in \mathbb{Z}^+$, let $\mathcal{M}(n, \mathcal{B}) = \mathcal{B} \otimes \mathcal{M}(n, \mathbb{C})$ be the C^*-algebra of $n \times n$ matrices with entries in $\mathcal{B}$. A linear map $\gamma : \mathcal{B} \to \mathcal{C}$ is said to be *n-positive* whenever $\gamma \otimes id_n : \mathcal{M}(n, \mathcal{B}) \to \mathcal{M}(n, \mathcal{C})$ is positive, and it is said to be *completely positive* (or to be a *CP-map*) whenever it is n-positive for all $n \in \mathbb{Z}^+$.

This notion was introduced by Stinespring (1955) and its mathematical meaning was sharpened by Størmer (1963), Arveson (1969) and Choi (1972); see also Nakamura, Takesaki & Umegaki (1960). Note that, already for $\mathcal{B} = \mathcal{C} = \mathcal{M}(2, \mathbb{C})$, the transposition map is positive, but is not 2-positive; this reflects the general fact that if every positive linear map from a C^*-algebra $\mathcal{B}$ to a C^*-algebra $\mathcal{C}$ is a CP-map, then at least one of these two C^*-algebras must be abelian. The converse is also true; in particular every state on a C^*-algebra is a CP-map. The notion of CP-map can in fact be considered to be the natural mathematical generalization of the notion of state in the sense that the GNS representation generalizes to CP-maps as follows. Let $\gamma : \mathcal{B} \to \mathcal{B}(\mathcal{H})$ be a CP-map; then there exists a triple $\{\mathcal{H}', \pi, V\}$ where $\mathcal{H}'$ is a Hilbert space, $\pi : \mathcal{B} \to \mathcal{B}(\mathcal{H}')$ is a representation, and $V : \mathcal{H} \to \mathcal{H}'$ is a bounded linear map, such that $\gamma[B] = V^*\pi(B)V \;\; \forall\, B \in \mathcal{B}$. Conversely, every map that can be written in this form is a CP-map.

These mathematical preliminaries having been recalled, we can move closer to our main purpose in this subsection. We first note from the above that: (i) the class of CP-maps is closed under compositions and positive linar combinations; in particular, we denote by $CP(\mathcal{B})$ the collection of all CP-maps of $\mathcal{B}$ into itself, equipped with the semi-group structure it inherits from composition; (ii) every automorphism α of $\mathcal{B}$ is a CP-map; (iii) every conditional expectation $\mathcal{E}$ is a CP-map.

We are now ready to describe a way in which CP-maps do naturally occur in the study of dynamical system, in a scheme that emerged from the work of Kraus (1971), Kossakowski *et al.* (1972–1977), Lindblad (1975, 1976), and Davies (1976), the latter being recommended for the wealth of sharp mathematical results it contains on this and related problems. Although some of the assumptions listed below should be relaxed for certain purposes, we choose to give the most stringent set of conditions in order to emphasize the compatibilities allowed by the theory.

We thus consider the following situation. Let ϕ be a state on a C^*-algebra $\mathcal{B}$ of quasilocal observables, and let us again denote by ϕ its unique normal extension to $\mathcal{M} = \pi_\phi(\mathcal{B})''$. Suppose ϕ is faithful on $\mathcal{M}$, and let $\sigma : \mathbb{R} \to \mathrm{Aut}(\mathcal{M})$ be the unique modular action associated to ϕ. Let $\alpha : \mathbb{R} \to \mathrm{Aut}(\mathcal{M})$, with $\phi \circ \alpha(t) = \phi \;\; \forall\, t \in \mathbb{R}$, be a dynamics in the conservative sense considered in the previous subsections. Let finally $\mathcal{N}$ be a von Neumann subalgebra of $\mathcal{M}$, corresponding to a partial description of the total system to which $\mathcal{M}$ refers. Suppose $\mathcal{N}$ is stable under $\sigma(\mathbb{R})$, and let $\mathcal{E} : \mathcal{M} \to \mathcal{N}$ be the unique faithful normal conditional expectation satisfying $\psi \circ \mathcal{E} = \phi$ where ψ denotes the restriction of ϕ to $\mathcal{N}$. With i denoting the natural embedding of $\mathcal{N}$ in $\mathcal{M}$, and t arbitrary in $\mathbb{R}$, let $\gamma(t) \equiv \mathcal{E} \circ \alpha(t) \circ i$. Upon combining the general results specifically mentioned above, we conclude that $\gamma(t) : \mathcal{N} \to \mathcal{N}$ are CP-maps.

Note that if, in addition to assuming that $\mathcal{N}$ is stable under $\sigma(\mathbb{R})$—so that Takesaki's condition on the existence of a conditional expectation be satisfied—we were to assume that the dynamics $\alpha(\mathbb{R})$ coincides with the modular action $\sigma(\mathbb{R})$, we would obtain that every $\gamma(t)$ is an automorphism; this would defeat our purpose in this subsection, which is to consider evolutions that are more general than the conservative actions we have grown accustomed to from the previous subsections. We will thus take the option of allowing $\alpha(\mathbb{R})$ to differ from $\sigma(\mathbb{R})$; another possible route would be to reject the assumption that $\mathcal{N}$ is stable under $\sigma(\mathbb{R})$, and proceed with Accardi's generalized conditional expectations.

We now formalize these motivating remarks. We say that a triple $\{\mathcal{N}, \psi, \gamma\}$ is a *generalized covariant W^*-dynamical system* (CWDS) if: $\mathcal{N}$ is a von Neumann algebra; ψ is a faithful normal state on $\mathcal{N}$; $\gamma : \mathbb{R}^+ \to CP(\mathcal{N})$ satisfies $\gamma(0) = id$; $\gamma(t)[I] = I \;\forall\, t \in \mathbb{R}^+$; $t \in \mathbb{R}^+ \mapsto < \chi; \gamma(t)[N] > \in \mathbb{C}$ is continuous for every normal state χ on $\mathcal{N}$ and every $N \in \mathcal{N}$; and $\psi \circ \gamma(t) = \psi \;\forall\, t \in \mathbb{R}^+$. A CWDS is said to be *conservative* if there exists an action $\alpha : \mathbb{R} \mapsto \mathrm{Aut}(\mathcal{N})$ with $\alpha(t) = \gamma(t) \;\forall\, t \in \mathbb{R}^+$; when this is not the case the CWDS is said to be *dissipative*. A CWDS is said to be *Markovian* if it satisfies the semigroup law $\gamma(s) \circ \gamma(t) = \gamma(s+t) \;\; \forall\, s,t \in \mathbb{R}^+$; clearly every conservative CWDS is Markovian, by definition of an action α. A CWDS is said to be *asymptotically stable* if ψ is a global attractor in the set S of all normal states on $\mathcal{N}$, i.e. if for every normal state χ on $\mathcal{N}$

$$w - \lim_{t \to \infty} \; \chi \circ \gamma(t) = \psi \tag{305}$$

Note that this condition is akin to the ergodic behaviour we encountered, in the conservative case, when the action α is norm asymptotically abelian; compare with Theorem 9.2.2. For dissipative CWDS that are Markovian, the general form of the generator L of $\gamma(\mathbb{R}^+)$ has been investigated—see Lindblad (1976), Gorini, Kossakowski & Sudarshan (1976), Davies (1976), Parravicini & Zecca (1977)—and conditions are known that ensure stability in the above sense—see Frigerio (1978); see also Davies (1970) and Spohn (1976). We further say that a CWDS is *reversible* if there exists $\nu : \mathbb{R}^+ \to CP(\mathcal{N})$ such that, for every $t \in \mathbb{R}^+$ and every $N_1, N_2 \in \mathcal{N} : < \psi; \gamma(t)[N_1]N_2 > = < \psi; N_1\nu(t)[N_2] >$; in connection with this condition, see Brattelli & Robinson (1976), Kossakowski *et al.* (1977) and Emch & Varilly (1979). This condition implies that $\gamma(\mathbb{R}^+)$ commutes with the modular action associated to ψ; in particular, the Bloch equation (304) generates a CWDS that is dissipative and asymptotically stable as soon as $\lambda > 0$, Markovian and reversible. More generally, every conservative CWDS $\{\mathcal{M}, \phi, \alpha\}$ is obviously reversible: take $\nu(t) = \alpha(-t)$; moreover, if $\mathcal{N}$ is a von Neumann subalgebra of $\mathcal{M}$, stable under the modular action associated to ϕ, let $\mathcal{E} : \mathcal{M} \to \mathcal{N}$ and $i : \mathcal{N} \to \mathcal{M}$ be the usual conditional expectation and embedding, $\gamma(t) \equiv \mathcal{E} \circ \alpha(t) \circ i \;\; \forall\, t \in \mathbb{R}^+$, and ψ be the restriction of ϕ to $\mathcal{N}$; then $\{\mathcal{N}, \psi, \gamma\}$ is a reversible CWDS. However, a CWDS obtained in this manner will not necessarily be Markovian or asymptotically stable, thus raising the question of whether the various structures just reviewed are compatible. While the discussion of the $x-y$ model presented in Subsection c can be recast in the present scheme to show that dissipative systems that are obtained in the manner just described can be asymptotically stable, it does not help decide whether the semigroup condition can possibly be also satisfied for dissipative systems obtained in this manner. This question is settled affirmatively by the following model.

We consider an infinite quantum lattice $\mathbb{Z}$ each site of which is occupied by a harmonic oscillator of fundamental frequency ω. Let ϕ be the canonical equilibrium state, at natural temperature β, for this non-interacting system, i.e. ϕ is described by the functional

(306) $$\phi : f \in \mathcal{T} \mapsto \exp\{-\Theta \parallel f \parallel^2 /4\} \in \mathbb{C}$$

with $\Theta = \coth(\beta\omega/2)$ and $\mathcal{T} = \mathcal{L}^2(\mathbb{R}, dx)$. Let $\{W(f) | f \in \mathcal{T}\}$ be the representation of the CCR associated to ϕ by the GNS construction, and $\mathcal{M}$ be the von Neumann algebra generated by this representation. Let $\{f_k | k \in \mathbb{Z}\}$ be an orthonormal basis in $\mathcal{T}$; we identify the algebra of local observables associated to the site $k \in \mathbb{Z}$ with

(307) $$\mathcal{B}_k \equiv \{W(zf_k) | z \in \mathbb{C}\}''$$

Since ϕ is a product state

(308) $$\phi = \bigotimes_{k \in \mathbb{Z}} \phi_k \quad \text{on} \quad \mathcal{B} = \bigotimes_{k \in \mathbb{Z}} \mathcal{B}_k$$

and its normal extension to $\mathcal{M}$, again denoted by ϕ, is faithful; $\mathcal{B}_k$ is stable under the modular group associated to ϕ and acting on $\mathcal{M}$. The faithful normal conditional expectation $\mathcal{E}_k : \mathcal{M} \to \mathcal{B}_k$ is explicitly given by:

(309) $$\begin{aligned} &\mathcal{E}_k[W(f)] = \phi((I - P_k)f)W(P_k f) \\ &\text{where } P_k : f \in \mathcal{T} \mapsto (f, f_k)f_k \in \mathcal{T} \end{aligned}$$

Consider now the action $\alpha : \mathbb{R} \to \mathrm{Aut}(\mathcal{M})$ defined by

(310) $$\begin{aligned} &\alpha(t)[W(f)] = W(\exp\{iVt\}f) \text{ with} \\ &(\exp\{iVt\}f)(x) = \exp(ixt)f(x) \end{aligned}$$

Compare with the modular action associated to ϕ, namely

(311) $$\begin{aligned} &\sigma(t)[W(f)] = W(\exp\{i\Omega t\}f) \text{ with} \\ &(\exp\{i\Omega t\}f)(x) = \exp(i\omega t)f(x) \end{aligned}$$

We now choose our orthonormal basis $\{f_k | k \in \mathbb{Z}\}$ in such a manner that

(312) $$f_0(x) = [\lambda/\pi(\lambda^2 + x^2)]^{1/2}$$

with $\lambda > 0$ arbitrary, but fixed. We now compute, for $z \in \mathbb{C}$:

$$\text{(313)} \quad \begin{array}{l} \gamma_0(t)[W(zf_0)] \equiv \mathcal{E}_0 \circ \alpha(t)[W(zf_0)] = \\ \exp\{-\Theta|z|^2[1-\exp(-2\lambda t)]/4\}W(\exp\{-\lambda t\}zf_0) \end{array}$$

With ϕ_0 denoting the restriction of ϕ to $\mathcal{B}_0$, the reader will verify that $\{\mathcal{B}_0, \phi_0, \gamma_0\}$ is a dissipative, Markovian, asymptotically stable, reversible, generalized covariant W^*-dynamical system, thus proving the compatibility of all the conditions introduced so far.

Some insight into the structure of this particular CWDS obtains from the following remark. For each $z \in \mathbb{C}$ with $|z| = 1$, let $\mathcal{X}_0(z) = \{W(\hat{x}zf_0)|\hat{x} \in \mathbb{R}\}''$. Note that $\mathcal{X}_0(i)$ [resp. $\mathcal{X}_0(1)$] is the maximal abelian von Neumann subalgebra associated to the position operator Q_0 [resp. the momentum operator P_0] for the harmonic oscillator attached to the site $k = 0$; since for each $z \in \mathbb{C}$ with $|z| = 1$ there exists $t \in \mathbb{R}$ such that $\mathcal{X}_0(z) = \sigma_0(t)[\mathcal{X}_0(i)]$, where $\sigma_0(\mathbb{R})$ is the modular action associated to ϕ_0, each of these $\mathcal{X}_0(z)$ is maximal abelian in $\mathcal{B}_0$, and corresponds to some operator, of the form $X_0(z) = a(z)P_0 + b(z)Q_0$, with absolutely continuous spectrum covering $\mathbb{R}$. Since moreover each $\mathcal{X}_0(z)$ is stable under $\gamma_0(\mathbb{R}^+)$—see (313)—one can describe this dynamics in classical terms by restricting it to $\mathcal{X}_0(z)$. In particular, for every normal state ψ_0 on $\mathcal{B}_0$, we can define a time-dependent probability measure $\psi_z(\cdot,t)d\mu_z(\cdot)$ on the spectrum $\mathbb{R}$ of $X_0(z)$ by:

$$\text{(314)} \quad <\psi_0; \gamma_0(t)[W(\hat{x}zf_0)]> = \int d\mu_z(x)\, \psi_z(x,t)\;\; \exp(-ix\hat{x})$$

where $d\mu_z(\cdot)$ is the probability measure defined by

$$\text{(315a)} \quad <\phi_0; W(\hat{x}zf_0)> = \int d\mu_z(x)\; \exp(-ix\hat{x})$$

From (306–308) one computes

$$\text{(315b)} \quad \begin{array}{l} d\mu_z(x) = (\beta\Omega_z^2/2\pi)^{1/2}\exp\{-\beta V_z(x)\}dx \\ \text{with} \quad V_z(x) \equiv \frac{1}{2}\Omega_z^2\, x^2 \\ \text{and} \quad \Omega_z^2 \equiv 2(\beta|z|^2\Theta)^{-1} \\ \text{where} \quad z \equiv \omega^{1/2}a(z) + i\omega^{-1/2}b(z) \end{array}$$

One then verifies that ψ_z satisfies the *diffusion equation*

$$\text{(316)} \quad \{\partial_t - D_z[\partial_z^2 - \beta V_z'(x)\partial_x]\}\, \psi_z(x,t) = 0$$

where the diffusion constant is given by $D_z = (\lambda\Omega_z^{-2})\beta^{-1} = \lambda|z|^2\Theta/2$, and the drift $-V_z'$ is computed from the effective potential V_z defined in (315b). For details, see Emch (1976a) where the relation is emphasized between this model and the diffusion model for an assembly of weakly coupled harmonic oscillators proposed by Ford, Kac & Mazur (1965) and studied further by Davies (1972).

This model calls for two remarks. We first want to mention that the asymptotic stability of the dissipative dynamical system $\{\mathcal{N}_0, \phi_0, \gamma_0\}$ is a reflection of an ergodic structure already present in the conservative dynamical system $\{\mathcal{M}, \phi, \alpha\}$ where it is embedded. Let indeed

$$\begin{aligned} \mathcal{T}_t &\equiv \overline{Span}\{\exp(iVs)f_0 \mid s \leq t\} \\ \mathcal{A}_t &\equiv \{W(f)|f \in \mathcal{T}_t\}'' = \alpha(t)[\mathcal{A}_0] \end{aligned} \tag{317}$$

Then $\{\mathcal{A}_t|t \in \mathbb{R}\}$ possesses the following four properties:

$$\mathcal{A}_s \subseteq \mathcal{A}_t \quad \forall\ s \leq t \tag{318a}$$

$$\mathcal{M} = \{\mathcal{A}_t|t \in \mathbb{R}\}'' \tag{318b}$$

$$\mathbb{C}I = \bigcap_{t\in\mathbb{R}} \mathcal{A}_t \tag{318c}$$

$$\sigma(s)[\mathcal{A}_t] = \mathcal{A}_t \quad \forall\ s,t \in \mathbb{R} \tag{318d}$$

These properties extend to the quantum realm the notion of K-system introduced in classical ergodic theory by Kolmogorov (1958); see also Sinai (1963), and for textbook presentations of the classical theory: Arnold & Avez (1968) or Cornfeld, Fomin & Sinai (1982). In the present context, a conservative CWDS $\{\mathcal{M}, \phi, \alpha\}$ is said to be a (generalized or) *extended Kolmogorov flow* whenever $\mathcal{M}$ contains a von Neumann subalgebra $\mathcal{A}_0$ such that $\{\mathcal{A}_t \equiv \alpha(t)[\mathcal{A}_0]|t \in \mathbb{R}\}$ satisfies (318); see Emch (1976b). Like their classical counterparts, which they encompass as particular cases—$\mathcal{M}$ is then abelian—the extended Kolmogorov flows enjoy very strong ergodic properties. For instance, the generator of the unitary group implementing $\alpha(\mathbb{R})$ has homogeneous Lebesgue spectrum (with infinite multiplicity for instance when ϕ is not a trace); from this, mixing properties follow as in the classical theory. An extension of the Kolmogorov dynamical entropy can also be defined in the present context, to measure quantitatively how much information is lost in the course of the evolution; this entropy can be shown to be infinite in the particular example detailed above: this is done by showing that one can embed in the dynamical system $\{\mathcal{M}, \phi, \alpha\}$ classical Bernoulli shifts of arbitrary large entropy. Other examples of extended Kolmogorov flows have been constructed; see e.g. Emch (1976b), Emch, Albeverio & Eckmann (1978), Evans (1979), Kümmerer & Schröder (1982). A further generalization of the notion of extended Kolmogorov flows has been proposed by Schröder (1983) under the name of *W^* -K-systems*: its main feature is to dispense with (318d) which is Takesaki's condition for the existence of a conditional expectation $\mathcal{E}_t$ from $\mathcal{M}$ onto $\mathcal{A}_t$; it is remarkable that much of the ergodic structure of $\{\mathcal{M}, \phi, \alpha\}$ remains even when the expectation hypothesis (318d) is violated; moreover W^* -K-systems have been produced for which the dynamics α coincide with the modular action σ, thus violating (318d) maximally. Hence (318d) is independent of (318 a–c), although all four conditions (318) are compatible with one another, as shown by the specific model we discussed in detail. This brings us to the second remark prompted by this model.

The dynamics $\alpha(\mathbb{R})$ on $(\mathcal{M}, \phi)$ that enters, via $\gamma_0(t) \equiv \mathcal{E}_0 \circ \alpha(t) \circ i$, in the diffusion equations (316) differs essentially from the modular action $\sigma(\mathbb{R})$ canonically associated to ϕ : $\sigma(\mathbb{R})$ is periodic, while there exist—for general reasons discussed in Emch (1976b)—two C^*-algebras $\mathcal{A}_+$ and $\mathcal{A}_-$, both norm dense in $\mathcal{M}$, such that for every $A_\pm \in \mathcal{A}_\pm$:

$$\text{(319)} \qquad \lim_{t\to\infty} \| [A_+, \alpha(t)[A_-]] \| = 0 \, .$$

This difference is an indication that ϕ and $\sigma(\mathbb{R})$ are not obtained by a straight thermodynamical limit of finite Hamiltonian systems; already from Ford, Kac & Mazur (1965) and Davies (1972), who start indeed from finite Hamiltonian systems, we learn that the semigroup $\gamma_0(\mathbb{R}^+)$, appearing in (313), obtains only in a "weak coupling limit". It is indeed here that, in addition to the thermodynamical limit and the choice of a reduced description $\{\mathcal{N}_0, \phi_0, \mathcal{E}_0\}$, a third ingredient has to be introduced into the conceptual picture we are forming for nonequilibrium statistical mechanics, and this is the limiting procedure proposed by van Hove (1955) for the theory of quantum many-body systems.

Some intuitive motivation for this limiting procedure can be gained from a heuristic consideration of an idealized classical model due to Lorentz (1905). In this model, very light particles bounce elastically against very heavy, static, randomly distributed, spherical obstacles of radius a; the particles do not interact with one another, and under their collisions with the obstacles, the latter are assumed to be recoilless; consequently, the speed $|v|$ of each particle is constant along its trajectory, and the particles evolve independently of one another. We can, therefore, concentrate on the one-particle dynamics, determined by its repeated collisions with the obstacles and its free evolution in between collisions. Its initial position x_0 and velocity v_0 at time t_0, together with the distribution law of the obstacles, determines then the probability $G(x, v, t; x_0, v_0, t_0)$ that it is found, at a later time t, in the position x with velocity v. Over a long time interval, the *mean free path* λ, i.e. the average length travelled by the particle between two successive collisions, is expected to approach the value given by $\pi a^2 \lambda \equiv (\Omega/N)$ where Ω is the total volume in which the system is enclosed, and N is the number of obstacles in Ω. Note that the fraction of the total volume, occupied by the obstacles is of the order of (Na^3/Ω), i.e. is of the order of (a/λ). As usual, one first let $\Omega, N \to \infty$ with $\rho \equiv (N/\Omega)$ fixed, to avoid having to consider the effect of the boundary of Ω. One then considers a limit where $(a/\lambda) \to 0$; this can be achieved in a number of ways: (i) keep λ fixed, and thus ρa^2 fixed, with $a \to 0$ and $\rho \to \infty$; (ii) keep a fixed, and thus $\rho\lambda$ fixed, with $\lambda \to \infty$ and $\rho \to 0$; or (iii) keep ρ fixed, and thus $a^2\lambda$ fixed, with $a \to 0$ and $\lambda \to \infty$. One can then argue see e.g. Martin (1979) or Spohn (1980)—that in the limit $(a/\lambda) \to 0$, $G(x, v, t; x_0, v_0, t_0)$ should satisfy the dissipative Boltzmann equation; these limiting procedures can be interpreted as rescallings of time and/or space, that allow the cumulative effect of many collisions to manifest itself over a macroscopic scale.

In quantum mechanical situations, the Hamiltonian for finite systems is often of the form $H_g(\Omega) = H_0(\Omega) + gV(\Omega)$ where $V(\Omega)$ represents the deviation from the evolution, generated by $H_0(\Omega)$, that does not affect the reduced description associated to the observables belonging to a subalgebra $\mathcal{N} \subseteq \{\exp[iH_0(\Omega)t] | t \in \mathbb{R}\}'$. The typical example is that of a system of N particles enclosed in a cubic volume Ω with periodic boundary conditions, where $H_0(\Omega) \equiv \sum_{k=1}^{N} (P_k^2/2m)$ is the free Hamiltonian and is thus diagonal in an orthonormal basis of eigenvectors of the momentum operators P_k, while $V(\Omega)$ describes an interaction that causes the momenta of the individual particles to change without changing the overall momentum of the total system.

The limiting procedures invoked in the classical case suggest that in the quantum case one should, in addition to taking the thermodynamical limit, consider a limiting process where $g \to 0$ and $t \to \infty$ while $g^2 t = \tau$ where τ is the time expected to appear in the dissipative equations describing the reduced evolution. Whereas the thermodynamical limit would produce an evolution $\alpha_g(\mathbb{R})$ and a canonical equilibrium state ϕ_g, this additional composite limit would result in an evolution $\alpha(\mathbb{R})$ that takes into account the long-time, cumulative effects of the interactions, while only the simple limit $g \to 0$ affects the state ϕ_g; this state would thus be expected to converge to a $\alpha(\mathbb{R})$-invariant state ϕ, whose modular group $\sigma(\mathbb{R})$ should, however, not be expected, in general, to coincide with the evolution $\alpha(\mathbb{R})$ responsible for the approach to equilibrium. We should therefore be prepared to consider conservative dynamical systems $\{\mathcal{M}, \phi, \alpha\}$ where α is not the modular action σ associated to ϕ; this is precisely the situation we encountered in (310–311).

To van Hove (1955) we owe to have shown, albeit at a level of mathematical rigour that younger generations of mathematical physicists would frown upon, that this idea of a combined limit $g \to 0$, $t \to \infty$ with $g^2 t = \tau$ fixed, can be used effectively to sort out, from an infinite perturbation expansion, those terms the contribution of which is essential to the appearance of dissipative phenomena governed by Markovian transport equations. This *van Hove limit*, with the justification for the selection of certain terms in the perturbation expansion, and the proof that these terms can be properly re-summed, were first controlled mathematically for a quantum version of the classical gas that we described in our motivating remarks; this quantum version of the model consists in non-relativistic particles moving, according to the laws of Hamiltonian quantum mechanics, through a three-dimensional lattice on which impurities are distributed in such a manner that they act on the particles as a static, translation-invariant, Gaussian random potential. The van Hove limit for this model results indeed in a Markov master equation, generating an evolution described by a dissipative semigroup; for details, see Martin & Emch (1975). To communicate the flavour of this result, we first define the one-particle Hamiltonian $H_g(\Omega) = H_0 + g\,V(\Omega)$ that we associate to every finite subset $\Omega \subset \mathbb{Z}^3$. Let $\Theta = [-\pi, \pi]^3$ be the dual space of $\mathbb{Z}^3$, and $f(\cdot)$ be the Fourier transform of $f_\bullet \in \ell^2(\mathbb{Z}^3)$; we take:

$$(320) \qquad (H_0 f)(\theta) = \omega(\theta) f(\theta) \text{ and } (V(\Omega)f)_n = \chi_n(\Omega)\Phi_n f_n$$

where $\omega(\theta) = \theta^2$, $\chi_\bullet(\Omega)$ is the characteristic function of Ω, and Φ is the Gaussian random field defined by

$$(321a) \qquad < \Phi_n > = 0 \quad , \quad < \Phi_n \Phi_m > = c_{|n-m|}$$

on which we impose the two conditions

$$(321b) \qquad \begin{aligned} \| c \|_1 &\equiv \sum_{n \in \mathbb{Z}^3} |c_n| < \infty \\ c(\theta) &\equiv (2\pi)^{-3/2} \sum_{n \in \mathbb{Z}^3} c_n \exp(in\theta) > 0 \;\; \forall\, \theta \in \Theta \end{aligned}$$

In the notation used by physicists, this model corresponds to a Hamiltonian, quadratic in the creation and annihilation operators, acting in Fock space, that can be formally written as $\tilde{H} = \tilde{H}_0 + g\tilde{V}$ with

$$(322) \qquad \begin{aligned} \tilde{H}_0 &\equiv \int_\Theta d\theta\; \omega(\theta)\; a^*(\theta) a(\theta) \\ \tilde{V} &\equiv (2\pi)^{-3/2} \iint_\Theta d\theta d\theta'\; \Phi(\theta - \theta')\; a^*(\theta) a(\theta') \end{aligned}$$

The cut-off, represented by $\chi(\Omega)$ in (320) is introduced in order to have well-defined self-adjoint operators to start the analysis with. Note that the interaction satisfies the following *van Hove diagonal singularity* ; for every operator of the form $(Af)(\theta) = A(\theta)f(\theta)$ we have:

$$(323) \qquad \begin{aligned} &\lim_{\Omega \to \mathbb{Z}^3} < V(\Omega) > (\theta, \theta') = 0 \\ &\lim_{\Omega \to \mathbb{Z}^3} < V(\Omega) A V(\Omega) > (\theta, \theta') = \\ &\qquad \delta(\theta - \theta') \int_\Theta d\theta'' A(\theta'')\, W(\theta'', \theta) \\ &\text{where } W(\theta'', \theta) \equiv (2\pi)^{-3/2} c(\theta'' - \theta) \end{aligned}$$

With $U_g(\Omega; \mathbb{R})$ [resp. $U_0(\mathbb{R})$] denoting the unitary group generated by $H_g(\Omega) = H_0 + g\,V(\Omega)$ [resp. H_0], one proves that there exists $T < \infty$ such that for all $\tau \in [0, T]$, and all $f, f' \in \mathcal{L}^2(\Theta)$:

$$(324a) \qquad \lim_{\substack{g \to 0 \\ g^2 t = \tau}} \; \lim_{\Omega \to \mathbb{Z}^3} < (U_0(-t) U_g(\Omega; t) f, f') > = (S(\tau) f, f')$$

where $\{S(\tau)|\tau \in \mathbb{R}^+\}$ is the contractive semigroup of multiplication operators given by

$$\begin{aligned} S(\tau) &= \exp\{-(\Gamma + i\Delta)\tau\} \quad \text{with} \\ \Gamma(\theta) &= \pi \int d\theta'\, W(\theta',\theta)\, \delta[\omega(\theta') - \omega(\theta)] \\ \Delta(\theta) &= P \int d\theta'\, W(\theta',\theta)\, [\omega(\theta') - \omega(\theta)]^{-1} \end{aligned} \tag{324b}$$

(where P denotes the principal part of the integral).

Spohn (1977) extended the domain where one can use the technical considerations that had first led to the mathematical solution of this particular model; see also Dell'Antonio (1982) for a discussion of the technical mathematical differences between the quantum case on the one hand, and on the other hand, the classical work of Papanicolaou & Varadhan (1973), Donsker & Varadhan (1975), and Kesten & Papanicolaou (1980). The rescalings involved when one starts with quantum Hamiltonian equations to derive hydrodynamical equations, say of the Vlasov-type, has been reinvestigated by Narnhofer & Sewell (1981); compare with Braun & Hepp (1977). Spohn (1980) has written a comprehensive review of the general area of Markovian limits in non-equilibrium statistical mechanics, both classical and quantum; see also Davies (1976). Three different models are discussed in detail by Martin (1979) to illustrate the physical premonitions that have actually been controlled in the mathematical theory of quantum irreversible processes.

To conclude this subsection on quantum non-equilibrium statistical mechanics, in which we have tried to hint at the conceptual links that exist between the quantum microscopic description and the classical macroscopic world, we want to mention briefly a specific application to the old, and most fundamental, problem of the quantum measurement process; this outline is based on Whitten-Wolfe & Emch (1976); for general reviews, see Hepp (1972), Reece (1973) and the commented collection of reprints edited by Wheeler & Zurek (1983).

Let $\{F_n\}$ be a partition of the identity in $\mathcal{B}(\mathcal{H})$, and consider the conditional expectation

$$\mathcal{E} : B \in \mathcal{B}(\mathcal{H}) \mapsto \sum_n F_n B F_n \;\in\; \{F_n\}' \tag{325}$$

The interpretation of the map $\phi \in S \mapsto \phi \circ \mathcal{E} \in S$ as a filtering process was discussed in the beginning of this section, under the name of *von Neumann measuring process*. Recall that $\{F_n\}$ was said to be compatible with a given state ϕ if $\phi \circ \mathcal{E} = \phi$, and that if $\mathcal{B}(\mathcal{H})$ had been replaced by an abelian algebra $\mathcal{B}$, with $\{F_n\} \subset \mathcal{B}$, then $\mathcal{E}$ would have reduced to the identity. The question we want to address now is whether it is possible to implement $\mathcal{E}$ as a dynamical process, i.e. whether it is possible to find a C^*-algebra $\mathcal{A}$, a state ψ on $\mathcal{A}$, and an action $\alpha : \mathbb{R} \to \mathrm{Aut}(\mathcal{B}(\mathcal{H}) \otimes \mathcal{A})$ such that, for every state ϕ on $\mathcal{B}(\mathcal{H})$, and every $B \in \mathcal{B}(\mathcal{H})$:

$$\lim_{t \to \infty} < \phi \otimes \psi ; \alpha(t)[B \otimes I] > = < \phi \circ \mathcal{E} ; B > \tag{326}$$

Moreover, if $\mathcal{A}$ is to be considered as the description of a measuring apparatus, there are a few more conditions that we would like this measuring process to satisfy, among which we single out the following three requirements: (i) $\alpha(\mathbb{R})$ should be given a Hamiltonian interpretation; (ii) some stability should hold against a sufficiently large class of perturbations of the initial state ψ of the apparatus; and (iii) as the process runs its course some information should be transferred from the microscopic level to a macroscopic level, where the experimenter can collect it with his own senses and have no problem interpreting a filtering process since then such a process does not change the state of the classical reduced description of the apparatus that is relevant to him.

These requirements can be made more precise with the help of a specific model, where their compatibility can be checked; we now present such a model.

We take for algebra of the system to be measured the algebra $\mathcal{B} = \mathcal{M}(2, \mathbb{C})$. The Pauli matrices τ^α (with $\alpha = x, y, z$) are defined as in (149); without loss of generality, we can assume the non-trivial partition $\{F_n\}$ to be given by $F_\pm = (I \pm \tau^z)/2$. Recall that we interpreted a state ϕ on $\mathcal{B}$ as a mode of preparation of the system described by its algebra of observable $\mathcal{B}$; what this means, in connection with the measuring process we are considering now, is that a collection—or beam—of individual systems has been prepared according to some prescription which we do not know, beyond the assertion that it can be summarized by associating to it a state ϕ on the algebra $\mathcal{B}$.

For the apparatus, we take in this model

$$\mathcal{A} = \bigotimes_{\omega \in \mathbb{Z}^+} \mathcal{A}_\omega \quad \text{with} \quad \mathcal{A}_\omega \simeq \mathcal{M}(2, \mathbb{C}) \tag{327}$$

where $\mathcal{A}$ is defined, as usual, as the C^*-inductive limit of the isotonic family of local algebras $\mathcal{A}(\Omega)$ associated to each finite subset $\Omega \subset \mathbb{Z}^+$. Our measuring apparatus is thus an infinite system, so that we have a sharp notion of what we should mean by "macroscopic". Specifically, we define first the reference state ψ on $\mathcal{A}$ by:

$$\begin{aligned} &\psi \equiv \bigotimes_{\omega \in \mathbb{Z}^+} \psi_\omega \text{ with } \psi_\omega : A \in \mathcal{A}_\omega \mapsto tr\, \rho_\omega A \in \mathbb{C} \\ &\rho_\omega = \exp\{\beta \varsigma_\omega \tau^z_\omega\} \,/\, tr[\exp\{\beta \varsigma_\omega \tau^z_\omega\}] \\ &\varsigma_\omega = \begin{cases} \varsigma_1 & \text{if } \omega \text{ is odd} \\ \varsigma_2 & \text{if } \omega \text{ is even} \end{cases} \end{aligned} \tag{328}$$

The real numbers β, ς_1 and ς_2 are kept fixed; they are arbitrary, except for the conditions $\beta > 0$ and $\varsigma_1 \neq \varsigma_2$. We denote by S_ψ the set of all normal states on the von Neumann algebra $\pi_\psi(\mathcal{A})''$, where π_ψ is the GNS representation associated to ψ. Note that for every $X \in \mathcal{A}$, the state

(329) $$\psi_X : A \in \mathcal{A} \mapsto < \psi; X^*AX > / < \psi; X^*X > \in \mathbb{C}$$

extends uniquely to a normal state on $\pi_\psi(\mathcal{A})''$, so that S_ψ contains in particular all the states that differ from ψ by a (quasi-)local perturbation. For every $N \in \mathbb{Z}^+$, let $\Omega = [1, 2N]$ and

(330a) $$M_1(\Omega) \equiv N^{-1} \sum_{n=1}^{N} \tau^z_{2n-1}$$
$$M_2(\Omega) \equiv N^{-1} \sum_{n=1}^{N} \tau^z_{2n}$$

(330b) $$M_-(\Omega) \equiv [\tanh(\beta\varsigma_1) - \tanh(\beta\varsigma_2)]^{-1}[M_1(\Omega) - M_2(\Omega)]$$
$$M_+(\Omega) \equiv I - M_-(\Omega)$$

The macroscopic observables $M_\pm$ are finally defined by

(330c) $$M_\pm \equiv w - op - \lim_{\Omega \to \mathbb{Z}^+} \pi_\psi\{M_\pm(\Omega)\}$$

and can be physically interpreted in terms of sublattice magnetizations.

The final step in the construction of the model is to specify the dynamics. For every $N \in \mathbb{Z}^+$, let again $\Omega = [1, 2N]$ and define now:

(331a) $$H(\Omega) \equiv F_+ \otimes H_0(\Omega) \in \mathcal{B} \otimes \mathcal{A}(\Omega)$$
$$H_0(\Omega) \equiv J \sum_{\omega=1}^{2N-1} (\tau^x_\omega \otimes \tau^x_{\omega+1} + \tau^y_\omega \otimes \tau^y_{\omega+1})$$

We denote by $\alpha(\Omega, \mathbb{R})$ the evolution defined on $\mathcal{B} \otimes \mathcal{A}(\Omega)$ by

(331b) $$\alpha(\Omega, t)[B \otimes A] \equiv \exp\{iH(\Omega)t\} \cdot A \otimes B \cdot \exp\{-iH(\Omega)t\}$$

One can then prove that, for every state ϕ on $\mathcal{B}$, every state $\psi' \in S_\psi$, and every $B \in \mathcal{B}$:

(332a) $$\lim_{t \to \infty} \lim_{\Omega \to \mathbb{Z}^+} < \phi \otimes \psi'; \alpha(\Omega, t)[B \otimes I] >$$
$$= \sum_{k=\pm} \lambda_k < \phi_k; B >$$

where

(332b) $$\lambda_\pm \equiv < \phi; F_\pm >$$
$$\phi_\pm : B \in \mathcal{B} \mapsto < \phi; F_\pm B F_\pm > / < \phi; F_\pm >$$

In this sense, we have satisfied our first set of requirements, namely to find a dynamics, implementing the process $\phi \mapsto \phi \circ \mathcal{E}$ in the sense of (326), interpretable in term of local Hamiltonian dynamics, and stable against a large class of perturbations of the initial state of the apparatus—recall (329).

Moreover, one can prove that for every state ϕ on $\mathcal{B}$, and every state $\psi' \in \mathcal{S}_\psi$:

$$\begin{aligned} \lim_{t\to\infty} \lim_{\Omega\to\mathbb{Z}^+} &< \phi \otimes \psi' ; \alpha(\Omega,t)[I \otimes M_\pm(\Omega)] > \\ &= \sum_{k=\pm} \lambda_k < \psi_k ; M_\pm > \end{aligned} \tag{333a}$$

where

$$< \psi_k ; M_j > = \delta_{kj} \quad \text{with zero dispersion} \tag{333b}$$

and λ_k determined by the state ϕ on $\mathcal{B}$ by (332b). In this sense, information has indeed been transferred from the microscopic level to the macroscopic level represented here by the macroscopic observables $M_\pm$. Again, this result is stable against the same class of perturbations of the initial state of the apparatus.

However artificial the interaction (331) may be, this model does prove that the conditions we wanted to impose on the von Neumann measuring process can be made precise enough so that one can decide that they are indeed compatible. It is true that the result of the measurement is only achieved asymptotically as time tends to infinity, in fact as $t^{-1/2}$ in this model. This can presumably be handled in the same manner as one handles the approach to equilibrium in statistical mechanics, or the approach to the free evolution in scattering theory: after a time, the actual state of the system and its final state cannot be distinguished within the arbitrary, but finite, precision of an actual experiment.

There are nevertheless serious problems that still remain to be solved.

The first problem is not so much a matter of conceptual compatibility, but one of making realistic contact with actual experiments.

The second problem is conceptual. We have so far adopted the traditional attitude that the state ϕ is a summary of the mode of preparation of the system. This is sufficient only as long as one limits one's attention to expectation values over a collection of systems prepared under identical conditions, at whatever level of description. The problem changes drastically if one asks what happens to individual systems, in particular in the course of a measurement process. One has then to face the fact that $\mathcal{B}$, the algebra of observables on a quantum system, being non-abelian implies that the convex set of the states on $\mathcal{B}$ is not a simplex. In particular, if $\mathcal{B}_1$ and $\mathcal{B}_2$ are two copies of the same non-abelian C^*-algebra $\mathcal{B}_0$, e.g. $\mathcal{M}(2,\mathbb{C})$, there are pure states on $\mathcal{B}_1 \otimes \mathcal{B}_2$, the restriction of which to neither $\mathcal{B}_1$ nor $\mathcal{B}_2$ is pure. When a state ϕ on $\mathcal{B}$, representing a beam of individual systems—e.g. particles or pairs of correlated particles—can be decomposed into different convex combinations of pure states, it is to be expected that *individual* measurements would be able to distinguish between these different decompositions. Even more acutely than with the von Neumann measurement process, the production of conceptually satisfactory dynamical models for this type of measurement process still belongs to the cutting edge of contemporary research on the empirical foundations of quantum mechanics. A stimulating query along this line was raised by Einstein, Podolsky & Rosen (1935); among the responses elicited by this so-called paradox, see in particular: Bell (1964); Clauser & Shimony (1978); Mermin (1981); Aspect (1983); and Wheeler & Zurek (1983).

Section 2. QUANTUM FIELD THEORY

The theory of relativistic quantum fields is a point of confluence for the main currents of 20th-Century physics; foremost amongst these, quantum theory and the theory of relativity are contributing the basic concepts. A synthesis between these two theories would fulfill a most natural metaphysical desire for a unification of our world views. There are also immediate physical circumstances that require this synthesis: it is indispensable to a consistent discussion of several situations that have persistently presented themselves in the laboratory, ever since the time of inception of these two theories in the first third of this Century; among these phenomena are those that involve interactions between matter and electromagnetic radiation, and in particular between electrons and photons. The most urgent theoretical challenge is therefore to produce a quantum theory, allowing for the creation and annihilation of particles, in an interaction scheme compatible with the four-dimensional Lorentz structure of the Minkowski space-time of special relativity; some astrophysical situations would further require that the synthesis be extended to a quantum formalism compatible with the tenets of general relativity.

The aim of this section is to outline the premises of some of the programmes that have been proposed to meet this challenge in the case of special relativity.

Each one of these programmes has led to the development of mathematical methods that are both very powerful and very subtly tuned; and each has contributed in turn to a better understanding of how difficult the problem really is. Since we will present here only the most elementary rudiments of the emerging theories, the reader interested in more detailed information should know that some of the best such information has been organized in full-length book form by Streater & Wightman (1964), Jost (1965), Simon (1974, 1979) and Glimm & Jaffe (1981).

a. Elementary quantum systems in special relativity

In this subsection we review the classification, by Wigner (1939), of the elementary quantum systems compatible with the special theory of relativity.

The fundamental symmetry group of the theory is taken to be the proper Poincaré group G, i.e. the connected component of the identity in the group of all transformations (5.1.2) of the Minkowski space $M^{3,1}$. Recall that every $g \in G$ can be identified as a pair (a, Λ) described by

$$g[x] = (a, \Lambda)\,[x] = \Lambda\,[x] + a \tag{1}$$

The resulting composition law

$$g_1 g_1 = (a_2, \Lambda_2)(a_1, \Lambda_1) = (a_2 + \Lambda_2\,[a_1], \Lambda_2\Lambda_1) \tag{2}$$

equips G with the structure of a semi-direct product of the group $T \equiv T^4$ of translations of $M^{3,1}$, by the group $H \equiv L^{3,1}$ of proper homogeneous transformations of $M^{3,1}$. The topology, defined on G by the condition that $g \mapsto g[x]$ be continuous, then equips G with the structure of a separable, connected, locally compact topological group; it is in fact a 10-dimensional Lie group. These properties are evidently inherited by T (which is 4-dimensional) and by H (which is 6-dimensional). Moreover, T is simply connected and abelian, while the (non-abelian) group H is doubly connected with universal covering group $\tilde{H} = SL(2, \mathbb{C})$. G is similarly doubly connected, and its universal covering group $\tilde{G}$ inherits from G the structure of a semi-direct product of T by $\tilde{H}$.

In line with the ideas described in Subsection 8.3.e, an elementary quantum system in special relativity is identified with a continuous action (see 8.3.142)

$$\alpha : g \in G \mapsto \alpha(g) \in \mathrm{Sym}\,(\mathcal{H}) \tag{3}$$

such that

$$\begin{array}{l} A \in \mathcal{B}_{sa}(\mathcal{H}) \ \text{ and } \ \alpha(g)[A] = A \ \forall\ g \in G \\ \text{imply } \ A = \lambda I \ \text{ with } \ \lambda \in \mathbb{R} \end{array} \tag{4}$$

As we saw in Subsection 8.3.e, the classification of these actions of G amounts to the classification of the irreducible, strongly continuous, unitary representations of $\tilde{G}$. The latter is what Wigner (1939) achieved. His proof was a judicious adaptation, to the Poincaré group, of the method of induced representations developed by Frobenius for discrete groups; this method was later generalized by Mackey (1955) to a much wider class of topological groups, of which the proper Poincaré group G, and its universal covering $\tilde{G}$, are particular representatives. We outline the method for $\tilde{G}$, with an emphasis on the fact that $\tilde{G}$ is a semi-direct product of two locally compact groups T and $\tilde{H}$ with T abelian; we omit, however, the measure theoretical trimmings involved in Mackey's general theory.

We first introduce the concept of induced representation. Let $L : \Gamma \to \mathcal{U}(\mathcal{H})$ be a strongly continuous, unitary representation of a closed subgroup Γ of $\tilde{G}$; μ be a quasi-invariant Borel measure on $\tilde{G}/\Gamma$; and $\mathcal{H}^L$ be the Hilbert space, the elements of which are the functions $\Psi : \tilde{G} \to \mathcal{H}$ that satisfy the following three conditions

(5a) $g \in \tilde{G} \mapsto (\Psi(g), \Phi) \in \mathbb{C}$ is Borel $\forall\, \Phi \in \mathcal{H}$

(5b) $\Psi(\gamma g) = L(\gamma)\, \Psi(g) \quad \forall\, (\gamma, g) \in \Gamma \times G$

(5c) $\int_{\tilde{G}/\Gamma} d\mu(\xi)\ \| \Psi(g) \|^2 < \infty$

Note that (5b) implies that $(\Psi(g), \Phi(g))$ is constant on each coset of Γ in $\tilde{G}$, so that (5c) does make sense, and an inner product on $\mathcal{H}^L$ is actually defined by

(5d) $$(\Psi, \Phi) \equiv \int_{\tilde{G}/\Gamma} d\mu(\xi)\, (\Psi(g), \Phi(g))$$

With ρ_g denoting the Radon-Nikodym derivative $d\mu_g/d\mu$, we define, for every $g \in \tilde{G}$, the operator $U^L(g)$ acting on $\mathcal{H}^L$

(6a) $$[U^L(g)\, \Psi](h) = [\rho_g(h)]^{1/2}\, \Psi(hg)$$

We clearly have that

(6b) $$U^L : g \in \tilde{G} \mapsto U^L(g) \in \mathcal{U}(\mathcal{H}^L)$$

is a strongly continuous, unitary representation of $\tilde{G}$ on $\mathcal{H}^L$; Mackey further showed that U^L is uniquely defined, up to unitary equivalence, by the equivalence class of μ. We will refer to U^L as the *representation* of $\tilde{G}$ *induced* by L.

Let now $\hat{T}$ denote the character group of T; its elements are the maps

(7) $$\chi : a \in T \mapsto \exp\{-i < p, a >\} \in \mathbb{C}$$
with $p \in M^{3,1}$ and $< p, a > \equiv g_{\mu\nu} p^\mu a^\nu$

Recall now that $\tilde{G}$ is a semi-direct product of T by $\tilde{H}$. Two characters χ_1 and χ_2 of T are said to be *equivalent* if there exists $\Lambda \in \tilde{H}$ such that $\chi_2 = [\chi_1]\Lambda$, where $[\cdot]\Lambda$ is the dual action of $\tilde{H}$ on $\hat{T}$ defined by

(8) $$[\chi]\Lambda : a \in T \mapsto \chi(\Lambda[a])$$

The *little group* $\tilde{H}_\chi$ of a character χ is defined by

(9) $$\tilde{H}_\chi = \{\Lambda \in \tilde{H} \mid [\chi]\Lambda = \chi\}$$

Note that if two characters are equivalent, then their little groups are conjugate to one another in $\tilde{H}$. We finally denote by $\tilde{G}_\chi$ the semi-direct product of T by $\tilde{H}_\chi$; and we remark that if L is a strongly continuous, unitary representation of $\tilde{H}_\chi$ acting on some Hilbert space $\mathcal{H}$, then

$$(\chi, L) : (a, \Lambda) \in \tilde{G}_\chi \mapsto \chi(a)L(\Lambda) \in \mathcal{U}(\mathcal{H}) \tag{10}$$

is a strongly continuous, unitary representation of $\tilde{G}_\chi$ acting on $\mathcal{H}$.

The *Frobenius-Wigner-Mackey theorem* then makes the following three assertions. Firstly, if L is an irreducible, strongly continuous, unitary representation of $\tilde{H}_\chi$ for some character χ of T, then the strongly continuous, unitary representation $U^{(\chi,L)}$ of G, induced from (χ, L) is also irreducible. Secondly, two irreducible representations $U^{(\chi_1,L_1)}$ and $U^{(\chi_2,L_2)}$ obtained in this manner are unitarily equivalent if and only if χ_1 and χ_2 are equivalent, *and*

$$\begin{array}{l} \Lambda \in \tilde{H}_{\chi_1} \mapsto L_{\chi_2}(\Lambda_0^{-1}\Lambda\Lambda_0) \in \mathcal{U}(\mathcal{H}_2) \\ \text{with} \quad \Lambda_0 \in \tilde{H} \quad \text{such that} \quad \chi_2 = [\chi_1]\Lambda_0 \end{array} \tag{11}$$

is unitarily equivalent to L_{χ_1}. Thirdly, up to unitary equivalence, all irreducible, strongly continuous, unitary representations of $\tilde{G}$ can be obtained in this manner.

The interest of this theorem is that: (i) it gives an immediate classification of the irreducible, strongly continuous, unitary representations of $\tilde{G}$, and thus a classification of those quantum systems that are elementary with respect to the proper Poincaré group; (ii) it gives an explicit procedure to construct a representative of each class of unitary equivalence.

The equivalence classes of characters are parametrized by the value of $g_{\mu\nu}p^\mu p^\nu = -m^2$, and when m is real, by the sign of p^4. We will restrict our attention to $m \geq 0$ and $p^4 > 0$, treating separately the cases $m > 0$ and $m = 0$.

In the first case, a typical representative is the character that corresponds to $p = m(0, 0, 0, 1)$. Its little group is $SU(2, \mathbb{C})$. The equivalence classes of irreducible, strongly continuous, unitary representations of $SU(2, \mathbb{C})$ have been described in Subsection 8.3.e. Recall that they are parametrized by dim $L = (2s + 1) \in \mathbb{Z}^+ \setminus \{0\}$. The corresponding elementary systems are particles of mass $m > 0$ and spin s ; e.g. scalar mesons ($s = 0$), electrons and protons ($s = 1/2$), etc. Note that whenever $(2s + 1)$ is odd, the representation $U^{(\chi,L)}$ is an ordinary representation of the proper Poincaré group G.

In the second case, a typical representative is the character that corresponds to $p = (1, 0, 0, 1)$. Its little group is isomorphic to the Euclidean group acting on $\mathbb{R}^2$. This group satisfies again the hypotheses under which the Frobenius-Wigner-Mackey theorem holds. Two types of representations then appear, leading respectively to particles with discrete "spin" s (again with $2s + 1 = n \in \mathbb{Z}^+ \setminus \{0\}$), and with "continuous spin"; only particles corresponding to the first type have been identified in nature, e.g. neutrinos ($s = 1/2$), photons ($s = 1$), etc.

The definition of elementary particle chosen in this subsection involves only the proper Poincaré group. One could naturally enlarge this symmetry group and search for the corresponding irreducible representations. The most immediate generalization is to include the space- and/or time-reflections; this is part of the folklore of theoretical physics, and the author first learned about it from an unpublished manuscript by Bargmann, Wigner & Wightman; what amounts to a brief summary of that part of this manuscript can be found in the appendix of Emch (1963). Several attempts have been made in the last twenty years to include some internal symmetries as well in the defining group of elementary systems; the reader should consult here a few of his high-energy physics friends, who will surely want to describe the latest favorite amongst the apparently limitless variety of ingenious schemes that have been proposed to achieve a fully satisfactory definition of elementary particles and of the fields that mediate their interactions.

b. The relativistic Fock spaces

The aim of this subsection is to generalize to the context of special relativity the non-relativistic Fock space techniques introduced in Section One.

The Fock spaces $\mathcal{F}_{m,s}$ to be constructed are to host any arbitrary, but finite, number of identical elementary particles of mass m and spin s. The one-particle subspace $\mathcal{F}^{(1)}_{m,s}$ of $\mathcal{F}_{m,s}$ is thus identified with the Hilbert space on which acts the irreducible, strongly continuous, unitary representation (m,s) of the universal covering group $\tilde{G}$ of the proper Poincaré group. Depending on whether $(2s+1)$ is odd (boson) or even (fermion) $\mathcal{F}^{(N)}_{m,s}$ is taken to be the completely symmetric or antisymmetric N-fold tensor product of $\mathcal{F}^{(1)}_{m,s}$ with itself. $\mathcal{F}_{m,s}$ is then defined to be

$$\mathcal{F}_{m,s} = \bigoplus_{N=0}^{\infty} \mathcal{F}^{(N)}_{m,s} \tag{12}$$

For the sake of simplicity, we shall concentrate on the case of a massive (i.e. $m > 0$), scalar (i.e. $s = 0$) meson; and we shall drop from the notation the indices m and s whenever they are not indispensable. The Frobenius-Wigner-Mackey theorem then leads us to realize $\mathcal{F}^{(1)}$ as the space of all Borel measurable functions Ψ on the "mass-hyperboloid"

$$H^3 \equiv \{p \in M^{3,1} \mid g_{\mu\nu} p^\mu p^\nu = -m^2\,;\; p^4 > 0\} \tag{13a}$$

with Ψ square-integrable with respect to the Lorentz-invariant measure defined on H^3 by

$$\mu(d^3p) \equiv (\| p \|^2 + m^2)^{-1/2} \; d^3p \tag{13b}$$

The representation $U^{(1)}$ of the proper Poincaré group G acting on $\mathcal{F}^{(1)}$, as constructed from the Frobenius-Wigner-Mackey theorem, extends naturally to a strongly continuous, unitary representation U of G acting on $\mathcal{F}$.

Let $S(\mathbb{R}^3)$ denote the Schwartz space of real-valued C^∞-functions of rapid decrease at infinity. For each $f \in S(\mathbb{R}^3)$ we then define the operators $a(f)$, $a^*(f)$, $\Phi(f)$ and $W(f)$ by a straightforward generalization of (10.1.217–223). We note that $\Psi_0 \in \mathcal{F}^{(0)}$ is in the domain of definition of every polynomial over $\mathbb{C}$ in the $\Phi(f)$'s. With $\mathcal{P}(\Phi)$ denoting the polynomial algebra, over $\mathbb{C}$, of the operators $\Phi(f)$, we have that $\mathcal{D} \equiv \mathcal{P}(\Phi)\Psi_0$ is dense in $\mathcal{F}$, and stable under $\Phi(f)$ and $U(g)$ for every $f \in S(\mathbb{R}^3)$ and every $g \in G$. Moreover, with $U_0(\mathbb{R})$ [resp. $U_0^{(1)}(\mathbb{R})$] denoting the restriction of U [resp. $U^{(1)}$] to the subgroup $\mathbb{R} \subset G$ of the time translations, we have, on $\mathcal{D}$:

(14a) $$\Phi_t(f) \equiv U_0(t)\Phi(f)U_0(t)^* = \Phi(U_0^{(1)}(t)f)$$

with

(14b) $$U_0^{(1)}(t) = \exp\{-iH_0\,t\}f$$

and

(14c) $$(H_0 f)(x) = [(-\Delta + m^2)^{1/2} f](x)$$

From this, we conclude that the operator-valued distribution defined by

(15) $$\Phi_t(f) = \int d^3x \; f(x)\; \Phi(x,t)$$

satisfies the Klein-Gordon equation

(16) $$[(-g^{\mu\nu}\partial_\mu\partial_\nu + m^2)\Phi](x,t) = 0$$

This manifestly Poincaré-invariant quantum equation establishes a suggestive link with classical relativistic field theory. The classical Lagrangean density

(17) $$\mathcal{L} = -\frac{1}{2}g^{\mu\nu}\partial_\mu\varphi \cdot \partial_\nu\varphi - \frac{1}{2}m^2\varphi^2$$

gives indeed the equation of motion

(18) $$[(-g^{\mu\nu}\partial_\mu\partial_\nu + m^2)\varphi](x,t) = 0$$

From the classical Lagrangean density (17), we can compute for future reference the momentum-energy tensor

(19) $$T_{\mu\nu} = \partial_\mu\varphi \cdot \partial_\nu\varphi + g_{\mu\nu}\mathcal{L}$$

and, in particular, the energy-density T_{00} for which we obtain the explicit form

(20) $$T_{00} = \frac{1}{2}(\partial_t\varphi)^2 + \frac{1}{2}\sum_{i=1}^{3}(\partial_i\varphi)^2 + \frac{1}{2}m^2\varphi^2$$

which is non-negative and vanishes only for the trivial field $\varphi \equiv 0$.

The Fock space objects which we constructed lend themselves to much combinatorics which is also very suggestive. Let $\Psi_0 \in \mathcal{F}^{(0)}$ with $\| \Psi_0 \| = 1$, and denote by $< A >_0$ the *vacuum expectation value* $(A\Psi_0, \Psi_0)$ of an operator A defined in $\mathcal{F}$. Upon noticing that

$$(21) \qquad < \Phi(f)^2 >_0 = \frac{1}{2} \| f \|^2$$

where $\| \cdot \|^2$ denotes the norm with respect to the scalar product in $\mathcal{F}^{(1)}$, we can transfer (10.1.229a) to the relativistic context of the present section, and obtain

$$(22a) \qquad < \exp\{i\Phi(f)\} >_0 = \exp\{-\frac{1}{2} < \Phi(f)^2 >_0\}$$

and thus, for all $n \in \mathbb{Z}^+$

$$(22b) \qquad < \Phi(f)^{2n+1} >_0 = 0$$

$$(22c) \qquad < \Phi(f)^{2n} >_0 = \frac{(2n)!}{2^n \, n!} < \phi(f)^2 >_0^n$$

Upon writing $f = \sum_i \lambda_i f_i$ with $\lambda_i \in \mathbb{R}$, and using the linearity of Φ, we further obtain

$$(22d) \qquad < \Phi(f_1) \ldots \Phi(f_{2n}) >_0 = \sum \prod < \Phi(f_{i_k}) \Phi(f_{j_k}) >_0$$

where the summation runs over the $(2n-1)!!$ pairings $(i_1, j_1) \ldots (i_n, j_n)$ of the indices $(1 \ldots 2n)$, and where the product runs, for each pairing, over all pairs occuring in that pairing. The relations (22) are evidently reminiscent of the situation encountered in classical probability theory when one deals with a collection of random variables that are jointly Gaussian. This Gaussian character is an essential reflection of the fact that the quantum field Φ has been obtained from a system of non-interacting particles.

The relations (22) further suggest introducing the so-called *Wick-ordering* $:\Phi(f)^n:$ defined recursively for $n = 1, 2, \ldots$ by

$$(23) \qquad \begin{aligned} \frac{\partial}{\partial \Phi(f)} :\Phi(f)^n: &= n :\Phi(f)^{n-1}: \\ < :\Phi(f)^n: >_0 &= 0 \\ :\Phi(f)^0: &= 1 \end{aligned}$$

This gives explicitly, for the Gaussian case considered here

$$(24) \qquad :\Phi(f)^n: = 2^{-n/2} \sum_{m=0}^{n} \frac{n!}{m!(n-m)!} a^*(f)^m \, a(f)^{n-m}$$

Hence : $\Phi(f)^n$: differs from $\Phi(f)^n$ by the fact that the creation operators $a^*(f)$ and the annihilation operators $a(f)$ appearing in the expansion of

$$\Phi(f)^n \equiv [\frac{1}{\sqrt{2}}\{a^*(f)+a(f)\}]^n \tag{25}$$

have been reordered in such a manner that all creation operators appear in (24) on the left of all annihilation operators. This property persists when the Wick ordering is extended by linearity to arbitrary polynomials in the $\Phi(f)$'s. For instance, this extension leads to

$$:\prod_{i=1}^{n}\Phi(f_i):\Psi_0 = 2^{-n/2}\prod_{i=1}^{n}a^*(f_i)\Psi_0 \tag{26}$$

and to the orthogonality relation

$$\begin{aligned} &< :\prod_{i=1}^{n}\Phi(f_i): \; :\prod_{j=1}^{m}\Phi(g_j): >_0 = \\ &\delta_{nm}\sum_{p\in P_n}\prod_{k=1}^{n} < \Phi(f_k)\Phi(g_{p(k)}) >_0 = \\ &2^{-n}\,\delta_{nm}\sum_{p\in P_n}\prod_{k=1}^{n}(f_k\,,\,g_{p(k)}) \end{aligned} \tag{27}$$

The Wick-ordering is a convenient computational tool that eliminates from the formalism undesirable vacuum expectation values. Its effect is most simply illustrated in the context of the quantum mechanics of a one-dimensional harmonic oscillator; it replaces the Hamiltonian

$$H = \frac{1}{2}(P^2+Q^2) = \frac{1}{2}(a^*a + aa^*) \tag{28a}$$

by the operator H from which the zero-point energy $< H >_0 = 1/2$ has been substracted, i.e.

$$:H: = H - < H >_0 = a^*a \tag{28b}$$

In the Fock space context, where $\dim[\mathcal{F}^{(1)}] = \infty$ reflects the infinite number of degrees of freedom of the field Φ, (23) implies immediately

$$:\Phi(f)^2: = \Phi(f)^2 - < \Phi(f)^2 >_0 \tag{29}$$

Physically, the Wick-ordering appears naturally in observable quantities. For instance, the free Hamiltonian H, defined as the generator of the evolution $U_0(\mathbb{R})$ on $\mathcal{F}$ (see 14), can be written as an integral over the 3-dimensional space-like hyperplane $\Sigma \equiv \{(x,0)|x \in \mathbb{R}^3\}$, namely

$$H = \int d^3x\, H_\Sigma(x) = \int d^3x\; a^*(x)\,[(-\Delta + m^2)a](x) \tag{30a}$$

where the energy-density on Σ can be rewritten, in terms of the field Φ and its canonical conjugate $\Pi = \partial_t\Phi$, in the form

$$H_\Sigma = \frac{1}{2} : \Pi^2 : + \frac{1}{2} : (\nabla\Phi)^2 : + \frac{1}{2}\, m^2 : \Phi^2 : \tag{30b}$$

which gives the quantum analog of the classical energy-density computed in (20). To this corresponds the quantum Lagrangean density

$$\mathcal{L} = -\frac{1}{2}\, g^{\mu\nu} : \partial_\mu\Phi\; \partial_\nu\Phi : - \frac{1}{2}\, m^2 : \Phi^2 : \tag{30c}$$

to be compared with (17).

c. Towards a general theory of quantum fields

While Fock space techniques do accomodate very well free quantum fields that are compatible with the Minkowski structure $M^{3,1}$ of special relativity, serious difficulties appear as soon as one tries to include non-trivial interactions, for instance by introducing polynomial terms of higher degree in the quadratic Lagrangean (30c). For these, as well as for all other relativistic interactions involving an infinite number of degrees of freedom, the difficulties manifest themselves by the occurrence of divergences that have plagued the investigations of quantum fields ever since the early 1930's.

The amazing success achieved in quantum electrodynamics by formal expansions notwithstanding, the disquieting status of the theory after some thirty years of efforts was dramatically summarized by Jost (1965): "We start out from equations which do not make sense. We apply certain prescriptions to their solutions and end up with a power series of which we do not know that it makes sense. The first few terms of this series, however, give the best predictions we know."

Many practitioners, in the theoretical physics community, have somehow learned to live with this endemic state of affairs. In the last twenty-five years, however, two main lines of attack have been deployed with the purpose to bring the symptoms under mathematical control; but it must be recognized that the disease itself, in its most virulent forms, has not been eradicated yet.

The first line of attack is a systematic reexamination of the basic tenets of the theory; it is sometimes referred to as "axiomatic quantum field theory" to distinguish it from the second line of attack, known as "constructive quantum field theory". The latter is a sustained effort to understand the inner workings of the situation, by building specific models that are exactly solvable, even if at the cost of dealing with a ficticious world where the dimension of space-time is different from the usual $d = 4$; in two and three space-time dimensions, much success has been achieved, leading to a vindication of the axiomatic schemes elaborated along the first line of attack. Whether the case $d = 4$ will similarly abide by the rules set forth by the axiomatists is an interesting question that is still open.

One sobering indication that $M^{3,1}$ might be very special indeed comes from a quite different quarter; it is the recognition of the serious challenge that the formulation of a relativistically covariant theory, including interacting systems, already offers in classical particle mechanics. This is illustrated by the Currie-Jordan-Sudarshan (1963) no-go theorem asserting, under assumptions one would otherwise be inclined to accept as quite general and natural, that particles may only move along straight lines; see several of the contributions collected in Llose (1982), and the references quoted therein.

In the remainder of this section, we only briefly indicate some of the most appealing axiomatic schemes that have been proposed for quantum field theory. For the detailed developments, and the actual implementation, of these schemes we refer again to the texts recommended at the end of the introduction to this section.

The basic object of the Wightman axioms in a triple $\{\mathcal{H}, U, \Phi\}$.

$\mathcal{H}$ is a Hilbert space, built over the field of the complex numbers for the purpose of hosting the description of a quantum system to be made precise now.

In line with the definition of symmetries presented in Subsection 8.3.e, and elaborated in Subsection a above, special relativity is introduced by assuming that

$$U : \tilde{G} \to \mathcal{U}(\mathcal{H}) \tag{31}$$

is a strongly continuous, unitary representation of the universal covering group $\tilde{G}$ of the proper Poincaré group G for the Minkowski space $M^{n,1}$; one naturally has in mind the case $d \equiv n + 1 = 4$, but it is sometimes technically advantageous to leave open the dimension d of space-time. Let $U(T^d)$ be the restriction of U to the group T^d of translations in $M^{n,1}$. There exists then a spectral family E on $\mathcal{H}$, with Lorentz-invariant support in $M^{n,1}$, such that

$$U(a) = \int \exp\{-i < p, a >\} \, dE(p) \quad \forall \, a \in T^d \tag{32}$$

where $< p, a >$ denotes the scalar product $g_{\mu\nu} p^\mu a^\nu$ in $M^{n,1}$. Two conditions are imposed on the joint spectrum of the generators $\{P^\mu | \mu = 1, \ldots, d\}$ of $U(T^d)$. The first condition is that there exists a unique (up to multiplication by $\lambda \in \mathbb{C}$) vector $\Psi_0 \in \mathcal{H}$ with $\Psi_0 \neq 0$, such that

$$U(a)\Psi_0 = \Psi_0 \quad \forall \, a \in T^d \tag{33}$$

Note that we have then, $\forall\, (a,\Lambda) \in \tilde{G}$:

$$U(a)U(\Lambda)\Psi_0 = U(\Lambda)U(\Lambda^{-1}[a])\Psi_0 = U(\Lambda)\Psi_0 \tag{34}$$

and thus

$$U(\Lambda)\Psi_0 = \lambda(\Lambda)\Psi_0 \quad \forall\, \Lambda \in \tilde{H} \tag{35}$$

In case $d = 4$: $\tilde{H} = SL(2, \mathbb{C})$; since the only one-dimensional, strongly continuous, unitary representation of this group is the identity representation, (33–35) read then

$$U(g)\Psi_0 = \Psi_0 \quad \forall\, g \in \tilde{G} \tag{36}$$

The second condition one imposes on the joint spectrum of $\{P^\mu | \mu = 1, \ldots, d\}$ is that the support of the spectral measure E defined in (32) satisfy

$$\text{supp}\ \ E \subseteq \{p \in M^{n,1} | -g_{\mu\nu}p^\mu p^\nu \geq 0\,,\ p^4 \geq 0\} \tag{37}$$

If one further assumes that the eigenvalue 0 is isolated, the fact that the support of E is a Lorentz-invariant closed subset of $M^{n,1}$ implies the existence of $m_0 > 0$ such that

$$\text{supp}\ \ E \subseteq \{0\} \bigcup \{p \in M^{n,1} | -g_{\mu\nu}p^\mu p^\nu \geq {m_0}^2; p^4 \geq m_0\} \tag{38}$$

The third member of the Wightman triplet $\{\mathcal{H}, U, \Phi\}$ is the quantum field Φ. It is introduced as a map

$$\Phi : f \in \mathcal{S}(\mathbb{R}^d) \mapsto \Phi(f) \in \mathcal{A}(\mathcal{H}) \tag{39}$$

where $\mathcal{S}(\mathbb{R}^d)$ is the Schwartz space of real-valued C^∞-functions of rapid decrease at infinity; and $\mathcal{A}(\mathcal{H})$ is a *-algebra of (unbounded, but densely defined) linear operators acting in $\mathcal{H}$. One assumes that there exists a dense domain $\mathcal{D} \subset \mathcal{H}$, contained in the domain of definition of $\Phi(f)\ \ \forall\, f \in \mathcal{S}(\mathbb{R}^d)$; all conditions stated here on $\Phi(f)$ are tacitly assumed to hold as well for $\Phi(f)^*$. One then assumes that $\forall\, (\Psi_1, \Psi_2) \in \mathcal{D} \times \mathcal{D}$, the maps

$$f \in \mathcal{S}(\mathbb{R}^d) \mapsto (\Phi(f)\Psi_1, \Psi_2) \in \mathbb{C} \tag{40}$$

are tempered distributions, i.e. are linear and continuous for the Schwartz topology on $\mathcal{S}(\mathbb{R}^d)$. One further assumes that $\mathcal{D}$ is stable under the action of each $\Phi(f)$, i.e.

$$\Phi(f)\mathcal{D} \subseteq \mathcal{D} \quad \forall\, f \in \mathcal{S}(\mathbb{R}^d) \tag{41a}$$

so that polynomials in the $\Phi(f)$'s are well-defined on $\mathcal{D}$; we write $\mathcal{P}(\Phi)$ for the polynomial algebra over $\mathbb{C}$ obtained in this manner from the $\Phi(f)$'s; we thus have

$$\mathcal{P}(\Phi)\mathcal{D} \subseteq \mathcal{D} \tag{41b}$$

Finally, one requires that the following conditions be satisfied, linking the above two structures—i.e. special relativity and quantum field.

Firstly, $\mathcal{D}$ is assumed to be stable under $U(\tilde{G})$ i.e.

$$U(g)\mathcal{D} \subseteq \mathcal{D} \quad \forall\, g \in \tilde{G} \tag{42}$$

Secondly, one assumes that, $\forall\, g \in \tilde{G}$ and $\forall\, f \in \mathcal{S}(\mathbb{R}^d)$:

$$U(g)\,\Phi(f)\,U(g)^* = \Phi(g[f]) \quad \text{on } \mathcal{D} \tag{43a}$$

where $g[f]$ is defined by

$$g[f]\,(x) = f(g^{-1}[x]) \quad \forall\, x \in M^{n,1} \tag{43b}$$

Thirdly, one assumes that the invariant vector Ψ_0 (see 33) belongs to $\mathcal{D}$; as a consequence of (41b), one has that $\mathcal{P}(\Phi)\Psi_0 \subseteq \mathcal{D}$. One then assumes that Ψ_0 is cyclic in $\mathcal{H}$ under $\mathcal{P}(\Phi)$, i.e.

$$\mathcal{P}(\Phi)\Psi_0 \quad \text{is dense in} \quad \mathcal{H} \tag{44}$$

Fourthly, a locality condition is required which we formulate here only in the boson case, i.e. when U is in fact a representation of G itself and not only of $\tilde{G}$. Two regions $\mathcal{O}_1$ and $\mathcal{O}_2$ in $M^{n,1}$ are said to be *causally disjoint* whenever

$$\{x_1 \in \mathcal{O}_1 \quad \text{and} \quad x_2 \in \mathcal{O}_2\} \ \text{implies} \ \{g_{\mu\nu}(x_1 - x_2)^\mu (x_1 - x_2)^\nu \geq 0\} \tag{45a}$$

The locality condition now reads

$$\begin{array}{l} [\Phi(f_1), \Phi(f_2)] = 0 \quad \text{on } \mathcal{D} \quad \text{whenever} \\ \text{supp}\,(f_1) \ \text{and supp}\,(f_2) \ \text{are causally disjoint} \end{array} \tag{45b}$$

where $[\,.\,,\,.\,]$ denotes the commutator, which is indeed well-defined on $\mathcal{D}$ since (41a) is assumed.

The list of the Wightman axioms is now complete. The reader will verify that all the axioms are satisfied for the free, massive, scalar field constructed on Fock space in Subsection b above. It is equally straightforward to check that the axioms are similarly satisfied for the other free, massive (i.e. $m > 0$), Bose (i.e. $2s + 1$ odd) fields constructed on their own Fock spaces. If one is willing to give up (38), one can include as well the free, massless (i.e. $m = 0$) Bose fields. The so-called generalized free fields are also controlled by these axioms; see e.g. Jost (1965). A minor modification of (45b), involving the substitution of anticommutators for the commutators, will also allow the free Fermi (i.e. $2s+1$ even) fields into the scheme.

Moreover, an axiomatization is not only interesting for what it explicitly says, but also for what it does not assume, i.e. for the distinction it proposes between what is essential and what is to be considered as incidental. The Wightman axioms leave open the choice of the Hilbert space $\mathcal{H}$ or, more precisely, the choice of the representations U of $\tilde{G}$ and Φ of $S(\mathbb{R}^d)$, as long as the conditions set forth by the axioms are satisfied. In particular, while the axioms postulate the existence of a unique vector Ψ_0 that is invariant under $U(T^d)$ [or $U(\tilde{G})$], this vector Ψ_0 is *not* assumed to satisfy the characteristic relation of the Fock space representation of the canonical commutation relations, namely

$$a(f)\Psi_0 = 0 \quad \forall\, f \in S(\mathbb{R}^d) \tag{46}$$

In contradistinction to the case of quantum systems with a finite number of degrees of freedom—where $S(\mathbb{R}^d)$ is replaced by a finite-dimensional Hilbert space—there is no von Neumann uniqueness theorem for quantum systems with infinitely many degrees of freedom. Not only does the proof of Proposition 8.3.1 fail to extend to this case, but an infinite maze of inequivalent irreducible representations of the canonical commutation [and anticommutation] relations has actually been constructed; see Gårding & Wightman (1954).

Nevertheless, the Wightman axioms imply that the so-called vacuum expectations

$$\varphi_0(f_1, \ldots, f_k) \equiv (\Phi(f_1)\ldots\Phi(f_k)\Psi_0, \Psi_0) \tag{47}$$

are well-defined. Moreover, the Schwartz nuclear theorem, together with the translation invariance of Ψ_0, implies that (47) defines, for every $k \in \mathbb{Z}^+$, a tempered distribution on $S(\mathbb{R}^{(k-1)d})$; these are known as the Wightman functions. The Wightman axioms can be reexpressed in terms of the Wightman functions only, thus making no explicit mention of an underlying Hilbert space. Now, given these Wightman functions, an argument akin to the GNS construction (see Theorem 9.1.5), and known as the Wightman reconstruction theorem, defines a cyclic representation in terms of covariant field operators acting on some Hilbert space, determined uniquely by the Wightman functions, and satisfying the Wightman axioms; see Wightman (1956), or Streater & Wightman (1964) and Jost (1965).

A very rich variation on the theme of the Wightman axioms has flourished out of the remark that the Wightman functions can be continued analytically to imaginary times. The resulting Schwinger functions, or alternatively the Euclidean Green's functions, retain enough information for their properties to be axiomatizable in such a way that one can recover from them the Wightman theory on Minkowski space. This Euclidean approach is encapsuled in the axioms proposed by Osterwalder & Schrader (1973, 1975); see also the version given in Glimm & Jaffe (1981). These axioms have provided mathematical physicists with a workable programme for the construction of Bose fields in interaction; the power of this approach derives in part from its making possible to use, refine, combine and generally strengthen ideas and methods stemming from both the theory of stochastic processes, or functional integration, and classical statistical mechanics. In spite of the technical difficulties encountered along the way, rigourous results have been obtained that are spectacular ... albeit restricted so far to space-time dimensions $d \neq 4$; e.g. the proof that a relativistic quantum field theory can harbour a non-trivial scattering matrix. The superrenormalizable models that have been solved allow to cloth with mathematical meaning some of the hitherto heuristic arguments found in the traditional physics literature. The reader is referred to Glimm & Jaffe (1981) for a recent, in depth, panorama of what has been achieved in bridging this gap; and to Simon (1974, 1979) for early presentations of some of the basic mathematical techniques.

The locality requirements to be satisfied by the theory are emphasized in the approach to the axiomatization of relativistic quantum fields advanced in the papers of Haag (1959), Haag & Schroer (1962), Araki (1962, 1963, 1964) and Haag & Kastler (1964).

In this approach one associates to every bounded (open) region $\mathcal{O} \subset M^{3,1}$ a C^*-algebra (with unit) $\mathcal{A}(\mathcal{O})$, subject to the following conditions. First, the usual isotony postulate:

$$\mathcal{O}_1 \subseteq \mathcal{O}_2 \quad \text{implies} \quad \mathcal{A}(\mathcal{O}_1) \subseteq \mathcal{A}(\mathcal{O}_2) \tag{48}$$

Second, the locality postulate:

$$\begin{aligned} &[A_1, A_2] = 0 \quad \forall\, (A_1, A_2) \in \mathcal{A}(\mathcal{O}_1) \times \mathcal{A}(\mathcal{O}_2) \\ &\text{whenever} \quad \mathcal{O}_1 \quad \text{and} \quad \mathcal{O}_2 \text{ are causally disjoint} \end{aligned} \tag{49}$$

Thirdly, let $\mathcal{A}$ denote the C^*-inductive limit of the net of C^*-algebras $\mathcal{A}(\mathcal{O})$, i.e. (see Subsection 9.2.a) $\mathcal{A}$ is the norm-closure of

$$\mathcal{A}_0 = \bigcup_{\mathcal{O} \subset M^{3,1}} \mathcal{A}(\mathcal{O}) \tag{50}$$

The compatibility with the tenets of special relativity is then completed by the postulate that an action α (see Section 9.2) of the Poincaré group G is given on $\mathcal{A}$ and satisfies

$$\alpha(g)[\mathcal{A}(\mathcal{O})] = \mathcal{A}(g[\mathcal{O}]) \tag{51a}$$

where, for every region $\mathcal{O} \subset M^{3,1}$, and every $g \in G$,

$$g[\mathcal{O}] \equiv \{g[x] \mid x \in \mathcal{O}\} \tag{51b}$$

The fourth, and last postulate in this scheme, is the technical condition that $\mathcal{A}$ be primitive, i.e. that $\mathcal{A}$ have a faithful irreducible representation.

In the view of Haag and his coworkers (loc.cit.), some emphasis was put on the fact that the postulates should be formulated independently of a concrete Hilbert space realization; this is the approach we just followed. If, however, a concrete realization is chosen, one may want to strengthen the postulates in two directions. Firstly, one may want to require that the action α be unitarily implemented; this is automatically the case as soon as the realization considered is obtained by the GNS construction from a G-invariant state on $\mathcal{A}$ (Proposition 9.2.7). Secondly, one may want to require that, in the realization considered, the local algebras $\mathcal{A}(\mathcal{O})$ be replaced by their weak-operator closures, i.e. that one frames the axioms in terms of local von Neumann algebras; this is the approach sucessfully taken by Araki (loc. cit.). One may then consider strengthening the locality condition (49) to read as the duality condition:

(52) $$\mathcal{A}(\mathcal{O}) = \mathcal{A}(\mathcal{O}')'$$

where $\mathcal{O}'$ denotes the causal complement of $\mathcal{O}$ in $M^{3,1}$, $\mathcal{A}(\mathcal{O}')$ is the global algebra relative to $\mathcal{O}'$, and $\mathcal{A}(\mathcal{O}')'$ is the commutant of $\mathcal{A}(\mathcal{O}')$. An interesting connection between this duality condition and the Takesaki theory of modular algebras is developed in Bisognano & Wichmann (1975, 1976).

From the original papers of Araki (loc.cit.) we know that these axioms are satisfied for the standard example of the free, massive scalar field on Fock space. The other models of relativistic quantum fields so far constructed (see the discussion following the statement of the Wightman axioms) seem to conform as well. The question of the general connections between the C^* (or W^*-) algebraic axioms of Araki-Haag-Kastler and the Wightman (or the Osterwalder-Schrader) axioms is the object of much contemporary research; for entries into literature on this, and related questions, see Doplicher, Haag & Roberts (1969–1974); Fredenhagen & Hertel (1981); Driessler, Summers & Wichmann (in preparation); and Glimm & Jaffe (1981). As the situation now stands, it appears that the Araki-Haag-Kastler axioms are more suited to a discussion of the conceptual relations between general properties of the systems under investigation [see e.g. Emch (1972)], while the Osterwalder-Schrader axioms reflect more closely the structures encountered in the explicit construction of mathematical models.

Finally, it would seem that, as long as a theory of quantum fields on the flat four-dimensional Minkowski space $M^{3,1}$ has not been brought under satisfactory control, one can at best dream of a mathematically consistent, unified formulation of the ideas of quantum theory and of general relativity: the programme of 20th-Century physics has not run its course yet ... but neither has the 20th Century itself.

BIBLIOGRAPHY

Abraham, D. B., Baruch, E., Gallavotti, G. & Martin-Löf, A. (1970): Thermalisation of a magnetic impurity in the isotropic XY model, *Phys. Rev. Lett.* **25**, 1449–1450; see also: Dynamics of a local perturbation in the XY-model, *Studies in Applied Mathematics* **50** (1971), 121–131.

Abraham, R. & Marsden, J. E. (1978): Foundations of mechanics, Benjamin, Reading, Mass.

Accardi, L. (1976): Non-relativistic quantum mechanics as a non-commutative Markov process, *Adv. Math.* **20**, 329–366.

Accardi, L. (1981): Topics in quantum probability, *Physics Rep.* **77**, 169–192.

Accardi, L. & Cecchini, C. (1982): Conditional expectations in von Neumann algebras and a theorem of Takesaki, *Journ. Funct. Analysis* **45**, 245–273.

Accardi, L. & Frigerio, A.: Quantum probability, Proc. 1982 Rome Conference, to appear.

Accardi, L., Frigerio, A. & Lewis, J. T. (1982): Quantum stochastic processes, *Publ. RIMS, Kyoto* **18**, 97–133.

Aharonov, Y. & Bohm, D. (1959): Significance of electromagnetic potentials in the quantum theory, *Phys. Rev.* **115**, 485–491. See also W. Bayh, *Zeits. f. Phys.* **169** (1962), 492–510.

Aizenman, M. (1981): Proof of the triviality of ${\varphi_d}^4$ field theory and some mean-field features of Ising models for $d > 4$, *Phys. Rev. Lett.* **47**, 1–4.

Aizeman, M., Goldstein, S., Gruber, C., Lebowitz, J. L. & Martin, Ph. (1977): On the equivalence between KMS-states and equilibrium states for classical systems, *Commun. Math. Phys.* **53**, 209–220.

Akhiezer, N. I. & Glazman, I. M. (1961): Theory of linear operators in Hilbert space, 2 Vols., Frederick Ungar, New York.

Albert, A. A. (1946): On Jordan algebras of linear transformations, *Trans. Am. Math. Soc.* **59**, 524–555.

Alfsen, E. M., Hanche-Olsen, H. & Shultz, F. (1980): State spaces of C^*-algebras, *Acta Math.* **144**, 267–305.

Alfsen, E. M. & Shultz, F. W. (1978): State spaces of Jordan algebras, *Acta Math.* **140**, 155–190.

Amrein, W. O. (1981): Non-relativistic quantum dynamics; Reidel, Dordrecht, Holland.

Amrein, W. O., Gorgescu, V. & Martin, Ph. A. (1974): Approche algébrique de la théorie non-relativiste de la diffusion à canaux multiples, *in* Physical reality & Mathematical description, C. P. Enz & J. Mehra, eds. Reidel, Dordrecht, Holland, pp. 255–276.

Amrein, W. O., Jauch, J. M. & Sinha, K. (1977): Scattering theory in quantum mechanics, W. A. Benjamin, Reading, MA.

Amrein, W. O., Martin, Ph. A. & Misra, B. (1970): On the asymptotic condition of scattering theory, *Helv. Phys. Acta* **43**, 313–344.

Anderson, J. (1979): Extensions, restrictions, and representations of states on C^*-algebras, *Trans. Amer. Math. Soc.* **249**, 303–329.

Andersson, S. I. (1976): Properties of equilibrium states on quasilocal C^*-algebras, *Fortschritte der Physik* **24**, 55–83.

Araki, H. (1962): Einführung in die axiomatische Quantenfeldtheorie, unpublished lecture notes ETH, Zurich.

Araki, H. (1963): A lattice of von Neumann algebras associated with the quantume theory of a free Bose field, *Journ. Math. Phys.* **4**, 1343–1362.

Araki, H. (1964): On the algebra of all local observables, *Progr. Theor. Phys.* **32**, 844–854; The type of von Neumann algebra associated with the free field, ibid. **32**, 956–965; Von Neumann algebras of local observables for the free scalar field, *Journ. Math. Phys.* **5**, 1–13.

Araki, H. (1969): Gibbs states of a one-dimensional quantum lattice, *Commun. Math. Phys.* **14**, 120–157.

Araki, H. (1973): Relative Hamiltonian for faithful normal states of a von Neumann algebra, *Publ. RIMS, Kyoto Univ.* **9**, 165–209.

Araki, H. (1974): On the equivalence of the KMS condition and the variational principle for quantum lattice systems, *Commun. Math. Phys.* **38**, 1–10.

Araki, H. (1975): On uniqueness of KMS states of one-dimensional quantum lattice systems, *Commun. Math. Phys.* **44**, 1–7.

Araki, H. (1978): Operator algebras and statistical mechanics, *in* Mathematical problems in theoretical physics, G. Dell'Antonio, S. Dopplicher & G. Jona-Lasinio, eds. Springer, Berlin.

Araki, H. (1980): On the characterization of the state space of quantum mechanics, *Commun. math. phys.* **75**, 1–24.

Araki, H. (1983): On the XY-model on two-sided infinite chain, preprint, Kyoto University, RIMS-435; see also: H. Araki & E. Baruch, On the dynamics and ergodic properties of the XY-model, *Journ. Stat. Phys.* **31**, 327–345.

Araki, H. & Jurzak, J. P. (1981): On a certain class of *-algebras of unbounded operators, *RIMS* **381**, Kyoto.

Araki, H. & Kishimoto, A. (1976): On clustering property, *RIMS* **198**, Kyoto.

Araki, H. & Raggio, G. A. (1982): A remark on transition probability, *Lett. Math. Phys.* **6**, 237–240; see also G. A. Raggio, *ibid* 233–236.

Araki, H. & Sewell, G. L. (1977): KMS conditions and local thermodynamical stability of quantum lattice systems, *Commun. Math. Phys.* **52**, 103–109.

Araki, H. & Woods, E. J. (1963): Representation of the canonical commutation relations describing a nonrelativistic infinite free Bose gas, *Journ. Math. Phys.* **4**, 637–662.

Arnal, D., Cortet, J. C., Flato, M. & Sternheimer, D. (1981): Star-products [...] *in* Field theory [...], E. Tirapegui, ed., Reidel, Dordrecht, Holland, pp. 85–111; see also: M. Flato & D. Sternheimer, Deformations of Poisson brackets [...] in Harmonic analysis [...], J. A. Wolf, M. Cahen & M. DeWilde, eds. Reidel, Dordrecht, Holland (1980), pp. 385–448.

Arnold, V. I. (1978): Mathematical methods of classical mechanics, Springer, New York.

Arnold, V. I. & Avez, A. (1968): Ergodic problems of classical mechanics, W. A. Benjamin, New York.

Artin, E. (1957): Geometric algebra, John-Wiley Interscience, New York.

Arveson, W. B. (1967): Analyticity in operator algebras, *Am. Journ. Math.* **89**, 578–642.

Arveson, W. B. (1969): Subalgebras of C^*-algebras, *Acta Math.* **113**, 141–224.

Ashtekar, A. & Magnon, A. (1975): Quantum fields in curved space-times, *Proc. R. Soc. London* **A 346**, 375–394.

Aspect, A. (1983): Trois tests expérimentaux des inégalités de Bell par mesure de corrélation de polarisation de photons, Thèse, Orsay.

Baer, R. (1952): Linear algebra and projective geometry, Academic Press, New York.

Balmer, J. J. (1885): Notiz über die Spectrallinien des Wasserstoffs, *Verhandl. der Naturforschenden Gesellschaft in Basel* **7**, 548–560; Zweite Notiz über die Spectrallinien des Wasserstoffs, *ibid.* **7**, 750–752; also *Wiedemannsche Annalen der Physik* **25**, 80–87.

Bardeen, J. L., Cooper, N. & Schrieffer, J. R. (1957): Theory of superconductivity, *Phys. Rev.* **108**, 1175–1204.

Bargmann, V. (1954): On unitary ray representations of continuous groups, *Ann. of Math.* **59**, 1–46.

Bargmann, V. (1964), Note on Wigner's theorem on symmetry operations, *Journ. Math. Phys.* **5**, 862–868.

Bargmann, V. & Wigner, E. P. (1948): Group theoretical discussion of relativistic wave equations, *Proc. Natl. Acad. Sci.* **34**, 211–223.

Becquerel, H. (1896): Sur les radiations émises par phosphorescence, *C. R. Acad. Sci. (Paris)* **122**, 420–421.

Bell, J. S. (1964): On the Einstein Podolsky Rosen paradox, *Physics* **1**, 195–200. On the problem of hidden variables in quantum mechanics, *Rev. Mod. Phys.* **38** (1966), 447–452.

Beltrami, E. (1865, 1868): Risoluzione del problema: Riportare i punti di una superficie supra un piano in modochele linee geodetiche vengano rappresentate da linee rette, *Annali di Mat.* **7**, 185–204. Opere matematiche Vol. 1, 262–280; Saggio di interpretazione della geometria non-euclidea, *Gior. di Math.* **6**, 284–312. Opere matematiche Vol. 1, 374–405.

Bernoulli, D. (1738): Hydrodynamica, Dulsecker, Argentorati [=Strassburg], Engl. transl. "Hydrodynamics" by T. Carmody & H. Kobus, Dover, 1968.

Bertin, G. & Radicati, L. A. (1976): The bifurcation from the Maclaurin to the Jacobi sequence as a second-order phase transition, *Astrophys. Journ.* **206**, 815–821.

Biggs, H. F. (1927): Wave mechanics, Oxford University Press, Oxford.

Birtwistle, G. (1928): The new quantum mechanics, Cambridge University Press, Cambridge.

Bisognano, J. J. & Wichmann, E. H. (1975, 1976): On the duality condition for a Hermitian scalar field, *Journ. Math. Phys.* **16** (1975) 985–1007; On the duality condition for quantum fields, *ibid.* **17** (1976), 303–321.

Bohr, N. (1913): On the constitution of atoms and molecules, *Phil. Mag.* **26**, 1–25; 476–502; 857–875. Repr. with introd. L. Rosenfeld, W. A. Benjamin, New York, 1963.

Boltzmann, L. (1866): Ueber die mechanische Bedeutung des zweiten Haupsatzes der Wärmetheorie, *Sitzungberichte, Akad. Wiss., Wien,* **53**, 195–220.

Boltzmann, L. (1871): Einige allgemeine Sätze über Wärmegleichgewicht, *Sitzungberichte, Akad. Wiss., Wien* **63**, 679–711.

Boltzmann (1877): Uber die Beziehung eines allgemeine mechanischen Satzes zum zweiten Hauptsatze der Wärmetheorie, *Sitzungberichte, Akad. Wiss., Wien* **75**, 67–73; Ueber die Beziehung zwischen des zweiten Hauptsatze der mechanischen Wärmetheorie und der Wahrscheinlichkeitsrechnung, respective den Satzen über das Wärmegleichgewicht, *Sitzungberichte, Akad. Wiss., Wien* 76, 373–435.

Bonola, R. (1938): Non-euclidean geometry, H. S. Carslaw, transl., Open Court, LaSalle, Ill.

Borchers, H. J. (1972): Algebraic aspects of Wightman field theory *in* Statistical mechanics and field theory, R. N. Sen & C. Weil, eds., Halsted Press, New York, 31–79.

Borchers, H. J. (1975): Algebraic aspects of Wightman quantum field theory, *in* Mathematical Problems in Theoretical Physics, H. Araki, ed., Lecture Notes in Physics **39**, Springer, New York, 283–292.

Borchers, H. J. & Hegerfeldt, G. C. (1972): The structure of space-time transformations, *Commun. Math. Phys.* 28, 259–266; Über ein Problem der Relativitätstheorie: Wann sind Punktabbildungen des $\mathbf{R}^n$ linear, *Nachr. Gött. Akad. Wiss. (1972)*, 1–25.

Born, M. (1926): Zur Quantenmechanik der Stossvorgänge *Zeitschr. f. Physik* 37, 863–867; *ibid*, **38**, 803–867; *Gött. Nachr.*, 146–160.

Born, M. (1955): Statistic interpretation of quantum mechanics, (Nobel laureate lecture), reprinted *in* M. Born (1969). *Science* **122**, 675–679.

Born, M. (1969): Physics in my generation, Pergamon Press, London & New York.

Born, M. (1971): The Born-Einstein letters, with commentaries by Max Born, I. Born, transl., Walker, New York.

Born, M., Heisenberg, W. & Jordan, P. (1925): Zur Quantenmechanik II, *Zeitschrift für Physik* **35**, 557–615.

Born, M. & Jordan, P. (1925): Zur Quantenmechanik, *Zeitschrift für Physik* **34**, 858–888.

Born, M. & Jordan, P. (1930): Elementare Quantenmechanik, Springer, Berlin.

Born, M. & Wiener, N. (1925–1926): A new formulation of the laws of quantization of periodic and aperiodic phenomena, *J. Math. and Physics (MIT)* **5**, 84–98; Eine neue Formulierung der Quantengesetze für periodische und nicht periodische Vorgänge, *Zeitschrift für Physik* **36** (1926), 174–187.

Bourbaki, N. (1960): Elements de mathématiques; groupes et algébres de Lie, Fasc. XXVI, Hermann, Paris.

Bradley, J. (1727): An account of a new discovered motion of the fixed stars, *Phil. Trans.* **35**, 627–661.

Brattelli, O. & Kastler, D. (1976): Relaxing the clustering condition in the derivation of the K.M.S. condition, *Commun. Math. Phys.* **46**, 37–42.

Brattelli, O., Kishimoto, A. & Robinson, D. W. (1978): Stability properties and the KMS condition, *Commun. Math. Phys.* **61**, 209–238.

Brattelli, O. & Robinson, D. W. (1979, 1980): Operator algebras and quantum statistical mechanics, 2 vols. Springer, New York.

Braun, H. & Koecher, M. (1966): Jordan-Algebren, Springer, Berlin.

Braun, W. & Hepp, K. (1977): The Vlasov dynamics and its fluctuations in the $1/N$ limit of interacting classical particles, *Commun. Math. Phys.* **56**, 101–113.

Brush, S. G. (1967): Note on the history of the FitzGerald-Lorentz contraction, *Isis* **58**, 230–232.

Brush, S. G. (1983): Statistical physics and the atomic theory of matter—from Boyle and Newton to Landau and Onsager. Princeton University Press, Princeton, N.J.

Buchdahl, H. A. (1966): The concepts of classical thermodynamics, Cambridge University Press, Cambridge.

Bunimovich, L. A. & Sinai, Ya. G. (1981): Statistical properties of Lorentz gas with periodic configuration of scatterers, *Commun. Math. Phys.* 78, 479–497.

Callen, H. B. (1960): Thermodynamics, John-Wiley, New York.

Cannon, J. T. (1973): Infinite volume limits of the canonical free Bose gas states on the Weyl algebra, *Commun. Math. Phys.* 29, 89–104.

Caratheodory, C. (1909): Untersuchungen über die Grundlagen der Thermodynamik, *Math. Ann.* 67, 355–386.

Carleman, T. (1933): Sur la théorie de l' équation intégro-différentielle de Boltzmann, *Acta Math.* 60, 91–146.

Cartan, H. & Eilenberg, S. (1956): Homological algebra. Princeton University Press, Princeton, N.J.

Cayley, A. (1859): Sixth memoir upon quantics, *Phil. Trans.* 149, 61–91; reprinted in Collected mathematical papers, vol. 2, 561–606.

Chernoff, P. R. (1981): Mathematical obstructions to quantization, *Hadronic Journ.* 4, 879–898.

Choi, M. D. (1972): Positive linear maps on C^*-algebras, *Can. J. Math.* 24, 520–529.

Clauser, J. F. & Shimony, A. (1978): Bell's theorem: experimental tests and implications, *Rep. Progr. Phys.* 41, 1881–1927.

Clausius, R. J. E. (1850): Ueber die bewegende Kraft der Wärme, und die Gesetze, welche sich daraus für die Wärmelehre selbst ableiten lassen, *Annalen der Physik (2)* 79, 368–397 and 500–524.

Clemence, G. M. (1943): The motion of Mercury 1765–1937, *Astron. Papers Am. Ephemeris* 11, *part I*; see also: The relativity effect in planetary motions, *Rev. Mod. Phys.* 19 (1947), 361–364.

Cohen, E. G. D. (1973): The generalization of the Boltzmann equation to higher densities; *in*: The Boltzmann equation, theory and applications, E. G. D. Cohen & W. Thirring, eds., Springer, Berlin.

Cook, J. (1957): Convergence of the Møller wave matrix, *J. Math. and Phys.* 36, 82–87.

Cornfeld, I. P., Fomin, S. V. & Sinai, Ya. G. (1982): Ergodic theory, Springer-Verlag, New York.

Courant, R. & Hilbert, D. (1924): Methoden der mathematischen Physik, Springer, Berlin; Methods of Mathematical Physics, J. Wiley-Interscience, New York, 1953.

Currie, D. G., Jordan, T. F. & Sudarshan, E. C. G. (1963): Relativistic invariance and Hamiltonian theories of interacting particles, *Rev. Mod. Phys.* 35, 350–375.

Davies, E. B. (1970): On the repeated measurement of continuous observables in quantum mechanics, *Journ. Funct. Analysis* 6, 318–346.

Davies, E. B. (1972): Diffusion for weakly coupled quantum oscillators, *Commun. Math. Phys.* 27, 309–325.

Davies, E. B. (1976): Quantum theory of open systems, Academic Press, London.

Davies, E. B. & Lewis, J. T. (1970): An operational approach to quantum probability, *Commun. Math. Phys.* 17, 239–260.

Davisson, C. J. & Germer, L. H. (1927): Diffraction of electrons by a crystal of nickel, *Phys. Rev.* 30, 705–740.

Day, M. M. (1957): Amenable semi-groups, *Illinois J. Math.* 1, 509–544.

de Broglie, L. (1924): Recherches sur la théorie des quanta, Thèses, Université de Paris, Masson & Cie, Paris; *Annales de Physique* 3 (1925), 22–128.

de Broglie, L. (1928): La mécanique ondulatoire, Gauthier–Villars, Paris.

Debye, P. (1912): Zur Theorie der spezifischen Wärme, *Annalen der Physik* **39**, 789–839.

Dell'Antonio, G. F. (1967): On the limits of sequences of normal states, *Comm. Pure and Appl. Math* **20**, 413–429.

Dell'Antonio, G. F. (1982): The van Hove limit in classical and quantum mechanics, *in* Stochastic processes in quantum theory and statistical physics, S. Albeverio, Ph. Combe & M. Sirugue-Collin, eds., Lecture Notes in Physics 173, Springer, Berlin.

Dell' Antonio, G. F., Doplicher, S. & Ruelle, D. (1966): A theorem on canonical commutation and anticommutation relations, *Commun. math. Phys.* **2**, 223–230.

Della Riccia, G. (1982): On the Lagrange representations of a system of Newton equations, *in* Dynamical systems, A. Avez, A. Blaquière and A. Marzollo, eds., Academic Press, New York.

Demoen, B., Vanheuverzwijn, P. & Verbeure, A. (1977): Completely positive maps on the *CCR*-algebra, *Lett. Math. Phys.* **2**, 161–166; also: *Rep. Math. Phys.* **15** (1979), 27–39.

Demoen, B., Vanheuverzwijn, P. & Verbeure, A. (1978): Energetically stable systems, *Journ. Math. Phys.* **19**, 2256–2259.

Descartes, R. (1637): La dioptrique, Jean Maire, Leyde.

Dicke, R. H. & Goldenberg, H. M. (1967): Solar oblateness and general relativity, *Phys. Rev. Letters* **18**, 313–316.

Dicke, R. H., Roll, P. G. & Krotkov, R. (1964): The equivalence of inertial and passive gravitational mass, *Ann. Phys. (NY)* **26**, 442–517.

Dimock, J. (1980): Algebras of local observables on a manifold, *Commun. Math. Phys.* **77**, 219–228.

Dirac, P. A. M. (1925, 1926): The fundamental equations of quantum mechanics, *Proc. Roy. Soc., London* **A109** (1925), 642–653; Quantum mechanics and a preliminary investigation of the hydrogen atom, *Proc. Roy. Soc., London* **A110** (1926), 561–579; The elimination of the nodes in quantum mechanics, *Proc. Roy. Soc., London* **A111** (1926), 281–305.

Dirac, P. A. M. (1930): The principles of quantum mechanics. Clarendon Press, Oxford, (see also 2nd ed. *ibid.* 1936).

Dixmier, J. (1950): Les fonctionelles linéaires sur l'ensemble des opérateurs bornés d'un espace de Hilbert, *Ann. of Math.* **51**, 387–408.

Dixmier, J. (1957): Les algèbres d'opérateurs dans l'espace Hilbertien, Gauthier-Villars, Paris.

Dixmier, J. (1964): Les C^*-algèbres et leurs représentations, Gauthier-Villars, Paris.

Dobrushin, R. L. (1968): The problem of uniqueness of a Gibbsian random field and the problem of phase transitions, *J. Funct. Anal. and Appl.* **2**, 44–57; see also: Gibbsian random fields for lattice systems with pairwise interactions, *ibid.* **2**, 31–43.

Dobrushin, R. L. (1968 D): The description of a random field by means of conditional probabilities and conditions on its regularity, *Theory of Prob. and Appl.* **13**, 197–224.

Dollard, J. (1964): Asymptotic convergence and the Coulomb interaction, *Journ. Math. Phys.* **5**, 729–738; see also: Adiabatic switching and the Schroedinger theory of scattering, *ibid.* **7** (1966), 802–810; Screening in the Schroedinger theory of scattering, *ibid.* **9** (1968) 620–624; and Scattering into cones I: potential scattering, *Commun. Math. Phys.* **12** (1969) 193–203.

Domb, C. & Green, M. S. (1972): Phase transitions and critical phenomena, Vol 1. Exact results, Academic Press, London.

Donsker, M. & Varadhan, S. (1975): Asymptotic evaluation of certain Markov process expectations for large time I, *Comm. Pure Applied Math.* **28**, 1–47.

Doplicher, S., Haag, R. & Roberts, J. (1969–1974): Fields, observables, and gauge transformations I, II, *Commun. Math. Phys.* **13** (1969) 1–23; **15** (1969) 173–200; Local observables and particle statistics I, II, *Commun. Math. Phys.* **23** (1971) 199–230; **35** (1974) 49–85.

Doplicher, S., Kadison, R. V., Kastler, D. & Robinson, D. W. (1967): Asymptotic abelian systems, *Commun. Math. Phys.* **6**, 101–120.

Dubin, D. A. & Sewell, G. L. (1970): Time translations in the algebraic formulation of statistical mechanics, *Journ. Math. Phys.* **11**, 2990–2998.

Dulong, P. L. & Petit, A. T. (1819): Recherches sur quelques points importants de la théorie de la chaleur, *Annales de Chimie et de Physique* **10**, 395–413.

Dunford, N. & Schwartz, J. T. (1958, 1963, 1971): Linear operators (3 vols.), Wiley-Interscience, New York.

Dyson, F. J. (1962): The threefold way, algebraic structure of symmetry groups and ensembles in quantum mechanics, *J. Math. Phys.* **3**, 1199–1215.

Eberlein, W. F. (1949): Abstract ergodic theorems and weak almost periodic functions, *Trans. Amer. Math. Soc.* **67**, 217–240.

Eckart, C. (1926): The solution of the problem of a single oscillator by a combination of Schroedinger's wave mechanics and Lanczos' field theory, *Proc. Nat. Acad. Sci.* **12**, 473–476. Operator calculus and the solutions of the equations of quantum dynamics, *Phys. Rev.* **28**, 711–726. See also Lanczos, K. (1926).

Eddington, A. S. (1924): A comparison of Whitehead's and Einstein's formulae, *Nature* **113**, 192.

Effros, E. G. & Hahn, F. (1968): Locally compact transformation groups and C^*-algebras, *Memoirs Am. Math. Soc.* **75**, 1–92.

Ehrenfest, P. (1911): Welche Züge der Lichquantenhypothese spielen in der Theorie der Wärmestrahlung eine wesentliche Rolle, *Annalen der Physik* **36**, 91–118.

Ehrenfest, P. (1913): Een mechanische theorema van Boltzmann en zijne betrekking tot de quanta teorie, *Verslag van de Gewoge Vergaderingen der Wissennatuurkundige Afdeeling Amsterdam* 22, 586–593; A mechanical theorem of Boltzmann and its relation to the theory of energy quanta, *Proceedings of the Amsterdam Academy* **16**, (1914), 591–597; Adiabatische Invarianten und Quantentheorie, *Annalen der Physik* **51**, (1916), 327–352; Adiabatic invariants and the theory of quanta, *Phil. Mag.* **33** (1917), 500–513.

Ehrenfest, P. & T. (1911): Begriffliche Grundlagen der statistischen Auffasung in der Mechanik, Enzyclopaedie der mathematische Wissenshaften, Vol. 4, pt. 32; Engl. transl.: Conceptual foundations of the statistical approach in mechanics, by M. J. Moravcsik, Cornell Univ. Press, Ithaca, 1959.

Einstein, A. (1905, PE): Ueber einen die Erzeugung und Verwandlung des Lichtes betreffenden heuristischen Gesichtpunkt, *Annalen der Physik* **17**, 132–148; Engl. transl. by A. B. Arons & M. B. Peppard, *Amer. J. Physics* **33**, (1965), 367–374.

Einstein, A. (1905, BM): Five papers on Brownian motion were published by Einstein between 1905 and 1908, see: A. Einstein, Investigations on the theory of the Brownian movement, R. Fürth, ed., A. D. Cooper, transl., Dover, New York, 1956.

Einstein, A. (1905, SR): Zur Elektrodynamik bewegter Körper, *Annalen der Phys.* **17**, 891–921; transl. in: Einstein (1952).

Einstein, A. (1907): Die Plancksche Theorie der Stahlung and die Theorie der spezifischen Wärme, *Annalen der Physik* **22**, 180–190.

Einstein, A. (1909): Zum gegenwärtigen Stand des Strahlungsproblems, *Phys. Zeits.* **10**, 185–193.

Einstein, A. (1911): Elementare Betrachtungen über die thermische Molekularbewegung in festen Körper, *Annalen der Physik* **35**, 679–694.

Einstein, A. (1916): Die Grundlage der allgemeinen Relativitätstheorie, *Annalen der Physik* **49**,769–822; transl. in: Einstein (1952).

Einstein, A. (1917): Kosmologische Betrachtungen zur allgemeinen Relativitätstheorie, *Sitz. Ber. Preuss. Akad. Wiss.* **1917 (II)**, 142–152, transl. in: Einstein (1952).

Einstein, A. (1921): Geometry and Experience, reprinted in Ideas and Opinions by Albert Einstein, based on Mein Weltbild, C. Seelig, ed. and other sources, with new translations and revisions by Sonja Bargmann; Bonanza Books, New York, 1954, 232–246.

Einstein, A. (1924–5): Quantentheorie das einatomigen idealen Gases, *Sitz. Ber. Preuss. Akad. Wiss* **1924**, 261–267; ibid. **1925**, 3–14.

Einstein, A. (1952): The principle of relativity, a collection of original memoirs ... by H. A. Lorentz, A. Einstein, H. Minkowski and H. Weyl, with notes by A. Sommerfeld, W. Perret & G. B. Jeffery, transl., Dover, New York.

Einstein, A. (1952 H): in a letter to R. S. Shankland, reprinted and retransl. *in* Holton (1973).

Einstein, A. (1956): The meaning of relativity [Stafford Little Lectures, May 1921], 5th ed., Princeton University Press, Princeton, N.J.

Einstein, A. & de Sitter, W. (1932): On the relation between the expansion and the mean density of the universe, *Proc. Nat. Acad. Sci.* **18**, 213–214.

Einstein, A., Podolsky, B. & Rosen, N. (1935): Can the quantum-mechanical description of physical reality be considered complete? *Phys. Rev.* **47**, 777–780.

Emch, G. G. (1963): Mécanique quantique quaternionienne et relativité restreinte, *Helv. Phys. Acta* **36**, 739–769 and 770–788.

Emch, G. G. (1966): Non-markovian model for the approach to equilibrium, *Journ. Math. Phys.* **7**, 1198–1206.

Emch, G. G. (1972): Algebraic methods in statistical mechanics and quantum field theory. Wiley-Interscience, New York.

Emch, G. G. (1976a): Nonequilibrium quantum statistical mechanics, *Acta Phys. Austr. Suppl.* **15**, 79–131.

Emch, G. G. (1976b): Generalized K-flows, *Commun. Math. Phys.* **49**, 191–215.

Emch, G. G. (1980): An algebraic approach for spontaneous symmetry breaking in quantum statistical mechanics, *in* Groups, systems and many-body physics, Proc. Intern. Summer School, Tübingen, 1977, P. Kramer & M. Dal Cin, eds., Vieweg u. Sohn, Braunschweig/Wiesbaden.

Emch, G. G. (1981): Prequantization and KMS structures, *Intern. Journ. Theor. Phys.* **20**, 891–904.

Emch, G. G. (1982): Quantum and classical mechanics on homogeneous Riemannian manifolds, *Journ. Math. Phys.* **23**, 1785–1791.

Emch, G. G. (1983): Geometric dequantization and the correspondence problem, *Intern. Journ. Theor. Phys.* **22**, 397–420.

Emch, G. G., Albeverio, S. & Eckmann, J.-P. (1978): Quasifree generalized K-flows, *Rep. Math. Phys.* **13**, 73–85.

Emch, G. G. & Guenin, M. (1966): Gauge invariant formulation of the BCS model, *Journ. Math. Phys.* **7**, 915–921.

Emch, G. G. & Knops, H. J. F. (1970): Pure thermodynamical phases as extremal KMS states, *Journ. Math. Phys.* **11**, 3008–3018.

Emch, G. G., Knops, H. J. F. & Verboven, E. J. (1970): Breaking of euclidean symmetry with an application to the theory of crystallization, *Journ. Math. Phys.* **11**, 1655–1668.

Emch, G. G. & Piron, C. (1963): Symmetry in quantum theory, *Journ. Math. Phys.* **4**, 469–473. The main result of this paper was announced by the authors *in* Note sur les symétries en théorie quantique, *Helv. Phys. Acta* **35** (1962), 542–3, together with its generalization to real and quaternionic Hilbert spaces.

Emch, G. G. & Radin, C. (1971): Relaxation of local thermal deviations from equilibrium, *Journ. Math. Phys.* **12**, 2043–2046.

Emch, G. G. & Sinha, K. B. (1979): Weak quantization in a nonperturbative model, *Journ. Math. Phys.* **20**, 1336–1340.

Emch, G. G. & Varilly, J. C. (1979): On the standard form of the Bloch equation, *Lett. Math. Phys.* **3**, 113–116.

Enss, V. Completeness of the three body quantum scattering, *in*: Bielefeld encounters in Physics and Mathematics III, Springer (to appear); see also *in*: Proc. VIIth Intern. Congr. Math. Phys., Boulder, Colorado (to appear).

Epstein, P. S. (1916): Zur Theorie des Starkeffektes, *Annalen der Physik* **50**, 489–520; Zur Quantentheorie, *Annalen der Physik* **51**, 168–188.

Erdélyi, A. (1953): Higher transcendental functions, McGraw-Hill, New York.

Ernst, M. H. (1983): Exact solutions of the nonlinear Boltzmann equation and related kinetic equations, *in* Lebowitz & Montroll (1983).

Evans, D. E. (1979): Completely positive quasifree maps on the CAR algebra, *Commun. Math. Phys.* **70**, 53–68.

Evans, G. C. (1910): Note on Kirchhoff's law, *Proc. Amer. Acad.* **46**, 95–106.

Ezawa, H. (1964): The representation of the canonical variables in the limit of infinite space volume: the case of the BCS model, *Journ. Math. Phys.* **5**, 1078–1090.

Faddeev, L. D. (1965): Mathematical aspects of the three-body problem in quantum scattering theory, Israel program of scien. transl, Jerusalem, Israel.

Fannes, M. & Verbeure, A. (1977): Correlation inequalities and equilibrium states, II, *Commun. Math. Phys.* **57**, 165–171.

Feinberg, M. & Lavine, R.: Foundations of the Clausius-Duhem inequality, to appear as an appendix in second edition of C. Truesdell (1969).

Fell, J. M. G. (1960): The dual spaces of C^*-algebras, *Trans. Am. Math. Soc.* **94**, 365–403.

Fermi, E. (1936): Thermodynamics, reprinted by Dover Publ., New York, 1956.

Feynman, R. P. (1953): The λ-transition in liquid Helium, *Phys. Rev.* **90**, 1116–1117; Atomic theory of the λ-transition in Helium, 91, 1291–1301; Atomic theory of liquid Helium near absolute zero, *ibid* **91**, 1301–1308.

Finkelstein, D. (1958): Past-future asymmetry of the gravitational field of a point particle, *Phys. Rev.* **110**, 965–967.

Finkelstein D., et al. (1959–1963): (with J. M. Jauch & D. Speiser) Notes on quaternion quantum mechanics, *CERN Reports 59-7, 59-9, 59-17 (1959)*; Zur frage der Ladungquantisierung, *Helv. Phys. Acta* **32** (1959), 258–260; Quaternionic representations of compact groups, *J. Math. Phys.* **4** (1963), 136–140; (with J. M. Jauch, S. Schiminovitch & D. Speiser) Foundations of quaternion quantum mechanics, *J. Math. Phys.* **3** (1962), 207–220; Principle of general Q-covariance, *J. Math. Phys.* **4** (1963), 788–796.

Finn, B. S. (1964): Laplace and the speed of sound, *Isis* **55**, 7–19.

Foias, C., Geher, L. & Sz-Nagy, B. (1960): On the permutability condition of quantum mechanics, *Acta Sci. Math. (Szeged)* **21**, 78–89.

Ford, G. W., Kac, M., & Mazur, P. (1965): Statistical mechanics of assemblies of coupled oscillators, *Journ. Math. Phys.* **6**, 504–515.

Fredenhagen, K. & Hertel, J. (1981): Local algebras of observables and pointlike localized fields, *Commun. Math. Phys.* **80**, 555–561.

Frenkel, J. (1929): Einführung in die Wellenmechanik, Springer, Berlin.

Friederich, W., Knipping, P. & von Laue, M. (1912): Interferenzerscheinungen bei Röntgenstrahlen (Münchener Berichte 1912, 303–322), *Annalen der Physik* **41** (1913), 971–1002.

Frigerio, A. (1978): Stationary states of quantum dynamical semi-groups, *Commun. Math. Phys.* **63**, 269–276.

Frigerio, A., Gorini, V. & Verri, M. (1978): Stability, detailed balance and KMS condition for quantum systems, *in* Mathematical problems in theoretical physics, G. F. Dell'Antonio, S. Doplicher & G. Jona-Lasinio, eds., Springer, Berlin.

Fronsdal, C. (1959): Completion and embedding of the Schwarzschild solution, *Phys. Rev.* **116**, 778–781.

Galileo (1632): Dialogue concerning the two chief world systems—Ptolemaic and Copernican, Drake, S. transl. (with a preface by A. Einstein), University of California Press, Berkeley and Los Angeles, 1967.

Galindo, A. & Sanchez del Rio, C. (1961): Intrinsic magnetic moment as a non-relativistic phenomenon, *Amer. J. Physics* **29**, 582–584.

Gallavotti, G. & Miracle-Sole, S. (1970): Absence of phase transitions in hard-core one-dimensional systems with long-range interactions, *Journ. Math. Phys.* **11**, 147–154; see also: A variational principle for the equilibrium of hard sphere systems, *Ann. Inst. Henri Poincaré* **18** (1968), 287–299.

Gallavotti, G. & Verboven, E. J. (1975): On the classical KMS boundary condition, *Nuovo Cim.* **28B**, 274–286.

Gårding, L. & Wightman, A. (1954): Representations of the anticommutation relations, *Proc. Natl. Acad. Sci.* **40** 617–621; Representations of the commutation relations, *ibid.* 622–626.

Gates, D. J. (1971): Exactly solvable cell model with a melting transition, *Journ. Math. Phys.* **12**, 766–778.

Gauss, C. F. (1816): This work was never fully published; whatever there is, can be gleaned from: Werke, Vol. 8, 157–268.

Gelfand, I. M. & Naimark, M. A. (1943): On the imbedding of normed rings into the ring of operators in Hilbert space, *Mat. Sborn. N.S.* **12** [54], 197–213.

Gibbs, J. W. (1873): Graphical methods in the thermodynamics of fluids and a method of geometrical representation of the thermodynamic properties of substances by means of surfaces, *Trans. Connecticut Acad. II*, 309–342, April–May, 1873; 382–404, Dec. 1873. Reprinted in: The Scientific Papers, vol. 1, 1–32, 33–54, Dover, New York, 1961.

Gibbs, J. W. (1878): On the equilibrium of heterogeneous substances, *Trans. Connecticut Acad. III*, 108–248, Oct 1874–May, 1876; 343–524, May 1877–July 1878 and *Amer. J. Science*, 3 ser. vol. XVI, 441–458, Dec. 1873. Reprinted in: The Scientific Papers, Vol. 1, 55–371, Dover, New York, 1961.

Gibbs, J. W. (1889): Rudolf Julius Emanuel Clausius, *Proc. Amer. Academy, new series, vol. XVI*, 458–465; reprinted in: The Scientific Papers, Vol. 2, 261–267, Dover, New York, 1961.

Gillispie, C. C. (1960): The edge of objectivity, Princeton University Press, Princeton, NJ.

Gleason, A. M. (1957): Measures on the closed subspaces of a Hilbert space. *Journ. of Mathematics and Mechanics* 6, 885–893.

Glimm, J. (1960): On a certain class of operator algebras, *Trans. Amer. Math. Soc.* **95**, 318–340.

Glimm, J. (1961): Type I C^*-algebras, *Ann. Math.* **73**, 572–612.

Glimm, J. & Jaffe, A. (1981): Quantum physics—a functional integral point of view, Springer, New York.

Golodez, V. Ya. (1972): Conditional expectations and modular automorphisms of von Neumann algebras, *Funct. Analysis & Appl.* 6, 68–69.

Gorini, V., Kossakowski, A. & Sudarshan, E. C. G. (1976): Completely positive dynamical semigroups on N-level systems, *Journ. Math. Phys.* **17**, 821–825.

Grad, H. (1958): Principles of the kinetic theory of gases, *in* Handbuch der Physik, Vol. 12, Springer, Berlin.

Greenberg, W., Polewczak, J. & Zweifel, P. F. (1983): Global existence proofs for the Boltzmann equation, *in* Lebowitz & Montroll (1983).

Greenleaf, F. P. (1969): Invariant means on topological groups, Van Nostrand-Reinhold, New York.

Groenewold, H. J. (1946): On the principles of elementary quantum mechanics, *Physica* **12**, 405–460.

Grossmann, A. & Huguenin, P. (1978): Group-theoretical aspects of the Wigner-Weyl isomorphism, *Helv. Phys. Acta* **51**, 252–261.

Grossman, A., Loupias, G., & Stein, E. M. (1968): An algebra of pseudodifferential operators and quantum mechanics in phase space, *Ann. Inst. Fourier, Grenoble* **18**, 343–368.

Gudder, S. & Marchand, J. P. (1977): Conditional expectations on von Neumann algebras: a new approach, *Rep. Math. Phys.* **12**, 317–329.

Guillemin, V. & Sternberg, S. (1977): Geometric asymptotics, *Mathematical Surveys #14*, AMS, Providence, R.I.

Guillemin, V. & Sternberg, S. (1984): Symplectic techniques in physics, Cambridge University Press, Cambridge.

Haag, R. (1959): Discussion des "axiomes" et des propriétés asymptotiques d'une théorie des champs locale avec particules composées, *in* Les problèmes mathématiques de la théorie quantique des champs, CNRS, Paris.

Haag, R. (1962): The mathematical structure of the Bardeen-Cooper-Schrieffer model, *Nuovo Cim.* **25**, 287–298.

Haag, R., Hugenholtz, N. & Winnink, M. (1967): On the equilibrium states in quantum statistical mechanics, *Commun. Math. Phys.* **5**, 215–236.

Haag, R., Kadison, R. V. & Kastler, D. (1970): Nets of C^*-algebras and classification of states, *Commun. Math. Phys.* **16**, 81–104.

Haag, R. & Kastler, D. (1964): An algebraic approach to quantum field theory, *Journ. Math. Phys.* **5**, 848–861.

Haag, R., Kastler, D. & Trych-Pohlmeyer, E. B. (1974): Stability and equilibrium states, *Commun. Math. Phys.* **38**, 173–193.

Haag, R. & Schroer, B. (1962): Postulates of quantum field theory *Journ. Math. Phys.* **3**, 248–256.

Haag, R. & Trych-Pohlmeyer, E. B. (1977): Stability properties of equilibrium states, *Commun. Math. Phys.* **56**, 213–224.

Hagedorn, G. A. (1980): Asymptotic completeness for two, three and four particle Schrödinger operators, *Trans. Amer. Math. Soc.* **258**, 1–75; see also: G. A. Hagedorn & P. Perry, *Comm. Pure Appl. Math.* **36** (1983), 213–232; and preprint, VPISU (1983).

Hamilton, W. R. (1833): On a general method of expressing the path of light, and of the planets, by the coefficients of a characteristic function, Dublin University Review 1833, 795–826; see also Essays on the theory of rays *Trans. Roy. Irish Academy* 15 (1828), 69–174; **16**, (1830), 1–61; **16**, (1830), 93–125; **17**, (1832), 1–144.

Hamilton, W. R. (1834, 1835): General method in dynamics, II. *Phil. Trans., 1834*, 247–308; I. *Phil. Trans., 1835*, 95–144; reprinted in: The Mathematical Papers, Vol. 2, 103–211, Cambridge University Press, 1940.

Hawking, S. W. (1982): The unpredictability of quantum gravity, *Commun. Math. Phys.* **87**, 395–415.

Hawking, S. W. & Ellis, G. F. R. (1973): The large scale structure of space-time, Cambridge Univ. Press, Cambridge.

Heckmann, O. (1931): Über die Metrik des sich ausdehnenden Universums, *Nachr. Ges. Wiss. Göttingen*, 127–130.

Heims, S. P. (1965): Approach to equilibrium, *Am. Journ. Phys.* **33**, 722–727.

Heisenberg (1925 M): Ueber quantentheoretische Kinematik und Mechanik, *Mathematischen Annalen* **95**, 694–705.

Heisenberg, W. (1925 P): Ueber quantentheoretische Umdeutung kinematischer und mechanischer Beziehungen, *Zeitschrift für Physik* **33**, 879–893.

Heisenberg, W. (1927): Ueber den anschaulichen Inhalt der Quantentheoretischen Kinematik und Mechanik, *Zeitschrift für Physik* **43**, 172–198.

Helmoltz, H. (1868): Ueber die Tatsachen, die der Geometrie zum Grunde liegen, *Nachrichten K. Ges. Wiss. Göttingen,* **9**.

Hepp, K. (1972): Quantum theory of measurement and macroscopic observables, *Helv. Phys. Acta* **45**, 237–248.

Hepp, K. (1974): The classical limit for quantum mechanical correlation functions, *Commun. Math. Phys.* **35**, 265–277.

Herglotz, G. (1911): Über die Mechanik des deformierbaren Körpers vom Standpunkte der Relativitätstheorie, *Ann. Phys.* **36**, 493–533.

Hertz, H. (1892): Über den Durchgang der Kathodenstrahlen durch dünne Metallschichten, *Ann. Phys. Chem.* **45**, 28–32.

Hewitt, E. & Ross, K. A. (1963): Abstract harmonic analysis, Springer, Berlin.

Hilbert, D. (1899): Grundlagen der Geometrie, (fünfte Auflage), Teubner, Leipzig, 1922; Engl. transl. of the tenth edition: Open Court, La Salle, IL, 1971.

Hilbert, D. (1900): Sur les problèmes futurs des mathématiques *in* Comptes Rendus du Deuxième Congrès International des Mathématiciens, Paris, 1900, Gauthier-Villars, Paris, 1902, 58–114, also (in German) *in* Nachr. Ges. Wiss. Göttingen, 253–297; also *in*: Gesammelte Abhandlungen, Vol. III; Engl. transl. *in* Amer. Math. Soc. Bull. 8 1901/2, 437–479.

Hilbert, D. (1912–14): I. Begründung der elementaren Strahlungstheorie, *Göttingen Nachrichten* (1912), 773–789; *Physik. Zeitschrift* 13 (1912), 1056–1064; II. Bemerkungen zur Begründungen der elementaren Strahlungstheorie, *Göttingen Nachrichten* (1913), 409–416; *Physik. Zeitschrift* 14 (1913), 592–595; III. Zur Begründung der elementaren Strahlungstheorie, *Göttingen Nachrichten,* (1914), 275–298;*Physik. Zeitschrift* 15 (1914), 878–889; the three papers are reprinted *in* Gesammelte Abhandlungen, Vol. III.

Hilbert, D. (1912): Grundzüge einer allgemeinen Theorie der linearen Integralgleichungen, B. G. Teubner, Leipzig u. Berlin; reprinted by Chelsea, New York (1953).

Hilbert, D., von Neumann, J. & Nordheim, L. (1927): Ueber die Grundlagen der Quantenmechanik, *Math. Annalen* 98, 1–30.

Hille, E. (1972): Methods in classical and functional analysis, Addison-Wesley, Reading, MA.

Hille, E. & Phillips, R. S. (1957): Functional analysis and semi-groups, American Mathematical Society, Providence, RI.

Hislop, P. D. & Longo, R. (1982): Modular structure of the local algebras associated with the free massless scalar field theory, *Commun. Math. Phys.* 84, 71–85.

Holton, G. (1973): Thematic origins of scientific thought, Kepler to Einstein; Harvard University Press, Cambridge, MA.

Huang, K. (1963): Statistical mechanics, John Wiley, New York.

Hugenholtz, N. M. (1967): On the factor type of equilibrium states in quantum statistical mechanics, *Commun. Math. Phys.* 6, 189–193.

Hugenholtz, N. M. & Wieringa, J. D. (1969): On locally normal states in quantum statistical mechanics, *Commun. Math. Phys.* 11, 183–197.

Ignatowski, W. v. (1911): Zur Elastizitätstheorie vom Standpunkte des Relativitätsprinzips, *Zeits. Phys.* 12, 164–169; see also 441–448.

Inonu, E. & Wigner, E. P. (1953): On the contraction of groups and their representations, *Proc. Nat. Acad. Sci. USA* 39, 510–524.

Isham, C. J., Penrose, R. & Sciama, D. W. (1981): Quantum gravity, Clarendon Press, Oxford.

Ishiwara, J. (1915): Die universelle Bedeutung des Wirkungsquantums, *Tokyo Sugaku Buturigakkawi Kizzi* 8, 106–116.

Ising, E. (1925): Beitrag zur Theorie des Ferromagnetismus, *Zeitschrift für Physik* 31, 253–258.

Israel, R. B. (1979): Convexity in the theory of lattice gases, Princeton University Press.

Ives, H. E. & Stilwell, G. R. (1941): Experimental study of the rate of a moving atomic clock, II, *J. Opt. Soc. Amer.* 31, 369–374.

Jacobi, C. G. J. (1866): Vorlesungen über Dynamik (R.F. A. Clebsch, ed.), G. Reimer, Berlin, reprinted *in* Gesammelte Werke, Vol. 8, Chelsea, New York, 1968.

Jacobs, K. (1960): Neuere Methoden und Ergebnisse der Ergoden-theorie, *Ergebnisse der Mathematik* 29, Springer, Berlin.

Jammer, M. (1966): The conceptual development of quantum mechanics, McGraw-Hill, New York; see also: The philosophy of quantum mechanics, John Wiley, New York, 1974.

Jauch, J. M. (1958): Theory of the scattering operator, *Helv. Phys. Acta* 31, 127–158 and 661–684.

Jauch, J. M. (1968): Foundations of quantum mechanics, Addison-Wesley, Reading, Mass.

Jauch, J. M. (1972): On a new foundation of equilibrium thermodynamics, *Found. Physics* 2, 327–332.

Jauch, J. M., Misra, B. & Gibson, A. G. (1968): On the asymptotic condition of scattering theory, *Helv. Phys. Acta* 41, 513–527.

Jeans, J. H. (1905): On the partition of energy between matter and ether, *Philosophical Magazine* 10, 91–98.

Jelinek, F. (1968): BCS-Spin model, its thermodynamic representations and automorphisms, *Commun. Math. Phys.* 9, 169–175.

Jordan, P. (1934): Ueber die Multiplikation quantenmechanischer Grossen II, *Zeitschrift f. Physik* 87, 505–512; see also: *ibid.* 80 (1933), 285–291, *Gött. Nachr.* (1932), 569–575; *ibid.* (1933), 209–217.

Jordan, P., von Neumann, J. & Wigner, E. P. (1934): On an algebraic generalization of the quantum mechanical formalism, Annals of Math. 35, 29–64.

Jordan, P. & Wigner, E. P. (1928): Ueber das Paulische Aequivalenzverbot, *Zeitschrift für Physik,* 47, 631–651.

Jordan, T. F. & Sudarshan, E. C. G. (1961): Lie group dynamical formalism and the relation between quantum mechanics and classical mechanics, *Rev. Mod. Phys.* 33, 515–524.

Jost, R. (1965): The general theory of quantized fields, Amer. Math. Soc., Providence, R.I.

Kac, M. (1959): Probability and related topics in physical sciences, Interscience Publ., New York.

Kac, M. (1968): Mathematical mechanisms of phase transitions (Notes by C. J. Thompson), *in* Brandeis Summer Institute 1966, M. Chrétien, E. P. Gross & S. Deser, eds., Gordon & Breach, New York; see also: Toward a unified view on mathematical theories of phase transitions, *in* Fundamental problems in statistical mechanics II, E. G. D. Cohen, ed., North-Holland, Amsterdam.

Kadison, R. V. (1951): Isometries of operator algebras, *Ann. Math.* 54, 325–338.

Kadison, R. V. (1965): Transformations of states in operator theory and dynamics, *Topology* 3, *Suppl.* 2, 177–198. See also Kadison (1951) and Størmer (1965).

Kadison, R. V. (1978): Similarity of operator algebras, *Acta Mathematica* 141, 147–163.

Kadison, R. V. & Ringrose, J. R. (1983): Fundamentals of the theory of operator algebras, Academic Press, New York.

Kadison, R. V. & Singer, I. M. (1959): Extensions of pure states, *Amer. Journ. of Math.* 81, 383–400.

Kahn, B. & Uhlenbeck, G. E. (1938): On the theory of condensation, *Physica* 5, 399–416.

Kangro, H. (1972): Planck's original papers in quantum mechanics, annotated by H. Kangro and transl. by D. ter Haar and S. G. Brush, John Wiley & Sons, New York.

Kastler, A. (1957): Optical methods of atomic orientation and of magnetic resonance, J. Opt. Soc. Am. 47, 460–465.

Kastler, D. (1976): C^*-algebras and their applications to statistical mechanics and quantum field theory, North-Holland, Amsterdam.

Kastler, D., Mebkhout, M., Loupias, G. & Michel, L. (1972): Central decomposition of invariant states with applications to the groups of time translations and of Euclidean transformations in algebraic field theory, *Commun. Math. Phys.* 27, 195–222.

Kato (1966): Perturbation theory for linear operators, Springer, New York (2nd ed., 1976).

Kennard, E. H. (1927–8): Zur Quantenmechanik einfacher Bewegungstypen, *Zeitschrift für Physik* 44, 326–352; Note on Heisenberg's indetermination principle, *Phys. Rev.* 31 (1928), 344–348.

Kesten, H. & Papanicolaou, G. (1980): A limit theorem for stochastic acceleration, *Commun. Math. Phys.* 78, 19–63.

Khinchin, A. I. (1957): Mathematical foundations of information theory, Silverman, R. A. & Friedman, M. D., transl., Dover Publications, New York.

Kirchhoff, G. (1859 I & II): I. Über die Fraunhoferschen Linien, *Monatsber. Akad. Wiss. Berlin, Oktober*, 662-665; II. Über den Zusammenhang zwischen Emission und Absorption von Licht und Wärme, *Monatsber. Akad. Wiss. Berlin, Dezember*, 783–787.

Kirchhoff, G. (1860): Über des Verhältniss zwischen dem Emissionsvermögen und dem Absorptionsvermögen der Körper für Wärme und Licht, *Annalen der Physik* 109, 275–301.

Kirchhoff, G. & Bunsen, R. (1860): Chemische Analyse durch Spectralbeobachtungen, *Oswalds Klassiker der exakten Wissenschaften* 72.

Klein, F. (1872, 1893): Vergleichende Betrachtungen über neuere geometrische Forschungen, *Math. Ann.* **43**, 63–100; reprinted in: Gesammelte mathematische Abhandlungen, Vol. 1, 460–497; Engl. transl. A comparative review of recent researches in geometry, *N. Y. Math. Soc. Bull.* 2 (1893), 215–249.

Klein, M. (1967): *in* Przibram, K. (1967).

Kline, M. (1972): Mathematical thought from ancient to modern times, Oxford Univ. Press, New York.

Kolmogorov, A. N. (1958): A new metric invariant of transitive systems and automorphisms of Lebesgue spaces, *Dokl. Akad. Nauk* **119**, 861–864.

Koopman, B. O. (1931): Hamiltonian systems and transformations in Hilbert spaces, *Proc. Natl. Acad. Sci.* **17**, 315–318.

Kossakowski, A. (1972): On quantum statistical mechanics of non-hamiltonian systems, *Rep. Math. Phys.* **3**, 247–274; On necessary and sufficient conditions for a generator of a quantum dynamical semigroup, *Bull. Acad. Polon. Sci. Ser. Math. Astro. et Phys.* **20**, 1021-1025; see also: Gorini, Kossakowski & Sudarshan (1976); Kossakowski, Frigerio, Gorini & Verri (1977).

Kossakowski, A., Frigerio, A., Gorini, V. & Verri, M. (1977): Quantum detailed balance and KMS condition, *Commun. Math. Phys.* **57**, 97–110.

Kostant, B. (1970): Quantization and unitary representations *in* Lecture Notes in Mathematics **170**, Springer, New York, 87–208. See also B. Kostant (1975).

Kostant, B. (1975): On the definition of quantization *in* Geométrie symplectique et physique mathématique, CNRS, Paris, 187–210.

Kramers, H. A. (1957): Quantum mechanics, D. ter Haar, transl., Interscience, New York.

Kraus, K. (1971): General state changes in quantum theory, *Ann. Phys.* **64**, 311–335.

Kruskal, M. D. (1960): Maximal extension of Schwarzschild metric, *Phys. Rev.* **119**, 1743–1745.

Kubo, R. (1957): Statistical mechanical theory of irreversible processes, *Journ. Phys. Soc. Japan* **12**, 570–586.

Kuhn, T. S. (1958): The caloric theory of adiabatic compression, *Isis* **49**, 132–140.

Kuhn, T. S. (1959): Energy conservation as an example of simultaneous discovery; reprinted in T. S. Kuhn, The essential tension, The University of Chicago Press, 1977.

Kuhn, W. (1925): Ueber die Gesamtstärke der von einem Zustande ausgehenden Absorptionslinien, *Zeitschrift für Physik* **33**, 408–412.

Kümmerer, B. & Schröder, W. (1982): A survey of Markov dilations for the spin-half relaxation and physical interpretation, *Semesterbericht Funktionalanalysis, Tübingen, Wintersemester* 1981/82, 187–213. see also: B. Kümmerer, Markov dilations of completely positive operators on W^*-algebras, *ibid* 175–186; B. Kümmerer, A non-commutative example of a continuous Markov dilation, *ibid.* **1982/83**, 61–91; B. Kümmerer & W. Schröder, A Markov dilation of a non-quasifree Bloch evolution, *Commun. Math. Phys.* **90**, 251–262; and W. Schröder (1983).

Lagrange, J. L. (1788): Mécanique analytique, La veuve Desaint, Paris.

Lamla, E. (1911): Hydrodynamics of the relativity principle, Dissertation, Berlin, Ann. Phys. (Leipzig) **37** (1912), 772–796.

Lanczos, J. (1926): Ueber eine feldmässige Darstellung der neuen Quantenmechanik, *Zeitschrift für Physik* **35**, 812–830.

Landau, L. D. & Lifshitz, E. M. (1958): Statistical physics, E. Peierls & R. F. Peierls, transl., Pergamon, London.

Landé, A. (1928): Optik, Mechanik und Wellenmechanik, *in* Handbuch der Physik **20**, Springer, Berlin.

Lanford, O. E. (1975): Time evolution of large classical systems, *in* Dynamical systems and applications, J. Moser, ed., Lecture notes in physics, **38**, Springer, Berlin; see also: O. E. Lanford's review in Lebowitz & Montroll (1983).

Lanford, O. E. & Robinson, D. W. (1968): Statistical mechanics of quantum spin systems III, *Commun. Math. Phys.* **9**, 327–338.

Lanford, O. E. & Ruelle, D. (1967): Integral representations of invariant states on B^*-algebras, *Journ. Math. Phys.* **8**, 1460–1463.

Lanford, O. E. & Ruelle, D. (1969): Observables at infinity and states with short range correlations in statistical mechanics, *Commun. Math. Phys.* **13**, 194–215.

Langevin, P. & de Broglie, M. (1912): La théorie du rayonnement et les quanta—Rapports et discussions de la réunion tenue à Bruxelles, 1911, Gauthier-Villars, Paris.

Lassner, G. (1975): Continuous representations of the test function algebra and the existence problem for quantum fields, *in* Mathematical Problems in Theoretical Physics, H. Araki, ed. Lecture Notes in Physics, **39**, Springer, New York, 297–302.

Lavine, R. B. (1965): The Weyl-transform Fourier analysis of operators in L^2-spaces. Ph.D. thesis, MIT (unpublished).

Lavine, R. B. (1970): Scattering theory for long range potentials, *Journ. Funct. Analysis* **5**, 368–382.

Lavine, R. B. (1974): Commutators and local decay, *in* Scattering theory in mathematical physics, J. A. Lavita & J. P. Marchand, eds., Reidel, Dordrecht, Holland.

Lavine, R. B. (1978): Spectral density and sojourn times, *in* Atomic scattering theory, J. Nuttal, ed., Univ. of Western Ontario Pres, London, Ontario.

Lebowitz, J. L. & Montroll, E. W. (1983): Nonequilibrium phenomena I: the Boltzmann equation, North-Holland, Amsterdam.

Lemaitre, G. (1925): Note on de Sitter's universe, *J. Math. and Physics (MIT)* **4**, 188-192.

Lemaitre, G. (1933): L'univers en expansion, *Ann. Soc. Sci. Bruxelles* **A53**, 51–85.

Lenz, W. (1920): Beitrag zum Verständnis der magnetischen Erscheinungen in fester Körpern, *Physikalische Zeitschrift* 21, 613–615.

Leverrier, U. J. (1859): Sur la théorie de Mercure et sur le mouvement du périhélie de cette planète, *C. R.* **49**, 378–383.

Lévy-Leblond, J.-M. (1965): Non-relativistic particles and wave equations, Thesis, Paris; Galilei group and non-relativistic quantum mechanics, Journ. Math. Phys. **4**, 776–788.

Lichnerowicz, A. (1977): Les variétés de Poisson et leurs algèbres de Lie associées, *J. Diff. Geom.* **12**, 253–300.

Lie, S. (1885,1893): Allgemeine Untersuchen über Differential-gleichungen, die eine kontinuierliche, endliche Gruppe gestatten, *Mathematische Annalen* **25**, 71–151; also: Theorie der Transformationsgruppen, Teubner, Leipzig, 1893.

Lieb, E. H. (1976): Stability of matter, *Rev. Mod. Phys.* **48**, 553–569.

Lieb, E. H. & Mattis, D. C. (1966): Mathematical physics in one dimension (exactly soluble models of interacting particles), Academic Press, New York.

Lindblad, G. (1975, 1976): Completely positive maps and entropy inequalities, *Commun. Math. Phys.* **40** (1975), 147–151; On the generators of quantum dynamical semigroups, *Commun. Math. Phys.* **48** (1976), 119-130.

Llosa, J. (1982): Relativistic action at a distance: classical and quantum aspects *Lectures Notes in Physics* **162**, Springer, New York.

London, F. (1938): The λ-phenomenon of liquid Helium and the Bose-Einstein degeneracy, *Nature* **141**, 643-644; On the Bose-Einstein condensation, *Phys. Rev.* **54**, 947–954.

Lorentz, H. A. (1905): Motion of electrons in metals, *Akad. Wetschap. Amsterdam* **7**, 438–453, 588–593, 684–691.

Lorentz, H. A. (1923): L'ancienne et la nouvelle mécanique *in* Le Livre du cinquantenaire de la Societé Française de Physique, Editions de la Revue d'Optique Théorique et Instrumentale, Paris, 1925; reprinted *in*: Collected papers, Vol. 7.

Lowe, I. J. & Norberg, R. E. (1957): Free-induction decay in solids, *Phys. Rev.* **107**, 46–61.

Lu, P.-C. (1982): Didactic remarks on the Sears-Kestin statement of the second law of thermodynamics, *Am. J. Phys.* **50**, 247–251.

Maasen, H. (1982): On a class of quantum Langevin equations and the question of approach to equilibrium, Proefschrift, Groningen.

Mach, E. (1883a): Die Mechanik und ihrer Entwickelung historisch-kritisch dargestellt, Leipzig. For the 5th ed. of the Engl. transl. from the 9th German ed., see: The Science of mechanics, Th. J. McCormack, transl., The Open Court Publ. Co., New York, 1942.

Mach, E. (1883b): On the concepts of electrostatics, *in*: Popular scientific lectures, Th. J. McCormack, transl., The Open Court Publ. Co., Chicago, IL, 1910.

Mackey, G. W. (1955): The theory of group representations, Lecture Notes, Univ. of Chicago; reprinted, with a 150-page updating appendix as: The theory of unitary group representations, Univ. of Chicago Press, Chicago, 1976.

Mackey, G. W. (1963): Mathematical foundations of quantum mechanics, W. A. Benjamin, New York.

Manuceau, J. & Trottin, J. C. (1969): On lattice spin systems, *Ann. Inst. Henri Poincaré* **A10**, 359–380.

Marle, C. M. (1982): Lie group actions on Poisson and canonical manifolds, *in* Dynamical systems ..., A. Avez, A. Blaquière and A. Marzollo, eds., Academic Press, New York.

Martin, N. G. & England, J. W. (1981): Mathematical theory of entropy, Encyclopedia of Mathematics and its applications, vol. 12, Addison-Wesley, Reading, MA.

Martin, P. C. & Schwinger, J. (1959): Theory of many-particle systems I, *Phys. Rev.* **115**, 1342-1373.

Martin, Ph. A. (1979): Modèles en mécanique statistique des processus irréversibles, Lecture Notes in Physics **103**, Springer, Berlin.

Martin, Ph. & Emch, G. G. (1975): A rigorous model sustaining van Hove's phenomenon, *Helv. Phys. Acta* **48**, 59–78.

Maxwell, J. C. (1860): Illustrations of the dynamical theory of gases. Part 1: On the motions and collisions of perfectly elastic spheres. Part 2: On the process of diffusion of two or more kinds of moving particles among one another. *Phil. Mag.* **19**, 19–32; **20**, 21–37.

Maxwell, J. C. (1865): A dynamical theory of the electromagnetic field, *Philosophical Trans.* **155**, 459–512; reprinted *in*: Scientific Papers, Vol. 1.

Maxwell, J. C. (1879): The electrical researches of the Honourable Henry Cavendish, F.R.S., written between 1771 and 1781, Cambridge Univ. Press, London.

Mermin, N. D. (1981): Quantum mysteries for anyone, *Journ. Philosophy* **7/81**, 397–408.

Michelson, A. A. (1891): On the application of interference methods to spectroscopic measurements, Phil. Mag. **31** (1891), 338–346; **34** (1892), 280–299.

Michelson, V. A. (1887): Essai théorique sur la distribution de l'énergie dans les spectra des solides, *Journal de Physique* 6, 467–480; (originally appeared in *Journal de la Societé Physico-Chimique Russe* 19, 79–92).

Millikan, R. A. (1910): A new modification of the cloud method of determining the elementary electrical charge and the most probable value of that charge, *Phil. Mag.* 19, 209–228; The isolation of an ion and a precision measurement of its charge and the correction of Stoke's law, *Phys. Rev.* 31, 92–93, *Science* 32, 436–448, *Phys. Zeitschr.* **11**, 1097–1109.

Millikan, R. A. (1916): A direct photoelectric determination of Planck's h, Phys. Rev. *7*, 355–388.

Minkowski, N. (1908): Space and time, Address delivered at the 80th Assembly of German Natural Scientists and Physicians, Köln, 21 Sept. 1908; reprinted *in*: Einstein (1952).

Minlos, R. & Sinai, Ja. G. (1967–8): The phenomenon of phase separation at low temperatures in some lattice models of a gas, I and II, *Math. USSR Sbornik* 2 (1967), 335–395, and *Trans. Moscow Math. Soc.* 19, (1968), 121–196; see also: Some new results on first order phase transitions in lattice gas models, *Trans. Moscow Math. Soc.* 17, (1967), 237–267.

Montgomery, D. & Zippin, L. (1955): Topological transformations groups, Interscience Publishers, New York.

Moser, J. (1968): Lectures on Hamiltonian systems, Memoirs Amer. Math. Soc. vol. **81**, Amer. Math. Soc., Providence, RI.

Moser, J. (1973): Stable and random motions in dynamical systems, with special emphasis on celestial mechanics, Ann. Math. Studies No. 77, Princeton University Press, Princeton, NJ.

Moser, J. (1975): Dynamical systems, theory and applications, Lecture notes in physics **38**, Springer, New York.

Moy, C. T. (1954): Characterizations of conditional expectation as a transformation of function spaces, *Pacific Journ. Math.* **4**, 47–64.

Moyal, J. E. (1949): Quantum mechanics as a statistical theory, *Proc. Cambridge Phil. Soc.* 45, 99–124.

Murray, F. J. & von Neumann, J. (1936): On rings of operators, *Ann. of Math.* **37**, 116–129. See also Murray & von Neumann (1937), von Neumann (1940); Murray & von Neumann (1943).

Murray, F. J. & von Neumann, J. (1937): On rings of operators II, *Trans. Amer. Math. Soc.* 41, 208–248.

Murray, F. J. & von Neumann, J. (1943): On rings of operators IV, *Ann. of Math.* **44**, 716–808.

Nakamura, M., Takesaki, M. & Umegaki, H. (1960): A remark on the expectations of operator algebras. *Kodai Math. Sem. Rep.* 12, 82–90.

Nakamura, M. & Turumaru, T. (1954): Expectations in an operator algebra, *Tohoku Math. Journ.* 6, 174–176.

Narnhofer, H. (1970): On Fermi lattice systems with quadratic hamiltonians, *Acta Phys. Austriaca* **31**, 349–353.

Narnhofer, H. & Sewell, G. L. (1981): Vlasov hydrodynamics of a quantum mechanical model, *Commun. Math. Phys.* 79, 9–24; see also: H. Narnhofer & G. L. Sewell (1980): Equilibrium states of gravitational systems, *Commun. Math. Phys.* 71, (1980), 1–28.

Nelson, E. (1967): Dynamical theories of Brownian motion, Princeton University Press, Princeton, NJ.

Newcomb, S. (1882): Discussion and results of observations on transits of Mercury from 1677 to 1881, *Washington, Astr. Pap. Ephemeris* 1, 363–487; see also *ibid.* 6 (586 pp.).

Newton, I. (1687): Philosophiae Naturalis Principia Mathematica, J. Streater, London.

Newton, I. (1704): Opticks, London, 1704, reprinted by Dover, NY, 1952.

Newton, R. G. (1966): Scattering theory of waves and particles, McGraw-Hill, New York.

Onsager, L. (1944): Crystal statistics. I. A two-dimensional model with an order-disorder transition, *Phys. Rev. (2)* 65, 117-149.

Osterwalder, K. & Schrader, R. (1973, 1975): Axioms for Euclidean Green's functions I, II *Comm. Math. Phys.* **31**, 83–112; **42** 281–305.

Paige, L. J. (1963): Jordan algebras, *in* Studies in modern algebra, A. A. Albert, ed., Prentice-Hall, Englewood Cliffs, NJ.

Pais, A. (1979): Einstein and the quantum theory, *Rev. Mod. Physics* **51**, 861–914.

Pais, A. (1982): Subtle is the Lord ... the science and the life of Albert Einstein, Clarendon Press, Oxford.

Papanicolaou, G. & Varadhan, S. (1973): A limit theorem with strong mixing in Banach space and two applications to stochastic differential equations, *Comm. Pure Applied Math* **26**, 497–524.

Parravicini, G. & Zecca, A. (1977): On the generator of completely positive dynamical semigroups of N-level systems, *Rep. Math. Phys.* **12**, 423–424.

Parry, W. (1981): Topics in ergodic theory, Cambridge University Press, Cambridge.

Paschen, F. (1916): Bohr's Helium Linien, *Annalen der Physik* **50**, 901–940.

Pauli, W. (1921, 1958): Relativitätstheorie, *in* Encyklopädie der mathematischen Wissenschaften, Vol. V 19, Teubner, Leipzig (1921); Engl. transl. by G. Field, with supplementary notes by Pauli, as: Theory of relativity, Pergamon Press, London, 1958.

Pauli, W. (1925): Über den Zusammenhang des Abschusses der Elektronengruppen im Atom mit der Komplexstruktur der Spektren, *Zeitschrift für Physik* **31**, 765–785.

Pauli, W. (1926): Über den Wasserstoffspectrum vom Standpunkt der neuen Quantenmechanik, *Zeitschrift für Physik* **36**, 336–363.

Pauli, W. (1933): Die allgemeinen Prinzipien der Wellenmechanik *in* Handbuch der Physik **24**, Springer, Berlin.

Pearson, D. B. (1975): An example in potential scattering illustrating the breakdown of asymptotic completeness, *Commun. Math. Phys.* **40**, 125–146; and General theory of potential scattering with absorption at local singularities, *Helv. Phys. Acta* **48**, 639–653.

Pedersen, G. K. (1979): C^*-algebras and their automorphism groups, Academic Press, New York.

Peierls, R. (1936): On Ising's model of ferromagnetism, *Proc. Camb. Phil. Soc.* **32**, 477–481.

Perelomov, A. M. (1972): Coherent states for arbitrary Lie group, *Commun. Math. Phys.* **26**, 222-236.

Perrin, J. (1895): Nouvelles propriétés des rayons cathodiques, *Comptes Rendus* **121**, 1130–1134.

Perry, P. A., Segal, I. & Simon, B. (1981): Spectral analysis of N-body Schrödinger operators, *Ann. Math.* **114**, 519–567.

Phelps, R. R. (1966): Lectures on Choquet's theorem, Van Nostrand, Princeton, NJ.

Pier, J. P. (1965): Sur une classe de groupes localement compacts remarquables du point de vue de l'analyse harmonique. Thèse, Nancy.

Piron, C. (1976): Foundations of quantum physics, Benjamin, New York.

Planck, M. (1900, A): Über irreversible Strahlungsvorgänge, *Annalen der Physik* 1, 69–122.

Planck, M. (1900, a & b): (a) Ueber eine Verbesserung der Wien'schen Spectralgleichung; (b) Zur Theorie des Gesetzes der Energieverteilung im Normalspectrum; *Verhandl. Deutschen Phys. Gesellschaft,* 2 (19 Okt.), 202–204; (14 Dez.) 237–245; reprinted *in*: Kangro (1972).

Poincaré, H. (1892): Les méthodes nouvelles de la mécanique céleste, Gauthier-Villars, Paris.

Poincaré, H. (1904): Létat actuel et l'avenir de la physique mathématique, Bulletin des Sciences Mathématiques, Première Partie 302–324; Engl. transl.: The Principles of mathematics—empirical physics, *Monist,* 15, (1905), 1–24.

Poincaré, H. (1905): Leçons de mécanique (professées à la Sorbonne), tome I, Gauthier-Villars, Paris.

Poincaré, H. (1913): Leçons sur les hypothèses cosmogoniques, 2ème ed., Hermann, Paris.

Poisson, S. D. (1809): Essai sur le calcul des variations, *Journ. Ec. Polytechn.* **8**, 266–344.

Pontrjagin, L. (1958): Topological groups, Princeton University Press, Princeton.

Porta, H. & Schwartz, J. T. (1967): Representations of the algebra of all operators in Hilbert space, and related analytic function algebras, *Comm. Pure & Appl. Math.* **20**, 457–492.

Powers, R. T. (1967): Representations of uniformly hyperfinite algebras and their associated von Neumann rings, *Ann. Math.* **86**, 138–171.

Powers, R. T. (1971, 1974): Self-adjoint algebras of unbounded operators I, II, *Commun. math. phys.* **21** (1971), 85–124; *Trans. Amer. Math. Soc.* **187** (1974), 261–293.

Powers, R. T. (1982): Algebras of unbounded operators, *Proc. Symp. in Pure Mathematics* **38**, R. V. Kadison, ed., Part 2, 389–406. See also R. T. Powers (1971, 1974).

Pringsheim, E. (1901, 1903): Einfache Herleitung des Kirckhoffschen Gesetzes, *Verhandl. Deutschen Phys. Gesellschaft* **3** (1901), 81–84; *Zeitschift für wissenschaftische Photographie* **1** (1903), 360–364.

Przibram, K. (1967): Letters on wave mechanics; Schroedinger, Planck, Einstein, Lorentz. K. Przibram, ed.; M. J. Klein, transl; Philosophical Library, New York.

Pukanszky, L. (1956): Some examples of factors, *Publ. Math. Debrecen* **4**, 135–156.

Pulvirenti, M. (1977): Stability, equilibrium and KMS condition for an infinite classical system, *Journ. Math. Phys.* **18**, 2099–2103.

Pusz, W. & Woronowics, S. L. (1978): Passive states and KMS states for general quantum systems, *Commun. Math. Phys.* **58**, 273–290.

Radin, C. (1979–): The infinite volume ground state of the Lennard-Jones potential (with C. Gardner), *Journ. Stat. Phys.* **20** (1979), 719–724; The symmetry of ground states under perturbations (with G. Hamrick), *Journ. Stat. Phys.* **21** (1979), 601–607; The ground state for sticky disks (with R. Heitmann), *Journ. Stat. Phys.* **22** (1980), 281–287; The crystal structure of the noble gases (with B. Borden), *Journ. Chem. Phys.* **75** (1981), 2012–2013. The ground state for soft disks, *Journ. Stat. Phys.* **26**, (1981), 365–373; Crystalline symmetry and surface tension, *Physica* **113A** (1982), 338–342; A first-order phase transition between crystal phases in the shift model. (with F. Nicolo), *Journ. Stat. Phys.* **28** (1982), 473–478; Periodicity of classical ground states (with L. Schulman), *Phys. Rev. Lett.* **51** (1983), 621–622.

Rayleigh (J. W. Strutt, Baron) (1900): Remarks upon the law of complete radiation, *Phil. Mag.* **49**, 539–540; reprinted *in*: Scientific Papers, Dover, New York, 1964.

Reece, G. (1973): The theory of measurement in quantum mechanics, *Intl. J. Theor. Phys.* **7**, 81–117.

Reed, M. & Simon, B. (1972): Methods of modern mathematical physics; Academic Press, New York.

Reed, M. & Simon, B. (1979): Methods of modern mathematical physics, vol. III, Scattering theory; Academic Press, New York.

Rickart, C. E. (1960): General theory of Banach algebras, D. van Nostrand, Princeton, N.J.

Rieffel, M. & van Daele, A. (1977): A bounded operator approach to Tomita-Takesaki theory, *Pacific J. Math.* **69**, 187–221.

Riemann, B. (1854): Ueber die Hypothesen, welche der Geometrie zum Grunde liegen. This Habilitationschrift was defended at Göttingen on June 10, 1854, but was only published, posthumously, in 1867; reprinted *in* Gesammelte mathematische Werke.

Riesz, F. & B. Sz-Nagy (1955): Leçons d'analyse functionnelle, Gauthier-Villars, Paris.

Robertson, H. P. (1928): On relativistic cosmology, *Phil. Mag. [7]* **5**, 835–848.

Robertson, H. P. (1929): On the foundations of relativistic cosmology, *Proc. Nat. Acad. Sci.* **15**, 822–829.

Robertson, H. P. (1929 U): The uncertainty principle, *Phys. Rev.* **34**, 163–164.

Robertson, H. P. (1933): Relativistic cosmology, *Rev. Mod. Phys.* **5**, 62–90.

Robinson, D. W. (1967, 1968): Statistical mechanics of quantum spin systems I, *Commun. Math. Phys.* **6**, 151–160; II, *ibid.* **7**, 337–348.

Robinson, D. W. (1973): Return to equilibrium, *Commun. Math. Phys.* **31**, 171–189.

Robinson, D. W. (1976): C^*-algebras and quantum statistical mechanics, *in* D. Kastler (1976), 235–252.

Roentgen, W. K. (1895): Ueber eine neue Art von Strahlen, *Sitz. Ber. d. Physik-Med. Ges. zu Würzburg* **137** (1895), 132–141; transl. as: On a new kind of rays, *Nature* **53** (1896), 274–276; see also: *Annalen der Physik* **64** (1898), 1–11; 12–17.

Ruelle, D. (1967): Quantum statistical mechanics and canonical commutation relations, *in* Cargèse Lectures in Theoretical Physics, F. Lurçat, ed. Gordon & Breach, New York.

Ruelle, D. (1968): Statistical mechanics of a one-dimensional lattice gas, *Commun.Math. Phys.* **9**, 267–278.

Ruelle, D. (1969): Statistical mechanics, rigorous results, W. A. Benjamin, New York.

Ruelle, D. (1970): Integral representation of states on a C^*-algebra, *J. Funct. Anal.* **6**, 116–151.

Rydberg, J. R. (1890): Recherches sur la constitution des spectres d'émission des éléments chimiques, *Kungliga Vetenskaps Akademiens Handlingar* **23** #11, 155pp.; On the structure of the line-spectra of the chemical elements, *Phil. Mag.* **29**, 331-337; Sur la constitution des spectres linéaires des éléments chimiques, *Comples Rendus* **110**, 394–397; Ueber den Bau der Linienspectren der chemischen Grundstoffe, *Zeitschrift für Physikalische Chemie* **5**, 227-232; Die neuen grundstaffe des Cleveitgases, *Annalen der Physik* **58** (1896), 674–679; La distribution des raies spectrales, *in* Rapports presentés au Congrès International de Physique, Vol. 2, 220–224, Gauthier-Villars, Paris, 1900.

Saccheri, G. G. (1733): Euclides vindicatus, P. A. Montano, Milano, Engl. transl. by G. B. Halsted, Open Court, Chicago, IL (1920).

Sachs, R. K. & Wu, H. (1977): General relativity for mathematicians, Springer, New York.

Sakai, S. (1957): On topological properties of W^*-algebras, *Proc. Japan Acad.* **33**, 439–444.

Sakai, S. (1965): On the central decomposition for positive functionals on C^*-algebras, *Trans. Amer. Math. Soc.* **118**, 406–419.

Sakai, S. (1971): C^*-algebras and W^*-algebras, Springer, New York.

Saletan, E. J. (1961): Contraction of Lie groups, *Journ. Math. Phys.* **2**, 1–21.

Sandage, A. (1961): The ability of the 200-inch telescope to discriminate between selected world models, *Astrophys. J.* **113**, 355-392; see also: (i) Observational Cosmology, *Observatory* 88 (1968), 91–106; (ii) Cosmology - A search for two numbers. *Physics Today, Feb. 1970.*

Schatten, R. (1950): A theory of cross-spaces, *Ann. of Math. Studies* **26**, Princeton, N.J.

Schatten, R. (1957): The space of completely continuous operators on Hilbert space, *Math. Ann.* **134**, 47–49.

Schrieffer, J. R. (1964): Theory of superconductivity, Benjamin, New York.

Schröder (1983): W^*-K-systems, Mathematicsches Institut, Tübingen.

Schroedinger, E. (1926): Quantisierung als Eigenwertproblem, I. *Annalen der Physik* **79**, 361–376; II. **79**, 489–527; III. **79**, 734–756; IV. **80**, 437–490. V. **81**, 109–139; reprinted *in* Abhandlungen zur Wellenmechanik, J. A. Barth, Leipzig. (See also 2nd ed. ibid. 1928). Engl. transl.: Collected papers on wave mechanics, J. F. Shearer & W. M. Deans, transl., Blackie & Son, London, Glasgow, 1928.

Schroedinger, E. (1926 N): Die stetige Uebergang von der Midro-zur Makromechanik, *Naturwissenschaften* **14**, 664–666.

Schultz, T. D., Mattis, D. C. & Lieb, E. H. (1964): Two-dimensional Ising model as a soluble problem of many fermions, *Rev. Mod. Phys.* **36**, 856–871.

Schwarzschild, K. (1916a): Ueber das Gravitationsfeld eines Massenpunktes nach der Einsteinschen Theorie, *Sitz. Ber. Preuss. Akad. Wiss. Berlin*, 186–196.

Schwarzschild, K. (1916b): Ueber das Gravitationsfeld einer Kugel aus inkompressibler Flüssigkeit nach der Einsteinschen Theorie, *Sitz. Ber. Preuss. Akad. Wiss. Berlin*, 424–434.

Schwarzschild, K. (1916c): Zur Quantenhypothese, *Sitz. Ber. Preuss. Akad. Wiss. Berlin*, 548–568.

Seelinger, H. (1906): Das Zodiakallicht und die empirischen Glieder in Bewegung der innern Planeten, *Bayer. Aka. Wiss. zu München* **36**, 595–622.

Segal, I. E. (1947a): Postulates for general quantum mechanics, *Annals of Math.* **48**, 930–948.

Segal, I. E. (1947b): Irreducible representations of operator algebras, *Bull. Amer. Math. Soc.* **53**, 73–88; see also I. E. Segal (1947a).

Segal, I. E. (1960): Quantization of nonlinear systems, *Journ. Math. Phys.* **1**, 468–488; see also I. E. Segal (1964), (1974).

Segal, I. E. (1963a): Transforms for operators and symplectic automorphisms over a locally compact abelian group, *Math. Scan.* **13**, 31–43.

Segal, I. E. (1963b): Mathematical problems of relativistic physics, Amer. Math. Soc. Publ., Providence, RI.

Segal, I. E. (1964): Explicit formal construction of nonlinear quantum fields, *Journ. Math. Phys.* **5**, 269–282.

Segal, I. E. (1974): Symplectic structures and the quantization problem for wave equations, *Symposia Mathematica* **14**, 9–117.

Serrin, J. (1978): The concepts of thermodynamics, *in* Contemporary developments in continuum mechanics, G. M. de La Penha & L. A. Medeiros, eds., North-Holland, Amsterdam.

Serrin, J. (1979): Conceptual analysis of the classical second laws of thermodynamics, *Arch. Rat. Mech. Analysis* **70**, 355–371.

Sewell, G. L. (1974): Relaxation, amplification and the KMS conditions, *Ann. Phys.* **85**, 336–377.

Sewell, G. L. (1977): KMS conditions and local thermodynamical stability of quantum lattice systems, II, *Commun. Math. Phys.* **55**, 53–61.

Sewell, G. L. (1980): Stability, equilibrium and metastability in statistical mechanics, *Physics Reports,* 57, 307–342.

Sewell, G. L. (1982): Quantum fields on manifolds: PCT and gravitationally-induced thermal states, *Ann. Phys.* 141, 201–224.

Shannon, C. E. (1948): A mathematical theory of communication, *Bell System Technical Journal* 27, 379–423; 623–656.

Siegert, A. J. F. & Vezetti, D. J. (1968): On the Ising model with long-range interaction, *Journ. Math. Phys.* 9, 2173–2193.

Simms, D. J. & Woodhouse, N. M. J. (1976): Lectures on geometric quantization, Lecture Notes in Physics 53, Springer, New York.

Simon, B. (1971): Quantum mechanics for Hamiltonians defined as quadratic forms, Princeton Univ. Press, Princeton, NJ.

Simon, B. (1974): The $P(\phi)_2$ Euclidean (quantum) field theory, Princeton University Press, Princeton, NJ.

Simon, B. (1979): Functional integration and quantum physics, Academic Press, New York.

Sinai, Ya. G. (1962, 1963): Probabilistic ideas in ergodic theory, International Congress of Mathematicians, Stockholm, 1962, 540–559 and *Amer. Math. Soc. Transl.* **(2) 31** (1963), 62–81.

Sinai, Ya. G. (1976): Introduction to ergodic theory, V. Scheffer, transl., Princeton University Press, Princeton, NJ.

Sniatycki, J. (1980): Geometric quantization and quantum mechanics, *Appl. Math. Ser.* **30**, Springer, New York.

Sommerfeld, A. (1915, 1916): Zur Theorie der Balmerschen Series, *Münchener Berichte 1915*, 425–458; Die Feinstruktur der Wasserstoff und wasserstoffähnlichen Linien,*Münchener Berichte 1915*, 459–500; Zur Quantentheorie der Spectrallinien, *Annalen der Physik* 51 (1916), 1–94, 125–167.

Sommerfeld, A. (1929): Atombau und Spektrallinien, Wellenmechanischer Ergänzungband, Vieweg, Braunschweig; see also (5. Auflage), 1931. Braunschweig.

Souriau, J. M. (1966): Quantification géométrique, *Commun. math. Phys.* 1, 374–398. See also J. M. Souriau (1970).

Souriau, J. M. (1970): Structure des systèmes dynamiques, Dunod, Paris.

Spohn, H. (1976): Approach to equilibrium for completely positive dynamical semi-groups of N-level systems, *Rep. Math. Phys.* **10**, 189–194; see also: An algebraic condition for the approach to equilibrium of an open N-level system, *Lett. Math. Phys.* 2, (1977), 33–38.

Spohn, H. (1977): Derivation of the transport equation for electrons moving through random impurities, *Journ. Stat. Phys.* 17, 385–412.

Spohn, H. (1980): Kinetic equations from Hamiltonian dynamics: Markovian limits, *Rev. Mod. Phys.* 53, 569–615.

Stern, O. & Gerlach, W. (1922): Der experimentelle Nachweis des magnetischen Moments des Silberatons, *Zeitschrift für Physik* 8, 110-111; Der experimentelle Nachweis der Richtungsquantelung im Magnetfeld, *Zeitschrift für Physik* 9, 349–355.

Stinespring, W. F. (1955): Positive functions on C^*-algebras, *Proc. Amer. Math. Soc.* **6**, 211–216.

Stone, M. H. (1932): Linear transformations in Hilbert space and their applications to analysis, Amer. Math. Soc., Providence, R.I.

Størmer, E. (1963): Positive linear maps of operator algebras, *Acta Math.* **110**, 233-278.

Størmer, E. (1965): On the Jordan structure of C^*-algebras, *Trans. Amer. Math. Soc.* **120**, 438–447.

Størmer, E. (1967): Types of von Neumann algebras associated with extremal invariant states, *Commun. Math. Phys.* **6**, 194–204.

Størmer, E. (1969): Asymptotically abelian systems, *in* Cargèse Lectures in Theoretical Physics, D. Kastler, ed., Gordon & Breach, New York.

Størmer, E. (1970): The even CAR-algebra, *Commun. Math. Phys.* **16**, 136–137.

Streater, R. F. (1966): Canonical quantization, *Commun. math. Phys.* **2**, 354–374.

Streater, R. F. & A. S. Wightman (1964), PCT, spin and statistics, and all that, W. A. Benjamin, New York.

Stueckelberg, E. C. G. (1960): Quantum theory in real Hilbert space, *Helv. Phys. Acta* **33** (1960), 727–752; Part II (with M. Guenin) *ibid.* **34** (1961), 621–628; Part III (with M. Guenin, C. Piron & H. Ruegg), *ibid.* **34** (1961), 675–698; Part IV (with M. Guenin) *ibid.* **35** (1962), 673–695.

Szekeres, G. (1960): On the singularities of a Riemannian manifold, *Publ. Mat. Debrecen* **7**, 285–301.

Takeda, Z. (1955): Inductive limit and infinite direct product of operator algebras, *Tohoku Math. J.* **7**, 68–86.

Takesaki, M. (1970a): Tomita's theory of modular Hilbert algebras and its applications, Lecture Notes in Mathematics 128, Springer, New York.

Takesaki, M. (1970b): Disjointness of the KMS-states of different temperatures, *Commun. Math. Phys.* **17**, 33-41.

Takesaki, M. (1972): Conditional expectations in von Neumann algebras, *Journ. Funct. Analysis* **9**, 306–321.

Takesaki, M. (1973): States and automorphisms of operator algebras—standard representations and KMS boundary condition, *in* Statistical Mechanics and Mathematical Problems, A. Lenard, ed., Lecture Notes in Physics 20, Springer, New York.

Takesaki, M. & Winnink, M. (1973): Local normality in quantum statistical mechanics, *Commun. Math. Phys.* **30**, 129–152.

Teichmueller, O. (1936): Operatoren im Wachsschen Raum, *Journal f. reine u. angewandte Mathematik* **174**, 73–124.

Thirring, W. (1968): On the mathematical structure of the BCS model, *Commun. Math. Phys.* **7**, 181–189.

Thirring, W. (1969): The mathematical structure of the BCS model and related models, *in* The many-body problem, L. M. Garrido, A. Cruz and T. W. Preist, eds. Plenum Press, London.

Thirring, W. (1983): Quantum mechanics of large systems, E. M. Harrel, transl., Springer, New York.

Thirring, W. & Wehrl, A. (1967): On the mathematical structure of the BCS model, *Commun. Math. Phys.* **4**, 303-314.

Thomas, W. (1925): Ueber die Zahl der Dispersionselektronen, die einem stationären Zustande zugeordnet sind, *Naturwissenschaften* **13**, 627.

Thompson, C. J., Siegert, A. J. F. & Vezzetti, D. J. (1970): On the Ising model with long-range interaction, II, Critical region analysis, *Journ. Math. Phys.* **11**, 1018–1028.

Thomson, G. P. (1927): Diffraction of cathode rays by a thin film (with A. Reid), *Nature* **119** (1927), 890; The diffraction of cathode rays by thin films of platinum, *Nature* **120** (1927), 802; Experiments on the diffraction of cathode rays, *Proc. Roy. Soc. London* **A117**

(1928), 600-609; see also: Early work in electron diffraction, *Amer. Phys.* 29 (1961), 821–825.

Thomson, J. J. (1897): Cathode rays, *Phil. Mag.* **44**, 293–316.

Thomson, W. (1848): On an absolute thermometric scale, founded on Carnot's theory of the motive power of heat, and calculated from Regnault's observations, *Phil. Mag.* **3:33**, 313–317.

Thomson, W. (1852): Dynamical theory of heat, *Phil. Mag.* **4:4**, 304–306; see also: Mathematical and physical papers, I, 174–175, Cambridge Univ. Press, 1882.

Tisza, L. (1966): Generalized thermodynamics, MIT Press, Cambridge, MA.

Tomita, M. (1967): Standard forms of von Neumann algebras *in* The *V*th Functional Analysis Symposium of the Mathematical Society of Japan, Sendai.

Tomiyama, J. (1957–1959): On the projection of norm one in W^*-algebras, *Proc. Japan Acad.* **33**, (1957), 608–612; II. *Tohoku Math. Journ.* **10**, (1958), 204–209; III, *ibid.* **11**, (1959), 125–129.

Truesdell, C. (1945): On a function which occurs in the theory of the structure of polymers, *Ann. Math.* **46**, 144–157.

Truesdell, C. (1969): Rational thermodynamics, Mc-Graw Hill, New York.

Truesdell, C. (1983): The influence of elasticity on analysis: the classic heritage, *Bull. AMS (new series)* **9**, 293–310.

Uhlenbeck, G. E. (1927): Over statistische methoden in de theorie der quanta, M. Nijhoff, 's Gravenhage.

Uhlenbeck, G. E. & Goudsmit, S. (1925, 1926): Ersetzung der Hypothese vom unmechanischen Zwang durch eine Forderung bezüglich des inneren Verhaltens jedes einzelnen Elektrons, *Die Naturwissenschaften* **13**, (1925), 953–954; Spinning electrons and the structure of spectra, *Nature* **117** (1926), 264–265.

Umegaki, H. (1954): Conditional expectation in an operator algebra, *Tohoku Math. Journ.* **6** (1954), 177–181; II, *ibid.* **8** (1956), 86–100; III, *Kodai Math. Sem. Rep.* **11** (1959), 51–64; IV, *ibid.* **14** (1962), 59–85.

Upmeier, H. (1981): Automorphism groups of Jordan C^*-algebras, *Math. Zeits.* **176**, 21–34; see also: Derivations of Jordan C^*-algebras *Math. Scan.* **46** (1980), 251–264.

van Hove, L. (1950): Sur l'intégrale de configuration pour les systèmes de particules à une dimension, *Physica* **16**, 137–143.

van Hove, L. (1951): Sur le problème des relations entre les transformations unitaires de la mécanique quantique et les transformations canoniques de la mécanique classique, *Mem. de l'Acad. Roy. de Belgique (cl.sc.)* **37**, 610–620; see also: Sur certaines représentations unitaires d'un groupe infini de tranformations, *ibid.* **36**, 1–102.

van Hove, L. (1955): Quantum-mechanical perturbations giving rise to a statistical transport equation, *Physica* **21**, 517–540.

von Helmoltz, H. (1868): Ueber die Tatsachen die der Geometrie zum Grunde liegen, *Nachr. Ges. Wiss. Göttingen* **15**, 193–221, reprinted *in*: Wissenschaftliche Abhandlungen, Vol. 2, 618–639.

von Hettner, G. (1922): Die bedeutung von Rubens Arbeiten für die plancksche Strahlungsformel, *Naturwissenschaften* **10**, 1033–1038.

von Laue, M. (1911): Das Relativätsprinzip, Vieweg u. Sohn, Braunschweig.

von Neumann, J. (1927): Wahrscheinlichkeitstheoretischer Aufbau der Quantenmechanik, *Göttinger Nachrichten*, 245–272.

von Neumann, J. (1931): Die Eindeutigkeit der Schroedingerschen Operatoren, *Math. Ann.* **104**, 570–578.

von Neumann, J. (1932): Grundlagen der Quantenmechanik, Springer, Berlin. Engl. transl. "Mathematical foundations of quantum mechanics", by R. T. Beyer, Princeton University Press, Princeton, NJ, 1955.

von Neumann (1936): On an algebraic generalization of the quantum mechanical formalism, *Mat. Sborn.* 1, 415–484.

von Neumann, J. (1940): On rings of operators III, *Ann. of Math.* 41, 94–161.

Walters, P. (1982): An introduction to ergodic theory, Springer, Berlin.

Wannier, G. H. (1966): Statistical Physics, John Wiley, New York.

Wehrl, A. (1978): General properties of entropy, *Rev. Mod. Phys.* 50, 221–260.

Weinberg, S. (1972): Gravitation and Cosmology, John Wiley & Sons, New York.

Wess, J. & Bagger, J. (1983): Supersymmetry & Supergravity, Princeton University Press, Princeton, NJ.

Weyl, H. (1918): Raum-Zeit-Materie, Springer, Berlin.

Weyl, H. (1927): Quantenmechanik und Gruppentheorie, *Zeitschrift für Physik* **46**, 1–46.

Weyl, H. (1928): Gruppentheorie und Quantenmechanik, Hirzel, Leipzig. (see also 2nd ed. ibid. 1931) Engl. transl.: The theory of groups and quantum mechanics (H. P. Robertson, transl.) Methuen, London, 1931, reissued by Dover, New York, 1949.

Wheeler, J. A. & Zurek, W. H. (1983): Quantum theory and measurement, Princeton University Press, Princeton, NJ.

Whitehead, A. N. (1925): Science and the modern world, Macmillan, New York.

Whittaker, E. T. & Watson, G. N. (1927): A course in modern analysis, Cambridge University Press, Cambridge.

Whitten-Wolfe, B. & Emch, G. G. (1976): A mechanical quantum measuring process, *Helv. Phys. Acta* 49, 45–55.

Wick, G. C., Wightman, A. S. & Wigner, E. P. (1952): Intrinsic parity of elementary particles, *Phys. Rev.* **88**, 101–105.

Wien, W. (1896): Über die Energieverteilung im Emissionspectrum eines schwarzen Körpers, *Annalen der Physik* 58, 662–669.

Wightman, A. S. (1956): Quantum field theory in terms of vacuum expectation values, *Phys. Rev.* 101, 860–866.

Wigner, E. P. (1931): Gruppentheorie und ihre Anwendung auf die Quantenmechanik der Atomspektren, Vieweg, Braunschweig. Engl. transl.: Group theory and its application to the quantum mechanics of atomic spectra (J. G. Griffin, transl.) Academic Press, New York, 1959.

Wigner, E. P. (1932): On the quantum correction for thermodynamical equilibrium, *Phys. Rev.* 40, 749–759.

Wigner, E. P. (1939): On unitary representations of the inhomogeneous Lorentz group, *Annals of Mathematics* 40, 149–204.

Wils, W. (1968): Désintégration centrale des formes positives sur les C^*-algèbres, *C. R. Acad. Sci. Paris* **267**, 810–812.

Wils, W. (1971): The ideal center of partially ordered vector spaces, *Acta Math.* **127**, 41–47.

Wilson, W. (1915): The quantum theory of radiation and line spectra, *Phil. Mag.* **29**, 795–802.

Winnink, M. (1972): Some general properties of thermodynamic states in an algebraic approach, *in* Statistical Mechanics and Field Theory, R. N. Sen & C. Weil, eds., Halsted Press, New York and Israel Universities Press, Jerusalem.

Wolf, J. A. (1974): Spaces of constant curvature, Publish or Perish, Inc., Boston, Mass.

Yaffe, L. G. (1982): Large N limits as classical mechanics, *Rev. Mod. Phys.* **54**, 407–435.

Yngvason, J. (1976): On the algebra of test functions for Wightman fields *in* C^*-algebras and their applications to statistical mechanics and quantum field theory, D. Kastler, ed. North-Holland, Amsterdam, 253–262.

Yngvason, J. (1981): Translationally invariant states and the spectrum ideal in the algebra of test functions, *Commun. math. Phys.* **81**, 401–418.

Yosida, K. (1971): Functional analysis, 3rd. ed. Springer, New York.